Plants, Pollutants and Remediation

Münir Öztürk • Muhammad Ashraf
Ahmet Aksoy • M. S. A. Ahmad
Khalid Rehman Hakeem
Editors

Plants, Pollutants and Remediation

Springer

Editors
Münir Öztürk
Botany Department and Center for Environmental Studies
Ege University Bornova
Izmir, Turkey

Ahmet Aksoy
Department of Biology
Akdeniz University
Antalya, Turkey

Khalid Rehman Hakeem
Faculty of Forestry
Universiti Putra Malaysia
Serdang, Selangor, Malaysia

Muhammad Ashraf
Pakistan Academy of Science
Islamabad, Pakistan

M. S. A. Ahmad
Botany Department
University of Agriculture
Faisalabad, Pakistan

ISBN 978-94-017-7193-1 ISBN 978-94-017-7194-8 (eBook)
DOI 10.1007/978-94-017-7194-8

Library of Congress Control Number: 2015958354

Springer Dordrecht Heidelberg New York London

Printed on acid-free paper

Springer Science+Business Media B.V. Dordrecht is part of Springer Science+Business Media (www.springer.com)

Foreword

Centuries ago, nature was dominant and human interference was negligible. Gradually population and industrialization increased, resulting in pollution of natural resources. Pollution is a glocal (global and local) problem. Of late, considerable efforts have been put in for decontaminating the polluted substrates such as air, water, and soil. In this regard, the subject of phytoremediation gained global momentum and has grown phenomenally.

Nature's cure using plant resources (phytoremediation) is a sustainable solution for environmental decontamination. As of now, about 25,000 articles have been published on various aspects of using biological resources for environmental cleanup starting with only 11 in 1989. The use of plants for the remediation of surface soils polluted and contaminated with toxic heavy metals is well established. The plant-based technologies are applicable to inorganic and organic contaminants and pollutants. A wide variety of technologies using plants and microbes to remediate or decontaminate soils, groundwaters, surface waters, or sediments, including air, are currently researched in various laboratories all over the world. These technologies have become attractive alternatives to conventional cleanup technologies due to relatively low capital costs and the inherently aesthetic nature. Biodiversity is the raw material for bioremediation and is an invaluable toolbox for wider application in the realm of geoenvironment and human health protection.

The industrial revolution, a feather in the cap of human civilization, has unwittingly rendered thousands of hectares of land tainted with the toxic by-products of many industries such as mining, batteries, and paints. Conventional remediation, which involves the physical removal and burial of contaminated soils, is neither feasible nor affordable. The growing awareness of the existence of a number of metal-accumulating plant species, called hyperaccumulators, that are endemic to metalliferous soils and can accumulate and tolerate high levels of heavy metals in the shoot is a major factor in the growing interest in phytoremediation. This technology, which uses plants with their extensive root systems and efficient uptake of a wide variety of molecules, offers a low-input affordable alternative to conventional remediation. The identification of several metal hyperaccumulator plant species demonstrates that the genetic potential exists for successful phytoremediation of contaminated soils.

Although extremely effective at accumulating metals, naturally occurring hyperaccumulators are less than ideal for phytoremediation due to their slow growth rate and low-to-the-ground rosette architecture, which makes them difficult to harvest. The transfer of these hyperaccumulating properties from the hyperaccumulators into a high-biomass-producing plant has been suggested as a potential avenue for making phytoremediation a commercial technology. Transgenic plants are used effectively for the remediation of soils containing a number of different xenobiotic contaminants. Progress in this area, however, is hindered by a lack of understanding of the basic physiological mechanisms involved in uptake into roots and translocation to aboveground tissues.

Recent research has focused on understanding the native molecular and physiological mechanisms of how plants remove pollutants from soils to aid in the creation of transgenic varieties optimized for soil remediation. Attempts made to cross these hyperaccumulators with their larger fast-growing relatives to produce desirable hybrids are not yet successful. To effectively accumulate a metal, a plant must be able to efficiently absorb, translocate through the xylem, unload into the shoot tissues, and finally sequester the metal into vacuoles. Many workers have speculated that the ability to hyperaccumulate metals could be the result of a broader change in the regulation of a response pathway. To be truly effective, plants used for the phytoremediation of metals need to be able to extract the toxic element from the soil and accumulate it in their aboveground tissues which can then be harvested and either composted or ashed to retrieve the extracted metals.

Soil, water, and air are the important natural resources that must be clean. Unfortunately, natural resources are polluted globally. Rapid industrialization and extraction of a large quantity of natural resources, including indiscriminate extraction of groundwater have resulted in environmental contamination and pollution. Large amounts of toxic wastes have been and are still dispersed in thousands of sites spread across the globe, resulting in varying degrees of contamination and pollution. Thus, every one of us is getting exposed to contamination from past and present industrial practices and emissions in natural resources (air, water, and soil) even in the most remote regions. The risk to human and environmental health is rising, and there is evidence that this cocktail of pollutants is a contributor to the global epidemic of cancers and other degenerative diseases. The challenge is to develop innovative and cost-effective solutions to decontaminate polluted environments from inorganic as well as organic pollutants.

Soil contamination with organics and inorganics is growing as a perennial problem all over the world. Its association with human health makes it a topic of more concern. Therefore, there is a need for research to evolve approaches and strategies for promoting sustainable technologies for environmental management which includes bioremediation. Currently the "gentle soil remediation options (GRO)" and the emerging "phytomanagement" practices highlight the use of bio-/phyto-/rhizoremediation-borne biomass as feedstock for "biorefinery" and ecosystem services.

In recent years, the number of studies evaluating GRO at a field level has been steeply on the rise. Most of the papers published are lab-scale and hydroponic experiments. This has received inadequate support from policy and decision makers who believed that phytoremediation is a temporary solution of transferring the pollutants and contaminants from one place to another. Often, scientists and academia are also subscribed to this feeling. Regulators have expressed apprehensions about phytoremediation due to the lack of contemporary knowledge of environmental sustainability. Thus, it is generally believed that pollution prevention by plants through phytoremediation strategy and approach is a temporary solution. Further, how to dispose of the contaminated photomaps is a puzzling question posed by environmental managers and regulators.

The move from greenhouse to field conditions requires incorporating agronomical and ecological knowledge into the remediation process. Agronomic practices such as crop selection, crop rotations/intercropping, planting density, fertilization, irrigation schemes (including chelator-supplemented water), bioaugmentation with microbial inoculants, and weed, pest, and herbivory management can be modified so as to suit both the characteristics of the contaminated soils and to meet the requirements of effective phytoremediating crops.

GRO can bring beneficial ecosystem services (e.g., habitat, C-storage, soil erosion, temperature regulation, etc.) and can also provide valuable sources of renewable biomass for the bio-based economy (e.g., bioenergy, biocatalysis and platform molecules for green chemicals, and ecomaterials). Harvested biomass can be burned/chemically converted for the energy sector and the recovery of accumulated metals (phytoextraction) or for the production of biomass suitable for the biorefinery industry (other GRO, e.g., phytostabilization). Some GRO-borne biomasses can be used as ecomaterials, notably in combination with plastics/biocomposites including geopolymers. Many economies are dependent on the supply of raw materials and trading values of metals such as copper, nickel, and zinc have been steadily on the rise. Metal-rich plant biomass has been used as an alternative to nonrenewable mineral materials to produce Lewis acid catalysts. GRO can offer a means of metal extraction, recovery, or recycling. For TE-contaminated soils, GRO are based on practices which decrease the labile ("bioavailable") pool and/or total contents of TE in the soil and include (in situ) contaminant stabilization ("inactivation") and plant-based (generally termed "phytoremediation") options. Phytoextraction aims to remove TEs from soils through their uptake and accumulation in plant parts that are removed by harvest. Here, bioavailable contaminant stripping (BCS) targets in particular the labile TE pool in the soil. Phytoextraction (or phytomining) can be carried out on metal-contaminated soils as well as low-grade ores or naturally metal-rich (serpentine) soils that cannot be economically utilized by traditional mining technology. Aided phytostabilization aims to establish a vegetation cover and progressively promote (in situ) inactivation of metal(loid)s by combining the use of TE-excluding plants and soil amendments. Although this technology does not lead to a cleanup of the soil, by altering TE speciation and mobility, it moderates potential negative environmental impacts and pollutant linkages.

I fervently believe that the chapters included in this book will contribute towards a broader understanding of pollution prevention by plants and the remediation strategies and approaches.

Department of Plant Sciences
University of Hyderabad,
Hyderabad, Telangana, India

Prof. Dr. M.N.V. Prasad

Preface

The hazards of environmental degradation transcend borders and are fully highlighted by different organizations. Environmental pollution is one of the most important and urgent problems faced by the global society, a problem which does not respect traditional political or geographical boundaries. Several steps have been taken by several organizations to look at the critical environmental and developmental challenges, arising from unprecedented pressures on the environment of planet Earth.

Nearly two decades have passed after the launch of Our Common Future, which defined sustainable development as a blueprint to address our environmental and developmental challenges. An evaluation of such issues is a social imperative of our time. We are experiencing rapid environmental change all around us, and many more problems like water shortages, land degradation, and biodiversity loss are on the horizon.

The problems related to the water resources are likely to grow wider. This will affect economic and social development as well as environmental sustainability. In order to overcome water scarcity, integrated water resource management will be of crucial importance. This will also be important for international peace and security and eradication of global poverty together with future developmental goals. The countries on individual basis will not be in a position to protect our environment. We on this planet are badly in need of a more coherent system of international environmental governance. We must move forward rapidly for the sake of current and future generations towards the global response to these challenges.

In the light of the statements given above, this book is being published at a time when there is a need for the pace of environmental degradation with a new sense of realism. The unprecedented environmental changes we face today are highlighted here. The book contains 19 chapters, which provide an overview of global social and economic trends, as well as the human dimensions of these changes. It highlights the challenges of environmental change, an outlook for the future, and policy options to address present and emerging environmental issues.

Environmental pollution endangers our biodiversity on one side and human health on the other. These pollutants, although generated in megacities and industrial areas, affect rural areas equally well through transport and dispersal. The relative distance from the pollutants in no way guarantees a lack of impact on our environment. Although several pollution control measures have been adopted, even then rapid development continues to produce significant impacts on global environmental quality. Many phytotoxic compounds, heavy metals, pesticides, and acidic precipitation highly affect our biodiversity. Many bioindicator species are used effectively to assess pollutant impacts; however, the knowledge and experience of the researchers is critical for an accurate evaluation.

Each chapter in this book includes information representing a compilation of material and references by internationally recognized experts. The main contribution is to provide a broad-based reference for pollutants in relation to our plant life. The need for producing this book was felt because environmental study is one of the most important and integral parts of life sciences. It describes the different components of the environment and their influence on plant as well as animal diversity. The degradation of the environment and its effects on our health have attained great importance in this branch of science. A global awareness on environment has been generated.

The editors have thus spent efforts to put together the fundamentals of existing knowledge in environmental perspectives and their remediation. Attempts have been made to include latest information available in this field. The environmental issues have been reviewed and efforts for protection as well as remediation outlined in different chapters. We were encouraged to undertake this editorial effort by the participation of a large number of scientists from all over the world together with Nobel laureates and other leading scientists. We express gratitude to all these participants who joined us. We hope this volume will be useful to all researchers as well as others concerned with our environment.

Bornova, Izmir, Turkey Münir Öztürk
Islamabad, Pakistan Muhammad Ashraf
Antalya, Turkey Ahmet Aksoy
Faisalabad, Pakistan M. S. A. Ahmad
Serdang, Selangor, Malaysia Khalid Rehman Hakeem

Acknowledgment

The foundation for editing this volume was laid during the "International Conference on Plants and Pollutants" held at Erciyes University. Our actively working colleagues from different parts of the world were kind enough to collaborate with us. We therefore take this opportunity to thank the contributors for their patience, full cooperation, and support.

The conference was sponsored by the IDB (Islamic Development Bank), Rectorate of Erciyes University (Rector Prof. Dr. Fahrettin Keleştemur), Provincial Grand Mayor Municipality of Kayseri (Mr. Mehmet Özhaseki), Trades Union of Kayseri (Mr. Hasan Ali Kilci), and Kayseri Directorate of the Ministry of Environment and Forests–Turkey. Our special thanks are due to them for the support and encouragement.

The motivation from the Nobel laureates Prof. Dr. Yuvan T Lee (Taiwan) and Prof. Dr. Ferid Murad (USA) as well as UNESCO laureate Prof. Dr. Atta-ur-Rehman (FRS) (Pakistan) inspired us greatly to work on this book. The editors would like to express their indebtedness and special thanks to all of them.

The success in the preparation of this volume depended largely on the encouragement from the Springer team who collaborated with us; therefore, our greatest appreciation goes to them.

Contents

Visible Injury, CO_2 Assimilation and PSII Photochemistry of *Eucaliptus* Plants in Response to Boron Stress

Cristina Nali, Alessandra Francini, Elisa Pellegrini, Stefano Loppi, and Giacomo Lorenzini

Contents

Abstract Boron is an essential element required for the normal growth of plants, but in high concentrations it is toxic, causing reduction of leaf area, induction of chlorotic and necrotic lesions in older leaves, delay in development and general inhibition of growth. To gain an insight into the role of photosynthetic mechanisms in the response to boron toxicity, physiological parameters were analyzed in seedlings of *Eucaliptus globulus* treated with 0.1 (control), 1 and 10 mg l^{-1} (excess) H_3BO_3 in nutrient solution during 12 weeks. After 42 days of treatment, plants grown in the excess of boron developed symptoms in the mature leaves, in form of marginal necrosis. At the end of treatment, CO_2 assimilation and stomatal conductance decreased (−71 % and −30 %, respectively, compared to control) when plants were supplied with 10 mg l^{-1} H_3BO_3; a reduction in growth (−30 % compared to control) and increase of B concentration in roots as a consequence of the treatment have been also observed.

C. Nali (✉) • A. Francini • E. Pellegrini • G. Lorenzini
Department of Agriculture, Food and Environment, University of Pisa, Via del Borghetto 80, 56124 Pisa, Italy
e-mail: cristina.nali@unipi.it

S. Loppi
Department of Life Sciences, University of Siena, Via P.A. Mattioli 4, 53100 Siena, Italy

M. Öztürk et al. (eds.), *Plants, Pollutants and Remediation*,
DOI 10.1007/978-94-017-7194-8_1

Keywords Boron excess • Photosynthesis • Chlorophyll *a* fluorescence • Nutrient uptake • Critical concentration

1 Introduction

Boron (B) is an essential element required for the normal growth of vascular plants. It is unique as a micronutrient in that the threshold between deficiency and toxicity is very narrow (Yau and Ryan 2008; Ozturk et al. 2010): it has long been known that the optimum B level for one species could be either toxic or insufficient for other species (Blevins and Lukaszewski 1998). The role of B in plant nutrition is little understood, which is surprising since on a molar basis the requirement for B is, at least for dicotyledons, higher than any other micronutrient (Marschner 1995); moreover, it has restricted mobility in many species and is freely mobile in others (Brown and Shelp 1997). B is implicated in three main processes: keeping cell wall structure, maintaining membrane function and supporting metabolic activities. However, in the absence of conclusive evidence, the primary role of B in plants remains elusive (Bolaños et al. 2004).

Boron toxicity is largely a local phenomenon, restricted to areas where soil or water supplies high amount of B (Aucejo et al. 1997). For this reason, worst cases of B toxicity occur in irrigated agricultural fields, re-vegetation projects and adjacent to industrial sites that emit B-laden aerosols (Sage et al. 1989). When the B concentration at root level is high, this element is accumulated in the leaf cell walls and may reach the cytoplasm, disturbing metabolism and resulting in the development of toxicity symptoms (Matoh 1997), generally in the older leaves in the form of marginal or interveinal chlorotic and/or necrotic lesions (Paull et al. 1992; Nable et al. 1997). As B concentrations in the roots remain relatively low compared to those in leaves even at very high levels of B supply (Nable et al. 1997), perhaps toxic concentrations do not occur in root tissues. The main concern is that B is mainly transported via the transpiration stream and B concentration typically decreases from older to younger leaves, and apical to basal leaf parts. Thus, stomatal movement may be affect B uptake behaviour. Increased stomatal resistance against the excessive B uptake was reported by several Authors (Alpaslan and Gunes 2001; Papadakis et al. 2004b; Gunes et al. 2006), no data being available on eucalypts.

Boron toxicity is an important disorder that causes negative physiological effects such as decreased leaf chlorophyll, inhibition of photosynthesis (Lovatt and Bates 1984), deposition of lignin and suberin (Ghanati et al. 2002), increased membrane leakiness. B excess inhibits photosynthesis by causing structural damage to thylakoids and thus decreasing CO_2 uptake. These effects disrupt photosynthetic transport of electrons, favoring a condition where molecular oxygen operates as an alternative acceptor for non-utilized electrons and light energy leading to generation of reactive oxygen species (ROS) (Molassiotis et al. 2006).

Although there are many reports in the literature relating to the development of leaf symptoms of B toxicity, the available information concerning the effects of B excess on CO_2 assimilation (Kamali and Childers 1967; Lovatt and Bates 1984; Sotiropoulos et al. 2002; Papadakis et al. 2004a; Han et al. 2009) and carbohydrate metabolism (Papadakis et al. 2004b; Cervilla et al. 2007) are scarce. Considering that in the eucalypts there is a certain sensitivity to B excess (Marcar et al. 1999; Poss et al. 1999) and the available information concerning the effects of B on photosynthesis, leaf anatomy and growth is scarce, we carried out this experiment in order to bridge some gaps.

2 Materials and Methods

2.1 *Plant Material, Growth Conditions and Treatments*

Uniform sized 1-year old *Eucaliptus globulus* were randomly assigned to 20-l pots (Ø 24 cm) filled with sand-vermiculite substrate (1:1, by vol.). Experiments were carried out in a greenhouse with natural daylight. During May-July, the minimum air temperature was 18 °C (night) and maximum 34 °C (day). Plants were irrigated with a modified Hoagland's solution (Hoagland and Arnon 1950). Three B concentrations were applied: 0.1 (control), 1 (B1) and 10 mg l^{-1} (B10, to induce B toxicity) as H_3BO_3. Each treatment solution was delivered to designated pots every 20 days (*ca.* 500 ml per pot). Daily irrigations were sufficiently frequent to avoid water stress.

2.2 *Gas Exchange and Chlorophyll* a *Analysis*

Measurements of leaf gas exchanges were carried out in mature leaves by an infrared gas-analyzer (CIRAS-1 PP-Systems) equipped with a Parkinson leaf chamber that controlled leaf temperature (25 °C), relative humidity (80 %), light (800 μmol m^{-2} s^{-1} PAR) and CO_2 concentration (350 ppm). Photosynthetic activity at saturation light level (A_{max}), stomatal conductance to water vapour (G_w) and apparent internal CO_2 concentration (C_i) were calculated according to the equations described in Von Caemmerer and Farquhar (1981) and related to one-sided leaf areas.

Modulated chlorophyll *a* fluorescence measurements were carried out with a PAM-2000 fluorometer (Walz) on dark-adapted leaves for 40 min using a dark leaf clip. Ground fluorescence, F_0 was determined using the measuring modulated light which was sufficiently low (<1 μmol m^{-2} s^{-1}) without inducing any significant variable fluorescence. The maximal fluorescence level, F_m, was determined by applying a saturating light pulse (0.8 s) at 8000 μmol m^{-2} s^{-1} in dark-adapted leaves; the variable fluorescence was calculated as $F_v = F_m - F_0$. The saturation pulse method was used for analysis of quenching components (qP and qN), as described by Schreiber et al. (1986).

Excitation pressure on PSII reflects the proportion of the primary stable quinone acceptor Q_A in the reduced state; it is calculated as $(1-qP)$. The actual quantum yield of PSII (Φ_{PSII}) was computed as $(F'_m-F_s)/F'_m$, where F_s is the steady-state fluorescence yield in the light-adapted state, as in Rohàček (2002). The apparent electron transport rate through PSII (ETR) was computed as $qP\times\Phi_{PSII}\times PFD\times 0.5\times0.84$ (Schreiber et al. 1986).

Chlorophylls were estimated non-destructively on intact parts of mature leaves with a SPAD meter (Minolta 502).

2.3 Boron Determination

After 12 weeks, plants were carefully removed from the pots and separated into leaves, stems and roots. All samples were washed with distilled water, then oven-dried at 60 °C for 4 days and weighed separately for dry mass determinations. The oven-dried samples were homogenised to a fine powder in a blender for subsequent analysis. About 300 mg of powder were mineralised with a 6:1 v:v mixture of ultrapure concentrated HNO_3 and H_2O_2 at 280 °C and a pressure of 0.55 MPa in a microwave digestion system (Milestone Ethos 900). Boron concentrations, expressed on a dry weight basis, were determined by inductively coupled plasma-mass spectrometry (ICP-MS, Perkin Elmer-Sciex Elan 6100). All analyses were carried out in triplicate in each of the three repeated experiments.

2.4 Statistical Analysis

Three repeated experiments were set up in a completely randomized design with seven replicate plants for each treatment. Data shown in tables and graphs represent the mean ± standard deviation. Analysis of variance (ANOVA) was applied in order to examine the effects of B treatment. Statistical analysis was conducted by using NCSS 2000 Statistical Analysis System software.

3 Results

Boron toxicity symptoms firstly appeared in leaves about 42 days after the beginning of the experiment only in B10 treatment. These symptoms occurred in the older leaves, as tip burn and marginal necrosis. However, it was observed that fresh (*data not shown*) and dry weight (DW) of these leaves were not significantly affected by B concentration in the nutrient solution (Fig. 1), while the roots growth decreased (−30 % compared to control). A significant increase of shoots DW (+21 %) was observed in plant treated with B1. Moreover, we observed that, the area between the

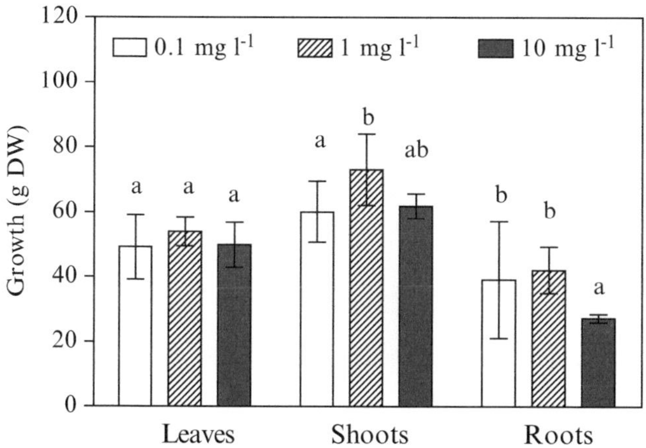

Fig. 1 Growth of leaves, shoots and roots of *Eucaliptus globulus* treated with 0.1 (control), 1 and 10 mg l^{-1} B. Values are means ± standard deviation. For each organ, different letters indicate values which statistically differ for $P \leq 0.05$

necrotic lesions remained green and apparently healthy: after 12 weeks of treatment total leaf chlorophyll ($a+b$) content was not changed by B supply (*data not shown*).

Boron concentration of all plant parts significantly increased as B concentrations in nutrient solution became higher, showing difference compared to control (Fig. 2). Higher B concentrations were found in leaves, while other organs had much lower contents, following the order leaves > stems > roots (Fig. 2). In particular, in the leaves, B concentration ranged in average between 29.4 (control) and 1624.4 mg kg^{-1} (B10).

After 12 weeks of treatment, gas exchange parameters are reported in Table 1. A_{max} of controls was higher than those of plants grown under B excess (−30 and −71 % in B1 and B10 plants, respectively). B stress induced a significantly decrease in G_w (−44 and −30 % in B1 and B10 plants, respectively) compared to control, as well as C_i that was reduced when B concentration became higher (−20 and −24 % in B1 and B10 plants, respectively).

The chlorophyll fluorescence parameters, F_v/F_m (which indicate the efficiency of excitation capture of PSII in the dark-adapted leaf) and F_v/F_0, significantly changed at the end of the experiment in leaves treated with the highest excess of B. This reduction corresponded to 4 % (F_v/F_m) and 19 % (F_v/F_0) (Table 2) and was essentially due to a concomitant decrease of F_0 and F_m. A significant increase of qN was observed in the B10 plants (+10 % compared to controls), while Φ_{PSII} decreased (−27 % when compared to controls). The reduction state of the primary stable quinone acceptor of PSII (Q_A) can be estimated as 1−qP: in leaves exposed to B10, the values increased (+11 % compared to control) (Table 2). A significant B-induced effect on ETR, whose decrease may be due to photoinhibition, was observed in B10 plants. The linear correlation between ETR and Ci showed positive slopes ($y=5.92x-930$, $R2=0.78$, $P=0.019$). Moreover, the positive association between

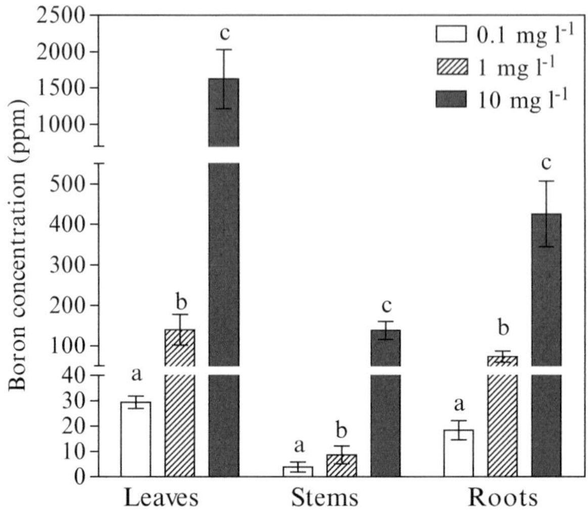

Fig. 2 Boron concentration in leaves, shoots and roots of *Eucaliptus globulus* treated with 0.1 (control), 1 and 10 mg l^{-1} B. Values are means ± standard deviation. For each organ, different letters indicate values which statistically differ for $P \leq 0.05$

Table 1 Effects of boron concentration in the nutrient solution on photosynthetic activity at saturation light level (A_{max}), stomatal conductance to water vapour (G_w) and apparent internal CO_2 concentration (C_i) of *Eucaliptus globulus*

	A_{max}	G_w	C_i
B (mg l^{-1})	(μmol CO_2 m^{-2} s^{-1})	(mmol H_2O m^{-2} s^{-1})	(ppm)
0.1	11.6±0.72 a	256±6.6 a	213±18.8 a
1	8.2±1.71 b	143±11.7 b	171±5.7 b
10	3.4±0.30 c	179±47.8 b	161±14.1 b

Data (means ± standard deviation) were captured after 42 weeks. For each parameter, different letters indicate values which statistically differ for $P \leq 0.05$

ETR and A_{max} suggested that B induced effects were well established ($y=9.81x+28.67$, $R^2=0.96$; $P=0.0004$).

Generally, at B1 concentration the parameters of chlorophyll *a* fluorescence did not change, with exception of $1-qP$, Φ_{PSII} and ETR values (−35, +24 and +51 %, respectively, in comparison to controls).

4 Discussion

Critical values for B toxicity have been established in many crops and trees showing a lot of difference between species. This is extremely true for accumulation of B in leaves: these tissues normally accumulate about from 40 to 100 mg B kg^{-1}

Table 2 Effects of boron concentration in the nutrient solution on chlorophyll *a* fluorescence parameters (arbitrary units; mean ± standard deviation) of *Eucaliptus globulus* leaves

B (mg l^{-1})	F_0	F_m	F_v/F_m	F_v/F_0	1-qP	qN	Φ_{PSII}	ETR
0.1	0.131 ± 0.015a	0.799 ± 0.0057a	0.835 ± 0.0020b	5.086 ± 0.114b	0.369 ± 0.0047b	0.653 ± 0.0042a	0.404 ± 0.0073b	85 ± 0.011b
1	0.123 ± 0.021a	0.791 ± 0.0015a	0.842 ± 0.0015b	5.451 ± 0.099b	0.238 ± 0.0020c	0.662 ± 0.0025a	0.503 ± 0.0060c	128 ± 0.004c
10	0.171 ± 0.030b	0.872 ± 0.0318b	0.804 ± 0.0095a	4.097 ± 0.127a	0.408 ± 0.0147a	0.715 ± 0.0133b	0.295 ± 0.0050a	58 ± 0.024a

For each treatment, different letters indicate values which statistically differ for $P \leq 0.05$. Abbreviations: F_0 minimal fluorescence, F_m maximal fluorescence, F_v/F_m variable and maximal fluorescence ratio, *1-qP* reduction state of Q_A, *qN* total nonphotochemical quenching, Φ_{PSII} actual quantum yield of PSII, *ETR* apparent electron transport rate through PSII

DW. However, the leaves can contain 250 mg kg^{-1} DW, when B in the soil approaches toxic levels, increasing up to 700–1000 mg kg^{-1} DW in extreme condition of B toxicity (Nable et al. 1997). In our study, injury became evident after 42 days of treatment in B10 plants and continued to the end of exposure (12 weeks), when leaf B concentration exceeded 1624 ± 407 mg kg^{-1} DW. This value is in accordance with those (1033 ± 828 mg kg^{-1} DW) found in *Eucalyptus* leaf tissue sampled in San Joaquin Valley of California when plants showed B incipient injury (Poss et al. 1999).

It was observed that B concentrations of all plant parts increased, by increasing B concentration in the nutrient solution. This observation is in accordance with those reported by other Authors studying *Eucaliptus* species (Poss et al. 1999). Much higher B concentrations were found in the leaves than in the other vegetative parts: in B10 treatment, the leaves contained up to four times B than roots. These data are in agreement with those reported for other species, where B was accumulated in leaves and low concentrations was found in woody stems and roots (Papadakis et al. 2004a). As already reported by Eaton (1944), these results suggest that B was transported to the leaves via the transpiration stream and the remobilization of B in phloem from leaves to the other organs was limited. Generally, phloem immobility could be considered as an internal tolerance mechanism to B excess. Moreover, we observed that good vegetative growth may continue in young leaves suggesting that the old ones are able to maintain enough photosynthetic leaf area explaining as our data of chlorophyll content resulted unchanged. In an affected leaf, phloem immobility keeps B away from metabolic sites, retaining it in the leaf margins, where despite suffering leaf burn, plants are still able to maintain enough healthy photosynthetic leaf area. The adequate photosynthetic area as well as the termination of the experiment before leaf abscission due to B toxicity, might probably explain because the total fresh and dry weight of leaves was not significantly affected by B supply (Papadakis et al. 2004a).

Although the B concentration in roots was lower compared to those found in leaves, significant reduction of growth of this organ was observed. Few studies confirm our results, explaining that the effect of B toxicity in roots is associated with abnormal cell division at the meristem level (Cervilla et al. 2009) and might be a result of the formation of hypodermis and the progressive deposition of suberin in cortical cell walls (Ghanati et al. 2005).

Researches about photosynthetic gas exchange responses under B toxicity give very different results. Sotiropoulos et al. (2002) found that B toxicity in kiwifruit induced a significant decrease of the photosynthetic rate and a significant increase of the intercellular CO_2 concentration, whereas stomatal conductance remained unaffected. Papadakis et al. (2004a) in orange plants showed that intercellular CO_2 concentration was not significantly affected by the increased B concentration in the nutrient solution, while at the same time both photosynthetic activity and stomatal conductance significantly decreased. Leaf stomatal resistance indicates the degree of stress in plants under adverse conditions and stomatal closure by reducing evaporation might play a restrictive role on the uptake of excessive B. Boron toxicity also may have damaged the ability of the stomata to open (Gunes et al. 1996). In the present study, the measured G_w showed the ability of *Eucalyptus* to close their stomata and to increase stomatal resistance under B excess. In Papadakis et al. (2003), higher stomatal resistance in B-tolerant *Citrus* genotype than that of B-sensitive *Citrus* genotype was reported.

Since PSII is believed to play an important role in the response of photosynthesis in higher plants under environmental stresses, the reduction of CO_2 assimilation by B excess should be reflected in the PSII behaviour. Our experiments have shown F_v/F_m decreased by increasing B concentration in the nutrient solution, as also observed by Papadakis et al. (2004a) in orange plants and by Guidi et al. (2009) in tomato leaves. Under non-stressed condition, C3-species had a theoretical value of F_v/F_m equal to 0.832 (Björkman and Demming 1987); this is true also for *E. globulus* and it is in agreement with other Authors (Rohàček 2002; Lee 2006). Its reduction means that treated plants were under stress conditions at the end of the experiments: molecular O_2 operates as an alternative acceptor for non-utilized electrons and light energy and, consequently, leads to generation of ROS.

The decrease of the F_v/F_0 is closely related to the structural damage of the thylakoid membranes that affect the photosynthetic transport of electrons (one of the probable reasons for the reduction of photosynthesis). Also the significant increase of F_0 observed suggests that this parameter is affected by environmental stresses that cause structural alterations in the pigment protein complexes of PSII or when the transfer from antennae to reaction centres is impeded (Bohlar-Nordenkampf et al. 1989). In addition, the increase in $1-qP$ and the decrease in F_v/F_m are known to be closely associated with photoinhibition (Ogren and Rosenqvist 1992). The higher level of $1-qP$ found in B10 concentration indicates that there was a greater excitation pressure on PSII centres and also suggests that a large proportion of PSII reaction centres are closed in severe B stressed leaves. The apparent electron transport rate through PSII was, in fact, reduced as effect of the photoinhibition.

Since Φ_{PSII} was significantly reduced, thus PSII reaction centres are unable to efficiently utilize the excitation energy which is dissipated as heat (Demmig-Adams et al. 1996). Highest B level induces an increase of photoprotective mechanism, qN showing higher values compared with controls. This indicates that thermal energy dissipation was activated trying to play a significant role in protecting plants from B excess.

These results clearly indicate that the B excess (10 mg l^{-1}) in *E. globulus*, even if in the presence of decrease of stomatal contuctance (and, thus, with reduced evaporation), leads to: (i) visible injury in old leaves; (ii) growth reduced in roots; (iii) increase in B concentration in all parts of plants following the order leaves > stems > roots; (iv) fall of photosynthetic activity, because of structural damage of the thylakoid membranes. Under these circumstances, *E. globulus* should be regarded as sensitive to B toxicity.

References

Alpaslan M, Gunes A (2001) Interactive effects of boron and salinity stress on the growth, membrane permeability and mineral composition of tomato and cucumber plants. Plant Soil 236:123–128

Aucejo A, Ferrer J, Gabaldon C, Marzal P, Seco A (1997) Diagnosis of boron, fluorine, lead, nickel and zinc toxicity in citrus plantations in Villarreal, Spain. Water Air Soil Pollut 94:349–360

Björkman PH, Demming B (1987) Photon yield of O_2 evolution and chlorophyll fluorescence characteristics among vascular plants of diverse origins. Planta 170:489–504

Blevins DG, Lukaszewski KM (1998) Boron in plant structure and function. Annu Rev Plant Physiol Plant Mol Biol 49:481–500

Bohlar-Nordenkampf HR, Long SP, Baker NR, Öquist G, Schreiber U, Lechner EG (1989) Chlorophyll fluorescence as a probe of the photosynthetic competence of leaves in the field: a review of current instrumentation. Funct Ecol 3:487–514

Bolaños I, Lukaszewski K, Bonilla I, Blevins D (2004) Why boron? Plant Physiol Biochem 42:907–912

Brown PH, Shelp BJ (1997) Boron mobility in plants. Plant Soil 193:83–101

Cervilla LM, Blasco B, Rios JJ, Romero L, Ruiz JM (2007) Oxidative stress and antioxidants in tomato (*Solanum lycopersicum*) plants subjected to boron toxicity. Ann Bot 100:747–756

Cervilla LM, Rosales MA, Rubio-Wilhelmi MM, Sànchez-Rodrìguez E, Blasco B, Rìos JJ, Romero L, Ruiz JM (2009) Involvement of lignification and membrane permeability in the tomato root response to boron toxicity. Plant Sci 176:545–552

Demmig-Adams B, Adams WW III, Barker DH, Logan BA, Bowling DR, Verhoeven AS (1996) Using chlorophyll fluorescence to assess the fraction of absorbed light allocated to thermal dissipation of excess excitation. Physiol Plant 98:253–264

Eaton FM (1944) Deficiency, toxicity and accumulation of boron in plants. J Agric Res 69:237–242

Ghanati F, Morita A, Yokota H (2002) Induction of suberin and increase of lignin content by excess boron in tobacco cells. Soil Sci Plant Nutr 48:357–364

Ghanati F, Morita A, Yokota H (2005) Deposition of suberin in roots of soybean induced by excess boron. Plant Sci 168:397–405

Guidi L, Degl'Innocenti E, Pardossi A, Carmassi G, Massa D (2009) Effects of boron and salinity stress on chlorophyll fluorescence parameters of tomato leaves. In: Book of abstracts of 2nd plant ecophysiology workshop "Chlorophyll fluorescence: from theory to (good) practice". Pisa, 25–26 May 2009, p 12

Gunes A, Inal A, Alpaslan A (1996) Effect of salinity on stomatal resistance, proline and mineral composition of pepper. J Plant Nutr 19:389–396

Gunes A, Soylemezoglu G, Inal A, Bagci EG, Coban S, Sahin O (2006) Antioxidant and stomatal responses of gravepine (*Vitis vinifera* L.) to boron toxicity. Sci Hortic 110:279–284

Han S, Tang N, Jiang HX, Yang LT, Li Y, Chen LS (2009) CO_2 assimilation, photosystem II photochemistry, carbohydrate metabolism and antioxidant system of citrus leaves in response to boron stress. Plant Sci 176:143–153

Hoagland DR, Arnon DI (1950) The water culture method for growing plants without soil. Calif Agric Exp Stat Circ 374:1–32

Kamali A, Childers N (1967) Effect of boron nutrition on peach anatomy. J Am Soc Hortic Sci 90:33–38

Lee SKD (2006) Hot pepper response to interactive effects of salinity and boron. Plant Soil Environ 52:227–233

Lovatt CJ, Bates LM (1984) Early effects of excess boron on photosynthesis and growth of *Cucurbita pepo*. J Exp Bot 35:297–305

Marcar NE, Guo J, Crawford DF (1999) Response of *Eucaliptus camaldulensis* Dhnh., E. *globulus* Labill. ssp. *globulus* and *E. grandis* W.hill to excess boron and sodium chloride. Plant Soil 208:251–257

Marschner H (1995) Mineral nutrition of higher plants. Academic, London
Matoh T (1997) Boron in plant cell walls. Plant Soil 193:59–70
Molassiotis A, Sotiropoulos T, Tanou G, Diamantidis G, Therious I (2006) Boron-induced oxidative damage and antioxidant and nucleolytic responses in shoot tip cultures of apple rootstock EM 9 (*Malus domestica* Borkh). Environ Exp Bot 56:54–62
Nable RO, Banuelos GS, Paull JG (1997) Boron toxicity. Plant Soil 193:181–198
Ogren E, Rosenqvist E (1992) On the significance of photoinhibition of photosynthesis in the field and its generality among species. Photosynth Res 33:63–71
Ozturk M, Sakcali S, Gucel S, Tombuloğlu H (2010) Boron and plants. In: Ashraf M et al (eds) Plant adaptation and phytoremediation, Part 2. Springer, the Netherlands, pp 275–311
Papadakis IE, Dimassi KN, Thcrios IN (2003) Response of two citrus genotypes to six boron concentration: concentration and distribution of nutrients, total absorption, and nutrient use efficiency. Aust J Agric Res 54:571–580
Papadakis IE, Dimassi KN, Bosabalidis AM, Therios IN, Patakas A, Giannakoula A (2004a) Effect of B excess on some physiological and anatomical parameters of "Navelina" orange plants grafted on two rootstocks. Environ Exp Bot 51:247–257
Papadakis IE, Dimassi KN, Bosabalidis AM, Therios IN, Patakas A, Giannakoula A (2004b) Boron toxicity in "Clementine" mandarin plants grafted on two rootstocks. Plant Sci 166:539–547
Paull JG, Nable RO, Rathjen AJ (1992) Physiological and genetic control of the tolerance of wheat to high contractions of boron and implications for plant breeding. Plant Soil 146:251–260
Poss JA, Grattan SR, Grieve CM, Shannon MC (1999) Characterization of leaf boron injury in salt-stressed *Eucalyptus* by image analysis. Plant Soil 206:237–245
Rohàček K (2002) Chlorophyll fluorescence parameters: the definitions, photosynthetic meaning, and mutual relationships. Photosynthetica 40:13–29
Sage RF, Ustin SL, Manning SJ (1989) Boron toxicity in the rare serpentine plant, *Streptanthus morrisonii*. Environ Pollut 61:77–93
Schreiber U, Schliwa U, Bilger W (1986) Continuous recording of photochemical and non-photochemical quenching with a new type of modulation fluorometer. Photosynth Res 10:51–62
Sotiropoulos TE, Therios TN, Dimassi KN, Bosabalidis A, Kofidis G (2002) Nutritional status, growth, CO_2 assimilation and leaf anatomical responses in two kiwifruit species under boron toxicity. J Plant Nutr 25:1249–1261
von Caemmerer S, Farquhar GD (1981) Some relationships between the biochemistry of photosynthesis and the gas exchange of leaves. Planta 153:376–387
Yau SK, Ryan J (2008) Boron toxicity tolerance in crops: a viable alterative to soil amelioration. Crop Sci 48:854–865

Vanadium in the Environment and Its Bioremediation

Tatsuya Ueki

Contents

Abstract Vanadium is an element with symbol V and atomic number 23. The vast majority of vanadium demand is from the steel industry, and the rest for titanium alloy and catalyst in chemical factory. Air pollution and water pollution by vanadium were recognized from early twentieth century. Increasing information on the toxicity and medicinal use enhanced the development of bioremediation of vanadium. In this chapter, the author would like to overview the history of pollution of vanadium, vanadium toxicity, bioaccumulation and bioremediation of vanadium.

Keywords Heavy metal • Bioremediation • Vanadium • Ascidians

T. Ueki (✉)
Molecular Physiology Laboratory, Graduate School of Science, Hiroshima University, 1-3-1 Kagamiyama, Higashi-Hiroshima, Hiroshima 739-8526, Japan

Marine Biological Laboratory, Graduate School of Science, Hiroshima University, 2445 Mukaishima, Onomichi, Hiroshima 722-0073, Japan
e-mail: ueki@hiroshima-u.ac.jp

M. Öztürk et al. (eds.), *Plants, Pollutants and Remediation*,
DOI 10.1007/978-94-017-7194-8_2

1 Introduction

Vanadium is an element with symbol V and atomic number 23. It is the 19th most abundant element in the earth's crust (0.015–0.016 %, 150–160 ppm) (Emsley 1998; WHO 2000). Metallic vanadium is not found in nature, but its compounds can be obtained as minerals such as vanadinite ($Pb_5(VO_4)_3Cl$) (Fig. 1), a lead vanadate ore from which vanadium was first discovered by a Mexican, Andrés Manuel del Río. In 1831, Nils Gabriel Sefström rediscovered this element and he called the element vanadium after Vanadis, an additional name of the Norse goddess Freyja, which represented beauty and fertility, because of beautifully colored chemical compounds of this element (Sefström 1831). Mine production including slag products increased year by year up to 75,000 tons in the world, about half of which is produced in China, followed by South Africa and Russia (Brown et al. 2014).

The vast majority (92 %) of vanadium demand is from the steel industry (Parles 2012). Vanadium is mainly used to produce high speed and high alloy tool steels. Vanadium is also used in the production of titanium alloys for aerospace and industrial purposes. Titanium alloys account for about 4 % of consumption in 2012 (Parles 2012). Vanadium pentoxide is used as a catalyst in sulfuric acid production and in the manufacture of ceramics. About 3 % of global vanadium consumption is in petrochemical, catalyst and pollution control applications as well as ceramic pigments, special glasses and other chemical industry applications.

In 2012, about 1 % of vanadium consumed was used in energy storage applications. Vanadium redox flow battery (Rychcik and Skyllas-Kazacos 1988) systems for grid energy storage applications and lithium battery systems incorporating vanadium for mobility applications are under development today with potential to have a significant impact on future vanadium demand (Parles 2012).

Fig. 1 Vanadinite, Mibladen Atlas Mountain, Morocco. *Dark orange color*

In this chapter, the author would like to overview the history of studies on pollution, toxicity, bioaccumulation and bioremediation of vanadium.

The readers may refer to a recent book on biological and biochemical aspects on vanadium edited by Dr. Michibata (2012). Bioinorganic and chemical topics can be found in a book by Dr. Rehder (2008).

2 Pollution of Vanadium

From early twentieth century, vanadium is regarded as a pollutant. Dutton was the first to describe vanadium poisoning, and produced a word "vanadiumism", which means a chronic intoxication caused by ingestion or absorption of some forms of vanadium, either industrially, medicinally, or accidentally (Dutton 1911). In his recognition, anemia is an early symptom, and the cough is a prominent and characteristic one. He also noted that some workers using vanadium are susceptible to tuberculosis. Anorexia, nausea and diarrhea indicated gastrointestinal involvement.

2.1 Air Pollution

Four principal oxides are known for vanadium: vanadium monooxide (VO), vanadium trioxide (V_2O_3), vanadium dioxide (VO_2) and vanadium pentoxide (V_2O_5), which ranges +2 to +5 oxidation states. Vanadium pentoxide dust is known to be one of hard metal irritants that affect the upper respiratory tract, producing tracheitis, bronchitis, pneumonia and pulmonary oedema (WHO 2014).

Experimental poisoning in animals indicated that accumulation does not occur and that acute and chronic symptoms are similar (Daniel and Lillie 1938). Studies in early 1900s on experimental administration of vanadium on animal models are well summarized in a review by Wyers (1946).

Stocks reported the relationship between atmospheric pollution in urban area and cancer, bronchitis and pneumonia (Stocks 1960). He especially noted the correlation between trace elements and lung cancer. Vanadium's action as respiratory irritant is significant.

Recent research on pollution of vanadium mainly focuses on the global movement of small particles. The United States of America and the European Union determined their own environmental baseline in 1971 and 1980, respectively, for PM10 and PM2.5. WHO first determined a guideline in Europe, and then extended it in 2005 as a global guideline (WHO 2005). In Japan, original guideline was first released in 1972, and the baseline for PM2.5 was determined in 2009.

Since vanadium is the major trace metal in fossil fuels (Filby and Branthaver 1987; Jacks 1976; Sundararaman et al. 1988), combustion of these materials provides an appreciable source of vanadium in the environment and can be a source for this heavy metal in particular materials in the air (Chen and Duce 1983; Duce and

Hoffman 1976; Weisel et al. 1984). Crude oil contains vanadium as high as 1580 ppm, and it varies depending on the source (Barwise 1990).

2.2 *Water Pollution*

Vanadium can either be dissolved in water as ions or may become adsorbed to particulate matter. The concentration of vanadium in water is largely dependent on geographical location and ranges from 0.2 to more than 100 ppb in freshwater, and from 0.2 to 29 ppb in seawater (WHO 2000). Typical average value of vanadium is recognized as 1.8 ppb (35 nM) (Cole et al. 1983; Collier 1984). Concentrations of vanadium in drinking water may range from about 0.2 to more than 100 μg/L (Nordberg et al. 2011). The concentration of vanadium in drinking-water depends significantly on geographical location.

World health organization (WHO) formulated no guidelines for vanadium in drinking water. US Environmental Protection Agency (EPA) did not formulate the limit, but designated vanadium as hazardous substances. The ministry of Health, Labor and Welfare, Japan, also does not determine the limit for vanadium in tap water, although those for other trace metal elements such as Fe, Zn, Al, Pb, Cd, Hg, Se, and Cr are determined.

A lot of toxicological studies on aquatic animals can be found in literatures for assessment of both acute and chronic toxicity on freshwater and marine fishes (e.g., Knudtson 1979; Perez-Benito 2006; Stendahl and Sprague 1982). A study using rainbow trout suggested that hardness did not exert a major effect, and the authors supposed that it was because vanadium is present in water as various anions (Stendahl and Sprague 1982).

3 Toxicity of Vanadium

The toxicity of vanadium, as vanadate anions, have been published from early twentieth century. Studies on rodent and avian models precedes the studies on aquatic animals, as mentioned in the previous section.

Moxon et al. published several papers on the toxicity of oxy anions including vanadate on rats (Franke and Moxon 1936, 1937; Moxon and DuBois 1939). It was revealed that arsenic and molybdenum were slightly toxic, tellurium and vanadium were moderately toxic, and selenium was very toxic as they were compared at the 50-ppm level.

Chicks were also used as test animals for vanadium toxicity by adding vanadium to the diets (Berg 1963, 1966; Hathcock et al. 1964; Nelson et al. 1962; Romoser et al. 1961). Nelson et al. showed that diets containing less than 20 ppm of vanadium were safe for young chicks (Nelson et al. 1962). Hathcock et al. examined the toxicity of vanadium with a diet added by 25-ppm vanadium for 2-weeks, which caused a significant decrease in growth rate and 90 % death in chicks (Hathcock et al. 1964).

Acute toxicity of vanadium compounds, both +5 and +4 oxidation states ($NaVO_3$ and $VOSO_4$), were examined by oral or intraperitoneal administration for rats and mice (Llobet and Domingo 1984), and LD_{50} (up to 14 days) were determined. The dose of vanadium was 39–845 mg/kg body weight. As a result, LD_{50} for V^{5+} was 2.2–3.0 times lower than that for V^{4+} after oral administration, as well for intraperitoneal administration where the factor was 1.2–1.9 times. Reproductive toxicity of vanadium was also examined in mice and rats (Elbetieha and Al-Hamood 1997; Jain et al. 2007; Llobet et al. 1993; Morgan and El-Tawil 2003) by using several different salts and compounds.

Later, more detailed studies using cell culture were conducted. Cytotoxic effects of vanadium on rabbit alveolar macrophages (RAM) was assessed *in vitro* with exposure to particulate forms of vanadium oxides in +5 or +4 oxidation state (V_2O_5, V_2O_3 and VO_2) (Waters et al. 1974). Cell viabilities after 20-h exposure were reduced to 50 % by 13–33 ppm vanadium, depending on chemical species.

Toxicity of vanadium is related to the production of reactive oxygen species (ROS) that cause several damages on nucleic acids, proteins and lipids. Exposure to air pollution particles also cause such damages (Kadiiska et al. 1997).

4 Bioaccumulation of Vanadium

Humans usually consume 10–60 μg of vanadium through foods daily. The mean vanadium concentration in the diet was reported to be 32 μg/kg and the mean daily intake was estimated to be 20 μg/day (WHO 2000). The human body is estimated to contain 50–200 μg of vanadium. In each organ, vanadium is present at very low concentrations (Underwood 2012). High levels of vanadium are found in marine organisms, such as ascidians and fan worms. On dry weight base, the vanadium level in a genus *Ascidia* reaches 4,000–20,000 ppm dry weight (Michibata et al. 1986). The fly agaric mushroom (*Amanita muscaria*) also contain relatively high levels of vanadium (120 ppm dry weight) (Michibata 2012). Comprehensive survey of vanadium levels in marine organisms suggested that around 20 ppm dry weight were found in sea weeds (Fukushima et al. 2009).

Approximately 100 years ago, the German physiological chemist Dr. Martin Henze discovered high levels of vanadium in the blood (coelomic) cells of the ascidian *Phallusia mammillata* collected from the Bay of Naples, Italy (Henze 1911). His discovery attracted the inter disciplinary attention of chemists, physiologists, and biochemists.

The greatest concentration was found in blood cells of the ascidian *Ascidia gemmata*, at up to 350 mM (Michibata et al. 1986, 1991), which is 10^7 times that in seawater (35 nM) (Cole et al. 1983; Collier 1984); this is believed to be the highest degree of accumulation of a metal in any living organism. Vanadium ions are mostly accumulated in the vacuole of signet ring cells, which are a type of blood (coelomic) cell and called "vanadocytes" (Michibata et al. 1987; Ueki et al. 2002).

Fig. 2 Amino acid sequences of the five Vanabins from *Ascidia sydneiensis samea* and the two from *Ascidia gemmata*. Conserved amino acid residues are *boxed*, and the 18 cysteines in the core region are *numbered*. Positively and negatively charged amino acids are shaded in *gray* (Reproduced from Samino et al. 2012)

Ongoing research during the last two decades has identified many proteins involved in the process of accumulating and reducing vanadium in vanadocytes, blood plasma, and the digestive tract of ascidians. Among the proteins identified so far, the vanadium-binding proteins (Vanabins) are most prominent.

Vanabins were first purified from blood cells of *Ascidia sydneiensis samea*, which contained 12.8 mM vanadium in the blood cells, by anion-exchange column chromatography (Kanda et al. 1997). The related proteins and genes were identified by ion exchange chromatography, metal ion affinity chromatography and a expressed sequence tag (EST) analyses from the same species (Ueki et al. 2003a; Yoshihara et al. 2005, 2008; Yamaguchi et al. 2004). In this species, the Vanabin family consists of at least five closely related proteins, Vanabins1–4 and VanabinP. All five Vanabins possess 18 cysteine residues, and the intervals between cysteines are well-conserved (Fig. 2).

A homology search of public DNA and protein databases, using both Vanabin1 and Vanabin2 amino acid sequences, revealed no proteins with striking similarities, other than those from two ascidian species, *Ciona intestinalis* and *A. gemmata*. We identified five Vanabins (CiVanabin1 to CiVanabin5) from *C. intestinalis* (Trivedi et al. 2003) and two Vanabins (AgVanabin1 and AgVanabin2) in *A. gemmata* (Fig. 2) (Samino et al. 2012). Thus, Vanabins appear to be ubiquitous among the vanadium-accumulating ascidians and may hold the key to resolving the mechanism underlying the highly selective and extremely high-level accumulation of vanadium ions.

More detailed review of the molecular mechanism of vanadium accumulation in ascidians can be found in publications from our research group (Michibata and Ueki 2010; Michibata et al. 2003, 2007; Ueki and Michibata 2011; Ueki et al. 2014).

The accumulation of vanadium is also revealed in the fan worms *Pseudopotamilla occelata* (Ishii et al. 1993) and *Perkinsiana littoralis* (Fattorini et al. 2010). In these fan worms, the concentration of vanadium is as high as 60 mM. Fan worms belong to the phylum Polychaeta, which is phylogenetically distant from ascidians (Chordata). Unlike the chordates, in fan worms, the highest level of vanadium is found not in blood (coelomic) cells but in the epithelial cells of the branchial crown.

5 Bioremediation of Vanadium

The decontamination of soil and water containing heavy metals from industrial activity is a troublesome problem. Natural or synthetic organic materials are useful to absorb heavy metals. Bioremediation strategies, using microorganisms or plants with metal-binding ability, have been proposed as attractive methods, because these are effective at low metal concentrations and are less expensive and more efficient than physicochemical methods of removing heavy metals.

5.1 *Organic Materials*

Efforts have been made from 1970s in order to process industrial waste waters by activated sludge. An early study succeeded in absorbing vanadium from a solution at the concentration of 30–40 mg/L, but it was not very efficient (Kunz et al. 1976). Metal sludge was also tested for removal of vanadium but, as compared with other heavy metal ions, vanadium removal efficiency was low (Namasivayam and Sangeetha 2007). One reason could be the behavior of vanadium in ambient environment as an anion (protonated forms of VO_4^{3-}) (Fig. 3) (Crans et al. 2004; Ueki et al. 2014).

Chitosan is very efficient at removing vanadium from dilute solutions (Guzman et al. 2002; Jansson-Charrier et al. 1996; Niu and Volesky 2003). Anionic metal complexes such anions as VO_4^{3-}, CrO_4^{2-}, SeO_4^{2-} are very effectively bound by biomass types like chitosan that contains abundant amine groups (Niu and Volesky 2003). In contrast, cationic form of vanadium (VO^{2+}) is also absorbed by chitosan (Jansson-Charrier et al. 1996). Adsorption of other cationic heavy metal ions such as Fe^{3+}, Cu^{2+} and Cd^{2+} using chitosan is also reported (Juang et al. 1999; Namdeo and Bajpai 2008; Prakash et al. 2012). Thus, chitosan is both effective for anions and cations.

By using plant materials, lead and vanadium were effeciently absorbed from a real industrial wastewater onto *Pinus sylvestris* sawdust (Kaczala et al. 2009). Removal of V^{3+} and Mo^{5+} from model wastewater using dried and re-hydrated biomass of a sea grass *Posidonia oceanica* is reported (Pennesi et al. 2013).

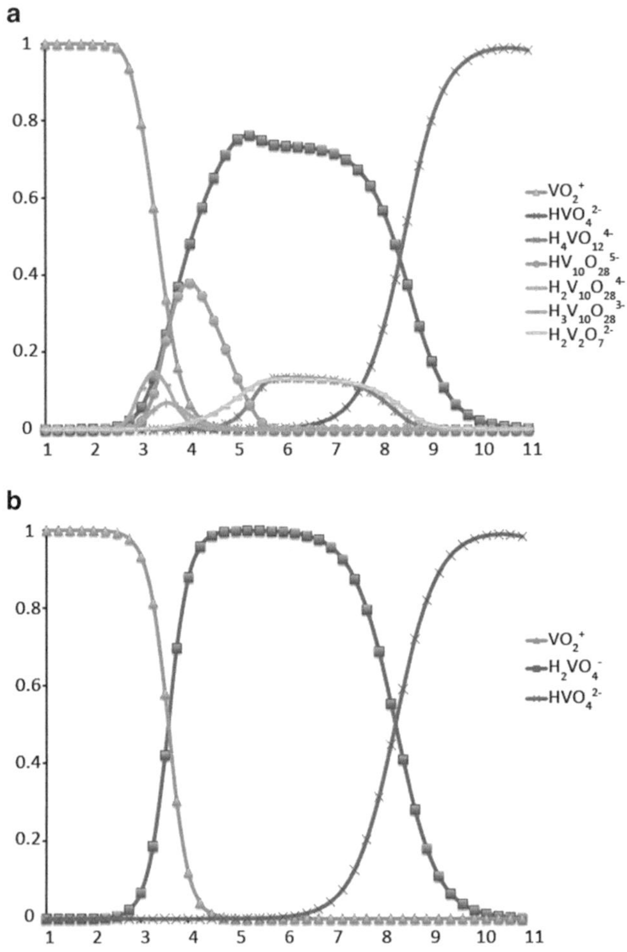

Fig. 3 Speciation diagram for aqueous vanadate solutions calculated by Visual MINTEQ ver. 3 based on MINTEQA2 (Allison et al. 1989). Data are normalized to total concentration and expressed as molar fraction x_V vs. pH. Ionic strength: 0.45 M. Vanadium concentration and species: 1 mM V^V (**a**) or 1 μM V^V (**b**). Temperature: 25 °C. Species that comprised less than 3 % are not shown. As compared with experimentally determined speciation diagrams (Rehder 2008), the ratios of polymeric vanadate species are relatively low (Modified from Ueki et al. 2014)

5.2 *Biotechnological Approaches*

To recover heavy metals, one possible approach is the biotechnological use of metal-binding peptides with the ability to bind heavy metals in various living organisms to improve the metal-binding abilities of microorganisms via heterologous expression.

Table 1 Bioremediation of heavy metals by *E. coli* cells expressing Vanabins from an ascidian *Ascidia sydneiensis samea*

	Vanabin1	Vanabin2	MBP	TB1
V(IV)	3.9	7.3	4.1	3.5
Cu(II)	876±215**	882±136**	87.5±22.4	43.2±20.9

Values are given as ng mg^{-1} dw.
***P*<0.005

Table 2 Bioremediation of heavy metals by *E. coli* cells expressing Ag Vanabins from an ascidian *Ascidia gemmata*

	AgVanabin1	AgVanabin2	MBP	TB1
Vanadium	6.25±0.48	10.12±0.95*	6.92±0.16	7.47±0.54
Iron	203,303±4,192	251,586±73,094	299,422±26,428	239,257±16,521
Copper	550.79±6.50	2360.91±462.05*	559.81±64.19	173.62±43.06
Cobalt	4.33±0.64	4.88±0.57	3.97±0.20	8.23±2.50
Nickel	25.04±1.08	26.64±1.53	25.12±1.97	20.46±1.16
Zinc	151.10±19.45	158.72±39.41	159.48±17.68	145.56±31.64

Values are given as ng mg^{-1} dw.
**P*<0.05

Many studies have focused on metallothioneins, which are small, cysteine-rich proteins that are widely distributed from prokaryotes to eukaryotes. When metallothioneins are expressed in the cytoplasm (He et al. 2014; Pazirandeh et al. 1995; Singh et al. 2008; Yoshida et al. 2002), periplasm (Mauro and Pazirandeh 2000; Pazirandeh et al. 1995, 1998) or outer membrane (Lin et al. 2010) of *Escherichia coli*, the cells remove heavy metal ions, such as Cd^{2+}, Hg^{2+}, Pb^{2+}, Cu^{2+} and As^{3+} from the culture media and accumulate them.

Phytochelatins(PCs) are also metal-binding cysteine-rich proteins found in plants and fungi, and heterologous expression of PC-synthase enhanced Cd^{2+}, Cu^{2+} and As^{3+} accumulation in bacteria (Sauge-Merle et al. 2003). Several studies have sought novel synthetic small peptides that enhance the bioaccumulation of specific metals (Kotrba et al. 1999; Mejáre et al. 1998; Samuelson et al. 2000).

It is well known that porin channels (exclusion size, 600 Da) exist on the outer membrane of gram-negative bacteria, including E. coli, and small molecules including heavy metal cations and anions can diffuse through this type of channel in a rather non-specific manner (Benz 1988; Benz et al. 1985; Nikaido and Rosenberg 1983).

In my research group, it was intended to express Vanabin genes in bacteria to construct bioremediation system for vanadium. First study was done by using two Vanabin genes from *Ascidia sydneiensis samea*. But unfortunately, *E. coli* cells expressing these Vanabins in the periplasm could not accumulate VO^{2+} significantly but absorption of Cu^{2+} was around 20-fold enhanced (Table 1) (Ueki et al. 2003b). A following study using two Vanabins from another ascidian species *A. gemmata* was performed. When AgVanabin2, was expressed in the periplasm of *E. coli*, absorption of both VO^{2+} and Cu^{2+} were enhanced significantly (Table 2) (Samino et al. 2012).

6 Conclusion and Future Prospective

From early twentieth century, vanadium is regarded as a pollutant, especially in air exhausted from industry. Waste water management for vanadium is also recognized. Bioremediation of vanadium is mainly intended to manage waste water, since natural water does not contain hazardous level of vanadium. Cost effective method is to use organic non-living materials such as chitosan. Biotechnological applications may provide much more specific method to remove vanadium, but it must need to improve both absorption activity and the cost-efficiency. Once these problems are solved, biotechnological methods may surpass the other technologies.

References

Allison JD, Brown DS, Novo-Gradac KJ (1989) MINTEQA2/PRODEFA2, A geochemical assessment model for environment system: version 3.0. US EPA, Office of Research and Development, Washington, DC, 20460

Barwise AJG (1990) Role of nickel and vanadium in petroleum classification. Energy Fuels 4:647–652

Benz R (1988) Structure and function of porins from gram-negative bacteria. Annu Rev Microbiol 42:359–393

Benz R, Schmid A, Hancock RE (1985) Ion selectivity of gram-negative bacterial porins. J Bacteriol 162:722–727

Berg LR (1963) Evidence of vanadium toxicity resulting from the use of certain commercial phosphorus supplements in chick rations. Poult Sci 42:766–769

Berg LR (1966) Effect of diet composition on vanadium toxicity for the chick. Poult Sci 45:1346–1352

Brown TJ, Idoine N, Raycraft ER, Shaw RA, Deady EA (2014) World mineral production 2008–12. British Geological Survey, pp 1–115

Chen L, Duce RA (1983) The sources of sulfate, vanadium and mineral matter in aerosol particles over Bermunda. Atmos Environ 17:2055–2064

Cole PC, Eckert J, Williams K (1983) The determination of dissolved and particulate vanadium in sea water by x-ray fluorescence spectrometry. Anal Chim Acta 153:61–67

Collier RW (1984) Particulate and dissolved vanadium in the North Pacific Ocean. Nature 309:441

Crans DC, Smee JJ, Gaidamauskas E, Yang L (2004) The chemistry and biochemistry of vanadium and the biological activities exerted by vanadium compounds. Chem Rev 104:849–902

Daniel EP, Lillie RD (1938) Experimental vanadium poisoning in the white rat. Pub Health Rep (1896–1970) 53:765–777

Duce RA, Hoffman GL (1976) Atmospheric vanadium transport to the ocean. Atmos Environ 10:989–996

Dutton WF (1911) Vanadiumism. J Am Med Assoc LVI:1648

Elbetieha A, Al-Hamood MH (1997) Long-term exposure of male and female mice to trivalent and hexavalent chromium compounds: effect on fertility. Toxicology 116:39–47

Emsley J (1998) The elements (Oxford chemistry guides). Oxford University Press, Oxford

Fattorini D, Notti A, Nigro M, Regoli F (2010) Hyperaccumulation of vanadium in the Antarctic polychaete *Perkinsiana littoralis* as a natural chemical defense against predation. Environ Sci Pollut Res 17:220–228

Filby RH, Branthaver JFE (1987) Metal complexes in fossil fuels: geochemistry, characterization, and processing. In: Presented at the ACS National meeting in New York. American Chemical Society, Washington, DC

Franke KW, Moxon AL (1936) A comparison of the minimum fatal doses of selenium, tellurium, arsenic and vanadium. J Pharmacol Exp Ther 58:454–459

Franke KW, Moxon AL (1937) The toxicity of orally ingested arsenic, selenium, tellurium, vanadium and molybdenum. J Pharmacol Exp Ther 61:89–102

Fukushima M, Suzuki H, Saito K, Chatt A (2009) Vanadium levels in marine organisms of Onagawa Bay in Japan. J Radioanal Nucl Chem 282:85–89

Guzman J, Saucedo I, Navarro R, Revilla J, Guibal E (2002) Vanadium interactions with chitosan: influence of polymer protonation and metal speciation. Langmuir 18:1567–1573

Hathcock JN, Hill CH, Matrone G (1964) Vanadium toxicity and distribution in chicks and rats. J Nutr 82:106–110

He Y, Ma W, Li Y, Liu J, Jing W, Wang L (2014) Expression of metallothionein of freshwater crab (*Sinopotamon henanense*) in *Escherichia coli* enhances tolerance and accumulation of zinc, copper and cadmium. Ecotoxicology 23:56–64

Henze M (1911) Untersuchungen über des Blut der Ascidien. I. Mitteilung. Die Vanadiumverbindung der Blutkörperochen. Hoppe- Seyler's. Z Physiol Chem 72:494–501

Ishii T, Nakai I, Numako C, Okoshi K (1993) Discovery of a new vanadium accumulator, the fan worm *Pseudopotamilla occelata*. Naturwissenschaften 80:268–270

Jacks G (1976) Vanadium in an area just outside Stockholm. Environ Pollut 11:289–295

Jain GC, Pareek H, Sharma S, Bhardwaj M, Khajja BS (2007) Reproductive toxicity of vanadyl sulphate in male rats. J Health Sci 53:137–141

Jansson-Charrier M, Guibal E, Roussy J, Delanghe B (1996) Vanadium (IV) sorption by chitosan: kinetics and equilibrium. Water Res 30:465–475

Juang RS, Wu FC, Tseng RL (1999) Adsorption removal of copper(II) using chitosan from simulated rinse solutions containing chelating agents. Water Res 30:2403–2409

Kaczala F, Marques M, Hogland W (2009) Lead and vanadium removal from a real industrial wastewater by gravitational settling/sedimentation and sorption onto *Pinus sylvestris* sawdust. Biores Technol 100:235–243

Kadiiska MB, Mason RP, Dreher KL, Costa DL, Ghio AJ (1997) *In vivo* evidence of free radical formation in the rat lung after exposure to an emission source air pollution particle. Chem Res Toxicol 10:1104–1108

Kanda T, Nose Y, Wuchiyama J, Uyama T, Moriyama Y, Michibata H (1997) Identification of a vanadium-associated protein from the vanadium-rich ascidian, *Ascidia sydneiensis samea*. Zool Sci 14:37–42

Knudtson BK (1979) Acute toxicity of vanadium to two species of freshwater fish. Bull Environ Contam Toxicol 23:95–99

Kotrba P, Dolecková L, de Lorenzo V, Ruml T (1999) Enhanced bioaccumulation of heavy metal ions by bacterial cells due to surface display of short metal binding peptides. Appl Environ Microbiol 65:1092–1098

Kunz RG, Giannelli JF, Stensel HD (1976) Vanadium removal from industrial wastewaters. J Water Pollut Control Fed 48:762–770

Lin K-H, Chien M-F, Hsieh J-L, Huang C-C (2010) Mercury resistance and accumulation in *Escherichia coli* with cell surface expression of fish metallothionein. Appl Microbiol Biotechnol 87:561–569

Llobet JM, Domingo JL (1984) Acute toxicity of vanadium compounds in rats and mice. Toxicol Lett 23:227–231

Llobet JM, Colomina MT, Sirvent JJ, Doningo JL (1993) Reproductive toxicity evaluation of vanadium in male mice. Toxicology 80:199–206

Mauro JM, Pazirandeh M (2000) Construction and expression of functional multi-domain polypeptides in *Escherichia coli*: expression of the *Neurospora crassa* metallothionein gene. Lett Appl Microbiol 30:161–166

Mejáre M, Ljung S, Bülow L (1998) Selection of cadmium specific hexapeptides and their expression as OmpA fusion proteins in *Escherichia coli*. Protein Eng 11:489–494

Michibata H (2012) Vanadium. Springer, Dordrecht

Michibata H, Ueki T (2010) Advances in research on the accumulation, redox behavior, and function of vanadium in ascidians. Biomol Concepts 1:97–107

Michibata H, Terada T, Anada N, Yamakawa K, Numakunai T (1986) The accumulation and distribution of vanadium, iron, and manganese in some solitary ascidians. Biol Bull 171:672–681

Michibata H, Hirata J, Uesaka M (1987) Separation of vanadocytes: determination and characterization of vanadium ion in the separated blood cells of the ascidian, *Ascidia ahodori*. J Exp Zool 244:33–38

Michibata H, Iwata Y, Hirata J (1991) Isolation of highly acidic and vanadium-containing blood cells from among several types of blood cell from Ascidiidae species by density-gradient centrifugation. J Exp Zool 257:306–313

Michibata H, Yamaguchi N, Uyama T, Ueki T (2003) Molecular biological approaches to the accumulation and reduction of vanadium by ascidians. Coord Chem Rev 237:41–51

Michibata H, Yoshinaga M, Yoshihara M, Kawakami N, Yamaguchi N, Ueki T (2007) Genes and proteins involved in vanadium accumulation by ascidians. In: ACS symposium series. American Chemical Society, Washington, DC, pp 264–280

Morgan AM, El-Tawil OS (2003) Effects of ammonium metavanadate on fertility and reproductive performance of adult male and female rats. Pharm Res 47:75–85

Moxon AL, DuBois KP (1939) The influence of arsenic and certain other elements on the toxicity of seleniferous grains three figures. J Nutr 18:447–457

Namasivayam C, Sangeetha D (2007) Removal of anions, heavy metals, organics and dyes from water by adsorption onto a new activated carbon from Jatropha Husk, an agro-industrial solid waste. Process Saf Environ Prot 85:181–184

Namdeo M, Bajpai SK (2008) Chitosan–magnetite nanocomposites (CMNs) as magnetic carrier particles for removal of Fe(III) from aqueous solutions. Colloid Surf A Physicochem Eng Asp 320:161–168

Nelson TS, Gillis MB, Peeler HT (1962) Studies of the effect of vanadium on chick growth. Poult Sci 41:519–522

Nikaido H, Rosenberg EY (1983) Porin channels in *Escherichia coli*: studies with liposomes reconstituted from purified proteins. J Bacteriol 153:241–252

Niu H, Volesky B (2003) Characteristics of anionic metal species biosorption with waste crab shells. Hydrometallurgy 71:209–215

Nordberg GF, Fowler BA, Nordberg M, Friberg L (2011) Handbook on the toxicology of metals. Academic, Waltham

Parles T (2012) Vanadium market fundamentals and implications. Presented at the metal bulletin 28th international ferroalloys conference. Berlin, Germany.

Pazirandeh M, Chrisey LA, Mauro JM, Campbell JR, Gaber BP (1995) Expression of the Neurospora crassa metallothionein gene in *Escherichia coli* and its effect on heavy-metal uptake. Appl Microbiol Biotechnol 43:1112–1117

Pazirandeh M, Wells BM, Ryan RL (1998) Development of bacterium-based heavy metal biosorbents: enhanced uptake of cadmium and mercury by Escherichia coli expressing a metal binding motif. Appl Environ Microbiol 64:4068–4072

Pennesi C, Totti C, Beolchini F (2013) Removal of vanadium(III) and molybdenum(V) from wastewater using *Posidonia oceanica* (Tracheophyta) biomass. PLoS One 8:e76870

Perez-Benito JF (2006) Effects of chromium(VI) and vanadium(V) on the lifespan of fish. J Trace Elem Med Biol 20:161–170

Prakash N, Sudha PN, Renganathan NG (2012) Copper and cadmium removal from synthetic industrial wastewater using chitosan and nylon 6 – Springer. Environ Sci Pollut Res 19: 2930–2941

Rehder D (2008) Bioinorganic vanadium chemistry. Wiley, Chichester

Romoser GL, Dudley WA, Machlin LJ, Loveless L (1961) Toxicity of vanadium and chromium for the growing chick. Poult Sci 40:1171–1173
Rychcik M, Skyllas-Kazacos M (1988) Characteristics of a new all-vanadium redox flow battery. J Power Sources 22:59–67
Samino S, Michibata H, Ueki T (2012) Identification of a novel Vanadium-binding protein by EST analysis on the most vanadium-rich ascidian, *Ascidia gemmata*. Mar Biotechnol 14:143–154
Samuelson P, Wernérus H, Svedberg M, Ståhl S (2000) Staphylococcal surface display of metal-binding polyhistidyl peptides. Appl Environ Microbiol 66:1243–1248
Sauge-Merle S, Cuine S, Carrier P, Lecomte-Pradines C, Luu DT, Peltier G (2003) Enhanced toxic metal accumulation in engineered bacterial cells expressing *Arabidopsis thaliana* phytochela tin synthase. Appl Environ Microbiol 69:490–494
Sefström NG (1831) Ueber das Vanadin, ein neues Metall, gefunden im Stangeneisen von Eckersholm, einer Eisenhütte, die ihr Erz von Taberg in Småland bezieht. Ann Phys Chem 97:43–49
Singh S, Mulchandani A, Chen W (2008) Highly selective and rapid arsenic removal by metabolically engineered *Escherichia coli* cells expressing *Fucus vesiculosus* metallothionein. Appl Environ Microbiol 74:2924–2927
Stendahl DH, Sprague JB (1982) Effects of water hardness and pH on vanadium lethality to rainbow trout. Water Res 16:1479–1488
Stocks P (1960) On the relations between atmospheric pollution in urban and rural localities and mortality from cancer, bronchitis and pneumonia, with particular reference to 3:4 benzopyrene, beryllium, molybdenum, vanadium and arsenic. Br J Cancer 14:397
Sundararaman P, Biggs WR, Reynolds JG (1988) Vanadylporphyrins, indicators of kerogen breakdown and generation of petroleum. Geochim Cosmochim Acta 52:2337–2341
Trivedi S, Ueki T, Yamaguchi N, Michibata H (2003) Novel vanadium-binding proteins (vanabins) identified in cDNA libraries and the genome of the ascidian *Ciona intestinalis*. Biochim Biophys Acta 1630:64–70
Ueki T, Michibata H (2011) Molecular mechanism of the transport and reduction pathway of vanadium in ascidians. Coord Chem Rev 255:2249–2257
Ueki T, Takemoto K, Fayard B, Salomé M, Yamamoto A, Kihara H, Susini J, Scippa S, Uyama T, Michibata H (2002) Scanning x-ray microscopy of living and freeze-dried blood cells in two vanadium-rich ascidian species, *Phallusia mammillata* and *Ascidia sydneiensis samea*. Zool Sci 19:27–35
Ueki T, Adachi T, Kawano S, Aoshima M, Yamaguchi N, Kanamori K, Michibata H (2003a) Vanadium-binding proteins (vanabins) from a vanadium-rich ascidian *Ascidia sydneiensis samea*. Biochim Biophys Acta 1626:43–50
Ueki T, Sakamoto Y, Yamaguchi N, Michibata H (2003b) Bioaccumulation of copper ions by Escherichia coli expressing vanabin genes from the vanadium-rich ascidian Ascidia sydneiensis samea. Appl Environ Microbiol 69:6442–6446
Ueki T, Yamaguchi N, Isago Y, Tanahashi H (2014) Vanadium accumulation in ascidians: a system overview. Coord Chem Rev 301–302:300–308. doi:10.1016/j.ccr.2014.09.007
Underwood E (2012) Trace elements in human and animal nutrition, 4th edn. Elsevier, Amsterdam
Waters MD, Gardner DE, Coffin DL (1974) Cytotoxic effects of vanadium on rabbit alveolar macrophages *in vitro*. Toxicol Appl Pharmacol 28:253–263
Weisel CP, Duce RA, Fasching JL, Heaton RW (1984) Estimates of the transport of trace metals from the ocean to the atmosphere. J Geophys Res 89:11607
WHO (2000) Vanadium air quality guidelines, WHO Regional Office for Europe, Copenhagen, Chapter 6.12, pp 1–9
WHO (2005) Air quality guidelines – global update 2005. WHO, Copenhagen
WHO (2014) Dust: definitions and concepts, Hazard prevention and control in the work environment: airborne dust (WHO, 1999). WHO, Chapter 1, pp 1–96
Wyers H (1946) Some toxic effects of vanadium pentoxide. Br J Ind Med 3:177–182

Yamaguchi N, Kamino K, Ueki T, Michibata H (2004) Expressed sequence tag analysis of vanadocytes in a vanadium-rich ascidian, Ascidia sydneiensis samea. Mar Biotechnol 6:165–174

Yoshida N, Kato T, Yoshida T, Ogawa K, Yamashita M, Murooka Y (2002) Bacterium-based heavy metal biosorbents: enhanced uptake of cadmium by *E. coli* expressing a metallothionein fused to beta-galactosidase. Biotechniques 32:551–558

Yoshihara M, Ueki T, Watanabe T, Yamaguchi N, Kamino K, Michibata H (2005) VanabinP, a novel vanadium-binding protein in the blood plasma of an ascidian, *Ascidia sydneiensis samea*. Biochim Biophys Acta 1730:206–214

Yoshihara M, Ueki T, Yamaguchi N, Kamino K, Michibata H (2008) Characterization of a novel vanadium-binding protein (VBP-129) from blood plasma of the vanadium-rich ascidian *Ascidia sydneiensis samea*. Biochim Biophys Acta 1780:256–263

Nitrogen Pollution, Plants and Human Health

Ovais Sareer, Samina Mazahar, W.M. Khanum Al Akbari, and Shahid Umar

Contents

Abstract Limited supply of reactive nitrogen compounds had been a key constraint to human development before the nineteenth century and mankind would rely heavily and entirely on fossil nitrogen and manure cycling for nitrogen needs. With the development of Haber-Bosch process mankind found a way to an almost inexhaustible supply of cheap reactive nitrogen. However humans could not foresee that the monolithic increase in the use of reactive nitrogen compounds aggravated by fossil fuel burning would culminate into a diverse array of environmental problems throughout the world. The learning to produce reactive nitrogen has come with huge costs to humans primarily due to mismanagement in its use. Agricultural practices followed worldwide generally have very low nitrogen use efficiency, which decreases further with increasing nitrogen inputs. This has led to increased losses of reactive nitrogen to the environment. These nitrogen losses have led to nitrate

O. Sareer • S. Mazahar • W.M.K. Al Akbari • S. Umar (✉)
Department of Botany, Faculty of Science, Jamia Hamdard, New Delhi 110062, India
e-mail: s_umar9@hotmail.com

M. Öztürk et al. (eds.), *Plants, Pollutants and Remediation*,
DOI 10.1007/978-94-017-7194-8_3

pollution of water courses and emissions of both ammonia and nitrous oxide to the atmosphere, with impacts on biodiversity and climate change. At the same time high-temperature combustion processes in industry and transport, which convert atmospheric di-nitrogen to nitrogen oxides, and the burning of fossil fuels has found humanity releasing reactive nitrogen directly into the atmosphere. These nitrogen compounds react with particles in the air we breathe and damage human health. Even with this brief listing of issues, it is evident that the unmortgaged human perturbation of the nitrogen cycle is having very dire consequences on the human health and environment that one could not have thought of. Therefore by assembling a handy summary on the issue of nitrogen, this chapter will serve as an essential reference for academicians, researchers and policy makers throughout the world.

Keywords Nitrate • Nitrogen • Pollution • Nitrogen use efficiency • Haber-Bosch • Human health

1 Introduction

Nitrogen the 14th element of the periodic table was discovered by Daniel Rutherford in 1772 and 18 years later named as "nitrogene" by Jean – Antonie Clad Chaptal. This inert diatomic gas is colorless, odorless and tasteless. Nitrogen gas is abundant in the earth's atmosphere and constitutes about 78.09 % by volume of it. It is the seventh most abundant element of our galaxy. Three centuries after its discovery the role of this element in biogeochemical processes and as an important nutrient are well established. Nitrogen being an important constituent of DNA (responsible for the blueprint of biodiversity), chlorophyll, amino acids, proteins and enzymes that drive the metabolic machinery of a cell, forms the very basis of life. It therefore is a key nutrient governing the growth and reproduction of living beings and concomitantly governs species diversity, composition, dynamics and functioning of almost all ecosystems. Going by its abundance on earth it is ironic that this element is the least available for life. Nitrogen at an approximate concentration of about 4×10^{21} g is more abundant than the combined mass of other four major elements. But almost 99 % of this is available in unusable forms to living beings the reason for which is the molecular nature of nitrogen, a form that cannot be used by most living beings. Breaking the triple bond of molecular nitrogen in order to make it available for living organisms requires a lot of energy and the bond can be cleaved only at very high temperature or by a group of some nitrogen fixing bacteria. Ionic forms of nitrogen (nitrate and nitrite) and ammonium ions are the chemically usable forms of nitrogen. The conversion of molecular nitrogen into these usable forms called as nitrogen fixation is done by certain nitrogen fixing bacteria in the rhizosphere and certain cyanobacteria in water. The ionic forms thus formed are used by plants to form proteins, amino acids and DNA. Herbivorous animals get there needed share of nitrogen from plants and carnivorous animals get it by feeding on herbivorous animals. After serving its purpose in living organisms bacteria and other organisms convert nitrogenous organic waste into ammonium ions. These inorganic forms of nitrogen are converted back to nitrogen ions by some specialized form of anaerobic bacteria, which is later released back to the atmosphere as nitrogen gas. Being the essential entity for life that it is increased availability of usable forms of this element usually boasts life directly by increasing the copiousness of primary producers.

The first step nitrogen fixation is the production of ammonia from atmospheric molecular nitrogen and the conversion of the former to biologically usable organic forms. This is mainly achieved in nature by two processes lightening and biological fixation. The latter is mediated by an enzyme nitrogenase present in the members of Rhizobium genus. Colonies of bacteria of this genus live in the root nodules formed in the rhizomes of leguminous plants. Ammonium produced after mineralization is biologically oxidized first to and then nitrate a process called as nitrification. Immobilization is the reverse of mineralization. Since all living organisms require nitrogen therefore bacteria compete for the available nitrogen. In this process nitrate and ammonium are taken up by the bacteria and become unavailable to the crops. Nitrate and nitrite are reduced back to molecular nitrogen by a group of facultative and anaerobic bacteria. This process is known as denitrification. Denitrifying bacteria occur in anaerobic places like swamps, wetlands and poorly drained soils. Under anaerobic conditions denitrification occurs at a very high rate. Under anaerobic conditions the other two processes i.e. nitrogen fixation and nitrification do not occur at all because these processes require oxygen. These three processes nitrogen fixation, nitrification and denitrification are very important considerations for global nitrogen pollution (Fig. 1).

Up until 1900, there was a concerning dearth of reactive nitrogen. Usable forms of nitrogen were required in agriculture as fertilizers in order to increase the yield to feed a very rapidly growing population. Besides nitrogen based explosives were needed as weaponry for the war which was a frequent occurrence during that time. However the solution came sooner rather than later. In 1909 Haber-Bosch process was developed by Fritz Haber. The process allowed the reaction of nitrogen and hydrogen gasses and the conversion of the former into ammonia. Industrial scale production of ammonia from nitrogen was first done in 1913 and the production immediately 20 tonnes per day and within the next year this production doubled the global nitrogen fixation leading to multiplied increase in food yield without which half of the humanity would not have been alive. With the use of nitrogen based fertilizers agriculture has become one of the most successful sectors in terms of productivity and has fulfilled the large growing demand for its output during the last 50 years. But there is a dark side of this agriculture also. It has negative effect on the environment by generating pollutants. The overwhelming production of reactive nitrogen and its use in agriculture has come with heavy costs. A lot of Nitrogen added to agricultural fields culminates into pollution and the earlier belief of values of advances in agricultural technology now appear as an overrepresentation of its actual value to the society.

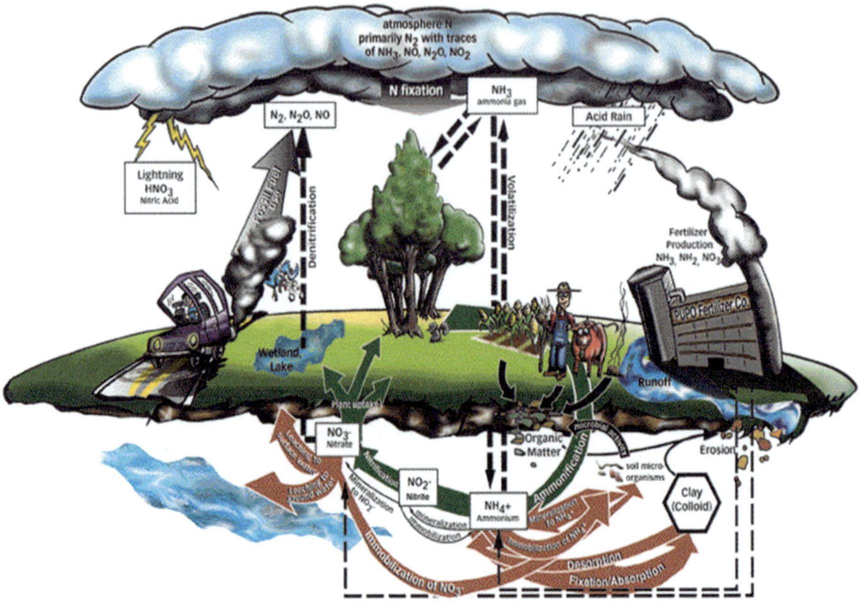

Fig. 1 Nitrogen cycle: Nitrogen cycle i.e. the movement of nitrogen along the atmosphere, land, water and living organisms, occurs in five phase's viz. (1) Nitrogen fixation (2) Mineralization (3) Nitrification (4) Immobilization and (5) Denitrification

High production levels of agriculture are being obtained by the application of plant nutrients such as Nitrogen, Phosphorus, Potassium, and Sulfur through fertilizers, manures, crop and animal residues, legumes etc. with the nutrients being added in excess to the requirements of the plants, pollution is bound to occur. Agriculturalists have thus turned into environmental degradators. These ill agricultural practices have led to pollution of ground water, soil and even our food which is a big concern. World total fertilizer consumption as of 2012 stands at a whopping 180.1 million metric tonnes (with a predicted annual growth rate 1.9 %) of which nitrogenous fertilizers alone comprise 109,928 thousand tonnes. Phosphorous and potassium have a share of 41,525 and 28,626 thousand tonnes respectively. More than half of the total fertilizers (82 million MT) is applied to cereals, equally distributed among three main cereals maize, wheat and rice. Fertilizer inputs to the other cereals represent about 4.6 % of the total fertilizer input. For oil crops fertilizer inputs have been estimated at 19 million MT (11 %) out of which 3.9 % are applied to soybean, 2.0 % for oil palm and 5.2 % for other oil seeds.. the rest of the fertilizers are shared by fiber crops, sugar crops, roots and tubers, fruits and vegetables grass lands and other miscellaneous crops. It is noteworthy to mention fertilizer inputs to vegetables comprised 9.3 % of the world total. This is amply demonstrated in the pie chart below.

Farmers are aware of the nitrogen status of the soils as nitrogen directly affects plant growth. Farmers apply nitrogen in the form of fertilizers and manure every time to increase the yield of their crops. Of the different forms of nitrogen available to the plants ammonium and nitrite being unstable readily accept oxygen and thus leave nitrate as the major form of nitrogen in the soils. However not all of this available nitrogen is taken up by the plants. The excess nitrate leaches out into ground water and rivers. The situation is aggravated in winter and spring months with most plants being dormant with no active uptake during these months. The presence of excess soil water during these months results in enormous nitrate leaching which peaks during these months. Nitrogenous compounds therefore tend to be added to the environment in ever increasing quantities. Figure 2 below shows this ever increasing addition of nitrogen to the environment.

Spatial patterns of total inorganic nitrogen deposition in 1860 (top) and early 1990 (middle) and 2050 (bottom) in units of kg nitrogen per square kilometer per year.

By doing so the global nitrogen cycle is getting altered the grave impacts of which on biodiversity, our environment, water quality and human health are beyond our imagination. As already discussed there are several sources that add reactive nitrogen to the environment but anthropogenic activities add more reactive nitrogen to the global nitrogen cycle than all the other sources combined the most heavily weighing among them being nitrogen fertilization. Figure 3 shows the increase in creation of reactive nitrogen species through various anthropogenic activities. As of

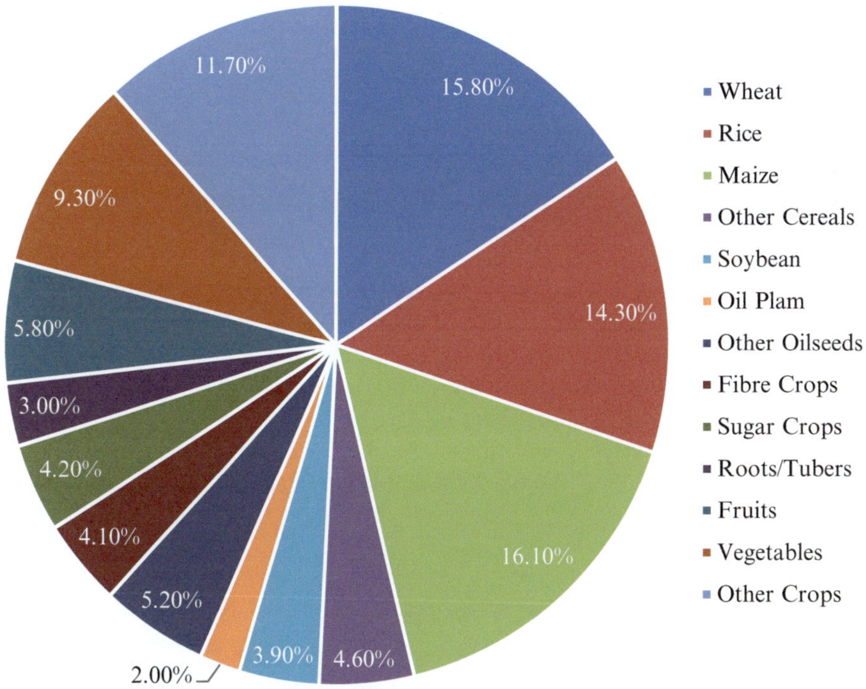

Fig. 2 Total fertilizer use by the crop at the global level (Adapted from IFA 2013)

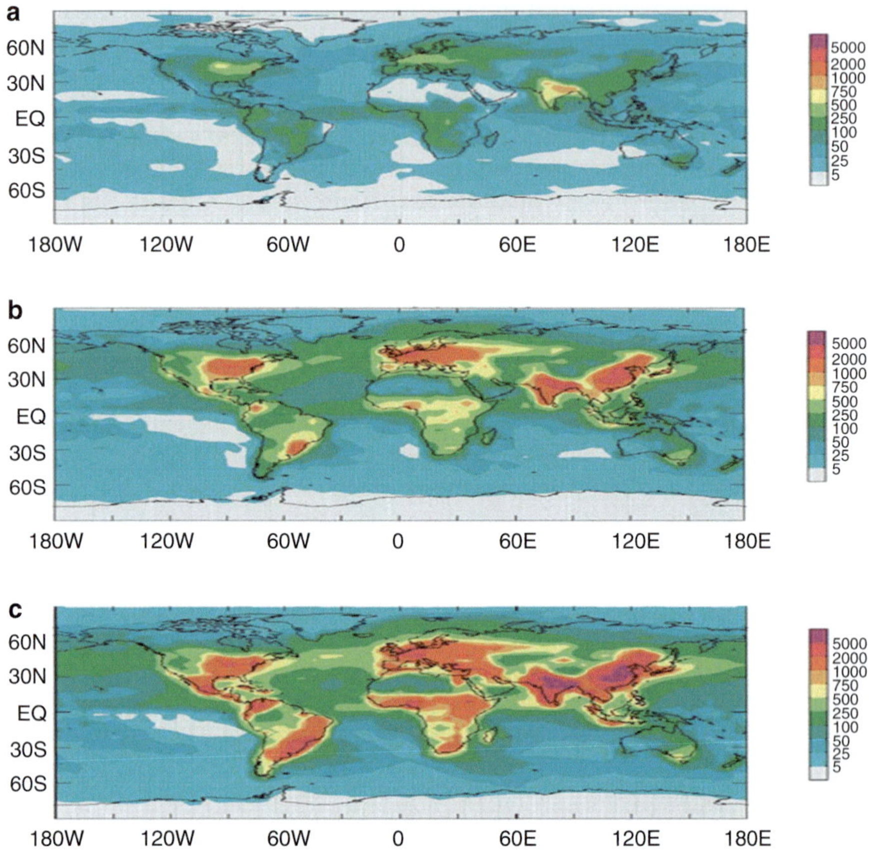

Fig. 3 Adapted from issue of nitrogen International Nitrogen Initiative (INI)

2000 about 100 Tg of reactive nitrogen were released from nitrogenous fertilizers spread on farm lands around the world. With the increased development of modern farming methods the rate at which nitrogen is being fixed has also increased. As of 2010 about 68.71 tonnes of nitrogen nutrients are being added per 100 ha of arable and permanent crop throughout the world.

Yet this reactive nitrogen is not all bad. The use of nitrogen fertilizer is critical to feeding the worlds hunger. The question that thus remains is how do we manage nitrogen responsibly?

2 Fertilizer and Nutrient Management

Modern sustainable agricultural practices are facing a complex challenge of increasing agricultural productivity by improving the plant nutrient management system, maintaining and increasing the soil fertility and limiting nutrient losses to the environment, to meet the food requirement of global population and to reduce the environmental pollution caused due to agricultural activity.

Nitrogenous fertilizers are the most important input in agricultural production. Today, the amount of nitrogenous fertilizers produced has reached a level of around 175 million tons. Many different forms of nitrogen fertilizers exist. Western fertilizer Handbook 2002 says that commercially at least eleven forms of nitrogen fertilizers are available everywhere around the globe. Some of the most commonly used nitrogen fertilizers are classified as:

1. Inorganic Nitrogenous Fertilizers

It includes chemical fertilizers with different forms and concentration of nitrogen in them.

(a) Nitrate Fertilizer (NO_3^-); these fertilizers get readily dissociated to release nitrate ions and are absorbed by plants. They are also free to be leached from the soil or be lost to the atmosphere through de-nitrification when soil become water saturated. Examples are: sodium nitrate ($NaNO_3$) having 16 % N and calcium nitrate (Ca $(NO_3)_2$ with 15.5 % N
(b) Ammonium Fertilizer (NH_4^+); these get adsorbed onto soil surface and are transformed to nitrate (NO_3^-) by nitrifying bacteria and further utilized by most of the crops. Some of these fertilizers are Ammonium Sulphate $(NH_4)_2SO_4$ with 20.6 % N, Ammonium Phosphate $NH_4(H_2PO_4)$ having 20 % N, Ammonium Chloride NH_4Cl with 25 % N, Anhydrous Ammonia NH_3 with 82 % N and Ammonia Solution NH_3 in water with 20–25 % N.
(c) Nitrate and Ammonium Fertilizers; It is a combined form of fertilizer containing both nitrate (NO_3^-) and ammonium ion (NH_4^+). The nitrate nitrogen gets readily available to plants whereas NH_4^+ becomes available after being transformed to nitrate by nitrifying bacteria. They are soluble in water and hence are found suitable for most of the crop plants. Example are: Ammonium nitrate NH_4NO_3 with 33–34 % N, Calcium Ammonium Nitrate (CAN) $Ca(NO_3)_2$ NH_4NO_3 with 25 % N, Ammonium Sulphate Nitrate (ASN) $(NH_4)_2SO_4NH_4NO_3$ with 26 % N.

2. Organic Nitrogenous Fertilizers; It includes decomposable plants and animal byproducts. These organic based fertilizers are slow in action but are available for plants for longer duration of time. Examples are: Compost, Rock phosphate, Bone meal (Elliott et al. 2005), Manure, Rock powder, Ash (Mikkelsen 2008), Fish emulsion (Rosen and Bierman 2005).
3. Slow Release Nitrogenous Fertilizers; These fertilizers help in better utilization of nitrogen by the crop plants as these fertilizers are released slowly and so are available for longer period of time. Some of the slow releasing nitrogenous fertilizers are Oxamide $H_2NCO{=}CONH_2$ having 31.8 % N, Urea-form (Urea+formaldehyde) with 38 % N, Crotonilidine diurea (CDU) (Urea+acetaldehyde) with 32 % N, Guany 1 urea (GU) with 37 % N and various others.

Several reports suggest that some forms of N are more effective for plant growth than others. All nitrogen materials are usually manufactured from ammonia. Such materials are less expensive, more concentrated, and just as plant available as the organics used in the past. The more readily available forms of nitrogen include anhydrous ammonia, ammonium nitrate, urea, non-pressure nitrogen solutions, ammonium sulfate, and ammonium phosphates. The required rates of application of these fertilizers depend on climatic and soil conditions. In warm and moist, soils with rapid decomposition rates, nitrogen released may be significant, but in cold and wet or very dry soils nitrogen is released very slowly. For these soils nitrogen should be applied sufficiently because an insufficient supply of nitrogen will result in losses in yield and quality, over usage, however, causes washing of excess unutilized nutrients away, culminating into the pollution of water bodies (Guler 2004; Atilgan et al. 2007; Ozturk et al. 2013). The import of fertilizers has increased greatly during the last two years, and these include primarily nitrogenous fertilizers. Although fertilizers are known for increasing crop yield and crop production, it also has a darker side which has a negative impact on the environment. Most of the plants are unable to utilize whole of the nitrogen supplied, the remaining nitrogen fertilizer leaches into the soil and water and hence pollutes lakes, rivers, aquifers and oceans one on the other. A major portion of the unabsorbed nitrogen fertilizer volatizes in the form of N_2O which is known as the major contributor of global greenhouse gases (GHGs). The N_2O levels in the atmosphere increase by 0.2–0.3 % annually globally (Bacon 1995; Abdalla et al. 1998). Asian countries including India and Pakistan consume about 48 % of the total fertilizer production in the world (Yilmaz 2004). It has also been observed in a survey that nitrogenous fertilizers that are used for farming account for one third of the total greenhouse gases produced due to agricultural activity. The leaching of nitrogenous fertilizers in water bodies create "Dead zones" which leads to death and decomposition of a large population of algal blooms leading to eutrophication. The natural decomposition of massive algal bloom depletes the oxygen of the water bodies thus increasing the Biological Oxygen Demand (BOD) creating a condition called Hypoxia which results in death of marine organisms. According to a report by United Nations Environment Program UNEP (2004) the dead zone is identified as one of the most significant Global Environmental threat faced by the world. The problem of nitrate leaching can be managed if a constant check is kept over the content of nitrate in soil and avoid overdoses of untimely nitrate (Addiscott et al. 1991) by ensuring that a little nitrate is present in the soil at all times. When any crop grows faster it utilizes the available nitrogen quickly and needs a good supply of nitrate fertilizer at this time. Once the plant growth ceases the requirement of nitrate fertilizer decreases and we need to check out that a little nitrate is available in the soil so that it is not sufficient enough to promote leaching. Recently it has been estimated that NO emissions from cropland ranges between 2.4 and 5.4 Tg N year^{-1} (Davidson and Kingerlee 1997; Yan et al. 2005; Yienger 1995). About 23.5 % of the global annual emission is due to fertilized cropland (12.6 Tg N year^{-1}) and ranks as the second source of atmospheric NH_3 (Bouwman et al. 1997). The major part of the emitted NO and NH3 from cropland is induced due to fertilizer application (Yan et al. 2003).

The sources of these pollutants include industrial waste, sewage disposal, detergent, manure and fertilizers, crop residues and organic matter that produces nitrate upon decomposition which at times remain unutilized by crops and contributes to nitrate leaching and hence pollutes the ground water. The harmful effects of these nitrogen influxes are fairly large and multiplex, ranging from eutrophication of terrestrial and aquatic systems to global acidification and stratospheric ozone loss. Of particular concern is the fact that chemical transformations of nitrogen along its transport pathway in the environment often lead to a cascade of effects. For example, an emitted molecule of nitrogen oxide can first cause photochemical smog and then, after it has been oxidized in the atmosphere to nitric acid and deposited on the ground, can lead to ecosystem acidification and eutrophication. Although there is still much to understand about the implications of nitrogen accumulation in the environment, there is also much to understand about how the increased availability of nitrogen interacts with other biogeochemical element cycles and how those interactions affect global climate change.

3 Absorption and Assimilation of Nitrates in Plant

Nitrate is the most important source of N absorbed by plants growing under field condition. Nitrate uptake is carried out by the root cells and further its reduction and assimilation occurs in plant tissues converting the inorganic nitrogen to organic nitrogen. Nitrate ions get absorbed from the soil solution via the epidermal and cortical cells of root across the plasma membrane. After reaching inside the root cells, nitrate gets reduced to ammonia (NH_4^+) by the enzymatic activity of nitrate reductase (NR) and nitrite reductase (NiR) and the $NH_4^+ - N$ is then assimilated via the GOGAT cycle into organic N (Lea et al. 1990). Kinetics of ionic absorption for various compound have been studied and has been found that nitrate gets absorbed at a rapid rate by plant roots at rate comparable to those of K^+, Rb^+, Cl^- and $H_2PO_4^-$. Rate of nitrate uptake in higher plants controls the rate of nitrate reduction and is not due to the fluctuation in the NR activity (Wilkinson and Crawford 1993) or by limiting the reducing power (Warner and Huffaker 1989). Thus, N assimilation in nitrate fed plants is regulated by the nitrate uptake. Rate of nitrate uptake in several crop plants such as Cowpea, Greengram and Soybean increases through the vegetative growth, maximum during the early reproductive stages and decreases during pod and seed development (Insand and Edwards 1998). The variation in the nitrate contents of plants has been attributed to the genetic variability of plants to absorb, reduce and assimilate nitrate (Olday et al. 1976; Hakeem et al. 2011a). For understanding the mechanism of unidirectional fluxes of nitrate, and for drawing a better conclusion on the understanding and accuracy of net uptake of nitrate and its depletion from the nearby medium nitrate influx and efflux have been studied successfully by using labelled nitrate as N^{13} (Oscarson 1987; Glass et al. 1992; King et al. 1992). Aslam et al. (1994)) also studied the influx-efflux experiments by using nitrate so as to inhibit nitrate uptake but employing the same transporter did not gave satisfactory results because the uptake system was distinct for both the ions. Different carriers have been predicted to be involved in the transportation of both anions (Ward et al. 1988; Galvan et al. 1996).

4 Accumulation of Nitrate in Plant

Nitrate accumulation in plant is a matter of global concern as all living beings whether human or animal depend upon plant for food and fodder. There are numerous sources in the environment that account for the total nitrate content in the human body. Whether it is water, food, either flesh or vegetable, milk and milk products, all are nitrate contributors. The concentration of nitrate increases with increasing level of food chain and keeps translocating in a food web with increasing magnitude. In recent years reduction of nitrate accumulation especially in fresh vegetable crops which possess high nutritional value and are consumed worldwide, has become an important task. Gaudreau et al. 1995 have reported nitrate content in lettuce leaves higher than 65×10^{-6} mol gram^{-1} fr. Wt. In cress shoots higher (110×10^{-6} mol gram^{-1} fr. Wt) concentrations have been reported (Quinche and Dvorak 1980). According to Maynard et al. (1976) 60 % of the total nitrogen may be in the form of nitrate.

Various plant species and even cultivars of the same species exhibit large variability in nitrate contents (Cantliffe 1973a, b; Barker et al. 1974; Maynard et al. 1976; Olday et al. 1976; Quinche and Dvorak 1980; Ostrem and Collins 1983; Blom-Zandstra and Eenink 1986; Reinink and Eenink 1988; Blom-Zandstra 1989; Hakeem et al. 2011a, b; Hakeem et al. 2012c). In case of higher plants, lowest shoot nitrate content have been reported in trees and shrubs (Mengel and Kirkby 1987; Bussi et al. 1997). Different plant parts have unequal distribution of nitrate. Leaf blades contain lower nitrate content as compared with the stems and petiole and older leaf tends to accumulate large nitrate as compared with the young leaves (Maynard et al. 1976; Ostrem and Collins 1983). High light intensity helps in the mobility and assimilation of nitrate and rather reduces nitrate accumulation. This study was supported through plant shading under natural light (Cantliffe 1973a, b; Blanc et al. 1980), during the growing period (Maynard et al. 1976; van der Boon et al. 1990), or experiments done in growth cabinets under control irradiance (Blom-Zandstra and Lampe 1985; Blom-Zandstra et al. 1988). Diurnal patterns were also studied in relation to nitrate content and it was found that nitrate content is lesser during the day period following an increase during the night hours (Maynard et al. 1976; Steingrover et al. 1986a, b; Delhon et al. 1995a, b; Cardenas-Navarro et al. 1998). Plant nitrate content gets lower by lowering the nitrate concentration in the nutrient solution or by using reduced compound such as ammonia or urea instead of nitrate (Gashaw and Mugwira 1981; van der Boon et al. 1990; Gunes et al. 1996; Santamaria and Elia 1997).

Nitrate content in plant is supposed to be an imbalance between the net absorption and assimilation rates (Maynard et al. 1976). Both nitrate uptake and assimilation system are genetically determined (Ferrario-Mery et al. 1997; Ourry et al. 1997) further explaining the variability of nitrate content among species and cultivars. It is also supposed that plant nitrate content may get stable through osmotic potential regulation (Blom-Zandstra and Lampe 1985; Steingrover et al. 1986a, b; McIntyre 1997). Thus nitrate accumulation is of great concern and need to be reduced by using N fertilizers in desirable amounts and limiting the nitrate availability to plant, increasing the light intensity, using green and farmyard manures. Harvesting vegetables at noon, removal of organ rich in nitrate content and cooking vegetable in water with low nitrate content removes at least 50 % of accumulated nitrate (Meah et al. 1994). Treatment of plants with varying doses of citric and salicylic acid also reduces nitrate accumulation as ascorbic acid and tocopherols inhibit the formation of N-nitrocompounds and hence prevent the conversion of nitrate to nitrite (Mowat et al. 1999).

5 Factors Affecting Nitrate Accumulation

Nitrate accumulation in plants is a multi-factorial process. It depends upon various exogenous factors such as light intensity, photoperiod, temperature, water supply and forms of nitrogen and its varying doses (Cantliffe 1973b; Maynard et al. 1976; Steingrover et al. 1986b; Hakeem et al. 2011a, b, 2012a, b, c) and various endogenous factors such as species and cultivars (Luo et al. 2006; Anjana et al. 2007; Hakeem et al. 2011a, b, 2012a, b, c). The uptake of nitrate is an energy driven active process involving low and high affinity transport system involving "active influx" and "passive efflux". When the uptake of nitrate is not in equilibrium with its assimilation and the uptake exceeds due to excess availability of nitrate in the surrounding medium, accumulation of nitrate occurs in the edible portion of plant tissues making them unfit for human and cattle consumption. So it is important to develop and effective nitrogen management in soil, plant and aquatic habitat.

5.1 Light Intensity

The light intensity has an inverse relation with the nitrate content. Higher the light intensity more will be the mobilization and utilization of the nitrate in plant parts and so lesser will be its accumulation. This is in accordance with the findings of Schuphan et al. (1967), Viets and Hageman (1971), and Wright and Davison (1964). Anjana et al. (2006, 2007); have reported the diurnal variation in the nitrate content of young three weeks old spinach leafy vegetable and it was found that nitrate concentration was 2025 and 2674 mg kg^{-1} in genotype IC 326869 at 1:00 pm and 5:00 pm and 641 and 998 mg kg-1 in IC 374686 at 1:00 pm and 5:00 pm. Harper and Paulsen (1968) found that the nitrate concentration of wheat (Triticum aestivum) leaf was lower and NR activity was found higher in blue light (380–470 nm) than in the red light (680–740 nm). Shading and low irradiance also effect the nitrate content in plants. Knipmeyer et al. (1962) studied the effect of shading in corn variety and the concentration of nitrate was negatively co-related with the hours of irradiance and was the same at both the times when shading was caused due to an artificial screen or due to the adjacent corn plant itself. Various experiments have revealed that long photoperiods decreased the N content to a considerable extent.

5.2 Temperature

Temperature also plays an important role in nitrate accumulation in plants, although temperature independently does not determine the extent of nitrate accumulation in plant. All the environmental factors such as light intensity, moisture content, photoperiod, nitrogen availability along with temperature are collectively responsible. Various studies have reported that nitrate accumulation may be due to an increase in temperature (Kretschmer 1958; Younis et al. 1965 or due to fall in temperature (Nightangle et al. 1930) or may remain unaffected. Mild water stress (Huffaker et al. 1970) along with the relatively high temperature of 30 °C or more, decrease the NR activity in corn (Younis et al. 1965). An interaction between temperature, nitrogen availability and light intensity was studied by Santamaria et al. (2001)) in relation to nitrate accumulation in rocket and it was found that under the influence of low light and high temperature the nitrate accumulation increases whereas under high light intensity nitrate content increases with an increase in temperature only in case when the nitrogen supply is high. Grzebelus and Baranski (2001) studied the influence of climate change on nitrate accumulation and it was observed that nitrate content was lower during the year with high rainfall. N accumulation was found higher in autumn winter as compared to spring season (Santamaria et al. 1999). N accumulation also changes with the change of season (Vieira et al. 1998). The above studies suggest that low light intensity and low temperature result in high nitrate accumulation because of an improper N assimilation.

5.3 Water Supply

Grzebelus and Baranski (2001) have shown that the nitrate content was lower during the year with highest rainfall and therefore a positive correlation exists between plant nitrate and water content. Condition of water stress or accumulation of salt in the soil can cause an increase in the nitrate content. The water potential of the soil may become negative under such conditions and may lead to nitrate accumulation by decreasing the rate of NR activity (Huffaker et al. 1970). As already discussed the regulation of nitrate content by osmotic potential may reduce the extent of nitrate accumulation (Blom-Zandstra and Lampe 1985; Steingrover et al. 1986a, b; McIntyre 1997).

5.4 Variability of Nitrogen Forms

As discussed in the previous section of nitrogenous fertilizers different forms of nitrogenous fertilizers are quite distinct from each other in their method of release in the soil and application to the plants. The varying doses of N supply to the plants are most important nutritional factor regulating N accumulation in vegetables

(Barker and Maynard 1971; Brown and Smith 1966, 1967; Lorenz and Weir 1974). Increasing the N application may increase the level of N accumulation if the assimilation rate is not in accordance with the N application. Increase in the nitrate content of vegetables has been reported with increase in the level of N nutrition (Arora and Luthra 1971; Barker and Maynard 1971; Barker et al. 1971; Brown and Smith 1966, 1967; Gately 1971; Hanway and Englehorn 1958; Peck et al. 1971; Regan et al. 1968; Schmidt et al. 1971; Trevino and Murray 1975). N sources which mineralizes slowly for example dried cow dung, tends to lesser accumulation of nitrate than materials which mineralizes quickly (Barker 1975; Nazaryuk et al. 2002). In the former case the N availability is slow and assimilation occurs at normal rate and so N accumulation is lesser. Ammonia based fertilizer or a mixture of nitrate and ammonium may materially reduce nitrate accumulation in plants (Inal and Tarakcioglu 2001; Santamaria et al. 2001). Various experiments are now being conducted in the hydroponic system dealing with the management of N fertilization strategies for raising low nitrate leafy vegetables (Andersen and Nielsen 1992; Liu and Li 1993; Gunes et al. 1994, 1996; Mozafar 1996; Chen and Gao 2002; Dong and Li 2003; Chen et al. 2007). However to have a quality yield production the hydroponics system may not be very successful as it implies the use of N reduced solution (Andersen and Nielsen 1992) and the pre-harvest transfer to N- free medium (Mozafar 1996). As nitrogen nutrition plays a pioneer role in agricultural production so it is thought that long term low or free N medium may not support high yields (Dong and Li 2003). More nitrates get accumulated in plants as the N fertilization level increases (Chen et al. 2004; Nazaryuk et al. 2002; Santamaria et al. 1998a, b; Anjana et al. 2007) whereas limiting the N availability reduces the nitrate content significantly (McCall and Willumsen 1999). There are different ways of increasing and decreasing the N fertilization. Manipulation in the nitrate content of plants can be done by stopping the N supply few days before harvesting in crop plants (Santamaria et al. 2001). Although for the effective control over nitrate accumulation it is best to apply nitrogen once at the start of the cropping cycle as the nitrate gets depleted from the plant and the soil as the plant reaches maturity (Vieira et al. 1998).

5.5 Genetic Factors

Different cultivars of a species, species itself and genotypes with different ploidy may exhibit difference in the nitrate accumulating capacity (Anjana et al. 2006; Grzebelus and Baranski 2001; Harada et al. 2003). The genetic makeup of the plant may be responsible for N variability. According to Harrison et al. (2004) the nitrate content in shoot is genetically controlled by several genes (QTLs). There might be several factors which determine the N variability. The response of different genotypes to the enzymes of nitrogen metabolic pathway and differential assimilation of nitrate may be an important reason for the same. The N uptake mechanism may vary among genotypes as this may involve an active transporter which may not be same in all the genotypes and so some genotypes may show a quick response and some may show a slow response to N uptake and hence exhibit variation in N accumulation. The variability in nitrate accumulation can be co-related with difference in the site of nitrate reductase enzyme (Andrews 1986), and difference in their photosynthetic capacity (Behr and Wiebe 1992). There is negative correlation between nitrate accumulation and sugar concentration (Blom-Zandstra and Lampe 1983) and dry matter content (Reinink et al. 1987), whereas a positive correlation is seen in both the parameters in different genotypes (Hakeem et al. 2011a, b). From the above finding a conclusion can be drawn that genotype with high carbohydrate content in vacuoles should have high dry matter content and so a little nitrate is needed for maintaining the osmoticum (Reinink et al. 1987). Further studies at molecular level suggest that a number of genes are involved in regulating the nitrate content in plant, e.g. genes encoding nitrate reductase (Scheible et al. 1997), glutamine synthetase and ferredoxin-dependent glutamate synthetase (Hausler et al. 1994), a putative anion channel, At CLC-a (Geelen et al. 2000). The nitrate assimilatory pathway involves the use of enzyme such as permeases, NR and NiR and is regulated by various endogenous and environmental stimuli including nitrate, glutamine (Gln), asparagines (Asn), light and sucrose (Crawford and Arst 1993; Oaks 1994; Sivasankar and Oaks 1996). Five quantitative trait loci have been identified for nitrate content in maize on dry matter basis, out of which one gene is known to encode for glutamine synthetase (Hirel et al. 2001). Similarly eight quantitative trait loci (QTLs) were identified for nitrate content on dry matter basis using Arabidopsis Bay-O and Shahdara recombinant inbred lines (RILs) (Loudet et al. 2003). An inhibition of nitrate uptake in maize seedling was studied by Lee et al. (1992), by using 10 mM exogenous Glutamine and it was found that increase in the concentration of Gln increased the concentration of Asparagin (Asn) which decreased N level inside the maize seedling. Further treating plant tissues with nutrient deprived of phosphate and sulphate also resulted in an increase concentration of Gln and Asn in plant tissue decreasing nitrate uptake (Karmoker et al. 1991; Rufty et al. 1993). Therefore different genes are involved in regulating the variability of nitrate levels in plants.

6 Nitrate Toxicity

Nitrate toxicity may be termed as a misnomer because it is the nitrite and not nitrate which is toxic to living beings. Nitrate absorbed from the soil through the plant roots is generally incorporated into plant tissues as amino acids and proteins. This reduction occurs in the actively growing green leaves and requires energy from sunlight, adequate water, nutrients and favorable temperature. When any such conditions are not prevalent, the plants are stressed and nitrate-to-protein conversion is disrupted and hence accumulation of nitrate occurs beyond the safe limits leading to nitrate toxicity. The problem arises when the conversion of nitrate to nitrite is faster than nitrite to ammonia. There is always a possibility of nitrate toxicity in both forage crops and green leafy vegetables whenever the normal growth conditions are disrupted by stress conditions such as drought, hail, frost, salinity etc. Rain or irrigating crops may release nitrates into water sources and hence drinking water may also get contaminated with nitrate so it can be rightly said that the nitrate concentration is an indicator of food & water quality. The toxic dose of ingested nitrate is between 2–5 g. WHO established the Acceptable Daily Intake (ADI) of nitrate as 0–3.7 mg/kg body weight and the permissible concentration of nitrate in drinking water as 50 mg/l or 50 ppm. A diet with 100 g of fresh vegetable with a nitrate concentration of 2500 mg/kg fr wt exceeds the ADI for nitrate by about 13 % (Anjana et al. 2007). Excess nitrate consumption may result in acute toxicity symptoms such as cancer of the digestive track, muscle tremors, cyanosis, gastroenteritis with abdominal pain, mental depression, rapid pulse, headache and weakness, blood in urine and faces. Recently the type-1 diabetes and high nitrate concentration in drinking water have been found to be correlated, although more facts are needed in this respect (Gupta et al. 2008). Fruits and vegetable account for 70 % of the total nitrate intake. Nitrate is found in most of the vegetables and exposure to nitrate is determined by the type of vegetable consumed and the concentration of nitrate in them rather than the absolute amount of vegetable consumed (Anonymous 2008). The nitrate content of several samples of leafy vegetable of spinach & chenopodium found in and around the local market of Delhi was as high as 4451 and 4293 mg/kg fr wt (Anjana et al. 2007) and this is far away from the limit set by ADI. According to Santamaria et al. (1999) various parts of a plant differ in their nitrate accumulating capacity. The decreasing trend of nitrate content in various vegetable organs is as follow: Petiole > leaf > Stem > root > inflorescence > tuber > bulb > fruit > seed (Santamaria et al. 1999). Further it was reported that vegetables that are consumed as only fruits and melons are low nitrate accumulators whereas a wholesome consumable vegetable along with root, stems and leaves are high accumulators of nitrate (Zhou et al. 2000). In spinach & parsley, petiole accumulate higher nitrate than leaf blade whereas in lettuce and 'head chicory' the outer leaves accumulate high nitrate than inner leaves (Santamaria et al. 1999, 2001). The petiole of the rocket leaf possessed more than double of the nitrate concentration than in the lamina of the rocket leaf (Elia et al. 2000). The most efficient way of decreasing nitrate in vegetable before it is ready to be sold is via making an appropriate selection of such genotypes on the basis of their relative levels of nitrate content and nitrate reductase activity (Gupta et al. 2008) as nitrate content are negatively correlated with the NRA. All plants contain some nitrate but excessively high levels are known to be present in forages that have undergone stress condition such as shade or low light intensity, unfavorable weather conditions, herbicide and insecticide application and various plant diseases. According to various literature, crop such as forage sorghum, grain sorghum, sudangrass, pearl millet, sudan-sorghum hybrid are notorious nitrate accumulator. For reducing the level of nitrate toxicity, it is important to avoid excess application of manure or nitrogenous fertilizers to the high nitrate accumulators. In stressed forage crop harvesting should be done after a week of favorable environmental condition so as to reduce the accumulated nitrate. Nitrate toxicity is a global problem and has been reported from a number of countries (Prakasa Rao and Puttanna 2006). Nitrate find its way into humans through two sources (1) Exogenous sources and (2) Endogenous sources.

7 Exogenous Sources

Increased intake of nitrate and nitrite by humans has resulted from the enrichment of biosphere with reactive nitrogen and consumption of more vegetables and (preserved) animal products. Nitrate and nitrite are added to food stuffs as preservatives to protect them from the growth of Clostridium botulinum (which causes botulism) or to enhance their color (characteristic pink colour of cured meat) (Food Safety Network 2010).

Erisman et al. (2008) reported that about half of the current global population is supported by nitrogen fertilizer. Greater dependency of farming practices on fertilizers also leads to increased human exposure as nitrate and nitrite forming salts are their key components. This arises from the consumption of crops and from nitrate contaminated drinking water due to agricultural land run off (Bryan and Grinsven 2013). Nitrate can be found at high concentrations, ranging from 200 to 2500 mg/kg, in vegetables and fruits (Van Duijvenboden and Matthijsen 1989). Vegetables constitute a major source of nitrate, providing over 85 % of the average daily human dietary intake (Gangolli et al. 1994; Hord et al. 2009). Many vegetables have been reported to contain high levels of nitrate, including lettuce, spinach, red beets, fennel, cabbage, parsley, carrots, celery, potatoes, cucumbers, radishes and leeks (Pennington 1998). The concentration of nitrite in vegetables and fruits is lower than that of nitrate, at less than 10 mg/kg, and it rarely exceeds 100 mg/kg (WHO 2007). However, vegetables that have been damaged, improperly stored, pickled or fermented may have nitrite levels up to 400 mg/kg (IARC 2010).

Nutrients that are not fixed in the soil or used by plants can leach into shallow groundwater or run off into surface water during rainfall events. Other sources of nitrate include animal waste from livestock operations and urban areas, and human waste from septic systems and municipal wastewater treatment plants, contamination from agro-based industries (Liebscher et al. 1992; WHO 2007; Harter 2009). Nitrate bearing rocks in geological formations as in some parts of Haryana, India also contribute to high nitrate levels in groundwater (Malik 2000).

8 Endogenous Sources

Varoius studies revealed that nitrates are also produced inside the body and constitute the endogenous sources. L- arginine- nitric oxide synthase is considered as the major source of endogenous nitrate and nitrite. The nitric oxide thus produced is oxidized to higher nitrogen oxides (Lundberg et al. 2011; Moncada and Higgs 1993;

Hibbs et al. 1987; Stuehr and Marletta 1985). Endogenous nitrate synthesis is also increased with gastrointestinal infections and diarrhea (WHO 1985, 1996).

9 Health Effects

Nitrate has always been considered as a toxic substance causing methemoglobinemia and cancers in humans for so many years. However, it is also associated with beneficial health effects, since nitrate represent an important alternative pathway to bioactive NO and its important physiological roles in vascular and immune function (Hakeem et al. 2012a, b, c). Methemoglobinemia is a side-effect of gastroenteritis and is not caused by nitrate but by nitric oxide, which is produced in a defensive reaction stimulated by gastroenteritis. The latter may be caused by a bacterium or a virus (Addiscott and Benjamin 2004). Nitrate per se is relatively nontoxic rather nitrite is the real culprit. With ingestion, nitrate is converted to nitrite in the saliva. Infants convert approximately double, or 10 % of ingested nitrate to nitrite compared to 5 % conversion in older children and adults. Thus, infants are more susceptible to nitrate toxicity than adults.

Nitrate toxicity can be classified as acute and chronic on the basis of dose and duration of exposure.

9.1 Acute (Short- Term) Toxic Effects

Acute toxicity may be caused due to intense exposure to lethal doses of nitrate both in humans and animals. Methemogloinemia or "baby blue syndrome" is the most common acute effect of nitrate toxicity. A broad range of oral lethal nitrate and nitrite doses to humans have been reported, likely due to the wide variability in individual sensitivity. For nitrate, human oral lethal doses range from 4 to 50 g (Mirvish 1991) and from 67 to 833 mg/kg bw (Boink et al. 1999). For nitrite, the estimated oral lethal dose for humans ranges from 1.6 to 9.5 g (Gowans 1990; Mirvish 1991) and from 33 to 250 mg/kg bw, the lower doses applying to children, the elderly and people with a deficiency in reduced nicotinamide adenine dinucleotide (NADH)–cytochrome b5– methaemoglobin reductase (Boink et al. 1999). In severe, untreated cases, brain damage and eventually death can result from suffocation due to lack of oxygen. Early symptoms of methemoglobinemia can include irritability, lack of energy, headache, dizziness, vomiting, diarrhea, labored breathing, and a blue-gray or pale purple coloration to areas around the eyes, mouth, lips, hands and feet. Infants up to 6 months of age are considered to be the most sensitive population. Not only do they convert a greater percentage of nitrate to nitrite, their hemoglobin is more easily converted to methemoglobin and they have less of the enzyme that changes methemoglobin back to its oxygen-carrying form. No cases of methemoglobinemia have been reported when water contained less than 10 ppm

(ppm) of nitrate nitrogen. The majority of cases involve exposure to levels in drinking water exceeding 50 ppm. Healthy adults do not develop methemoglobinemia at nitrate levels in drinking water that place infants at risk. Pregnant women are more sensitive to the effects of nitrate due to a natural increase in methemoglobin levels in blood during the later stage of pregnancy beginning around the 30th week. At increased risk are those individuals with rare conditions, which are generally passed on hereditarily, who have higher than normal levels of methemoglobin in their blood. Individuals with digestive difficulties due to reduced stomach acidity are also at higher risk. Boiling water that has elevated nitrates should be avoided, since this only results in increasing the concentration of nitrate as the water evaporates.

Acute nitrate intoxication is accompanied by serious gastroenteritis with abdominal pain, blood in urine and feces. Presence of bluish discoloration referred to as cyanosis, along with dyspepsia, mental depression, headache and weakness are other prominent symptoms (Fassett 1973).

9.2 *Chronic (Long- Term) Effects*

They arise from long term exposure to non-lethal doses of nitrate. These are briefly discussed below.

9.2.1 Methemoglobinemia or (Blue Baby Syndrome)

Current epidemiological evidences about nitrate and methemoglobinemia are quite confusing. While some studies have shown nitrate to be a major cause of the disease, some studies related to nitrate and nitrite exposure on human, both in children and adults, have not produced methemoglobinemia. Nitrate cannot be concluded as the sole causative agent for methemoglobinemia infants exposed to 175–700 mg nitrate per day did not experience methemoglobin levels above 7.5 % (Cornblath and Hartmann 1948). Dejam et al. (2007) in recent nitrite infusion studies of up to 110 μg/kg body wt/min for 5 min induced methemoglobin concentrations of only 3.2 %. Thus, methemoglobinemia is believed to be caused by other factors like bacterial infections in addition to elevated nitrate water (L'Hirondel et al. 2006; Powlson et al. 2008). Fan and Steinberg (1996) review also reported the possibility that methemoglobinemia may be associated with both the presence of nitrate and bacterial contamination of drinking water, favoring the conversion of nitrate to nitrite and the occurrence of diarrhea, which, in infants, could increase the risk of developing methemoglobinemia. Endogenous nitrite formation due to bacterial-contaminated water as the cause of many cases of methemoglobinemia is reported (Avery 1999): No exposure-response relationships between levels of nitrate in drinking water and methemoglobinemia were found in a review conducted for the WHO (Fewtrell 2004).

9.2.2 Cancer

Nitrate acts as a "procarcinogen" and is not carcinogenic as such. Under acidic conditions, nitrates are reduced to nitrites (NO^{2-}), which in turn may combine with nitrosatable amino compounds (amines or amides) to form potentially carcinogenic N-Nitroso compounds (nitrosamines), (Oshima and Miyoshi 2010). The close association between nitrate contamination of drinking water and increased cancer rates is strongly supported by various physiological studies (Weisenberg et al. 1982; NAS 1977; Møller et al. 1989). Endogenously formed N-nitroso compounds are important in human cancer have been reported (Michaud et al. 2004; Mirvish 1995). Szaleczky et al. (2000) reported that endogenously formed nitrogen and oxygen free radicals may be involved in causing cancer in humans. Cancer of the relevant target organ might be increased by ingestion of larger amounts of nitrate (NAS 1981). Moreover, products of nitric oxide, generated by macrophages during inflammation, can react with water at neutral pH to form nitrite and nitrate and with amines to form nitrosamines (Mirvish 1995). Hence, the International Agency for Research on Cancer (IARC 1978; Fraser et al. 1980) has classified nitrate/nitrite as possibly carcinogenic to humans.

9.2.3 Gastric Cancer

Increasing rates of stomach cancer is associated with increasing nitrate intake and have been documented (Mirvish 1983; Armijo et al. 1981; Cuello et al. 1976; Xu et al. 1992; Reed et al. 1981, 1983). There is an increased risk of gastric cancer under low gastric acidity and lends support to the hypothesis that N-nitroso compounds may be involved in its development. In Italy in a case control study, Palli et al. (2001) found that the highest risk of gastric cancer was among those with a higher nitrite and a lower antioxidant intake, subgroups of the population that would be expected to have higher rates of endogenous nitrosation. A positive association between esophageal and/or stomach cancer with nitrite intake in the diet as well as a significant interaction with vitamin C was seen in two case-control studies (Mayne et al. 2001; Rogers et al. 1995). Some studies conducted in Slovakia, Spain and Hungary found positive correlations between stomach cancer incidence or mortality and historical measurements of drinking water nitrate concentrations near or above 10 mg NO3-N/L – equivalent to 44 mg NO3 -/L (Morales-Suarez-Varela et al. 1995; Sandor et al. 2001; Gulis et al. 2002). In a matched case–control study, Yang et al. (1998) reported association between gastric cancer mortality and nitrate levels in municipal supplies in Taiwan.

9.2.4 Nitrate and Non-Hodgkins Lymphoma (NHL)

Weisenburger (1990) showed association between nitrate contamination and non-Hodgkins lymphoma in epidemiological study of nitrate in well water in Nebraska. The authors concluded that "these findings suggest that NHL in eastern Nebraska may be related to the use of pesticides and nitrogen fertilizers". Ward et al. (2010) found no association between processed meat intake and an increased risk of NHL, but rather found an association with plant based sources (baked good and cereals) which could not be explained. No association was seen in an earlier dietary study by Ward et al. (1996).

9.2.5 Nitrate and Colorectal Cancer

De Roos et al. (2003) found that dietary nitrite intake was positively associated with colon and rectum cancers, with 50–70 % increased risk at levels in the highest quartile; this increased risk was associated primarily with nitrite intake from animal sources rather than vegetables. They also found out that exposure to nitrate concentrations above 5 mg NO3-N/L for more than 10 years was associated with increased colon cancer risk among subgroups with low vitamin C intake (OR = 2.0, 95 % CI = 1.2–3.3) and high meat intake (OR = 2.2, 95 % CI = 1.4–3.6). These patterns were not observed for rectal cancer.

9.2.6 Nitrate and Urinary Bladder Cancer

Risk of bladder cancer and high nitrate ingestion has been pointed out in many recent studies. Weyer et al. (2001) in a recent epidemiological study promulgated that drinking water nitrate levels were associated with a significant elevation in bladder cancer risk. High levels of gastric juice nitrites and elevated urine levels of nitrosamines have been reported in patients with chronic atrophic gastritis or high intragastric pH levels (Farinati et al. 1996, 1989; Jaskiewicz et al. 1990; Vermeer et al. 2001). Chiu et al. (2007) observed a positive association between bladder cancer mortality and nitrate in drinking water in a case control study at levels ≤2.86 mg NO3-N/L (equivalent to 13 mg NO3-/L), and in a cohort study by Weyer et al. (2001) at levels >2.46 mg NO3-N/L (equivalent to 11 mg NO3-/L) in drinking water.

9.2.7 Respiratory System

A correlation among drinking water nitrate concentration, high methemoglobin levels and pathological changes in bronchi and lung parenchyma have been reported in animal studies (Shuval and Gruener 1972; Gruener and Shuval 1970). Gupta et al. (2000), reported a high percentage (40–82 %) of cases of acute respiratory tract infection with history of recurrence in children drinking high nitrate water. These findings were similar with an animal experiment on rabbits (Gupta et al. 1999b).

9.2.8 NOx, Tobacco and Malignancy

20 % of cancer deaths worldwide is caused by tobacco use. The International Agency for Research on Cancer predicts ten million tobacco-related deaths annually by 2020, of which 70 % will occur in the developing world (IARC 2004). Sleiman et al. (2010) reported that residual nicotine from tobacco smoke sorbed to indoor surfaces reacts with ambient nitrous acid (HONO) to form carcinogenic tobacco-specific nitrosamines (TSNAs).

9.2.9 Cardiovascular System

A profound effect of excess nitrate consumption on human health is manifested on the cardiovascular system. Malberg (1978) reported that ingestion of high nitrate induce early onset of hypertension. Blood pressure was found to increase by high nitrate levels in drinking water and salt in school children Pomeranz et al. (2000). High nitrate consumption also causes inflammation and degeneration of cardiac musculature (Shuval and Gruener 1972; Gruener and Shuval 1970; Gupta et al. 1999). Endothelial dysfunction is also reported to be linked with nitrate toxicity (Deanfield et al. 2005). Nitric oxide is dangerous and may be the cause of Ischemia in conditions of inadequate oxygen (Garg and Bryan 2009).

9.2.10 Gastro Intestinal System

Recurrent diarrhea in children has been reported in children up to 8 years of age with increased consumption of nitrate rich water Gupta et al. (2001). Increased Cytochrome b5 reductase activity following high nitrate ingestion is associated with stomatitis (Gupta et al. 1999a). Gastro-intestinal malignancies are also common effect of nitrate toxicity in humans.

9.2.11 Abortions

Expecting women or who are planning for pregnancy should not consume water containing high nitrate as birth related complications and deaths have been reported. Major complications include spontaneous abortions, death, still birth, low birth weight and slow weight gain (Fewtrell 2004, Committee on nitrate accumulation, 1972).

9.2.12 Birth Defects- Malformations

Malformations related to increased nitrate consumption and exposure have been shown in a number of studies. Animal studies have indicated that there is trans placental transfer of N-Nitroso compounds to the fetus (Shuval and Gruener 1972) and this fetal exposure can cause cancer later in life (Druckrey et al. 1966). An Australian study reported an increased risk for central nervous system malformations in infants whose mothers consumed drinking water from private wells with nitrate levels at 26 ppm. A California study found an increased risk for neural tube defects (anencephaly) in babies of women who consumed drinking water with nitrate levels >10 ppm during pregnancy. Croen et al. (2001) found that exposure to nitrate in groundwater at concentrations above the 45 mg/L maximum contaminant level was associated with increased risk for anencephaly (Odds ratio 4.0; confidence interval 1.0–15.4).

9.2.13 Diabetes

A few studies revealed an association between intake of nitrate and insulin-dependent type 1 diabetes mellitus (IDDM). An increased risk of developing insulin dependent diabetes in residents of counties whose water supplies had nitrate levels between 0.77 and 8.2 ppm, compared to counties with water nitrate levels <0.77 ppm has been reported in a Colarado study (Kostraba et al. 1992). The production of free oxide radicals following high nitrate ingestion have been further supported by studies conducted by Gupta et al. (1999a, 2000).

9.2.14 Effect of Nitrate on Thyroid Function and Morphology

Thyroid gland is also affected by high nitrate intake. High nitrate intake leads to competitive inhibition of thyroidal iodide uptake. This leads to decreased thyroid hormone secretion (triiodothyronine [T3], thyroxine [T4]) and increased levels of thyroid stimulating hormone (TSH) and ultimately causes goiter. Studies on children in Slovakia showed that nitrate exposed children had larger thyroid glands and increased frequency of thyroid disorder (Radikova et al. 2008; Tajtakova et al. 2006). Children exposed to water pollution by nitrates have a higher relative risk of goiter (Vladeva et al. 2000). Long-term nitrate exposure via drinking water may result in an increase in thyroid gland weight (Eskiocak 1995; Eskiocak et al. 2005; Zaki et al. 2004).

9.2.15 NOx (Nitrite/Nitrate) and Nephrotic Syndrome

Balat et al. (2000) obtained children with MCNS had increased urinary nitrite excretion in comparison with healthy controls. Greater number of apoptotic T cells was present in patients with nephrotic syndrome than in children in remission from nephrotic syndrome and in controls Zachwieja et al. (2002). Under high NO concentration, the indirect effects mediated by reactive nitrogen species prevail, which induce cell toxicity by nitrosating DNA and tyrosine residues and inducing lipid peroxidation (Davis et al. 2001).

9.2.16 Adrenal Gland

Adrenal gland is reported to be affected by excessive intake of nitrate. Nitrite induces hypertrophy of the adrenal zona glomerulosa by reducing blood pressure and stimulating the rennin angiotensin axis (Vleeming et al. 1997; Til et al. 1988).

9.2.17 Immunity

Human immune system is another common target impacted by nitrate ingestion. Ustyugova et al. (2002) found out that nitrate had no effect on lymphocyte growth, but nitrite decreases proliferation of lymphocytes. Animal studies also reported an immune suppression due to high nitrate ingestion (Porter et al. 1999).

9.2.18 Health Benefits

The discovery of mammalian endogenous nitric oxide generation and the ability of its oxidation products nitrate and nitrite to recycle back to bioactive NO led to the emergence of a novel field of research. This explores a potentially beneficial role of these anions in physiology, nutrition and therapeutics. Hence, in moderation, emerging evidence indicates that nitrate can play an important and beneficial role in human physiology (Lundberg et al. 2008).

9.2.19 Cardiovascular Effects of Nitrate

Nitric oxide NO is the most important molecule in regulating blood pressure and maintaining vascular homeostasis. A diet rich in fruits and vegetables has a positive influence on blood pressure (Appel et al. 1997). Lundberg et al. (2006) speculate that the blood-pressure-lowering effects of certain diets may be due to the nitrate content. A recent food sample survey indicated that a Dietary Approach to Stop Hypertension (DASH) diet exceeds the ADI for nitrate by >500 % and may account for the modest blood pressure effects (Hord et al. 2009). A significant decrease in systolic and diastolic blood pressure within 3 h of ingestion of 0.5 L of beetroot juice (a rich source of nitrate) has been demonstrated by Webb et al. (2008). These data were consistent with earlier results from Lundberg's group showing a significant blood pressure lowering effect from 3 days supplementation with sodium nitrate (Larsen et al. 2006). Although the reduction efficacy from nitrate to NO is very inefficient, it is clear that a diet rich in nitrate can provide a source of bioactive NO (Byran and van Grinsven 2013). Nitrate and nitrite also emerge to play key role on endothelial function. NO synthesized by endothelial nitric oxide synthase (eNOS) is essential for vascular homeostasis by maintaining vessels in their relaxed state. Defects of endothelial NO function, referred to as endothelial dysfunction, not only is associated with all major cardiovascular risk factors such as hyperlipidemia, diabetes, hypertension, and severity of atherosclerosis, but also has a profound predictive value for future atherosclerotic disease progression (Bugiardini et al. 2004; Halcox et al. 2002; Schachinger et al. 2000). Nitrite has been shown to predict exercise capacity in humans (Rassaf et al. 2007). Plasma nitrite levels increase in response to exercise in healthy individuals, whereby in aged patients with endothelial dysfunction, there is no increase in nitrite from exercise (Lauer et al. 2008). A short-term (3-day) dietary supplementation with sodium nitrate results in improved muscular efficiency and a reduction in oxygen consumption during submaximal exercise in healthy subjects and enhance tolerance to high-intensity exercise in humans (Bailey et al. 2009; Larsen et al. 2007).

9.2.20 Nitrate Effects on Host Defense

Duncan et al. (1995) found out that nitrite produced on the dorsal surface of the tongue by bacterial reduction is further reduced to NO under acidic conditions and results in the destruction of acid-producing organisms. Post intake of a nitrate-rich meal results in increased production of NO along with other reactive nitrogen species quantities than that required to produce vasodilatation suggesting a role other than modulation of gastric blood flow (McKnight et al. 1997). Addition of nitrite in concentrations found in human saliva results in much more effective killing of pathogens like (Candida albinos, Escherichia coli H7:O157, Salmonella, Shigella) than in acid alone (Benjamin et al. 1994; Dykhuizen et al. 1996). Helicobacter pylori, a bacteria well adapted to colonize the human stomach, is even susceptible to acidified nitrite (Dykhuizen et al. 1998). Acidification of nitrite derived from dietary nitrate appears to play a pivotal role in protection against ingested pathogens and is supported by the reduced levels of post-prandial salivary nitrite found in subjects treated with the broad-spectrum antibiotic amoxicillin (Dougall et al. 1995). Benjamin et al. (1997) propounded that licking of human skin results in production of NO as salivary nitrite is reduced on the acidic skin surface. A number of common skin pathogens are effectively killed by acidified nitrite (Weller et al. 2001).

10 Conclusion

The nitrogen cycle has undergone a big change during the last century with the invention of the Haber-Bosch process. It has changed from one mastered by nature to one mainly determined by human activity. Growing human demand of food and energy production has terminated into substantive increment in the quantity of reactive nitrogen into the environmental reservoirs. These increments bear both positive and negative effects on our environment. The first and the foremost positive effect of reactive nitrogen brought into agro ecosystems is that it has helped to feed human hunger. Increased concentrations of reactive nitrogen in the atmosphere, has both lineal and non-lineal impact on human and ecosystem health on a global basis. The production of molecular nitrogen being quite small in comparison to reactive nitrogen inputs into the agro ecosystems, Nitrogen continues to accumulate in its reactive forms, resulting in the loss of biodiversity and productivity. Transfer and cycling of reactive nitrogen from the atmosphere, agro ecosystems, forests, and grasslands into the wetlands, streams and rivers is increasing rapidly which has resulted in numerous effects, including acidification, eutrophication, and human health problems. Therefore managing the reactive forms of nitrogen and reducing its inflows into the agro ecosystems through the judicious use of fertilizers while sustaining productivity is of utmost importance.

References

Abdalla MI, Abdalla MMA, Abdel-Aal SA, Farrag IA (1998) Effect of nitrogen fertilization on the yield and quality of some salad crops. Hortsci Abstr 33:2

Addiscott TM, Benjamin N (2004) Nitrate and human health. Soil Use Manag 20:98–104

Addiscott TM, Whitmore AP, Powlson DS (1991) Farming, fertilizers and the nitrate problem. CAB, Wallingford

Andersen L, Nielsen NE (1992) A new cultivation method for the production of vegetables with low content of nitrate. Sci Hortic 49:167–171

Andrews M (1986) The partitioning of nitrate assimilation between root and shoot of higher plants: mini review. Plant Cell Environ 9:511–519

Anjana US, Iqbal M, Abrol YP (2006) Are nitrate concentrations in leafy vegetables within safe limits? In: Proceedings of the workshop on nitrogen in environment. Industry and Agriculture, New Delhi, India, pp 81–84

Anjana, Umar S, Iqbal M (2007) Nitrate accumulation in plants, factors affecting the process, and human health implications. A review. Agron Sustain Dev 26:45–57

Anjana, Umar S, Iqbal M, Abrol YP (2007) Are nitrate concentrations in leafy vegetables within sale limit? Curr Sci 92:355–360

Anonymous (2008) Risk to consumer from nitrate found in vegetables is minimal. Horticult Week 36

Appel LJ, Moore TJ et al (1997) A clinical trial of the effects of dietary patterns on blood pressure. DASH Collaborative Research Group. N Engl J Med 336(16):1117–1124

Armijo R, Orellana M, Medina E, Coulson AH, Sayre JW, Detels R (1981) Epidemiology of stomach cancer in Chile I-Case control study. Int J Epidemiol 10:53–56

Arora SK, Luthra YP (1971) Nitrogen metabolism of leaves during growth ofPhaseolus aureus L. as affected by S, P and N application. Plant Soil 34(1):283–291

Aslam M, Travis RL, Huffaker RC (1994) Stimulation of nitrate and nitrite efflux by ammonium in barley (Hordeum vulgare L.) seedlings. Plant Physiol 106:1293–1301

Atilgan A, Coskan A, Saltuk B, Erkan M (2007) The level of chemical and organic fertilizer usage and potential environmental impacts in greenhouses in Antalya region. Ekoloji 15:37–47

Avery AA (1999) Infantile methemoglobinemia: reexamining the role of drinking water nitrates. Environ Health Perspect 107(7):583–586

Bacon PE (1995) Nitrogen fertilization in the environment. Marcel Dekker, New York

Bailey SJ, Winyard P et al (2009) Dietary nitrate supplementation reduces the O_2 cost of low-intensity exercise and enhances tolerance to high-intensity exercise in humans. J Appl Physiol 107(4):1144–1155

Balat A, Cekmen M, Yürekli M, Gülcan H, Kutlu O, Türköz Y, Yologlu S (2000) Adrenomedullin and nitrite levels in children with minimal change nephrotic syndrome. Pediatr Nephrol 15:70–73

Barker AV (1975) Organic vs. inorganic nutrition and horticultural crop quality. Hortscience 10:50–53

Barker AV, Maynard DN (1971) Nutritional factors affecting nitrate accumulation in spinach. Commun Soil Sci Plant Anal 2:471–478
Barker AV, Peck NH, Mac Donald GE (1971) Nitrate accumulation in vegetables. I. Spinach grown in upland soil. Agron J 63:126–129
Barker AV, Maynard DN, Mills HA (1974) Variations in nitrate accumulation among spinach cultivars. J Am Soc Hortic Sci 99:132–134
Behr U, Wiebe HJ (1992) Relation between photosynthesis and nitrate content of lettuce cultivars. Sci Hortic 49:175–179
Benjamin N, O'Driscoll F et al (1994) Stomach NO synthesis. Nature 368(6471):502
Benjamin N., Pattullo S., Weller R., Smith L., & Ormerod, A. (1997) Wound licking and nitric oxide. The Lancet, 349(9067), 1776.
Blanc D, Otto C, Mars S (1980) Etude de l'accumulation des nitrates dans la racine de carotte. Conse'quences agronomique. Annales de la Nutrition et de l'Alimentation 34:963–968
Blom-Zandstra M (1989) Nitrate accumulation in vegetables and its relationship to quality. Ann Appl Biol 115:553–561
Blom-Zandstra M, Eenink AH (1986) Nitrate concentration and reduction in different genotypes of lettuce. J Am Soc Hortic Sci 111:908–911
Blom-Zandstra M, Lampe JEM (1983) The effect of chloride and sulphate salts on the nitrate content in lettuce plants. J Plant Nutr 6:611–628
Blom-Zandstra M, Lampe JEM (1985) The role of nitrate the osmoregulation of lettuce (Lactuca sativa L.) grown at in different light intensities. J Exp Bot 36:1043–1052
Blom-Zandstra M, Lampe JEM, Ammerlaan FHM (1988) C and N utilization of two lettuce genotypes during growth under non-varying light conditions and after changing the light intensity. Physiol Plant 74:147–153
Boink ABTJ, Dormans JAMA, Speijers GJA, Vleeming W (1999) Effects of nitrates and nitrites in experimental animals. In: Wilson WS, Ball AS, Hinton RH (eds) Managing risks of nitrates to humans and the environment. Royal Society of Chemistry, Cambridge, pp 317–326
Bouwman AF, Lee DS, Asman WAH, Dentener FJ, Hoek KWVD (1997) A global high-resolution emission inventory for ammonia. Glob Biogeochem Cycles 11:561–587
Brown JR, Smith GE (1966) Soil fertilization and nitrate accumulation in vegetables. Agron J 58:209–212
Brown JR, Smith GE (1967) Nitrate accumulation in vegetable crops as influenced by soil fertility practices. Research bulletin/Agricultural experiment station. University of Missouri (920) 1-63
Bugiardini R, Manfrini O et al (2004) Endothelial function predicts future development of coronary artery disease: a study of women with chest pain and normal coronary angiograms. Circulation 109(21):2518–2523
Bussi C, Gojon A, Passama L (1997) In situ nitrate reductase activity in leaves of adult peach trees. J Hortic Sci 72:347–353
Byran NS, van Grinsven H (2013) The role of nitratet in human health. Adv Agron 119:153–181
Cantliffe DJ (1973a) Nitrate accumulation in table beets and spinach as affected by nitrogen, phosphorous and potassium nutrition and light intensity. Agron J 65:563–565
Cantliffe DJ (1973b) Nitrate accumulation in spinach cultivars and plant introductions. Can J Plant Sci 53:365–367
Cardenas-Navarro R, Adamowicz S, Robin P (1998) Diurnal nitrate uptake in young tomato (Lycopersicon esculentum Mill.) plants: test of a feedback based model. J Exp Bot 49:721–730
Chen GL, Gao XR (2002) Effect of partial replacement of nitrate by amino acid and urea on nitrate content of non-heading Chinese cabbage and lettuce in hydroponics. Sci Agric Sin 35:187–191
Chen BM, Wang ZH, Li SX, Wang GX, Song HX, Wang XN (2004) Effects of nitrate supply on plant growth, nitrate accumulation, metabolic nitrate concentration and nitrate reductase activity in three leafy vegetables. Plant Sci 167:635–643
Chen YL, Sun SR, Wang JQ, Gao XS (2007) Study on nutritional effects of organic-N on solution cultured lettuce. J Henan Agric Univ 41:506–510

Chiu HF, Tsai SS, Yang CY (2007) Nitrate in drinking water and risk of death from bladder cancer: anecological case–control study in Taiwan. J Toxicol Environ Health A 70(12):1000–1004

Cornblath M, Hartmann AF (1948) Methemoglobinaemia in young infants. J Pediatr 33:421–425

Crawford NM, Arst HN (1993) The molecular genetics of nitrate assimilation in fungi and plants. Annu Rev Genet 27:115–146

Croen LA, Todoroff K, Shaw GM (2001) Maternal exposure to nitrate from drinking water and diet and risk for neural tube defects. Am J Epidemiol 153:325–331

Cuello C, Correa P, Haenszel W, Cordillo G, Brown C, Archer M, Tannenbaum S (1976) Gastric cancer in Colombia I. Cancer risk and suspect environmental agents. J Natl Cancer Inst 57:1015–1020

Davidson EA, Kingerlee W (1997) A global inventory of nitric oxide emissions from soils. Nutr Cycl Agroecosyst 48:37–50

Davis KL, Martin E, Turko MF (2001) Novel effects of nitric oxide. Annu Rev Pharmacol Toxicol 41:203–236

Deanfield J, Donald A, Ferri C, Giannattasio C, Halcox J, Halligan S, Lerman A, Mancia G, Oliver JJ, Pessina AC, Rizzoni D, Rossi GP, Salvetti A, Schiffrin EL, Taddei S, Webb DJ (2005) Working Group on Endothelin and Endothelial Factors of the European Society of Hypertension (2005) Endothelial function and dysfunction. Part I: methodological issues for assessment in the different vascular beds: a statement by the working group on endothelin and endothelial factors of the European Society of Hypertension. J Hypertens 23:7–17

Dejam A, Hunter CJ et al (2007) Nitrite infusion in humans and nonhuman primates:endocrine effects, pharmacokinetics, and tolerance formation. Circulation 116(16):1821–1831

Delhon P, Gojon A, Tillard P, Passama L (1995a) Diurnal regulation of NO_3 uptake in soybean plants. I. Changes in NO_3 influx, efflux, and N utilization in the plant during the day/night cycle. J Exp Bot 46:1585–1594

Delhon P, Gojon A, Tillard P, Passama L (1995b) Diurnal regulation of NO_3 uptake in soybean plants. II. Relationship with accumulation of NO3 and asparagine in the roots. J Exp Bot 46:1595–1602

De Roos AJ, Ward MH et al (2003) Nitrate in public water supplies andthe risk of colon and rectum cancers. Epidemiology 14:640–649

Dong XY, Li SJ (2003) Effect of nutrient management on nitrate accumulation of pakchoi under solution culture. J Plant Nutr Fertil Sci 9:447–451

Dougall HT, Smith L et al (1995) The effect of amoxycillin on salivary nitrite concentrations: an important mechanism of adverse reactions? Br J Clin Pharmacol 39(4):460–462

Druckrey H, Ivankovic S, Preussmann R (1966) Teratogenic and carcinogenic effects in the offspring after single injection of ethylnitrosourea in pregnant rats. Nature 210:1378–1379

Dykhuizen RS, Frazer R et al (1996) Antimicrobial effect of acidified nitrite on gut pathogens: importance of dietary nitrate in host defense. Antimicrob Agents Chemother 40(6): 1422–1425

Dykhuizen RS, Fraser A et al (1998) Helicobacter pylori is killed by nitrite under acidic conditions. Gut 42(3):334–337

Elia A, Conversa G, Gonnella M (2000) Dosi di azoto, produzione e accumulo di nitrati di lattuga allevata in idrocoltura. Atti V Giornate Scientifiche SOI, Sirmione, pp 229–230

Elliott AL, Davis JG, Waskom RM, Self JR, Christensen DK (2005) Phosphorus fertilizers for organic farming systems. Colorado State University. http://www.ext.colostate.edu/pubs/crops/00569.html

Erisman JW, Galloway JN et al (2008) How a century of ammonia synthesis changed the world. Nat Geosci 1:636–639

Eskiocak S (1995) The effects of administration chronic nitrate by drinking water on thyroid hormones, thyroidal radioiodine uptake and mass. Thesis in biochemistry speciality, Samsun, Turkey

Eskiocak S, Dundar C, Basoglu T, Altaner S (2005) The effects of taking chronic nitrate by drinking water on thyroid functions and morphology. Clin Exp Med 5:66–71

Fan AM, Steinberg VE (1996) Health implications of nitrate and nitrite in drinking water: an update on methemoglobinemia occurrence and reproductive and development toxicity. Regul Toxicol Pharmacol 23(1 Pt 1):35–43

Farinati F, Della Libera G et al (1996) Gastric antioxidant, nitrites, and mucosal lipoperoxidation in chronic gastritis and Helicobacter pylori infection. J Clin Gastroenterol 22:275–281

Farinati F, Lima V et al (1989) Mutagenic activity in gastric juice andurine of subjects with chronic atrophic gastritis, gastric epithelial dysplasia and gastriccancer. Cancer Lett 48:169–175

Fassett DW (1973) Nitrates and nitrites. In: Toxicants occurring naturally in foods. National Academy of Sciences, Washington, DC, pp 7–25

Ferrario-Mery S, Murchie E, Hirel B, Galtier N, Quick WP, Foyer CH (1997) Manipulation of the pathways of sucrose biosynthesis and nitrogen assimilation in transformed plants to improve photosynthesis and productivity. In: Foyer CH, Quick WP (eds) A molecular approach to primary metabolism in higher plants. Taylor & Francis, London, pp 125–153

Fewtrell L (2004) Drinking-water nitrate, methemoglobinemia, and global burden of disease: a discussion. Environ Health Perspect 112(14):1371–1374

Food Safety Network (2010) Nitrites/nitrates. In: Preservatives. Food safety network. University of Guelph, Guelph, Ontario, 3 pp. Available at www.uoguelph.ca/foodsafetynetwork/preservatives

Fraser P, Chilvers C, Beral V, Hill MJ (1980) Nitrate and human cancer: a review of the evidence. Int J Epidemiol 9:3–11

Galvan A, Quesada A, Fernandez E (1996) Nitrate and nitrite are transporter by different specific transport system and by a bispecific transporter in Chlamydomonas reinhardtii. J Biol Chem 271:2088–2092

Gangolli SD, Van Den Brandt PA, Feron VJ, Janzowsky C, Koeman JH, Speijers GJA, Spiegelhalder B, Walker R, Wishnok JS (1994) Nitrate, nitrite and N-nitroso compounds. Eur J Pharmacol 292(1):1–38

Garg HK, Bryan NS (2009) Dietary sources of nitrite as a modulator of ischemia/reperfusion injury. Kidney Int 75:1140–1144

Gashaw L, Mugwira LM (1981) Ammonium-N and nitrate-N effects on the growth and mineral compositions of triticale, wheat, and rye. Agron J 73:47–51

Gately TF (1971) Effects of Nitrogen on potato yields and on the total-N and nitrate-N content of the tops. Potato Res 14 84–90

Gaudreau L, Charbonneau J, Vezina LP, Gosselin A (1995) Effects of photoperiod and photosynthetic photon flux on nitrate content and nitrate reductase activity in greenhouse-grown lettuce. J Plant Nutr 18:437–453

Geelen D, Lurin C, Bouchez D, Frachisse JM, Lelievre F, Courtil B, Barbier-Brygoo H, Maurel C (2000) Disruption of putative anion channel gene AtCLC-a in Arabidopsis suggests a role in the regulation of nitrate content. Plant J 21:259–267

Glass ADM, Shaff JE, Kochian LV (1992) Studies of the uptake of nitrate in barley. Iv electrophysiology. Plant Physiol 99:456–463

Gowans WJ (1990) Fatal methemoglobinemia in a dental nurse. A case of sodium nitrite poisoning. Br J Gen Pract 40(340):470–471

Gruener N, Shuval HI (1970) Health aspects of nitrates in drinking water. In: Developments in water quality research. Ann Arbor Humphrey Science Publishers, Ann Arbor/London, pp 89–106

Grzebelus D, Baranski R (2001) Identification of accessions showing low nitrate accumulation in a germplasm collection of garden beet. Acta Hortic 563:253–255

Guler S (2004) Dunya'da ve Turkiye'de Gubre Tuketiminde Yasanan Gelismeler. In: Turkiye 3. Ulusal Gubre Kongresi, Karaman and Brohi (eds) Tarim-Sanayi-Cevre, 11–13 Ekim 2004, Tokat, pp 47–54

Gulis G, Czompolyova M, Cerhan JR (2002) An ecologic study of nitrate in municipal drinking water and cancer incidence in Trnava District. Slovakia Environ Res 88(3):182–187

Gunes E, Post WHK, Kirkby EA, Aktas M (1994) Influence of partial replacement of nitrate by amino acid nitrogen or urea in the nutrient medium on nitrate accumulation in NFT grown winter lettuce. J Plant Nutr 17:1929–1938

Gunes A, Inal A, Aktas M (1996) Reducing nitrate content of NFT grown winter onion plant by partial replacement of NO3- with amino acid in nutrient solution. Sci Hortic 65:203–208
Gupta SK, Gupta RC, Gupta AB, Seth AK, Bassin JK, Gupta A, Sharma ML (2001) Recurrent diarrhoea in areas with high nitrate in drinking water. Arch Environ Health 56:369–373
Gupta SK, Gupta RC, Seth AK, Gupta AB, Bassin JK, Gupta A (2000) Methaemoglobinaemiain areas with high nitrate concentration in drinking water. Natl Med J India 13:58–61
Gupta SK, Gupta RC, Seth AK, Gupta AB, Bassin JK, Gupta DK, Sharma S (1999a) Epidemiological evaluation of recurrent stomatitis nitrates in drinking water and cytochromeb5 reductase activity. Am J Gastroenterol 94:1808–1812
Gupta SK, Gupta RC, Seth AK, Gupta AB, Sharma ML, Gupta A (1999b) Toxicological effectsof nitrate ingestion on cardiorespiratory tissues in rabbit. South Asian J Prev Cardiol 2:101–105
Gupta SK, Gupta RC, Chhabra SK, Eskiocak S, Gupta AB, Gupta R (2008) Health issue related to nitrogen pollution in water and air. Curr Sci 94(11):1460–1477
Hakeem KR, Ahmad A, Iqbal M, Gucel S, Ozturk M (2011a) Nitrogen efficient rice genotype can reduce nitrate pollution. Environ Sci Pollut Res 18:1184–1193
Hakeem KR, Ahmad A, Iqbal M, Gucel S, Ozturk M (2011b) Nitrogen-efficient cultivars can reduce nitrate pollution. Environ Sci Pollut Res 18:1184–1193
Hakeem KR, Chandna R, Ahmad A, Iqbal M (2012a) Physiological and molecular analysis of applied nitrogen in rice (*Oryza sativa* L.) genotypes. Rice Sci 19(3):213–222
Hakeem KR, Chandna R, Ahmad A, Iqbal M (2012b) Physiological studies and proteomic analysis for differentially expressed proteins and their possible role in the root of N-efficient rice (*Oryza sativa* L.). Mol Breed 32(4):785–798
Hakeem KR, Chandna R, Ahmad A, Iqbal M (2012c) Reactive nitrogen inflows and nitrogen use efficiency in agriculture, an environment perspective. In: Ahmad P, Prasad MNV (eds) Environmental adaptations and stress tolerance of plants in the era of climate change. Springer, New York, pp 217–232
Halcox JP, Schenke WH et al (2002) Prognostic value of coronary vascular endothelial dysfunction. Circulation 106(6):653–658
Hanway JJ, Englehorn AJ (1958) Nitrate accumulation in some Iower crop plants. Agron J 50:331–334
Harada H, Yoshimura Y, Sunaga Y, Hatanaka T, Sugita S (2003) Breeding of Italian ryegrass (Lolium multiflorum Lam.) for a low nitrate concentration by seedling test. Euphytica 129:201–209
Harper JE, Paulsen GM (1968) Influence of intensity, quality, and duration of light on nitrogen reduction and assimilation in wheat. Crop Sci 8:537–539
Harrison J, Hirel B, Limani AM (2004) Variation in nitrate uptake and assimilation between two ecotypes of Lotus japonicus and their recombinant inbred lines. Physiol Plantarum 120:124–131
Harter T (2009) Agricultural impacts on groundwater nitrate. Hydrology 8(4):22–23
Hausler RE, Blackwell RD, Lea PJ, Leegood RC (1994) Control of photosynthesis in barley leaves with reduced activities of glutamine synthase. Planta 194:406–417
Hibbs JB Jr, Taintor RR, Vavrin Z (1987) Macrophage cytotoxicity: role for L-arginine deiminase and imino nitrogen oxidation to nitrite. Science 235:473–476
Hirel B, Bertin P, Quillere I (2001) Towards a better understanding of the genetic and physiological basis for nitrogen use efficiency in maize. Plant Physiol 125:1258–1270
Hord NG, Tang Y et al (2009) Food sources of nitrates and nitrites: the physiologic context for potential health benefits. Am J Clin Nutr 90(1):1–10
Huffaker RC, Radin T, Kleinhoff GE, Cox EL. (1970) Effects of Mild Water Stress on Enzymes of Nitrate Assimilation and of the Carbosylative Phase of Photosynthesis in Barley. Crop. Sci., 10:471-473.
IARC (1978) IARC monographs on the evaluation of carcinogenic risks of chemicals to man. Some N-nitroso compounds, vol 17. International Agency for Research on Cancer, 150 cours Albert Thomas, 69372 Lyon Cedex 08, France

IARC (2010) Ingested nitrate and nitrite and cyanobacterial peptide toxins. International Agency for Research on Cancer, Lyon, 464 pp (IARC Monographs on the Evaluation of Carcinogenic Risks to Humans, Volume 94)
IARC (2004) Tobacco Smoke and Involuntary Smoking,IARC Monographs International Agency for Research on Cancer, Lyon, France. Vol 83
IFA (2013) Patrick Heffer. Assessment of fertilizer use by crop at the global level 2010–2010/11. International Fertilizer Industry Association (IFA) - 28, rue Marbeuf - 75008 Paris - France
Inal A, Tarakcioglu C (2001) Effects of nitrogen forms on growth, nitrate accumulation, membrane permeability, and nitrogen use efficiency of hydroponically grown bunch onion under boron deficiency and toxicity. J Plant Nutr 24:1521–1534
INI. The issues of nitrogen. The International Nitrogen Initiative. www.initrogen.org
Jaskiewicz K, Van Helden PD, Wiid IJ, Steenkamp HJ, Van Wyk MJ (1990) Chronicatrophic gastritis, gastric pH, nitrites and micronutrient levels in a population at risk forgastric carcinoma. Anticancer Res 10:833–836
Karmoker JL, Clarkson DT, Saker LR, Rooney JM, Purves JV (1991) Sulphate deprivation depresses the transport of nitrogen to the xylem and the hydraulic conductivity of barley (Hordenm vulgare L.) roots. Planta 185:269–278
King BJ, Siddiqi MY, Glass ADM (1992) studies of the uptake of nitrate in barley- V. Estimation of root cytoplasmic nitrate concentration using nitrate reductase activity-implications for nitrate influx. Plant Physiol 99:1582–1589
Knipmeyer JW, Hageman RH, Earley EB, Sief RD (1962) Effect of light intensity on certain metabolites of the corn plant (Zea mays L.). Crop Sci 1:1–5
Kostraba JN, Gay EC, Rewers M, Hamman RF (1992) Nitrate levels in community drinking waters and risk of IDDM. An ecological analysis. Diabetes Care 15:1505–1508
Kretschmer AE Jr (1958) Nitrate accumulation in everglades forages. Agron J 50:314–316
L'Hirondel JL, Avery AA et al (2006) Dietary nitrate: where is the risk? Environ Health Perspect 114(8):A458–A459, author reply A459–461
Larsen FJ, Ekblom B, Sahlin K, Lundberg JO, Weitzberg E (2006) Effects of dietary nitrate on blood pressure in healthy volunteers. N Engl J Med 355:2792–2793
Larsen FJ, Weitzberg E, Lundberg JO, Ekblom B (2007) Effects of dietary nitrate on oxygen cost during exercise. Acta Physiol 191:59–66
Lauer T, Heiss C et al (2008) Age-dependent endothelial dysfunction is associated with failure to increase plasma nitrite in response to exercise. Basic Res Cardiol 103(3):291–297
Lea PJ, Robinson SA, Stewart GA (1990) The enzymology and metabolism of glutamine, glutamate and asparagine. In: Miflin BJ, Lea PJ (eds) The biochemistry of plants. Intermediary metabolism, vol 16. Academic, San Diego, pp 121–157
Lee RB, Purves JB, Ratcliffe RG, Saker LR (1992) Nitrogen assimilation and the control of ammonium and nitrate absorption by maize roots. J Exp Bot 43:1385–1396
Liebscher H, Hii B, McNaughton D (1992) Nitrate and pesticide contamination of ground water in the Abbotsford aquifer, southwestern British Columbia. Inland Waters Directorate, Environment Canada, Vancouver, BC
Liu GQ, Li SJ (1993) Seasonal fluctuation of amide-nitrogen effects on yield and nitrate accumulation of nonheading Chinese cabbage in liquid culture. J Nanjing Agric Univ 16:111–113
Lorenz OA, Weir BL (1974) Nitrate accumulation in vegetables. In: White PL, Robbins D (eds) Environmental quality and food supply. Futura Publications, Mt. Kisco, New York, pp 93–105
Loudet O, Chaillow S, Krapp A, Daniel-Vedele F (2003) Quantitative trait loci analysis of water and anion contents in interaction with nitrogen availability in Arabidopsis thaliana. Genetics 163:711–722
Lundberg JO, Carlström M, Larsen FJ, Weiitzberg E (2011) Roles of dietary inorganic nitrate in cardiovascular health and disease. Cardiovasc Res 89:525–532
Lundberg JO, Weitzberg E, Gladwin MT (2008) The nitrate–nitrite–nitric oxide pathway in physiology and therapeutics. Nat Rev Drug Discov 7(2):156–167

Luo JK, Jia SB, Chen LJ, Shen W (2006) The mechanism of nitrate accumulation in pakchoi (Brassica compestris L.ssp. Chinensis (L.)). Plant Soil 282:291–300
Malberg JW, Savage EP, Osteryoung J (1978) Nitrates in drinking water and the early onset of hypertension. Environ Pollut 15:155–160
Malik RPS (2000) Agriculture and water quality in India towards sustainable management. In: Water quality management and control of water pollution. Water reports 21. FAO, Rome, pp 73–85
Maynard DN, Barker AV, Minotti PL, Peck NH (1976) Nitrate accumulation in vegetables. Adv Agron 28:71–118
Mayne ST, Risch HA, Dubrow R, Chow WH, Gammon MD, Vaughan TL, Farrow DC, Schoenberg JB, Stanford JL, Ahsan H, West AB, Rotterdam H, Blot WJ, Fraumeni JF Jr (2001) Nutrient intake and risk of subtypes of esophageal and gastric cancer. Cancer Epidemiol. Biomark Prev 10(10):1055–1062
McCall D, Willumsen J (1999) Effects of nitrogen availability and supplementary light on the nitrate content of soil grown lettuce. J Hortic Sci Biotech 74:458–463
McIntyre GI (1997) The role of nitrate in the osmotic and nutritional control of plant development. Aust J Plant Physiol 24:103–118
McKnight GM, Smith LM et al (1997) Chemical synthesis of nitric oxide in the stomach from dietary nitrate in humans. Gut 40(2):211–214
Meah MN, Harrison N, Davies A (1994) Nitrate and nitrite in foods and the diet. Food Addit Contam 11:519–532
Mengel K, Kirkby EA (1987) Principles of plant nutrition. International Potash Institute, Worblaufen-Bern
Michaud DS, Mysliwiec PA, Aldoori W, Willett WC, Giovannucci E (2004) Peptic ulcer disease and the risk of bladder cancer in a prospective study of male health professionals. Cancer Epidemiol Biomarkers Prev 13:250–254
Mikkelsen R (2008) Managing potassium for organic crop production. Better Crops 92. http://ucanr.org/sites/nm/files/76654.pdf
Mirvish SS (1991) The significance for human health of nitrate, nitrite, and n-nitroso compounds. In: Bogardi I, Kuzelka RD, Ennenga WG (eds) Nitrate contamination: exposure, consequence, and control. NATO ASI series, vol 30. Springer. Berlin, pp 253–266
Mirvish SS (1995) Role of N-nitroso compounds (NOC) and N-nitrosation in etiology of gastric, esophageal, nasopharyngeal and bladder cancer and contribution to cancer of known exposures to NOC. Cancer Lett 93:17–48
Mirvish SS. (1983) The etiology of gastric cancer: Intragastric nitrosamide formation and otherthe-ories. J Natl Cancer Inst. 71: 631–647.
Møller H, Landt J, Pedersen E, Jensen P, Autrup H, Jensen OM (1989) Endogenous nitrosation in relation to nitrate exposure from drinking water and diet in a Danish rural population. Cancer Res 49:3117–3121
Moncada S, Higgs A (1993) The L-arginine-nitric oxide pathway. N Engl J Med 329:2002–2012
Morales-Suarez-Varela MM, Llopis-Gonzalez A, Tejerizo-Perez ML (1995) Impact of nitrates in drinking water on cancer mortality in Valencia. Spain Eur J Epidemiol 11(1):15–21
Mowat C, Carswell A, Wirz A, McColl KEL (1999) Omeprazole and dietary nitrate independently affect levels of vitamin C and nitrite in gastric juice. Gastroenterology 116:813–822
Mozafar A (1996) Decreasing the NO3- and increasing the vitamin C contents in spinach by a nitrogen deprivation method. Plant Foods Hum Nutr 49:155–162
NAS (National Academy of Sciences) (1977) Drinking water and health. Safe Drinking Water Committee, Washington, DC
NAS National Academy of Sciences (1981) The health effects of nitrate, nitrite, and Nnitrosocompounds. Part 1 of a two-part study by the Committee on Nitrite and Alternative CuringAgents in Food. Report by the US National Research Council, National Academy of Sciences. National Academy Press, Washington, D.C.
Nazaryuk VM, Klenova MI, Kalimullina FR (2002) Ecoagrochemical approaches to the problem of nitrate pollution in agroecosystems. Russ J Ecol 33:392–397

Nightangle GT, Schermerhorn LG, Robbins WR (1930) Some effects of potassium deficiency on the histological structure and nitrogenous and carbohydrate constituents of plants. Natl J Agric Exp Stn Bull (499) 1–36

Oaks A (1994) Primary nitrogen assimilation in higher plants and its regulation. Can J Bot 72:739–750

Olday FC, Barker AV, Maynard DN (1976) A physiological basis for different pattern of nitrate accumulation into spinach cultivars. J Am Soc Hortic Sci 101(3):217–219

Oscarson B (1987) Short-term studies of NO_3- uptake in Pisum using $13NO_{3,}$ Planta 87:1–6

Oshima H, Miyoshi N (2010) Ingested nitrate and nitrite: beneficial to human health? Genes Environ 32(3):43–52

Ostrem JA, Collins GB (1983) Genetic variation for nitrate concentration in Nicotiana tabacum L. J Hered 74:431–434

Ourry A, Gordon AJ, Macduff JH (1997) Nitrogen uptake and assimilation in roots and root nodules. In: Foyer CH, Quick WP (eds) A molecular approach to primary metabolism in higher plants. Taylor & Francis, London, pp 237–253

Ozturk M, Guce S, Sakcali S, Baslar S (2013) Nitrate and edible plants in the Mediterranean Region of Turkey: an overview. In: Umar S et al (eds) Nitrate in leafy vegetables-toxicity and safety measures. I.K. International Publishing House Pvt. Ltd, New Delhi/Bangalore, pp 17–51

Palli D, Saieva C, Coppi C, Del Giudice G, Magagnotti C, Nesi G, Orsi F, Airoldi L (2001) O6-alkylguanines, dietary N-nitroso compounds, and their precursors in gastric cancer. Nutr Cancer 39(1):42–49

Peck NH, Barker AV, MacDonald GE, Shallenberger RS (1971) Nitrate accumulation in vegetables II. Table beets grown in upland soils. Agron J 63:130–132

Pennington JAT (1998) Dietary exposure models for nitrates and nitrites. Food Control 9(6):385–395

Pomeranz A, Korzets Z, Vanunu D, Krystal H, Wolach B (2000) Elevated salt and nitrate levels in drinking water cause an increase of blood pressure in schoolchildren. Kidney Blood Press Res 23:400–403

Porter WP, Jaeger JW, Carlson IH (1999) Endocrine, immune, and behavioral effects of aldicarb (carbamate), atrazine (triazine) and nitrate (fertilizer) mixtures at groundwater concentrations. Toxicol Ind Health 15:133–150

Powlson DS, Addiscott TM et al (2008) When does nitrate become a risk for humans? J Environ Qual 37(2):291–295

Prakasa Rao EVS, Puttanna K (2006) Strategies for combating nitrate pollution. Curr Sci 91:1335–1339

Quinche JP, Dvorak V (1980) Le dosage des nitrates dans lele'gumes, les plantes condimentaires et les terres par ionome'trieset par chromatographie gaz-liquide. Revue Suisse deViticulture, Arboriculture et Horticulture 12:7–20

Radikova Z, Tajtakova M, Kocan A, Trnovec T, Sebokov E, Klimes I, Langer P (2008) Possible effects of environmental nitrates and toxic organochlorines on human thyroid in highly polluted areas in Slovakia. Thyroid 18(3):353–362

Rassaf T, Lauer T et al (2007) Nitric oxide synthase-derived plasma nitrite predicts exercise capacity. Br J Sports Med 41(10):669–673

Reed PI, Smith PL, Haines K, House FR, Walters CL (1981) Gastric juice N nitrosamines in health and gastroduodenal disease. Lancet 2:550–552

Reed PI, Summers K, Smith PLR, Walters CL, Bartholomew BA, Hill MJ et al (1983) Effect of ascorbic acid treatment on gastric juice nitrite and N-nitroso compound concentration in achlorhydric subjects. Gut 24:1492–1493

Regan WS, Lambeth VN, Brown JR, Blevins DG (1968) Fertilization interrelationships in yield, nitrate and oxalic acid content of spinach. Proc Am Soc Hortic Sci. 93:485–492

Reinink K, Eenink AH (1988) Genotypical differences in nitrate accumulation in shoots and roots of lettuce. Sci Hortic 37:13–24

Reinink K, Groenwold R, Bootsma A (1987) Genotypical differences in nitrate content in Lactuca sativa L. and related species and correlation with dry matter content. Euphytica 36:11–18

Rogers MAM, Vaughan TL, Davis S, Thomas DB (1995) Consumption of nitrate, nitrite, and nitrosodimethylamine and the risk of upper aerodigestive tract cancer. Cancer Epidemiol Biomark Prev 4(1):29–36
Rosen CJ, Bierman PM (2005) Maintaining soil fertility in an organic fruit and vegetable crops system. University of MN Extension. http://www.extension.umn.edu/distribution/horticulture/M1191.html
Rufty TW Jr, Israel DW, Volk RJ, Qiu J, Sa T (1993) Phosphate regulation of nitrate assimilation in soybean. J Exp Bot 44:879–891
Sandor J, Kiss I, Farkas O, Ember I (2001) Association between gastric cancer mortality and nitrate content of drinking water: ecological study on small area inequalities. Eur J Epidemiol 17(5):443–447
Santamaria P, Elia A (1997) Producing nitrate-free endive heads: effect of nitrogen form on growth, yield, and ion composition of endive. J Am Soc Hortic Sci 122:140–145
Santamaria P, Elia A, Parente A, Serio F (1998a) Nitrate and ammonium nutrition in chicory and rocket salad plants. J Plant Nutr 21:1779–1789
Santamaria P, Elia A, Parente A, Serio F (1998b) Fertilization strategies for lowering nitrate content in leafy vegetables: chicory and rocket. J Plant Nutr 21:1791–1803
Santamaria P, Elia A, Serio F, Todaro E (1999) A survey of nitrate and oxalate content in retail fresh vegetables. J Sci Food Agric 79:1882–1888
Santamaria P, Elia A, Gonnella M, Parente A, Serio F (2001) Ways of reducing rocket salad nitrate content. Acta Hortic 548:529–537
Schachinger V, Britten MB et al (2000) Prognostic impact of coronary vasodilator dysfunction on adverse long-term outcome of coronary heart disease. Circulation 101(16):1899–1906
Scheible WR, Lauerer M, Schulze ED, Caboche M, Stitt M (1997) Accumulation of nitrate in the shoot acts as a signal to regulate shoot-root allocation in tobacco. Plant J 11:671–691
Schmidt DR, MacDonald HA, Brockman FE (1971) Oxalate and nitrate contents of four tropical leafy vegetables grown at two soil fertility levels. Agron J 63:559–561
Schuphan W, Bengtsson B, Bosund I, Hylmo B (1967) Nitrate accumulation in spinach. Qual Plant Mater Veg 14:317–330
Shuval HI, Gruener N (1972) Epidemiological and toxicological aspects of nitrates and nitrites in the environment. Am J Public Health 62:1045
Sivasankar S, Oaks A (1996) Nitrate assimilation in higher plants: the effect of metabolites and light. Plant Physiol Biochem 34:609–620
Sleiman M, Gundel LA, Pankow JF, Jacob P, Singer BC and Destaillats H. (2010) Formation of carcinogens indoors by surface-mediated reactions of nicotine with nitrous acid, leadingto potential third hand smoke hazards. Proc Natl Acad Sci, 107: 6576–6581
Steingrover E, Ratering P, Siesling J (1986a) Daily changes in uptake, reduction and storage of nitrate in spinach grown at low light intensity. Physiol Plant 66:550–556
Steingrover E, Siesling L, Ratering P (1986b) Effect of one night with "low light" on uptake, reduction and storage of nitrate in spinach. Physiol Plant 66:557–562
Stuehr DJ, Marletta MA (1985) Mammalian nitrate biosynthesis: mouse macrophages produce nitrite and nitrate in response to Escherichia coli lipopolysaccharide. Proc Natl Acad Sci U S A 82:7738–7742
Szaleczky E, Prónai L, Nakazawa H, Tulassay Z (2000) Evidence of in vivo peroxynitrite formation in patients with colorectal carcinoma, higher plasma nitrate/nitrite levels, and lower protection against oxygen free radicals. J Clin Gastroenterol 30:47–51
Tajtakova M, Semanova Z, Tomkova Z, Szokeova E, Majoros J, Radikova Z, Sebokova E, Klimes I, Langer P (2006) Increased thyroid volume and frequency of thyroid disorders signs in schoolchildren from nitrate polluted area. Chemosphere 62(4):559–564
Til HP, Falke HE, Kuper CF, Willems MI (1988) Evaluation of the oral toxicity of potassium nitrite in a 13-week drinking-water study in rats. Food Chem Toxicol 26:851–859
Trevino IC, Murray GA (1975) Nitrogen effects on growth, seed yield and protein of seven pea cultivars. Crop Sci 15:500–502

UNEP, United Nation Environmental programme (2004). An overview of our changing environment, GEO Year book 2004/05 1-96. DOI: www.unep.org/yearbook/2004/pdf/geo_yearbook_2004.pdf

Ustyugova IV, Zeman C, Dhanwada K, Beltz LA (2002) Nitrates/nitrites alter human lymphocyteproliferation and cytokine production. Arch Environ Contam Toxicol 43:270–276

Van der Boon J, Steenhuizen JW, Steingrover EG (1990) Growth and nitrate concentration of lettuce as affected by total nitrogen and chloride concentration, NH_4/NO_3 ratio and temperature of the recirculating nutrient solution. J Hortic Sci 65:309–321

Van Duijvenboden W, Matthijsen AJCM (1989) Integrated criteria document nitrate. National Institute for Public Health and the Environment (RIVM) Report no. 758473012 Rijksintituut voor de Volksgezondheid en Milieuhygiene (National Institute of Public Health and Environmental Protection): Bilthoven, The Netherlands, 1989

Vermeer IT, Engels LG, Pachen DM, Dallinga JW, Kleinjans JC, van Maanen JM (2001) Intragastricvolatile N-nitrosamines, nitrite, pH, and Helicobacter pylori during long-termtreatment with omeprazole. Gastroenterology 121:517–525

Vieira IS, Vasconselos EP, Monteiro AA (1998) Nitrate accumulation, yield and leaf quality of turnip greens in response to nitrogen fertilisation. Nutr Cycl Agroecosys 51:249–258

Viets FG, Jr, Hageman RH (1971) Factors affecting the accumulation of nitrate in soil, water and plants. U.S. Department of Agriculture. Agricultural Handbook 413, 1–63

Vladeva S, Gatseva P, Gopina G (2000) Comparative analysis of results from studies of goitre in children from Bulgarian villages with nitrate pollution of drinking water in 1995 and 1998. Cent Eur J Public Health 8:179–181

Vleeming W, van de Kuil A, te Biesebeek JD, Meulenbelt J, Boink AB (1997) Effect of nitrite on blood pressure in anaesthetized and free-moving rats. Food Chem Toxicol 35(6):615–619

Ward MR, Tischner R, Huffaker RC (1988) Inhibition of nitrate transport by anti-nitrate reductase lgG fragments and the identification of plasma membrane associated nitrate reductase in roots of barley seedlings. Plant Physiol 88:1141–1145

Ward MH, Mark SD, Cantor KP, Weisenburger DD (1996) Drinking water nitrate and the risk of non-Hodgkin's lymphoma. Epidemiology 7:465–470

Ward PS, Patel J, Wise DR, Abdel-Wahab O, Bennett BD, Coller HA, Cross JR, Fantin VR, Hedvat CV, Perl AE, Rabinowitz JD, Carroll M, Su SM, Sharp KA, Thompson CB (2010) The common feature of leukemia-associated IDH1 and IDH2 mutations is a neomorphic enzyme activity converting α-ketoglutarate to 2-hydroxyglutarate. Cancer Cell 17(3):225–234

Warner RH, Huffaker RC (1989) Nitrate transport is independent of NADH and NAD(P)H nitrate reductases in barley seedlings. Plant Physiol 91:947–953

Webb AJ, Patel N et al (2008) Acute blood pressure lowering, vasoprotective, and antiplatelet properties of dietary nitrate via bioconversion to nitrite. Hypertension 51(3):784–790

Weisenberg E, Grauer F, Baron Y, Rajstein J (1982) An assessment of the implication of nitrate-rich water on salivary nitrate. Ecotoxicol Environ Saf 6:97–101

Weisenburger DD. (1990) Environmental epidemiology of non-Hodgkin's lymphoma in eastern Nebraska. Am J Indust Med.;18:303–5

Weller R, Price RJ et al (2001) Antimicrobial effect of acidified nitrite on dermatophyte fungi, candida and bacterial skin pathogens. J Appl Microbiol 90(4):648–652

Western Fertilizer Handbook (2002) Ninth Edition. California Plant Health Association, 1801 I Street, Suite 200, Sacramento, CA, 95814

Weyer PJ, Cerhan JR, Kross BC, Hallberg GR, Kantamneni J, Breuer G, Jones MP, Zheng W, Lynch CF (2001) Municipal drinking water nitrate level and cancer risk in olderwomen: the Iowa Women's Health Study. Epidemiology 12:327–338

WHO (1985) Health hazards from nitrate in drinking-water. In: Report on a WHO meeting, Copenhagen, 5–9 March 1984. Copenhagen, WHO Regional Office for Europe. Environmental health series no. 1, Copenhagen, pp 23–40

WHO (1996) Toxicological evaluation of certain food additives and contaminants. Prepared by the Forty-Fourth Meeting of the Joint FAO/WHO Expert Committee on Food Additives (JECFA).

World Health Organization, International Programme on Chemical Safety (WHO Food Additives Series 35), Geneva
WHO (2007) Nitrate and nitrite in drinking-water. Background document for development of WHO guidelines for drinking-water quality. World Health Organization, Geneva (WHO/SDE/WSH/07.01/16)
Wilkinson JQ, Crawford NM (1993) Identification and characterization of a chlorate-resistant mutant of Arabidopsis thaliana with mutations in both nitrate reductase structural genes NIAl and NIAZ. Mol Gen Genet 239:289–297
Wright MJ, Davison KL (1964) Nitrate accumulation in crops and nitrate poisoning in animals. Adv Agron 16:197–247
Xu G, Song P, Reed PI (1992) The relationship between gastric mucosal changes and nitrate intake via drinking water in a high risk population for gastric cancer in Moping county. China Eur J Cancer Prev 1:437–443
Yang CY., Cheng MF., Tsai SS., & Hsieh YL. (1998) Calcium, magnesium, and nitrate in drinking water and gastric cancer mortality. Cancer Science, 89(2), 124-130.
Yan XY, Akimoto H, Ohara T (2003) Estimation of nitrous oxide, nitric oxide and ammonia emissions from croplands in East, Southeast and South Asia. Glob Chang Biol 9:1080–1096
Yan XY, Ohara T, Akimoto H (2005) Statistical modeling of global soil NOx emissions. Global Biogeochem Cycles 19:GB3019, doi:3010.1029/2004GB002276
Yienger JJ (1995) Empirical model of global soil biogenic NOx emissions. J Geophys Res 100:11447–11464
Yilmaz H (2004) Turkiye'de Kimyasal Gubre Uretim, Tuketim ve Dis Ticaretindeki Gelismeler. In: Turkiye 3. Ulusal Gubre Kongresi, Tarim-Sanayi-Cevre, 11–13 Ekim, Tokat, pp 35–46
Younis MA, Pauli AW, Mitchell HL, Stickler FC (1965) Temperature and its interaction with light and moisture in nitrogen metabolism of corn (Zea mays L.) seedlings. Crop Sci 5:321–326
Zachwieja J, Dworacki G, Bobkowski W, Dobrowolska-Zachwieja A, Zaniew M, Maciejewski J (2002) Increased apoptosis of peripheral blood lymphocytes in children with nephrotic syndrome. Pediatr Nephrol 17:197–200
Zaki A, Ait CA, Talibi A, Derouiche AF, Aboussaouira T, Zarrouck K, Chait A, Himmi T (2004) Impact of nitrate intake in drinking water on the thyroid gland activity in male rat. Toxicol Lett 147:27–33
Zhou ZY, Wang MJ, Wang JS (2000) Nitrate and nitrite contamination in vegetables in China. Food Rev Int 16:61–76

Evaluation of Wild Halophytes of Aralo-Caspian Flora Towards Soil Restoration and Food Security Improvement

Esmira Alirzayeva, Valida Ali-zade, Tamilla Shirvani, and Kristina Toderich

Contents

Abstract Nowadays, due to an increase in the salinization induced by climate change and desertification process in the dryland ecosystems, evaluation of potential use of halophytes for soil remediation, forage and food supply, medicinal and other purposes is gaining wide attention and acceptability. Documentation of valuable knowledge about these plants is assuming urgent priority.

Desert halophytes of Aralo-Caspian flora safely survive under extreme saline edaphic and arid climate conditions by development of specific ion translocation and salt-accumulation mechanisms by developing specific salt/storage cells and salt excretion though salt-producing trichomes/hairs for removal, degradation and immobilization of wide spectrum of soil pollutants. Innovative remediation strategies are based on domestication of wild halophytes which are able to grow and produce seeds successfully under dry and saline environments, can accumulate high shoot and root biomass, have deep root systems and are distinguished by efficient

E. Alirzayeva (✉) • V. Ali-zade • T. Shirvani
Institute of Botany of Azerbaijan National Academy of Sciences, Baku, Azerbaijan
e-mail: hh.esmal@hotmail.com

K. Toderich
International Center for Biosaline Agriculture for Central Asia and Caucasus (ICBA-CAC), Tashkent, Uzbekistan

M. Öztürk et al. (eds.), *Plants, Pollutants and Remediation*,
DOI 10.1007/978-94-017-7194-8_4

nutrient uptake mechanisms, resistance to salinity and water deficiency. Different mechanisms and strategies for the sequestration and regulation of salt ion concentrations in the plant tissues are operating in the stem and leaf of succulent halophytes and in the recreto-and pseudohalophytes of Aralo-Caspian flora. The significant levels of sclerification of perianth segments simultaneously with the thickening of pericarp and spermoderma epidermis are known to protect embryo from pollutants. Peculiarities of reproduction mechanisms and CO_2 fixation pathways also play an important role in vital function of halophytes crops under harsh desert conditions. In this paper new integrated approaches for utilization of wild halophyte taxa in remediation of soils contaminated by toxic salts/heavy metals, stabilization of ecosystem function and sources for functionality of biologically active substances (primary and secondary metabolites) will be discussed.

Keywords Salinity • Contamination • Planttolerance • Halophytes • Kur-Araz lowland • Kyzylkum Desert • Aralo-Caspian sub-province

1 Introduction

Soil degradation is one of the most significant and strength edaphic stresses limiting the plant growth and development as well as declining the crop productivity, thereby represents a threat for sustainable development of agriculture and is the serious desertification factor (Qadir and Oster 2004; Dagar et al. 2006; Chedlly et al. 2008; Ahmad and Sharma 2008; Ondrasek et al. 2011; Lokhande and Suprasanna 2012). Drought, all types of erosion, non-appropriate temperature, anthropogenic activities, deficit of nutrients and excess of pollutants (salts, heavy metals, organic wastes etc.) result in deterioration of soil properties and decrease the soil fertility. Particularly, salinization becomes more dangerous, when plants simultaneously undergo metal toxic stress (Ozturk et al. 2008a, b; Shafi et al. 2010; Altay and Öztürk 2012; Khan et al. 2013; Dağhan and Ozturk 2015; Hasanuzzaman et al. 2015; Hakeem et al. 2013; 2015). Under the ongoing climate change arid and semi-arid regions are more vulnerable to land degradation and biodiversity loss.

A steady growing of salts and heavy metal (HM) contents in environment puts more actual a question on selection of new plant genetic resources with the specific evolutionarily developed morphological, structural and metabolic protective mechanisms of salt-tolerance. Due to these peculiarities they are capable not only to survive and reproduce under these extreme conditions, but also highly to accumulate and remove the pollutants in their harvestable aboveground organs/biomass.

About 1 % of the remarkable species of terrestrial plants named halophytes are able to grow and fulfill their whole life cycle at the salt concentration much exceeding the norm for living (Guvensen et al. 2006; Rozema and Flowers 2008; Ozturk et al. 2014). Halophytes are also considered to be HM tolerant and their accumulator plants (Manousaki and Kalogerakis 2009; 2011a; Khan et al. 2014).

The problem of soil degradation in Aralo-Caspian lowlands is on the agenda. Salinization is one of the major ecological and agricultural problems in arid and semi-arid areas of Central Asia. Continuous use of the major rivers of Central Asia (Amu Darya, Zarafshan and Syr Darya) for irrigation with purpose to produce and export of cotton, oil and minerals during Soviet period has resulted in rising water tables, waterlogging and the saline lands in the whole Aral Sea Basin (Toderich and Tsukatani 2007; Toderich et al. 2013b).

Historical intensive exploitation of petroleum resources, development of the relevant industries as well as heavy and irrational agro-technical approaches (irrigation, fertilization etc.) lead to a sharpening of the situation, consequently, soil salinization and HM contamination in Caspian coast of Azerbaijan.

Significance of the territories for the economy of these regions dictates an improving of a quality of the marginal lands and searching of alternative sustainable technologies for their utilization. Considering the high cost, durability and laboriousness of the various agro-technical, chemical and meliorative approaches there is a critical necessity in the development more effective and environmental-friendly biological technologies for the soil restoration. In this aspect, an alternative economically and socially accepted strategy can be based on the appropriate selection of plants successfully growing and producing seeds under dry and saline environments, possessing a great metal/salt removal potential without affecting on soil fertility. This localized management named a phytorehabilitation already successfully found an application in number regions of the world (Rabhi et al. 2009; Schwitzguébel et al. 2009; Ashraf et al. 2010; Toderich et al. 2010a, b; Hasanuzzaman et al. 2013; Sabir et al. 2015) can be introduced to practical management to enhance economic productivities of Aralo-Caspian lowlands.

The purpose of the present work is to explore the strained environments of Aralo-Caspian lowlands on example of arid areas in Uzbekistan and Azerbaijan, to evaluate wild halophytic resources for their rational use in the soil restoration and improvement of food security in these regions. New integrated approaches for utilization of natural halophyte taxa in remediation of soils contaminated by toxic salts/HM and stabilization of ecosystem function are discussed.

2 Soil Salinization (Global Context)

Salinity and waterlogging as two manifestations of the salt-affected soils are a serious threat to crop production in the world. Due to an increase in natural as well as anthropogenic impacts, last decades the salinization of the dryland ecosystems spreads in more than 100 countries independence on climate conditions and reaches the global level (Squires and Glenn 2009; Hasanuzzaman et al. 2013). On a global scale, about one-third of all agricultural lands are becoming saline (Munns et al. 1999).

About one billion hectares of world lands is highly affected by different types of salinization and contamination (Qadir et al. 2000; Squires and Glenn 2009; Ozturk et al. 2010a, b). Natural or primary salt-affected soils are classified as two types: saline soils contain mainly sodium chloride and sodium sulfate, more rarely other neutral salts, while the second type are sodic or alkaline, clay soils (Squires and Glenn 2009). Major cations in soil extract include Na^+, Ca^{2+}, Mg^{2+} and major anions include Cl^-, SO_4^{2-}, HCO_3^-, CO_3^{2-} and NO_3^-.

Human-induced or secondary salinization being the result of change in the hydrologic balance of the soil occupies much smaller area (approximately 77 Mha), nonetheless represents even more serious problem, mainly to cropland (Munns et al. 1999; Squires and Glenn 2009). However, on the physical parameters soils are grouped into three categories: saline, sodic and saline-sodic soils. Saline soils distinguish with electrical conductivity (EC_e) greater than 4 dS m^{-1} and sodium absorption ratio (SAR) of less than 13, saline-sodic – with EC_e greater than 4 dS m^{-1} and SAR of 13 or higher, sodic – with EC_e lower than 4 dS m^{-1} and SAR of 13 or higher (Qadir et al. 2008).

Salinization is the great problem for agriculture, whereas the irrigation results in the accumulation of water soluble salts (e.g. NaCl, Na_2CO_3 and $CaCl_2$) to above normal concentrations in the rooting zone of arable land (Ozturk et al. 2006a, b, 2009; 2011a, b, 2012; Chandna et al. 2013), since high rates of evaporation and transpiration draw soluble salts from deep layer of the soil profile (Rozema and Flowers 2008).

Today, 20 % of the world's cultivated and nearly half of all irrigated lands are secondary salt-affected soils (Rhoades and Loveday 1990; Flowers and Flowers 2005). In Azerbaijan this type of soils occupy over 1,5 million ha constituting a significant proportion of agro-ecosystems in a number of irrigated agricultural regions (Mammadov et al. 2010). Kur-Araz lowland which is considered to be a more problematic region of Azerbaijan in this respect covers 46.7 % (665,7 th. ha) of all irrigated lands of this country. The salt-affected and waterlogging lands make about 65 % of territories in the southern part of Uzbekistan and more than 90 % in the deltas of Amudarya and Syrdarya Rivers.

2.1 *Salinization Problem in Kur-Araz Lowland*

Kur-Araz lowland with an area of about 2,2 mln hectares is a part of Aralo-Caspian lowland. It occupies a central part of Azerbaijan separating the Greater and Lesser Caucasus and washed by Caspian Sea in the east (Fig. 1).

Large and small rivers cut this lowland into the some major natural zones, each zone is named due to their historical and cultural detachment: Mugan plain occupies 478.4 th. ha, Salyan plain – 149 th. ha, Garabakh plain – 324.7 th. ha, Mil plain – 368.7 th. ha, Shirvan plain – 859.7 th. ha. Each of these plains is characterized by peculiarities of their relief, soil types, vegetation and economic utilization (Azizov 2006; Soil Atlas of Azerbaijan Republic 2007).

Kur-Araz lowland is considered by some authors to be the typical arid zone with aridity coefficient of 3–3.5 (Nabiyeva 2004), but other researches allocate it to semi-

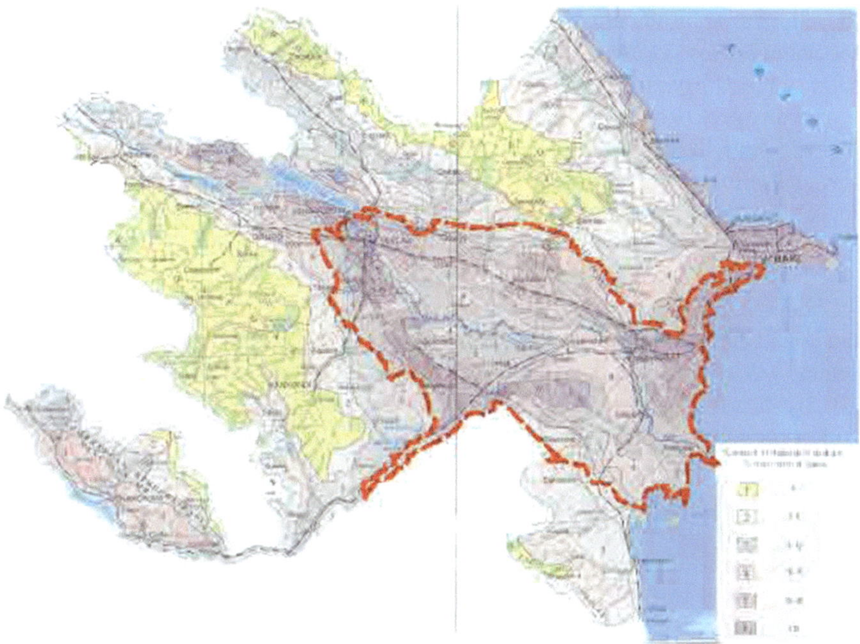

Fig. 1 Map on the degree of soil digestible sodium in Azerbaijan (Soil Atlas of Azerbaijan Republic 2007). *Red outlined* Kur-Araz lowland (by A.Jafarova)

arid zone (Shukurov et al. 2008). However, a strong degree of desertification in some sites of lowland is observed (Budagov et al. 2003).

It is the most plain area in Azerbaijan with the dry climate, prolonged hot summer and short comparable mild winter. The average temperature of July is 25–28 °C, January – 1.3–3.6 °C, precipitation is 200–400 mm. The portion of surface evaporation is for 3–3.5 folds the amount of atmospheric precipitation. The annual evaporation is 100–120 mm in Northeastern and 900 mm in Southeastern parts of lowland (Hajiyev and Rahimov 1971). Total radiation throughout lowland ranges 130–133 kkal cm^2 annually (Nabiyeva 2004).

This region is known as the biggest irrigated agricultural territory and primary source of forage, cotton and grain in Azerbaijan, the region of powerful electric and oil energy resources. Although Kur-Araz lowland is of importance in agricultural industry of the country it directly faces with the multiplicity of ecological problems, in particular, surplus of salts and HM as well as a desertification (Azizov 2006; Alirzayeva et al. 2006a).

Currently, a surface of this lowland is essentially affected by both natural impact and anthropogenic activities. Salinity of soils is caused by the flowing close to surface mineralized ground water and a high evaporation of excess of moisture.

Kur river, being a most powerful artery of runoff of Greater Caucasus with its annual effluent of 18 km^3 containing about 4,4 mln. ton of salts, plays great role in the salinization of lowland soils. Surface of this lowland is essentially affected by the activities also of Araz and other rivers which bring a huge amount of alluvial sand and mud from mountains.

Salinity degree of ground water is also raised, nearby the rivers and channels it is characterized by low level of mineralization (0.5–1.0 g/l), while in the center of lowland salinity is increasing up to 10–50 g/l, in exceptional cases up to 100 g/l and more. The depth of ground water varies as at average 10 m, but in the some areas – 2–3 m, while in the foothills – 80–90 m (Abduyev 1961; Azizov 2006; Soil Atlas of Azerbaijan Republic 2007).

Mineralization degree of sulfate waters is higher than carbonated ones and they are found on the edges of the lowland as well as in Shirvan plain. Chlorinated ground waters are distributed in the center and south parts of the lowland (Heydarova 2013). Salts from underground runoff from Caucasus Mountain enter into the subsoil of rock forming the Kur-Araz lowland. This runoff is confirmed by the significant contents of salts in the deep of horizons of rocks and the increase of mineralization of groundwater with depth (Azizov 2006).

Besides, an intensive irrigation of agricultural lands in the preceding years also leads to a rising of ground water levels and consequently an increasing the salt content on the soil surface. As a result, the significant territories of this massif were converted into marginal degraded lands out of agricultural use and soils here have turned into unfertile (Hajiyev 2002) (Fig. 2).

Consequently, along with sierozem predominance on the territory of Kur-Araz lowland, area of brown semi-desert, saline and solonetzic soils increased here. 60–70 % of whole territory of the lowland suffers from different degrees of salinity (Table 1).

Based on the images of "LANDSAT-TM" satellite, it was revealed that strongly saline soils are widely distributed in Shirvan plain, but moderate salinity covers most parts of the plain (Heydarova 2013).

Solonchaks in this lowland differ by high alkaline reactions and surplus of Mg and Na contents (Mammadov 2005). The content of chlorine and sulfate ions in soils of lowland is variable and ranges from 0.035 to 1.2 % (Cl^-) and from 0.05 to 1 % (SO_4^{2-}) on different sites depending on the zones. These values are low at the foothills, while they are high in the central part of the lowland and coastal areas. Sulphated saline soils are distributed in the top and middle cones of flowing of rivers in Shirvan plain, along Kur river territories in Aghdash and Ujar regions, also near Mughan village. Sulfates are found to be the main components of the saline soils here (Heydarova 2013). Ground waters are mainly in shallow at the depth of 1–2 m in these regions and very saline (Naghiyev and Heydarova 2010).

In majority of soils the digestible portion of Na comprises 10 % of the total alkaline base content. There are areas where the absorbed sodium makes up 20 or more of the total Na (Fig. 1). The absorptive capacity of the soil in Kur-Araz lowland varies between 10 and 40 mg equivalents (per 100 g) depending on the texture of the soil. Among the digestible bases calcium ions have the advantage, accounting 50 % and sometimes 80 % of the sum of bases, the second place is occupied by magnesium, the content of which is 30–40 % of the total absorbed bases. Total alkalinity

Fig. 2 Marginal land in Shirvan plain of Kur-Araz lowland of Azerbaijan at the beginning of summer

Table 1 Soil salinization degree in Kur-Araz lowland

The degree of soil salinity,%	Kur-Araz lowland											
	Garabakh plain		Mil plain		Shirvan plain		Mughan plain		Salyan plain		Total area	
	Area, th. ha	%	Area, th. ha	%	Area, th. ha	%	Area, th. ha	%	Area, th. ha	%	Area, th. ha	%
<0,25	144,2	44,4	135,7	36,8	137,6	16,0	117,1	24,5	30,1	20,2	564,7	25,9
0,25–0,5	84,4	26,0	65,6	17,8	184,8	21,5	49,8	10,4	21,7	14,6	406,3	18,6
0,5–1,0	33,8	10,4	71,5	19,4	180,5	21,0	78,6	16,4	27,3	18,3	391,7	18,0
1,0 – 2,0	33,1	10,2	57,9	15,7	233,0	27,2	137,3	28,7	25,2	16,9	487,3	22,3
>2,0	29,2	9,0	38,0	10,3	123,0	14,3	95,6	20,0	44,7	30,0	331,3	15,2
Total	324,7	100	368,7	100	859,7	100	478,4	100	149,0	100	2181,3	100

Soil Atlas of Azerbaijan Republic (2007)

of soils distributed in the central part of the lowland ranges from 0.01 to 0.1 %. The widespread grading of alkalinity is 0.02–0.08 %. The highest alkalinity observed in soils of Garabakh plain. Percentage of sodium makes up 20 or more from the sum of absorbed bases here (Azizov 2006).

Besides salinization, a degradation and deterioration of soil structure as well as HM contamination of this lowland are also conditioned by the agricultural and oil-power industry activities (Alirzayeva et al. 2006a).

2.2 *Salinization Problem in Kyzylkum*

Population growth and development of agricultural industries result in increasing of pressures on rangelands in arid and semi-arid zones. In spite of the wide use of crop residues and grains in livestock production (Nordblom et al. 1997; Gintzburger et al. 2003), overgrazing of the good rangelands still is the major factor leading to deterioration of natural vegetation of desert areas. Erratic cropping in low rainfall zones and cutting of shrubs by the local population for firewood even more aggravate the situation. All these factors pose a threat for useful, endemic or rare desert plant species and lead to the reduction of productivities of pastures. Above mentioned have induced the disintegration of rural infrastructure in arid zones of the Zarafshan River Basin and the Kyzylkum Desert, which led to the migration of local population from the native areas to neighboring cities or countries (Toderich and Tsukatani 2007).

Effect of drought and salinity on food security in Central Asia can be higher than in other areas. Crop production becomes less sustainable under rapid increase of secondary salinization here. Besides salinization, contamination by HM and chemical compounds released by agriculture, uranium, oil and gas industries has been frequently reported for the Kyzylkum sandy desert (Goldshtein 1997; Goldshtein et al. 2000; Tsukatani and Katayama 2001; Toderich et al. 2004; Aparin et al. 2006). As a result, the former highly productive livestock system has deteriorated and livelihoods of the people have dramatically declined.

Marginal and natural rangelands in the lower reaches of the Zarafshan River Valley and Kyzylkum Desert, including Karakata are affected by aridity and salinity (Fig. 3)

Temperature in the Kanimekh agro-ecological zone, Navoi region (Central Kyzylkum, Uzbekistan) in average is about 1–4 °C in winter and 30–32 °C in summer periods (Fig. 4). The soil type of this area is silt-sandy loam, throughout the profile to a depth of 60 cm. The soil is highly saline near the surface and in the lower layers.

Salt-affected lands in Aral Sea Basin demonstrate the most characteristic features of natural continental terrestrial salinization, sodication and alkalinization. Low organic matter (<1.0 %), high salt contents and poor water holding capacity render these soils unproductive. The organic matter of soil was less than 1 %, while the cation exchange capacity varied between 10 mg eq/100 g soil and 35 mg eq/100 g soil. Soil fertility of the desert saline soils is characterized as rather low. Total nitrogen and phosphorus ranged between 0.7–5.5 mg kg^{-1} and 10.0–18.26 mg kg^{-1}, respectively.

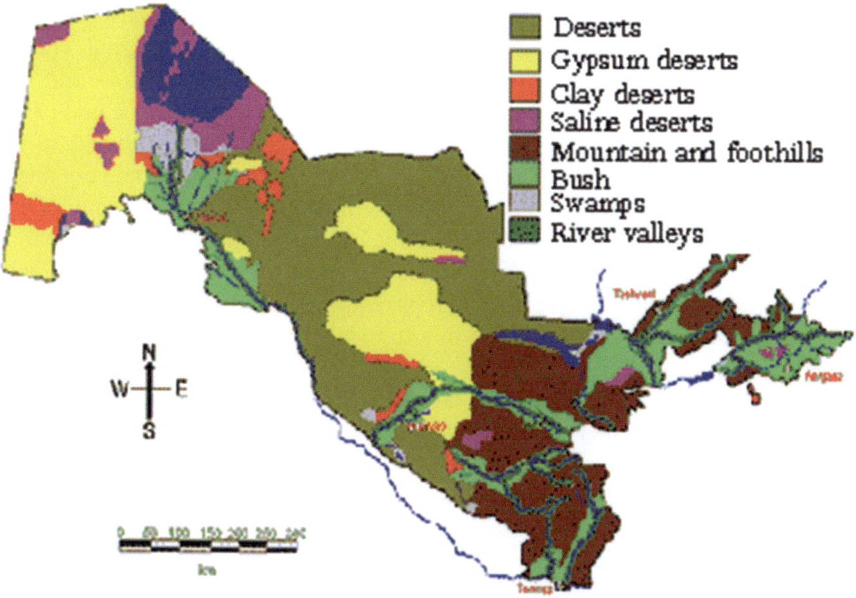

Fig. 3 Map of distribution of main desert types in Uzbekistan

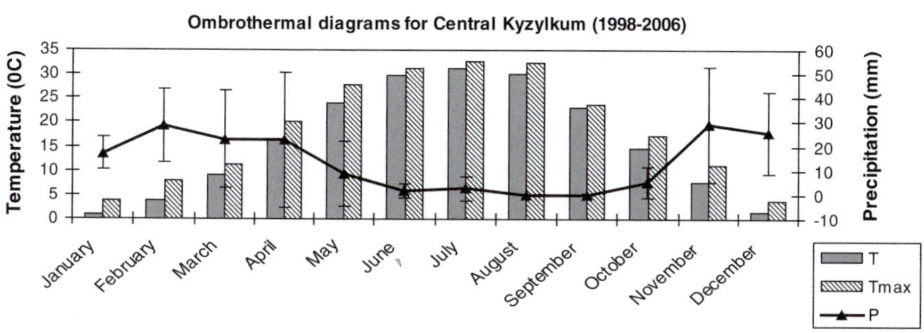

Fig. 4 Climatic conditions for the Kanimekh agroecological zone (Central Kyzylkums, Uzbekistan), *T* mean temperature, *Tmax* maximum temperature, *P* mean precipitation

The predominant salinity type is sulfate-chloride. Main type of mineralization of soils is calcium-sodium and sulfate. High evaporation rates dry the ponds in summer inducing the soil and groundwater (GW) salinization. The poor natural drainage system of marginal cropping irrigated lands has also caused an increase in the salt content at the superficial crust and groundwater that leads to secondary salinization of the soils. Drainage water is about 3.5 time saline than irrigation water. Element composition for the Kyzylkesek water was found as follows: Sr>Ba>Ti>Mn>Cr>Ni>Cu>Mo>Pb (Toderich et al. 2009). GW salinity varies in the range of 2.0–8.2 g l^{-1}. GW table fluctuates from 0.5 to 2.5 m during May-July at the dry solonchaks and the selected experimental agricultural plots and up to 8.0–20.0 m in the virgin desert degraded rangeland area. The GW table under the plantations ranged from 1.2–2.2 m below the soil surface during the observation period. The EC values of the GW varied between 0.4 and 4.2 dS m^{-1} under plantations and was 0.4–2.8 dS m^{-1} in the open field. EC values in the beginning and at the end of the vegetation season indicated the slight to medium salinity of soil, although at the upper 40 cm horizon at some points this value was over 25 dS m^{-1}. The very intensive processes of soil salinization occur in the area located in the vicinity of the artesian wells. Average EC of the irrigation water (artesian hot spring) varies between 8.3 and 18.1 dS m^{-1}, pH is between 7.3 and 8.1. Taking into account these features, the cultivation of agricultural crops requires high inputs of chemical fertilizers and costly leaching practice to free root zone from salinity (Toderich et al. 2009).

The salt affected marginal lands in the plains and studied saline depression were classified into three habitat types: **A** – with shallow water table and moderate salinity (marginal old agricultural lands), **B** – with relatively deep water table and high salinity, and **C** – transitional habitats in which salinity and water table are no controlling factors (virgin *Artemisia perennial* pastures). The most important factor of zonal vegetation was soil salinity and the salt tolerance limits of species. In general, the distinction of each ecological vegetation group differs according to relief, floristic composition and the ion concentration (halo-accumulation plant ability) in the aboveground dry biomass. It was found that the most important direct source of soil salinization, as is seen in the Table 3, was shallow groundwater level calculated through the soil moisture content and soil salinity based on the content of ion Na^+ in the upper soil profiles.

Thus, dryland salinity and associated water quality are recognized to be among most severe natural resource degradation problems in the marginal desert belt of Aral Sea Basin. Access to irrigation water in this region has drastically decreased in the last years, which caused additional obstacles to rangelands productivity and agricultural production (Toderich et al. 2013a). Replacement of deep-rooted, perennial native vegetation with shallow-rooted, annual agricultural crops and halophytic pastures has resulted in increased recharge causing shallow saline water tables leading to dryland salinity and loss of plant diversity. This results in greater amounts of water entering a GW system, water table rise and the concentration of naturally occurring salts near the soil surface. Slight changes in temperature or soil moisture and dissolved salts regime therefore substantially alter the composition, distribution and abundance of species. Increased frequency of climatic extremes and changes in soil salinity induce changes in plant functional group composition with invasion of non-native annual plant, which significantly reduce productivity in arid ecosystems. Therefore, functioning of these arid systems depends to a high degree on plant diversity (Toderich et al. 2013b).

Toxic metal pollution of waters and soils is a major environmental problem in the industrial zones of Kyzylkum desert (Uzbekistan). For this area very little information is available on the accumulation of toxic ions, which derive from both in soils and waters, and pass through plants translocation into the food chain. There are also limited numbers of species being able to be established on strained conditions.

3 Structural and Functional Mechanisms of Plant Tolerance to Salt Stress

Soil salt contents and composition are the major determinant in the growth and productivity of plants. Plants regulating salt content in their tissues vary on the response reactions to salinity stress, often similar to other abiotic strained conditions they display common stress-tolerance pathways being sensitive or tolerant (Flowers et al. 2010). They use different mechanisms to overcome high salt concentrations and to grow safely in saline conditions (Glenn et al. 1999; Zhu 2001; Hameed et al. 2010; Dikilitas and Karakas 2010).

Most plants are salt-sensitive (glycophytes) a growth of which is affected by the elevated root-zone salinity. Salinity causes direct and indirect stresses on plant tissues. Primarily affecting on the water availability of plants (Qadir and Oster 2004) and resulting in ion specific effects, that is excessive accumulation of toxic ions (Na^+ and Cl^-) (Sun et al. 2009) salinity induces hyperosmotic stress in them (Marcelis and Van Hooijdonk 1999; Zhu 2001). Salt-induced secondary stress causes over-production of ROS (reactive oxygen species) (Sun et al. 2009). The most common effects of salinity display in inhibition of photosynthesis, ion imbalance, changes in metabolic activities, mainly in decline of enzyme activities and hormonal ratio, disturbance in solute accumulation, so, in a seed forming of plants etc. (Seemann and Critchley 1985; Dikilitas and Karakas 2010; Ahmad and Prasad 2012; Waskiewicz et al. 2013; Geissler et al. 2013). As a consequence, high concentrations of salt reduce nutrient uptake by plants (Gorham 1992, 1994; Grattan and Grieve 1994; Hasanuzzaman et al. 2013) and disrupt ion distribution (Zhu 2001). Nutrient availability is known to be reduced in plants by negative effects of sodium in the saline soils and by high pH in sodic soils (Grattan and Grieve 1994; Curtain and Naidu 1998; Arshad et al. 2011). High concentrations of salts lead to decrease in seed germination and seedlings vigor.

Salt tolerance is a multigenic trait which is governed by various morphological and physiological factors (Lokhande and Suprasanna 2012; Hasanuzzaman et al. 2013). To achieve a salt tolerance, three interconnected aspects of plant activities are important: (i) detoxification, damage of plants must be prevented or alleviated; (ii) homeostasis, homeostatic conditions must be re-established in the new, stressful environment; (iii) growth regulation, growth must resume, albeit at a reduced rate. Slower growth is an adaptive feature for plant survival under salinity allowing them to rely on multiple resources (Zhu 2001).

Complex molecular responses are involved in the tolerance of plants to salinity stress. These mechanisms include an osmoregulation through accumulation of compatible solutes (polyols, sugars, proline, glycinbetaine and other osmoprotectants) to maintain a sufficient amount of water content and production of stress proteins (Hasegawa et al. 2000; Zhu 2001; Yokoi et al. 2002; Lokhande and Suprasanna 2012). Latter play role in the protection of cellular structure by scavenging of ROS, in particular, by antioxidant defense system (Zhu et al. 2004), which includes antioxidant compounds (tocopherol and carotenoids) and enzymes (superoxide dismutase (SOD), catalase (CAT), peroxidase (POD) and many others (Garratt et al. 2002).

Salt tolerance is also associated with the uptake of potassium which is affected by high external sodium due to their chemical similarity (Seaman 2005).

There are two main mechanisms in plant adaption to saline environments: salt accumulation is the capability of plants to preserve a normal metabolic activity even in the presence of high intracellular salt levels; and salt avoidance – to prevent the penetration of salt ions into their sensitive cells (Manousaki and Kalogerakis 2011b). Perennial deep-rooted plants, avoiding the highly salinity in the soil solution near the surface, access water mostly from the regions of the root-zone with the lowest salinity. Whereas more shallow-rooted annuals complete their life cycle during period when the salinity of the soil solution near the surface is lower than in the subsoil (Bennett et al. 2009).

The plant avoiding high ion concentration in soil and salt exclusion at root level, compartmentalization of toxic ions in vacuole and ion excretion through specialized cells on the stem and leaves (Toderich et al. 2010b) are plant adaptation mechanisms in saline soils.

Salt tolerant plants that survive to complete their life cycle at the salt concentrations much exceeding the norm for living are called halophytes. Halophyte species are of special interest since they can also tolerate environments characterized by stresses as chilling, freezing, heat and drought (Shevyakova et al. 2003; Kholodova et al. 2010; Manousaki and Kalogerakis 2011a, b).

Halophytes are also considered to distinguish by their better natural adaption to HM contamination and may be able to accumulate them in large amount compared to salt-sensitive crop plants commonly chosen for phytoremediation purposes (Jordan et al. 2002; Manousaki and Kalogerakis 2009; 2011a; Kholodova et al. 2010). It was suggested that tolerance to salt and HM may partly demonstrate the common physiological mechanisms (Przymusinski et al. 2004). Studies by Thomas et al. (1998) on the halophyte *Mesembryanthemum crystallinum* suggest that halophyte copes with copper metal exposure through distinct genetic mechanisms. Individual environmental stresses such as NaCl and copper overlap to some extent and one explanation may be that several integrated mechanical and chemical signals are responsible for stress-related responses (Manousaki and Kalogerakis 2011b).

Drought resistant mechanisms may indirectly contribute to accumulation of HM by halophytes and their tolerance, since HM stress is also considered to be responsible for secondary water stress in plants (Poschenrieder et al. 1989; Boyd and Martens 1998; Macnair 2003).

Halophytes have different mechanisms and strategies for the sequestration and regulation of the salt ion concentration in the plant tissues. Some halophytes have salt excretion organs (salt glands, salt bladders or thichomes) in their leaves as a secondary tolerance mechanism (Glenn et al. 1999; Lefevre et al. 2009; Toderich et al. 2010a, b; Eid and Eisa 2010). High concentration of various ions can be accumulated in the vacuoles of bladder-trichome terminal cells which are frequently developed on the adaxial surface of epidermal cells of leaves or bract/bracteoles. Increasing of sclerification, availability of pigments and tracheid-like cells holding moisture, abundance of crystals in the fruit tepals, perianth tissues also promote the protection of embryo from unfavorable desert conditions (Toderich et al. 2012). It was revealed that an important strategy common to other halophytes also are vesicular hairs for salt excretion (Grigore and Toma 2007).

Although not all halophytes have these regulating organs, salt accumulating glands are mostly common in the *Poaceae*, *Tamariccacea*, *Chenopodiaceae* and *Frankenaciaceae* and many species from these families are known to have grandular structures (see Manousaki and Kalogerakis 2011a, b).

To prevent water loss halophytes develop some morphological features such as increased succulence of the leaves and stems which is considered as another important adaptive phenomenon in co-called "obligatory halophytes" as *Suaeda maritima*, *Salicornia europaea*, *Camphorosma annua*, *Bassia hirsuta* and *Petrosimonia oppositifolia*. Tracheoidioblastes in the structure of tissues of *Salicornia europaea* are possible involved in the maintenance and dynamics of water inside the plant. Some specific arrangements of foliar tissues were found as atriplicoid type Kranz anatomy in *Atriplex tatarica*, the kochioid type anatomy in *Petrosimonia oppositifolia* and *Camphorosma annua*, "sympegmoid" structure in *Suaeda maritima* and *Salicornia europaea* (Grigore and Toma 2007). A thick cuticle on leaves, a reduced number of stomata or sunken stomata, altered stomatal distribution and rolled leaves are also adaptive mechanisms of halophytes to salinization (Flowers et al. 1986; Cruz and Cuartero 1990; Dikilitas 2003; Ashraf and Foolad 2007; Lokhande and Suprasanna 2012). These mechanisms preventing water losses thereby might reduce the uptake and toxic effect of excessive of ions and improve salinity tolerance (Flowers et al. 1991).

Diversities in sexual reproduction mechanisms and CO_2 fixation pathways are also important factors regarding reproduction and survival under saline and technogenic contaminated desert environments (Toderich et al. 2010a, b).

4 Ecological Halophyte Groups and Their Characterization

Halophytes generally use similar salt tolerance effectors and regulators and regulatory pathways that have been found in glycophytes, but that subtle differences in regulation account for large variations in tolerance or sensitivity (Zhu 2000). Such plants are reported to tolerate salt concentrations around 200–250 up to 1000 mM NaCl that abolish 99 % of other plant species (Munns and Tester 2008; Flowers and Colmer 2008; Flowers et al. 2010).

Plant salt tolerance has been revealed for more than 200 years ago (Flowers et al. 1986) and there have been several definitions on "halophyte" constitutes. Any definition must include a threshold value of salt concentration and salt concentration used in the definition counts the number of halophytes. Based on this parameter Aronson listed about 1550, but Menzel and Lieth 2600 salt-tolerant plant species at about 80 mM NaCl (Flowers et al. 2010).

Halophytes were variously classified as euhalophytes (true halophytes, salt accumulators), psuedohalophytes or glycohalophytes (salt avoiders) and crinohalophytes (salt excretors) (Nagalevskii 2001; Zhao et al. 2002; Hameed et al. 2010).

The presence of sodium ions in halophytes has been reported to be partitioned in the cell vacuole and thereby depleting the cytoplasmic sodium level conferring the plants to function as normal. As above mentioned, the most salt resistant plants are the salt accumulating halophytes with succulent structure like *Salicornia*, *Salsola*, *Suaeda* and concentration of Na^+ and Cl^- in vacuoles generally exceeds the external concentration and is equal to or greater than that of seawater (Flowers 1985). Another group plants like *Atriplex* and *Halimione* consists of the salt localizing halophytes, in which the salt is localized in special hairs covering the stem and the leaves (Krüger and Peinemann 1996). 50 % or higher of the salt entering the leaf of a crinohalophytes is found can be excreted (Glenn et al. 1999).

Halophytes are not uniformly distributed across the taxa of higher plants and also bryophytes and pteridophytes. *Caryophyllales* order distinguishes by the greatest number of halophytes (Flowers et al. 2010). Large percentage of halophytes belongs to *Chenopodiaceae* and *Poaceae* families. According to Le Houerou (1996) there are as many as 6000 species of terrestrial and tidal halophytes in the world, the largest proportion of which belongs to *Chenopodiaceae. Poaceae*, *Fabaceae* and *Asteraceae* families also distinguish by numbers of halophytes, although they represent fewer than 5 % of their species (Glenn et al. 1999).

More than 140 species of the *Salsola*, which includes shrubs, semishrubs, perennial and annuals herbs, are known to be tolerant to water, heat and salt stresses have been described in Central Asia (Botcszantsev 1969; Freitag 1997). They colonize marginal and desert lands and constitute up to 45 % of all *Chenopodiacea* species here. They have developed adaptive features that give them more competitive advantage over sensitive glycophyte species (Toderich et al. 2012).

The Mediterranean flora includes about 700 species of halophytes, 70 % of them are perennials and 30 % – annuals (Le Houerou 1996). 430 halophyte species are identified for China (Zhao et al. 2002). More than 70 species of halophytes are counted only in *Chenopodiaceae* family from flora of Azerbaijan (Movsumova 2011).

4.1 Halophytes of Kur-Araz Lowland

Kur-Araz lowland is considered as a winter pasture with a dominating of the salt resistant grasses and brushes widely used by cattle. Dry steppes and semi-deserts with ephemeras, saltworts and wormwoods, saltwort-wormwood associations on sierozem and grey-brown podzolic soils, with cereal-wormwood and bearded associations on dark grey-brown soils are generally prevalent in this region (Movsumova 2003). *Salsola*, *Suaeda*, *Kalidium* etc. genus are typical representatives of clayey deserts, solonchaks and solonetzic soils forming pure and different associations mixed with grasses and wormwoods predominated on the desert landscapes (Movsumova 2011).

More than 1,215 plant species (27 % of the total flora of Azerbaijan numbering 4,500 species) are found here. A reason of the comparative poverty of species composition of flora in this lowland is that, firstly, the Caspian Sea relatively recently moved away from these lands and secondly, soils here are mainly covered by agricultural crops (Musayev and Fataliyev 2004).

Dominant halophytes and other attendant plants in Kur-Araz lowland are species from *Salsola* (Moq) Iljin, *Halostachys* (Biob). C.A. Mey, *Suaeda* (C.A.Mey) Mog, *Artemisia* L., *Lagonychium* Bieb., *Glycyrrhiza* L. etc. (Table 2). *Salsoleta nodulosae* occupies a large area on saline soils here (Movsumova 2003; Gurbanov 2009). Among 16 *Artemisia* species distributed in Azerbaijan only five species (*A.annua, A.arenaria, A.scoparia, A.szovitsiana, A.fragrans*) are described in Kur-Araz lowland at the altitude from −28 m to 100–150 m. All these *Artemisia* species forming pure phytoceonosis are the edificators. In the floristical composition of *Artemisia* also take part *Salsola dendroides, S.ericodes, S.crassa, S.nodulosa, Halocnemum strobilaceum, Kalidium caspicum, Suaeda dendroides, Kochia prostrata, Capparis spinosa, Glycyrrhysa glabra, Convolvulus persicus, Limonium meyeri, Botriochlos ischaemum, Stipa szovitsiana, Festuca sulcata, Agropyron cristatum, Achillea nobilis, Teucrium polium* (Melikov and Huseynova 2004). The *Artemisia* species are the fodder storage stocks in these areas and are considered a best forage for animals due to the comparatively high contents of crude protein (12.7 %) and fat (9.2 %) (Movsumova 2005) (Fig. 5).

Artemisia fragrans L. is the widespread plant in all the type of polluted areas, is found to vigorously grow on soils polluted by HM, petroleum hydrocarbons as well as saline conditions in Azerbaijan (Alirzayeva et al. 2006a; Ali-zade et al. 2010, 2011; Alirzayeva et al. 2011; Alirzayeva and Neumann 2013). This plant is dominating in wide areas of Kur-Araz lowland, also found in dry slopes (Shukurov et al. 2008; Gurbanov 2009). Due to rich contents of secondary metabolites this plant is valuable as medicinal resource and is the main important component of winter pastures (Alirzayeva et al. 2006b).

Halostachys caspica (Bieb) C. A. Mey being succulent shrub is found to accumulate 30 % of the salts. Plants with the height of 70–150 cm are distributed in the following communities: *Halostachys caspica + Petrosimonia brachiata, Halostachys caspica + Suaeda dendroides* and form a formation with 29 species of flowering plants in the Kur-Araz lowland (Movsumova 2005).

Kallidium capsicum (L.) Ung. Stemb. distributed in very salinized soils is indicator of salty subsurface water basins. This plant is one of the predominant plants of the solonchak desert and semi-desert of Azerbaijan. This is important component of the winter pastures for the sheep and cattle. Its ash contains 19 % of potash and 4 % of soda (Shukurov et al. 2008).

Kochia prostrata L. is the semi-shrub plant, the root system of which operates on a soil depth of 220–280 cm. This plant possesses ecological plasticity and distinguishes by ability to adapt to different types of soils. Containing 12.4–11.8 % of protein and 1.5–2.5 % of fat this plant is a valuable fodder source. Being widespread plant in the deserts and semi-deserts of the Kur-Araz lowland and resistant to various extreme factors it is perspective to create the pastures here (Melikov and Bayramova 2004).

Table 2 Basic information on some dominant plant species from saline areas of Kur-Araz lowland

Species	Family	Habit	Life form	Height	Biologically active substances (Mehtiyeva 2011; Mehtiyeva and Zeynalova 2013)
Alhagi pseudalhagi	*Fabaceae*	Herb	Perennial	30–70 sm	Steroids, essential oils, coumarins, alkaloids, tannins, vitamin C, K, B2, B1, organic acids, vitamin K, carotene, flavonoids, anthocyanins
Artemisia fragrans	*Asteraceae*	Herb	Perennial	30–40 sm	Essential oils, alkaloids, flavonoids, carbohydrates, organic acids, rubber sesquiterpenoids
Artemisia scoparia	*Asteraceae*	Herb	Biennial	30–70 sm	Essential oils, rubber, phenol carboxylic acids, flavonoids, organic acids, coumarins, tannins, lactones
Atriplex tatarica	*Chenopodiaceae*	Semishrub	Annual	100 sm	Alkaloids, coumarins, flavonoids, saponins
Carthamus oxyacanthus	*Asteraceae*	Herb	annual	14–40 sm	Sesquiterpenoids, flavonoids, anthocyanins, carotenoids
Climacoptera crassa	*Chenopodiaceae*	Herb	Annual	10–40 sm	Organic acids (citric acid), alkaloids, ash, fatty oil
Cirsium arvense	*Asteraceae*	Herb	Perennial	50–60 sm	Glycosides, fatty oil, alkaloids, carbohydrates, vitamin C, essential oils, rubber, flavonoids, steroids, coumarins
Glycyrrhiza glabra	*Fabaceae*	Herb	Perennial	60–80 sm	Alkaloids, essential oils, resins, steroids, phenol carboxylic acids, coumarins, tannins, flavonoids, vitamin C, carotene, anthocyanins
Halocnemum strobilaceum	*Chenopodiaceae*	Shrub	Perennial	10–30 sm	Alkaloids, organic acids (oxalic acid, citric acid), saponins

(continued)

Table 2 (continued)

Species	Family	Habit	Life form	Height	Biologically active substances (Mehtiyeva 2011; Mehtiyeva and Zeynalova 2013)
Halostachys caspica	*Chenopodiaceae*	Shrub	Annual	3,5 m	Betaine, diphenylamine, sitosterol, benzoic acids
Kallidium capsicum	*Chenopodiaceae*	Semishrub		10–70 sm	Alkaloids, carotenoids (capsanthin, capsorubin, β-carotene, lutein, zeaxanthin etc.), fats, vitamins A, C and others; volatile oils
Kochia prostrata	*Chenopodiaceae*	Semishrub	Annual	30–60 sm	Saponins, alkaloids, tannins, flavonoids, coumarins
Lagonychium farctum	*Fabaceae*	Shrub	Perennial	50–60 sm	Tannins, flavonoids, steroids
Limonium meyeri	*Limoniaceae*	Herb	Perennial	40–100 sm	Tannins, carbohydrates
Medicago minima	*Fabaceae*	Herb	Annual	5–25 sm	Saponins in the hydrolyzate
Medicago orbicularis	*Fabaceae*	Herb	Annual	7–40 sm	Saponins in the hydrolyzate
Rumex tuberosus	*Polygonaceae*	Herbs	Perennial	50–100 sm	Vitamin C, tannins, anthraquinones
Salicornia europaea	*Chenopodiaceae*	Herb	Annual	5–40 sm	Alkaloids, flavonoids, tannins, anthocyanins, steroids, carbohydrates, phenolcarboxylic acid
Salsola dendroides	*Chenopodiaceae*	Semishrub	Annual	Up to 1 m	Organic acids, alkaloids, saponins
Salsola nodulosa	*Chenopodiaceae*	Shrub	Perennial	30 sm	Alkaloids, flavonoids, tannins, fenolcarboxilic acids, Vitamin C, K, carotins, coumarins, laktons, organic acids
Suaeda confusa	*Chenopodiaceae*	Shrub	Annual	10–40 sm	Alkaloids, fatty oil

(continued)

Table 2 (continued)

Species	Family	Habit	Life form	Height	Biologically active substances (Mehtiyeva 2011; Mehtiyeva and Zeynalova 2013)
Suaeda altissima	*Chenopodiaceae*	Shrub	Annual	30–200 sm	Alkaloids, fatty oils
Silybum marianum	*Asteraceae*	Herb	Biennial	90–100 sm	Rubber, fatty oil, organic acids, alkaloids, flavonoids, carbohydrates, steroids, vitamin K, resins, saponins, essential oils, terpenoids, anthocyanins, carotenoids
Tamarix ramosissima	*Tamaricaceae*	Shrub and small tree	Perennial	2–6 (8) m	Alkaloids, tannins, steroids, coumarins, flavonoids, anthocyanins, phenolcarboxylic acids, tannins

Fig. 5 Photo of some plants dominated in Kur-Araz lowland

Limonium meyeri (Boiss) O. Kuntze is distributed in chal-meadows, salty-meadows and semi-desert habitats in this lowland. This plant possesses a wide range importance being used for medicinal purpose, as decorative plant as well as in mode of life (Shukurov et al. 2008).

Salsola dendroides Pall. Illustr. is the indicator of deteriorated and salty soils as well as subsurface water basins. It is very important component of the winter pastures, sheep and cattle eat it voraciously after hard winter (Shukurov et al. 2008). This species is edificator of plant community which includes 100 species. It forms the following associations in the Kur-Araz lowland: *S. dendroides* (pure), *S. dendroides + Suaeda microphilla*, *S. dendroides + Alhagi pseudoalhagi + Artemisia meyeriana*. A deep root system of *S. dendroides* reaches the ground water (Movsumova 2005).

Salsola nodulosa (Moq.) Iljin is endemic species for Caucasus. This plant possesses a strong root system that reaches deep soil layers in the Kur-Araz lowland. It is main important component of the winter pastures and valuable fodder for lifestock. Containing alkaloid this plant has medical importance, too (Movsumova 2003; Shukurov et al. 2008).

Suaeda microphilla Pall. Illustr. is widespread plant in the degraded lands of Kur-Araz lowland. This plant is considered as the high quality fodder in the winter pastures (Shukurov et al. 2008; Gurbanov 2009).

Tamarix ramosissima Ldb. is found in saline areas of Kur-Araz lowland and known as the indicator of salinization. This plant is considered as producer of black dye in its all parts (Shukurov et al. 2008).

4.2 Halophytic Flora of Kyzylkum Desert

Lower reaches of Zarafshan River Valley (old agricultural zone) and Kyzylkum Deserts are considered as the waste marginal lands and natural rangelands affected by aridity and salinity. Survey of the regions revealed wild halophytes representing 19 taxonomical families with more than 380 species of different groups of salt tolerant plants (Akjigitova 1982; Goldshtein et al. 2000; Shamsutdinov et al. 2000; Gintzburger et al 2003; Toderich and Tsukatani 2007).

These regions of Uzbekistan are distinguished by the highest numbers of representatives from *Chenopodiaceae* (33 %), *Asteraceae* (20 %), *Poaceae* (11 %), *Fabaceae* and *Brassicaceae* (about 11 %) families. Species belonging to *Polygonaceae*, *Plumbaginaceae*, *Zygophyllaceae*, *Cyperaceae* take smaller part (3–5 %), whereas, *Eleagnaceae*, *Plantagainaceae* and *Frankeniaceae* cover less than 1.0 % of rangelands halophytic pastures (Table 3). The area is also concerned due to a high plant endemism which makes up about 3.4 % of the total species here (Toderich and Tsukatani 2007).

Many populations of local halophyte species from flora of Uzbekistan are small and sometimes fragmented and are frequently not able to complete their life cycles, to reproduce under these conditions. At the same time, overgrazing leads to nearly disappearing of germplasm and an irreversible loss of biodiversity resources.

Table 3 Description of halophytic plant communities in the Karakata salt depression, Central Kyzylkum

	Biotope/ecological groups	Description of plant communities	Water table depth (m)	Improvement practice	Soil salinity level as per sodium content (Na+, mg.eq/100 g)	Halophytic pasture yield (t ha−1)a	Soil t ha−1 moisture (%)
1	Sandy and gray-brown *desert soils*/aboriginal psammophytic pastures (xerophyte)	*Ferula assa-foetida, Aellenia subaphylla, Ammothamnus Lehmannii, Astragalus villosissimus, Artemisia diffusa, Salsola praecox, Turnefortia sp., Calligonum leocacladum, Stipa sp., Ammodedron Connoliy*	20–28	No	Low (0.3–0.5)	0.7	2.8
2	Sagebrush with ephemers (Artemisia spp.) sandy desert (xerohalophyte)	*Artemisia diffusa, Haloxylon aphyllum, Peganum harmala, Salsola sp.,Climacoptera lanata*	20–18	No	Low (0.2–1.2)	2.48	3.3
3	Haloxylon forest (haloxerophyte)	*Peganum harmala, Haloxylon aphyllum, Alhagi pseudalhagi, Salsola sp., Sueda sp., Climacoptera lanata*	4–6	No	Low-moderate (3.5–7.0)	1.5	7.9
4	4.1. Desert salt affected soil improved through an agro-silvi – pastoral model	*Aboriginal strips of halophytes with Climacoptera lanata mixed with moderately salt tolerant tree-shrubs-traditional crops (Sorghum, Pennisetum-Amaranthus etc. crops)*	1.5–2.8	Without irrigation and fertilizers (control)	moderately (7.9–12.0)	4.7	8.1

	4.2. Desert salt affected soil improved through an agro-silvi – pastoral model	*Aboriginal strips of halophytes with Climacoptera lanata mixed with moderately salt tolerant tree-shrubs-traditional crops (Sorghum, Pennisetum-Amaranthus etc. crops)*	0.9–2.0	Irrigation with saline water, without fertilizers	Moderately (7.9–12.0)	21.8	8.1
	4.3. Desert salt affected soil improved through an agro-silvi – pastoral model	Aboriginal strips of halophytes with *Climacoptera lanata* mixed with moderately salt tolerant tree-shrubs-traditional crops (*Sorghum, Pennisetum-Amaranthus* etc. crops)	0.9–2.0	Irrigation with saline water and fertilizers	Moderately (7.9–12.0)	43.6	8.1
5	Wet solonchak/hyperhalophytes (pure stands)	*Tamarix hispida, Salicornia europea, Sueda sp., Climacoptera lanata, Alleropus litoralis, Halostachys caspia, Halimochnemis strobilaceum*	0.5–0.9	No	High (17.5–34.0)	11.5	8.9

aCalculated for *Climacoptera lanata* under different agropastoral management practices

According to Toderich and Tsukatani (2007) a new concept (based on soil characteristics, water table level and mineral composition of aboveground biomass, morphological and reproductive traits and carbon discrimination values) for classification of the following ecological groups of halophytes from flora of Uzbekistan was developed.

Group I Hyperhalophytes: This group of plants is characterized by the presence of highly saline water (full-strength sea water up to 100 dS/m). The water table varies from 0.5 to 1.5 m depth, with solonchak-alkaline and solonetz soils. Plants are succulent and have both C_3 and C_4 types of photosynthesis. The main species of this group includes *Climacoptera longistylosa, C.kasakaroum, C.bucharica, C.crassa, C.subcrassa, C.transoxana, C.ferganica, C aralensis, C.turcomanica, C.turgaica, C.itricata, C.turkestanica, Plantago coronopus*; *Halostachys belangerana; Halocnemum strobilaceum*; *Petrosimonia crassifolia; P. litwinowi; Halocnemis varia; Halogeton glomeratus; Salicornia* spp.

Group II Hydro-halophytes: Species within this group that has C_3 photosynthesis occur in wet to standing water, varying from freshwater to brackish water marshes, ditches, around seeps and springs. They can tolerate up to 10,000 ppm of salts and more. The main species include *Phragmites australis*; *P. communis*; *Hippophae ramnoides; Populus diversifolia; Elaeagnus angustifolia; Arundo* spp.; *Typha* spp.

Group III Euhalophytes: This group of plants is a good indicator of superficial distribution (1–2 m) of underground water and mainly consists of salt-accumulator and salt-excluder plants. Such species occur in wet sandy areas at the edge of salt flats, marshes and salt deserts; wet marshes saline soils on the margin of lakes, tugai, salted desert depression, takyrs. The vegetation mainly comprises of *Suaeda*, annual *Salsola* species; *Aeluropus repens; A. littoralis; A. villosus; Poa littoralis*; *Tamarix hispida, T. androsovii, T. rasimossisima; Dactylis littoralis; Kochia scoparia* which tolerate full-strength sea water and they mostly have C_4 photosynthesis.

Group IV Halo-xerophytes: The water table is at >4 m depth having different types of soils (from sandy to clayey), grey with gypsum content, alkaline meadow salt-marshes and sandy desert soils. *Haloxylon aphyllum; Ephedra strobilaceae; Halothamnus subaphylla; Campharosma lessingii; Kochia scoparia; Alhagi pseudoalhagi; Lycium turcomanicum; L. ruthenicum; Ceratoides ewersmanniana; Anabasis annua; A.salsa; A.aphylla; A.eriopoda; A.ferganica;* perennial *Salsola* spp.; *Zygophyllum* spp.; some *Calligonum* spp. are the major contributors of this group.

Group V Halogemimezophytes: Where the water table varies between 1.5 and 2.5 m in depth, this group inhabits steppes, semi-desert and desert zones mostly on solonetz-alkaline soils, lake shores and river banks. *Cynadon dactylon; Limonium gmelinii; Salsola arbuscula; Karelinia caspica; Zygophyllum fabago; Halimodendron halimodedron; Agropyron desertorum; Eremopyrum orientalis; Psylliostachys suvorovii; Atriplex tatarica; Bassia hyssopifolia; Glycyrrhiza* *glabra; Limonium otolepis; Frenkenia* spp. are the major dominants of the group. It includes both C_3 and C_4 plants.

Group VI Halo-gemipetrophytes: The water table varies between 1.5 and 4.0 m where the plants grow on stony skeletal saline substrate. It is a small group of plants mostly shrubs, semi-shrubs and semi-shrublets: *Haloxylon ammodendron; Salsola arbusculiformis* and some species of *Atraphaxis, Nanophyton, Anabasis.*

Group VII Metallo-halophytes: Most hyperaccumulators of HM and salt ions grow slowly and have smaller biomass with delayed flowering and low seed quality (Toderich et al. 2004). Most distinguishable species within flora of Uzbekistan are *Artemisia diffusa, A.halophyla; Karellinia caspica; Allysum desertorum; Tamarix hispida; Frankenia* spp. and they translocate and accumulate toxic metals from soils into the roots and aboveground biomass (shoots, leaves etc.).

Salicornia, Halostachys, Halimocnemis, Climacoptera genus, i.e. hyperhalophytes are widespread near artesian hot spring with a high salt mineral content. The conventional salt tolerant crops (sorghum, pearl millet, fodder beet, safflower etc.) occupy an intermediate place between hyperhalophytes and xero-halophytes (Toderich and Tsukatani 2007).

Kyzylkum vegetation contains only a restricted number of species, mainly from *Salsola* (both annuals and perennial), tamarisk and grasses graminous, which have metal/salts removal abilities and also can survive and reproduce under these contaminated environments. However, typical halophytes such as *Suaeda salsa, Halostachys caspica, Salicornia europea* and many species of annual chenopods show insignificant tolerance to heavy metals toxicity.

Plants accumulating high concentrations of salts are known to be toxic for livestock and cause different types of diseases and physiological disorders (Escarre et al. 2000). Halophytic forage species growing under Kyzylkum desert conditions differ on their chemical compositions, especially on nutrient contents. The total mineral ions were found to be highest in *Halocnemum strobilaceum* (Pall.), then in *Climacoptera lanata* Pall., *Suaeda salsa* (L.) Pall., *Tamarix hispida* Willd., *T. laxa* Willd. and *Haloxylon aphyllum* (Minkw.) Iljin., *Salsola* spp., *Zygophyllum* spp. plants. While *Alhagi pseudoalhagi, Artemisia diffusa, Poaceae spp.* distinguish by minimum concentration of mineral ions and are considered as more suitable fodder plants than other salt tolerant species. Due to gradually accumulation of K^+ they as well as some salt tolerant graminous fodder grass can be categorized as relatively more palatable forage plants. The highest concentration of K^+ was found in *Kochia scoparia, Atriplex nitens* and *Suaeda arcuata* growing in Kyzylkum desert. Naturally growing plants, e.g. *Halocnemum strobilaceum, Tamarix hispida, Climacoptera lanata, C. bucharica, C. turkestanica, Halostachys caspia, Halothamnus subaphylla* demonstrated Na^+ concentrations in near the critical limit for livestock. Due to exclusion of the excess salts through salt glands abundant on the surface of epidermis low amount of mineral ions were revealed in annuals, e.g. *Salsola* spp. and grasses like *Bromus tectorum* L., *Aeluropus litoralis* (Goun) Parl, *Eremopyrum orientale* (L), *Agropyron desertorum* (Fisch.) Schult (Toderich and Tsukatani 2007; Toderich et al. 2010a, b).

Species of *Artemisia* have good ability to translocate through plant tissues a significant high quantity of sodium, chlorides and other mineral ions. *Tamarix hispida* (a salt extraction hyperhalophyte) demonstrated a remarkably high Fe and Co levels in the aerial dry matter. It was also revealed that native desert metallohalophytes tend to accumulate the highest ions concentrations (primarily sodium, chlorates and oxalates) in the epidermal and subepidermal tissues, including various glandular structures on leaves, bracteoles and perianth segments. Some naturally highly adapted metallohalophytes develop a cellular mechanism to partition toxic salts into vacuoles or to exclude salt at the root zone, so it does not impact on cell metabolism and division. A high concentration of various ions can accumulate in the vacuoles of bladder-trichomes terminal cells (Toderich et al. 2010b).

5 Use of Halophytes in Phytorehabilitation Technologies for Improvement of Productivity and Ecosystem Resilience and Function

The strategy of the phytorehabilitation and desalinization of saline lands with the use of halophytes is expanding and developing, as this is more cost effective approach and could be successfully used than genetic and biochemical ones. The ability of halophytic plants to remove toxic ions varies significantly between species and ecotypes. By use of these plants land reclamation and rehabilitation can be achieved. Boyko (1966) was the first to suggest that halophytes could be used to desalinate soil and water. A number of untraditional native and exotic salt tolerant plant species and valuable crops suitable for reclamation of salt-affected and dry lands were shown to be very useful in preliminary cultivation trials (Nechayeva 1989; Shamsutdinov and Shamsutdinov 2007; Toderich et al. 2010a, b). Na^+ and Cl^- hyperaccumulating halophyte species such as *Suaeda maritima, S. portulacastrum, S. fruticosa, S. salsa, S. calceoliformis, Kalidium folium, Sesuvium portulacastrum, Arthrocnemum indicum, Atriplex nummularia* and *A. prostrata* are found to accumulate high concentrations of salt in their aboveground tissues and consenquently, to upgrade saline soils by harvesting the plants on a regular basis (Zhao 1991; Ravindran et al. 2007; Manousaki and Kalogerakis 2011a, b).

Investigation of some *Suaeda* (C.A.Mey) Mog species revealed their capacity to accumulate salt ions and to decrease the soil salt content in their habitat, consequently, to increase the area of land available for cultivation and the yield of crops grown on the marginal saline soil (Sharma and Gupta 1986; Zhao 1991). *S.salsa* normally growing in highly saline soils 4.5 % reduced Na content of the soil at depth 20–30 cm at a density of 15 plants m^{-2} and 6.7 % with density of 30 plants m^{-2} (Zhao 1991). As salt accumulating halophytes plant *S.maritima* also was found to exhibit greater accumulation of salts in their tissues compartmentalizing the toxic Na^+ in their vacuoles. During 4 months sodium chloride was removed by this plant in amount of 504 kg from 1 ha saline land (Ravindran et al. 2007). Studies with naturally growing perennial halophyte *S.fruticosa* reported that more than 1088.6 kg of salt can be removed from 1 acre by a single harvest of the aerial parts in the fall each year (Chaudhri et al. 1964). These *Suaeda* species can be considered to be successfully used for accumulation of NaCl in highly salinized areas for crop production after a few repeated cultivation and harvest.

Similar tendency regarding accumulation of salt ions and HM has been reported from many other halophytes associated with salt tolerance. C_4 perennial native shrub of Mediterranean basin *Atriplex halimus* L. is known as a tolerant plant to drought (Le Houerou 2000), salinity (Bajji et al. 1998) and HM stress (Lutts et al. 2004). This plant with high biomass production, deep root system and favorable crude protein content is dominant livestock fodder reserves in arid and semi-arid countries (Ortíz-Dorda et al. 2005; Osman et al. 2006). *A.halimus* also distinguishes by accumulation capacity of HM, namely Pb, Zn and Cd in salt affected soils, as salinity is known to change the bioavailability of metals in soil and be a key factor in the translocation of metal from roots to the aerial parts of the plant (Jordan et al. 2002; Manousaki and Kalogerakis 2009; Nedjimia and Daoudb 2009). Seedlings of this species displayed significant values of Cd and Zn accumulation in its aerial parts (830 and 440 mg kg^{-1}, respectively), when were exposed to treatment by these metals and the rate of their translocation was increased with exposure time (Lutts et al. 2004). Most of Cd taken up by a newly revealed Cd-hyperaccumulator highly salt-tolerant halophyte *A.halimus* subsp *schweinfurthii* was found to be retained in roots (606.51 mg g^{-1} DW) after 15 days at 400 μM $CdCl_2$ (Nedjimia and Daoudb 2009) and soil salinity was improved for 40 % by this plant during 1 year experiment (Ould Ahmed 2012). Another species of *Atriplex* genus *A.nummularia* can grow on land of marginal quality and accumulate significantly more Cu and Pb in the shoots than glycophyte species in the same conditions (Jordan et al. 2002; Eid and Eisa 2010). The salinity-induced threefold increase in the Cd concentration was also observed in *A.hortensis* (Lopez-Chuken and Young 2005). *Atriplex* spp. are suggested to be an effective plants for phytoextraction and phytostabilization of HM in saline soils (Lutts et al. 2004; Nedjimia and Daoudb 2009), phytoremediation and phytorehabilitation of degraded lands like sand dunes, saline/alkaline soils, marginal sites with low fertility and poor soil structure (Le Houerou 1992; Ortíz-Dorda et al. 2005; Eid and Eisa 2010).

Resistance mechanisms to HM may be explained by precipitation of Cd in oxalate crystals in the stems in *A.halimus* (Lutts et al. 2004), while *A.nummularia* (Eid and Eisa 2010) and *Tamarix aphylla* L. and *T. smymensis* Bunge (Manousaki and Kalogerakis 2011a) have salt glands through which sodium and chloride ions as well as Cd, Zn, Cu and Pb are excreted from their leaf tissues onto the leaf surface as a possible detoxification mechanism against the metal burden.

Investigation of phytoextraction of Cd and Pb from soil and their excretion by the leaves of salt cedar *Tamarix smymensis* Bunge being widespread salt tolerant plant in the Mediterranean region showed that increased soil salinity results in an increase of the cadmium uptake and excretion by plant without visible signs of metal toxicity. This resistance mechanism of *T.smymensis* is an unique advantage that may change current phytoextraction practices in salt-affected soils (Manousaki et al. 2008; Kadukova et al. 2008).

Artemisia L. species are plants tolerant to a wide range of different stress factors, they are known as halophytes being tolerant to salinity (Ishikawa and Kachi 2000), semi-desert plants surviving drought (Evans et al. 1992) and high temperature (Wen et al. 2005) and various organic and inorganic pollutants (Ali-zade et al. 2010, 2011; Alirzayeva and Neumann 2013). A number of *Artemisia* L. species/ecotypes are widespread and vigorously grow at various locations in different geographical regions in the world heavily exposed to environmental pollutants and distinguish by their HM hyperaccumulation capacity (Morishita and Boratynski 1992; Kim et al. 2003; Takeda et al. 2005; Alirzayeva et al. 2006a, b). Results obtained in the previous studies(Alirzayeva et al. 2006a; 2011) have shown that some species of halophyte plants *Artemisia* L. (*A. scoparia, A. fragrans, A.caucasica, A. szovitsiana, A. arenaria*) capable of growing under conditions of strong salt-affected soils markedly differ in their HM accumulation capacity, depending on the locations. All the *Artemisia* L. species tested growing at different distances from a polluting source were found to be metal accumulator plants. Among them *A.scoparia* showed the highest capacity to accumulate Cd, Cu, Pb and Zn in their shoots, and the high values of the bio-accumulation factors for all HM tested also revealed. All these *Artemisia* species were found to accumulate HM in their organs higher than the investigated attendant plants (*Argusia sibirica, Salsola dendroides, Climacoptera crassa*) from the same contaminated sites. *Artemisia* species (especially *A.scoparia*) growing in contaminated soils with their large biomass and long roots reaching 60–140 cm depth (Abduyev 1963), could be promising for use in the restoration of degraded soils (Alirzayeva et al. 2006a; Ali-zade et al. 2010, 2011).

Artemisia argyi L., *Limonium bicolor* K., *Melilotus suaveolens* L. and *Salsola collina* P. being the halophyte plants are shown by Sakai et al. (2012) to be effective for amelioration the salt-affected soils and confirmed the good correlation between growth of halophytes and Na and K content in them.

Diversities in sexual reproduction mechanisms and CO_2 fixation pathways are important factors for tree-like Asiatic *Salsola* L. species, regarding reproduction and survival under saline and technogenic contaminated desert environments. Phenotypic plasticity in the sexual expression of flower organs affects interspecies and intra-population genetic structure both for annual and perennial species. Salt gland secretion products on the surface of fruit tepals and perianth might protect reproductive organs and provide a strategy for germplasm conservation. Small and isolated populations of the annual species with a predominantly wind-crossed mode of reproduction exhibit low levels of genetic variation and in return the ability of plants to adapt to frequent droughts and climate changing conditions. Asiatic *Salsola* L. complex is an example of evolutionary convergence of ecological, structural and physiological mechanisms, which are determined genetically, to adapt to harsh desert environments. Stems, fruits and leaves of these plants can be used as a year-round feed for camels and a summer feed for sheeps and goats. So, appropriate conservation and sustainable utilization measures require protecting natural habitats of Asiatic *Salsola* L. species (Toderich et al. 2010a, b; 2012). C_4 species with NADP-ME photosynthesis, *S.arbuscula* and *Salsoloid* type of Kranz-anatomy both in leaves and reproductive organs form an unique "functional reproductive plant group" that is very resistant to the severe stresses in arid, contaminated habitats (Toderich et al. 2008).

Milić et al. (2012) indicate that halophyte species *Salicornia europaea, Suaeda maritima* and *Salsola sada* may accumulate significant amount of salt ions from saline soils and therefore remediate land to the point where native plants can invade and become established or the site can be returned to agricultural productivity.

Salt tolerant trees and shrubs species e.g. *Populus*, *Haloxylon*, *Salix*, *Elaeagnus*, *Morus*, *Atriplex*, *Berberis*, *Hippophae* established on good deep soils demonstrated good potential as part of the arid fodder production system. Fodder shrubs were associated with cereal farming system, including rangeland species alone, or mixed with different salt tolerant traditional fodder crops, such as *Sorghum bicolor* and *Pennicetum glaucum*. As part of the desert land revegetation, saltbushes *Atriplex canescens*, *A. undulata* and *A.nitens* were recently introduced to the saline sandy desert zones of Kyzylkum. Better plant growth, accumulation of green biomass and consequently yield of both fresh and dry matter were observed for *Kochia scoparia*, *Climacoptera lanata*, *Atriplex nitens* both in pure soil and mixed artificial agrophytocenous. *Kochia scoparia* and *Agropyron desertorum* were revealed to produce up to 1.3 t DM/ha in association with wild growing *Alhagi pseudoalhagi*. Ion contents of evaluated wild native halophytes were relatively low and hence these species could be recognized as alternative forages, both in pure halophytic pastures and/or in mixed grass stands. They can be recommended to farmers for cultivation and creating a livestock grazing system (Toderich et al. 2010a, b).

Plants used for phytodesalination purposes can be also utilized and have several post-harvest applications as edible, medicinal and industrial materials, thus they can have two simultaneous benefits (Zorrig et al. 2012; Lokhande and Suprasanna 2012). Some of these plants are considered as valuable materials for fodder and protein meals, bioenergy and biofuel production, extraction of oils and biological active substances, curative tools and seed production (Swingle et al. 1996; Anwar et al. 2002; Wu and Sessa 2004; Rogers et al. 2005; Masters et al. 2007).

Considering the great and diversified value of native halophytic and salt tolerant plants, their reproduction is rational for further domestication and large scale utilization of these resources in the restoration of marginal lands and improvement of livelihood of population. Suitable agro-technologies are needed to multiply seeds and/or reproductive plant materials, establish them within natural plant communities and introduce them wherever they are desired in other ecosystems (Toderich et al. 2007). These approaches are successfully found the application in arid and desert areas of Central Asia and use of this environmentally and economically viable method was revealed to increase the fodder productivity and sheep capacity of natural pastures in Uzbekistan and Turkmenistan (Nechayeva 1989; Shamsutdinov and Shamsutdinov 2007, 2010). Since flowering and seed maturating of most of halophytes are late they can be recommended for livestock feed during autumn and winter seasons (Toderich et al. 2010a, b).

6 Conclusions and Recommendations

Results of this literature review and own experiments revealed that halophytes of Aralo-Caspian lowlands can selectively remove specific toxic ions from soils. Careful selection and field trial, however, are necessary to produce optimally useful germplasm for strained sites. The mobilization and introduction of phytogenetic resources both native and very-carefully-trailed, domesticated salt-tolerant plants as well as study of their morphological, physiological and reproductive properties, cultivation with testing of alternative technologies will optimize the selection of halophytic arid plants and reveal the novel models for prediction of rangeland productivity and will provide a sustainable development of saline/sandy deserts ecosystems.

The use of halophytes from the Aralo-Caspian floras to reclaim soils could represent both a practical and economical viable strategy. Domestication of halophytes by introducing into a biosaline agriculture production presents a feasible alternative in the remediation of highly saline abandoned soils. Even though the scientific technology for molecular transforming plants is established, unfortunately plants that are well adapted to desert environments have not yet been transformed. This shortage of high quality plants greatly hinders work to mass propagate, to restore and maintain desert plant ecosystems.

However, the suitable strategies and integrated (water and ecosystem) approach are required to be developed for the regeneration, management and sustainable development of the degraded natural desert and semi-desert areas. Further studies should include: (1) revelation of proper approach for the selection and introduction of indigenous halophytic and salt-tolerant plants well-adapted to the local strained conditions in Aralo-Caspian regions; (2) investigation of plants suitability from morphological, anatomical-structural and physiological-biochemical aspects for prediction improving of saline prone lands productivity; (3) evaluation of ecological, agricultural nutritional or medicinal potential of the selected plant species, value chains analysis, their economic cost benefit and market demand.

The study of plant cellular mechanisms, enzymatic and genetic analyses to select the optimal genotypes from the local populations of Aralo-Caspian flora and their use in the genetic improvement to increase the biomass and seed production of species distinguished by the resistance to salinity, water use efficiency and pollution as well as by the essential nutrient uptake as food valuation are following steps of the research.

Selection and development of suitable innovative remediation and agro-management strategies for domestication of wild halophytes successfully producing seeds under marginal environments for stabilization of ecosystem function, renewal of plant-energy and sources for functionality of biologically active substances should become essential measures. The expedient realization of these measures will allow recommending: corresponding wild native metallo-halophytic species/ecotypes; timetable and techniques for seed collection and planting; optimal and low-cost agronomic practices for provision and amplification of yield viability and quality with purpose to remediate and rehabilitate the degraded lands as well as improvement their fertility. Consequently, increase of food security can be achieved by the enlargement of the area of lands suitable for agricultural and livestock use owing to revitalizing of unused marginal lands. These characteristics may offer a new and valuable source of income to local populations.

Acknowledgement This work was supported by the Science Development Foundation under the President of the Republic of Azerbaijan- Grant No EIF- 2012-2(6)-39/18/3 and partially by financial support of ICBA Funds.

References

Abduyev MR (1961) Solonetz soils of Azerbaijan and ways of their improvement. Azerbaijan Government Publication, Baku, 67 p (in Azerbaijaniani)

Abduyev MR (1963) Root system of wormwood plants under lowland conditions of Azerbaijan. Proc Acad Sci Azerbaijan SSR 5:15–20 (in Russian)

Ahmad P, Prasad MNV (2012) Abiotic stress responses in plants: metabolism, productivity and sustainability, 1st edn. Springer, New York, XIX, 551 p

Ahmad P, Sharma S (2008) Salt stress and phyto-biochemical responses of plants. Plant Soil Environ 54:89–99

Akjigitova NI (1982) Halophytic vegetation of Middle Asia and its indicator properties. "FAN" Publisher, Tashkent, 201 p

Alirzayeva E, Neumann G (2013) Multiple tolerance of *Artemisia fragrans* to environmental stress factors. International conference on "Environmental changes and conservation of plant diversity", Baku, p 33

Alirzayeva EG, Shirvani TS, Yazici MA, Alverdiyeva SM, Shukurov ES, Ozturk L, Ali-zade VM, Cakmak I (2006a) Heavy metal accumulation in *Artemisia* and foliaceous lichen species from the Azerbaijan flora. For Snow Lands Res 80:339–348

Alirzayeva EG, Shirvani TS, Yazici MA, Aleskerova AN, Ali-zade VM, Cakmak I (2006b) Heavy metal accumulation in some medicinal wormwood species from Azerbaijan. Materials of International Conference, dedicated to 75 years of formation of All-Russian Res Inst of Med and Arom Plants M XII, pp 231–235

Alirzayeva E, Shirvani T, Alizade V, Romheld V, Cakmak I (2011) Chapter 6: Genetic capacity of some medicinal plants to accumulate heavy metals. In: Altaf Ahmad et al (eds) Medicinal plants in changing environment. Capital Publication, New Delhi, pp 72–97

Ali-zade VM, Alirzayeva EG, Shirvani TS (2010) Plant resistance to anthropogenic toxicants: approaches to phytoremediation. In: Ashraf M, Ozturk M, Ahmad MSA (eds) Plant adaptation and phytoremediation. Springer, Dordrecht Heidelberg London New York, pp 173–191

Ali-zade VM, Shirvani TS, Alirzayeva EG (2011) Plant resistance to toxicity of metals and petroleum hydrocarbons. Approaches to phytoremediation. Elm, Baku, 280 p (in Russian)

Altay A, Öztürk M (2012) Halophytic plant diversity of Milleyha (Samandağ-Hatay) freshwater ecosystem, land degradation due to human disturbances & conservation measures. Pak J Bot 44:37–50

Anwar F, Bhanger MI, Nasir MKA, Ismail S (2002) Analytical characterization of *Salicornia bigelovii* seed oil cultivated in Pakistan. Agric Food Chem 50:4210–4214

Aparin VB, Kawabata Y, Ko S, Kunio S, Masahiro N, Masayoshi Y (2006) Evaluation of geoecological status and anthropogenic impact on the central Kyzylkum Desert (Uzbekistan). J Arid Land Stud 15(4):219–222

Arshad M, Murtaza G, Ali MA, Shafig M, Dumat C, Ahmed N (2011) Wheat growth and phytoavailability of copper and zinc as affected by soil texture in salin-sodic conditions. Pak J Bot 43(5):2433–2439

Ashraf M, Foolad MR (2007) Roles of glycine betaine and proline in improving plant abiotic stress resistance. Environ Exp Bot 59:206–216

Ashraf M, Ozturk M, Ahmad MSA (2010) Toxins and their phytoremediation. In: Ashraf M, Ozturk M, Ahmad MSA (eds) Plant adaptation and phytoremediation. Springer, Dordrecht Heidelberg London New York, pp 1–32

Azizov GZ (2006) Water-salt balance of meliorated soils of Kur-Araz lowland and scientific analysis of their results. "Elm", Baku, 260 p (in Azerbaijani)

Bajji M, Kinet JM, Lutts S (1998) Salt stress effects on roots and leaves of *Atriplex halimus* L. and their corresponding callus cultures. Plant Sci 137:131–142

Bennett SJ, Barrett-Lennard EG, Colmer TD (2009) Salinity and waterlogging as constraints to saltland pasture production: a review. Agric Ecosys Environ 129:349–360

Botcszantsev VP (1969) Genus *Salsola* L.-concise history of its development and dispersal. Bot J 54:989–1001

Boyd RS, Martens SN (1998) The significance of metal hyperaccumulation for biotic interactions. Chemoecology 8:1–7

Boyko H (1966) Basic ecological principles of plant growing by irrigation with highly saline or seawater. In: Boyko H (ed) Salinity and aridity: new approaches to old problems. DW Junk Publishers, The Hauge

Budagov B, Mammadov R, Ismatova Kh, Mikailov A (2003) Current state and reasons of desertification of coastal zone of Caspian Sea. "Man and Biosphere" (MAB, UNESCO) Azerbaijan National Committee, vol 2, pp 16–34 (in Russian)

Chandna R, Azooz MM, Ahmad P (2013) Recent advances of metabolomics to reveal plant response during salt stress. In: Ahmad P et al (eds) Salt Stress in Plants: Omics, Signaling and Adaptation. Springer, New York, pp 1–14

Chaudhri II, Shah BH, Naqvi N, Mallick IA (1964) Investigations on the role of *Suaeda fruticosa* Forsk in the reclamation of saline and alkaline soils in west Pakistan plains. Plant Soil 21:1–7

Chedlly A, Ozturk M, Ashraf M, Grignon C (eds) (2008) Biosaline agriculture and high salinity tolerance. Birkhauser Verlag (Springer), Basel, 367 pp

Cruz V, Cuartero J (1990) Effects of salinity at several developmental stages of six genotypes of tomato (*Lycopersicon* spp.). In: Cuartero J, Gomez-Guillamon ML, Fernandez-Munoz R (eds) Eucarpia tomato 90, Proceedings of the XIth Eucarpia meeting on tomato genetics and breeding, Malaga, Spain, pp 81–86

Curtain D, Naidu R (1998) Fertility constraints to plant production. In: Sumner ME, Naidu R (eds) Sodic soils. Distribution, management and environmental consequences. Oxford University Press, New York, pp 107–123

Dagar JC, Tomar OS, Kumar Y, Bhagwan H, Yadavand RK, Tyagi NK (2006) Performance of some under-explored crops under saline irrigation in a semiarid climate in northwest India. Land Degrad Dev 17:285–299

Dağhan H, Ozturk M (2015) Soil pollution in Turkey and remediation methods. In: Hakeem S, Ozturk M (eds) Soil remediation and plants: prospects and challenges. Academic/Elsevier, New York, pp 287–312

Dikilitas M (2003) Effect of salinity and its interactions with *Verticillium albo-atrum* on the disease development in tomato (*Lycopersicon esculentum* Mill.) and lucerne (*Medicago sativa* and *M. media*) plants. Ph.D thesis, University of Wales Swansea

Dikilitas M, Karakas S (2010) Salts as potential environmental pollutants, their types, effects on plants and approaches for their phytoremediation. In: Ashraf M, Ozturk M, Ahmad MSA (eds) Plant adaptation and phytoremediation. Springer, Dordrecht, pp 357–383

Eid MA, Eisa SS (2010) The use of some halophytic plants to reduce Zn, Cu and Ni in soil. Aust J Basic Appl Sci 4:1590–1596

Escarre J, Lefebvre C, Gruber W, Leblanc M, Lepart J, Riviere Y, Delay B (2000) Zinc and cadmium hyperaccumulation by *Thlaspi caerulescens* from metalliferous and nonmetalliferous sites in the Mediterranean area: implications for phytoremediation. New Phytol 145:429–437

Evans DR, Black RA, Loescher WH, Fellows RJ (1992) Osmotic relations of the drought-tolerant shrub *Artemisia tridentata* in response to water stress. Plant Cell Environ 15(1):49–59
Flowers TJ (1985) Physiology of halophytes. Plant Soil 89:41–56
Flowers TJ, Colmer TD (2008) Salinity tolerance in halophytes. New Phytol 179:945–963
Flowers TJ, Flowers SA (2005) Why does salinity pose such a difficult problem for plant breeders? Agric Water Manage 78:15–24
Flowers TJ, Hajibagheri MA, Clipson NJW (1986) Halophytes. Q Rev Biol 61:313–337
Flowers TJ, Flowers SA, Hajibagheri MA, Yeo AR (1991) Ion accumulation in the cell wall of rice plants growing under saline conditions: evidence for the Oertli hypothesis. Plant Cell Environ 14:319–325
Flowers TJ, Galal HK, Bromham L (2010) Evolution of halophytes: multiple origins of salt tolerance in land plants. Funct Plant Biol 37:604–612
Freitag H (1997) *Salsola* L. (*Chenopodiaceae*). In: Rechinger KN (ed) Flora Iranica, vol 173. Akadesche Druck-u Verlagsanstalt, Graz, pp 154–255
Garratt LC, Janagoudar BS, Lowe KC, Anthony P, Power JB, Davey MR (2002) Salinity tolerance and antioxidant status in cotton cultures. Free Radic Biol Med 33:502–511
Geissler N, Huchzermeyer B, Koyro HW (2013) Effect of salt stress on phytosynthesis under ambient and elevated atmospheric CO_2 concentration. In: Ahmad P et al (eds) Salt Stress in Plants. Signalings, Omics and Adaptations. Springer, New York, pp 377–414
Gintzburger G, Toderich K, Mardonov B, Mahmudov M (2003) Rangelands of the arid and semi arid zones in Uzbekistan. CIRAD/ICARDA, Montpellier, 420 p
Glenn EP, Brown JJ, Blumwald E (1999) Salt tolerance and crop potential of halophytes. Crit Rev Plant Sci 18:227–255
Goldshtein RI (1997) Ecological situation in the Kyzylkums in connection with its industrial development. Bull SCST Repub Uzbekistan N 3–4:70–75
Goldshtein RI, Tsukatani T, Toderich KN, Ashurmertov OA, Aparin VB (2000) Ecological state and conservation of arid salt- affected lands using Asiatic *Salsola* species. Abstracts book of International seminars of saline agriculture, Islamabad, pp 10–12
Gorham J (1992) Salt tolerance of plants. Sci Prog 76:273–285
Gorham J (1994) Salt tolerance in the *Triticeae*: K/Na discrimination in some perennial wheat grasses and their amphiploids with wheat. J Exp Bot 45:441–447
Grattan SR, Grieve CM (1994) Mineral nutrient acquisition and response by plants grown in saline environments. In: Pessarakli M (ed) Handbook of plant and crop stress. Marcel Dekker, New York, pp 203–226
Grigore MN, Toma C (2007) Histo-anatomical strategies of *Chenopodiaceae* halophytes: Adaptive, ecological and evolutionary implications. WSEAS Trans Biol Biomed 4(12):204–218
Gurbanov EM (2009) New taxons of vegetation in Azerbaijan. Proc Baku State Univ 3:107–116
Guvensen A, Gork G, Ozturk M (2006) An overview of the halophytes in Turkey. In: Khan et al (eds) Sabkha ecosystems vol. II, West & Central Asia. Springer, Dordrecht, pp 9–30
Hajiyev VD (2002) Reconstruction and protection of Kur-Araz lowland deserted ecosystem. 6-th Baku International Congress "Energy, Ecology, Economy", Baku, pp 403–404
Hajiyev OA, Rahimov VA (1971) Climatic level of administrative regions of Azerbaijan. "Elm", Baku, p 269 (in Azerbaijani)
Hakeem KR, Chandna R, Rehman R, Tahir I, Sabir M, Iqbal M (2013) Unraveling salt stress in plants through proteomics In: Ahmad P, Azooz MM, Prasad MNV (eds) Salt Stress in Plants: Signalling, Omics and Adaptations. Springer, New York, pp 47–61
Hakeem KR, Sabir M, Ozturk M, Mermut A (2015) Soil remediation and plants: prospects and challenges. Academic/Elsevier, New York, 724 pp
Hameed M, Ashraf M, Ahmad MSA, Naz N (2010) Structural and functional adaptations in plants for salinity tolerance. In: Ashraf M, Ozturk M, Ahmad MSA (eds) Plant Adaptation and Phytoremediation. Springer, Dordrecht Heidelberg London New York, pp 151–170
Hasanuzzaman M, Nahar K, Fujita M, Ahmad P, Chandna R, Prasad MNV, Ozturk M (2013) Enhancing Plant Productivity under Salt Stress- Relevance of Poly-omics. In: Ahmad P et al (eds) Salt Stress in Plants: Signalling, Omics and Adaptations. Springer, New York, pp 113–156

Hasanuzzaman M, Nahar K, Hakeem KR, Ozturk M, Fujita M (2015) Arsenic Toxicity in Plants and Possible Remediation. In: Hakeem S, Ozturk M (eds) Soil remediation and plants: prospects and challenges. Academic/Elsevier, New York, pp 433–502

Hasegawa PM, Bressan RA, Zhu J-K, Bohnert HJ (2000) Plant cellular and molecular responses to high salinity. Annu Rev Plant Physiol Plant Mol Biol 51:463–499

Heydarova RM (2013) The study of sulfate saline soils and subsoil grounds on the basis of processing of Space Images. Proc Azerbaijan Nat Acad Sci Biol Med Sci 68(1):90–94 (in Azerbaijani)

Ishikawa SI, Kachi N (2000) Differential salt tolerance of two *Artemisia* species growing in contrasting coastal habitats. Plant Cell Environ 5(1):49–59

Jordan FL, Robin-Abbott M, Maier RM, Glenn EP (2002) A comparison of chelator-facilitated metal uptake by a halophyte and a glycophyte. Environ Toxicol Chem 21:2698–2704

Kadukova J, Manousaki E, Kalogerakis N (2008) Pb and Cd accumulation and phytoexcretion by salt cedar *(Tamarix smyrnensis* Bunge*)*. Int J Phytorem 10:31–46

Khan F, Hakeem KR, Sidiqqui TO, Ahmad A (2013) RAPD markers associated with salt tolerance in soybean genotypes under salt stress. Appl Biochem Biotech 170(2):257–272

Khan MA, Boer B, Ozturk M, Al Abdessalaam T, Clüsener-Godt M (eds) (2014) Sabkha ecosystems: volume IV: cash crop halophyte and biodiversity conservation. Tasks for vegetation science. Springer, Dordrecht

Kholodova V, Volkov K, Kuznetsov Vl (2010) Plants under heavy metal stress in saline environments. In: Soil heavy metals. Soil biology, vol 19. Springer, Heidelberg Dordrecht London New York, pp 163–183

Kim JG, Cho NH, Kim NB et al (2003) Bioremediation method of heavy metal contaminated soils. Patent no KR 2003079062 A 20031010 (in Korean)

Krüger HR, Peinemann N (1996) Coastal plain halophytes and their relation to soil ionic composition. Vegetation 122:143–150

Le Houerou HN (1992) The role of saltbushes (*Atriplex* species) in arid land rehabilitation in the Mediterranean basis: a review. Agrofor Syst 18:107–148

Le Houerou HN (1996) Forage halophytes in the Mediterranean basin. In: Chakour-Allah R, Malcolm CV, Hamdy A (eds) Halophytes and biosaline agriculture. Marcel Dekker, New York, pp 115–136

Le Houerou HN (2000) Utilization of fodder trees and shrubs in the arid and semiarid zones of West Asia and North Africa. Arid Soil Res Rehab 14:101–135

Lefevre I, Marchal G, Meerts P, Correal E, Lutts S (2009) Chloride salinity reduces cadmium accumulation by the Mediterranean halophyte species *Atriplex halimus* L. Environ Exp Bot 65:142–152

Lokhande VH, Suprasanna P (2012) Prospects of halophytes in understanding and managing abiotic stress tolerance. In: Ahmad P, Prasad MNV (eds) Environmental adaptations and stress tolerance of plants in the era of climate change, XVI. Springer, New York, pp 29–56

Lopez-Chuken UJ, Young SD (2005) Plant screening of halophyte species for cadmium phytoremediation. Z Naturforsch 60:236–243

Lutts S, Lefevre I, Delperee C, Kivits S, Dechamps C, Robledo A, Correal E (2004) Heavy metal accumulation in the halophyte species Mediterranean saltbush. J Environ Qual 33:1271–1279

Macnair MR (2003) The hyperaccumulation of metals by plants. Adv Bot Res 40:63–105

Mammadov GSh (2005) Assessment and management of ecological status of Azerbaijan soils. In: Alekperov UK (ed) Proceedings of the "Man and Biosphere" (MAB), Azerbaijan National Commitee for UNESCO, vol 3. Elm Publish, Baku, pp 64–77 (in Azerbaijaniani)

Mammadov G, Khalilov M, Mammadova S (2010) Agroecology. Elm Publish, Baku, 551 p (in Azerbaijani)

Manousaki E, Kalogerakis N (2009) Phytoextraction of Pb and Cd by the Mediterranean saltbush (*Atriplex halimus* L.): metal uptake in relation to salinity. Environ Sci Pollut Res 16:844–854

Manousaki E, Kalogerakis N (2011a) Halophytes present new opportunities in phytoremediation of heavy metals and saline soils. Ind Eng Chem Res 50:656–660

Manousaki E, Kalogerakis N (2011b) Halophytes- an emerging trend in phytoremediation. Int J Phytoremed 13:959–969
Manousaki E, Kadukova J, Papadantonakis N, Kalogerakis N (2008) Phytoextraction and phyto-excretion of Cd by *Tamarix smyrnensis* growing on contaminated nonsaline and saline soils. Environ Res 106:326–332
Marcelis LFM, Van Hooijdonk J (1999) Effect of salinity on growth, water use and nutrient use in radish (*Raphanus sativus* L.). Plant Soil Hague 215:57–64
Masters DA, Benes SE, Norman HC (2007) Biosaline agriculture for forage and livestock production. Agric Ecosys Environ 119:234–248
Mehtiyeva NP (2011) Biodiversity of medicinal flora of Azerbaijan. Baku, 186 p
Mehtiyeva N, Zeynalova S (2013) Medicinal and aromatic plants of Azerbaijan. In: Medicinal and Aromatic Plants of the World Encyclopedia of Life Support Systems (EOLSS), developed under the Auspices of the UNESCO, Eolss Publishers, Oxford
Melikov RV, Bayramova KK (2004) Forage Kochia. Trans Bot Inst Azerbaijan Nat Acad Sci XXV:297–298
Melikov RV, Huseynova AD (2004) Tipology of wormwood groves in Kur-Araz lowland. Trans Bot Inst Azerbaijan Nat Acad Sci XXV:279–282
Milić D, Luković J, Zorić L, Vasin J, Ninkov J, Zeremski T, Milić S (2012) Halophytes relations to soil ionic composition. J Serb Chem Soc 77:1–14
Morishita T, Boratynski JK (1992) Accumulation of cadmium and other metals in organs of plants growing around metal smelters in Japan. Soil Sci Plant Nutr 38:781–785
Movsumova FG (2003) Biodiversity of solonchak desert vegetation of Kur-Araz lowland of Azerbaijan, In: Alekperov UK (ed) Proceedings of the "Man and Biosphere" (MAB), Azerbaijani National Committee for UNESCO, vol 2. Elm Publish, Baku, pp 163–174 (in Azerbaijaniani)
Movsumova FG (2005) Flora and vegetation of saltwort deserts of Nakhichevan AR. "Shams", Baku, 134 p (in Azerbaijaniani)
Movsumova FG (2011) Distribution, ecological-phytocoenotic peculiarities and useful features of species of *Chenopodiaceae* family in Azerbaijan. In: Proceedings of the "Man and Biosphere" (MAB, UNESCO), Annual edition, vol 7. Azerbaijan National Committee, Bakupp 311–318
Munns R, Tester M (2008) Mechanisms of salinity tolerance. Annu Rev Plant Biol 59:651–681
Munns R, Cramer GR, Ball MC (1999) Interactions between rising CO_2, soil salinity and plant growth. In: Luo Y, Seemann JR (eds) Carbon dioxide and environmental stress. Academic, New York, 139–167
Musayev SG, Fataliyev RA (2004) Flora of Azerbaijan: new data. Trans Inst Bot ANAS XXV:16–22 (in Russian)
Nabiyeva FH (2004) Climatic factors of desertification in Kur-Araz lowland. Trans Inst Bot ANAS XXV:337–339
Nagalevskii VY (2001) Halophytes of the Northern Caucasus. Kuban State University, Krasnodar (in Russian)
Naghiyev PY, Heydarova RM (2010) The study of saline soils in Shirvan plain on the base of digitalized processing of space images. In: Proceedings of National Aerospace Agency, p 12 (In Azerbaijani)
Nechayeva NT (1989) Ecological foundations of conservation and enrichment of rangelands of arid zones in USSR. Prob Desert Reclam 2:3–12
Nedjimia B, Daoudb Y (2009) Cadmium accumulation in *Atriplex halimus* subsp. *schweinfurthii* and its influence on growth, proline, root hydraulic conductivity and nutrient uptake. Flora 204:316–332
Nordblom T, Shomo F, Gintzburger G (1997) Food and feed prospects for resources in Central Asia. In: Demment M (ed) Central Asia regional livestock assessment workshop. Tashkent, Uzbekistan. Management Entity, Small Ruminant CRSP, University of California. USA, 226 p
Ondrasek G, Rengel Z, Veres S (2011) Soil salinisation and salt stress in crop production. In: Shanker A, Venkateswarlu B (eds) Abiotic stress in plants – mechanisms and adaptations. Springer, Dordrecht, pp 171–190

Ortíz-Dorda J, Martinez-Mora C, Correal E, Simon B, Cenis JL (2005) Genetic structure of *Atriplex halimus* populations in the Mediterranean Basin. Ann Bot 95:827–834
Osman AE, Bahhady F, Hassan N, Ghassali F, Al Ibrahim T (2006) Livestock production and economic implications from augmenting degraded rangeland with *Atriplex halimus* and *Salsola vermiculata* in northwest Syria. J Arid Environ 65:474–490
Ould Ahmed BA (2012) Soil salinity control through halophytes in arid and semiarid area in Mauritania. J Arid Land Stud 22:21–24
Ozturk M, Baslar S, Dogan Y, Sakcali S (2006a) Alleviation of salinity stress in the seeds of some *Brassica* species. In: Khan MA, Weber DJ (eds) Biosaline agriculture and salinity tolerance. Kluwer/Springer, Dordrecht, pp 145–156
Ozturk M, Guvensen A, Gork G, Gork C (2006b) An overview of the coastal zone plant diversity and management strategies in the Mediterranean region of Turkey. In: Ozturk M et al (eds) Biosaline Agriculture and Salinity Tolerance in Plants. Birkhauser Verlag, Basel, pp 89–100
Ozturk M, Yucel E, Gucel S, Sakcali S, Aksoy A (2008a) Plants as biomonitors of trace elements pollution in soil. In: Prasad MNV (ed) Trace elements: environmental contamination, nutritional benefits and health implications, Chapter 28. Wiley, New York, pp 723–744
Ozturk M, Guvensen A, Gork G (2008b) Halophyte plant diversity in the Irano-Turanian phytogeographical region of Turkey In: Abdelly C et al (eds) Biosaline agriculture& salinity tolerance. Birkhauser Verlag, Basel, pp 141–155
Ozturk M, Gucel S, Sakcali S, Dogan Y, Baslar S (2009) Effects of temperature and salinity on germination and seedling growth of *Daucus carota* cv. *nantes* and *Capsicum annuum* cv. *sivri* and flooding on *Capsicum annuum* cv. *sivri* Chapter 6, In: Ashraf M et al (eds) Salinity and water stress: improving crop efficiency. Tasks for vegetation science, vol 44. Springer, New York, pp 51–64
Ozturk M, Mermut A, Celik A (2010a) Land degradation, urbanisation, land use & environment, NAM S. & T., Delhi-India, 445 pp
Ozturk M, Sakcali S, Gucel S, Tombuloğlu H (2010b) Boron and plants. In: Ashraf M et al (eds) Plant adaptation & phytoremediation, Part 2. Springer, New York,pp 275–311
Ozturk M, Gucel S, Guvensen A, Kadis C, Kounnamas C (2011a) Halophyte plant diversity, coastal habitat types and their conservation status in Cyprus. Sabkha Ecosyst Tasks Veg Sci Springer 46:99–111
Ozturk M, Turkyilmaz B, Gucel S, Guvensen A (2011b) Proline accumulation in some coastal zone plants of the Aegean Region of Turkey. In: Muscolo TJ, Flowers, A (eds) Proceedings of the European COST action FA0901. Eur J Plant Sci Biotechnol 5(Special Issue 2):54–56
Ozturk M, Memon AR, Gucel S, Sakcali MS (2012) Brassicas in Turkey and their possible role in the phytoremediation of degraded habitats. In: Anjum NA et al (eds) The plant family Brassicaceae: contribution towards phytoremediation. Environmental pollution book series, vol 21. Springer, New York, pp 265–288
Ozturk M, Altay V, Gucel S, Guvensen A (2014) Halophytes in the East Mediterranean-their medicinal and other economical values. In: Khan MA et al (eds) Sabkha ecosystems vol IV cash crop halophytes biodiversity & conservation. Tasks for vegetation science series, Springer, New York, pp: 247–272
Poschenrieder C, Gunse B, Barcelo J (1989) Influence of cadmium on water relations, stomatal resistance, and abscisic-acid content in expanding bean leaves. Plant Physiol 90:1365–1371
Przymusinski R, Rucinska R, Gwozdz EA (2004) Increased accumulation of pathogenesis-related proteins in response of lupine roots to various abiotic stresses. Environ Exp Bot 52:53–61
Qadir M, Oster JD (2004) Crop and irrigation management strategies for saline-sodic soils and waters aimed at environmentally sustainable agriculture. Sci Total Environ 323:1–19
Qadir M, Ghafoor A, Murtaza G (2000) Amelioration strategies for saline soils: a review. Land Degrad Dev 11:501–521
Qadir M, Tubeileh A, Akhtar J, Larbi A, Minhas PS, Khan MA (2008) Productivity enhancement of salt-affected environments through crop diversification. Land Degrad Dev 19:429–453
Rabhi M, Barhoumi Z, Atia A, Lakhdar A, Hafsi C, Hajji S, Abdelly C, Smaoui A (2009) Evaluation of the capacity of three halophytes to desalinize their rhizosphere as grown on saline soils under non-leaching conditions. Afr J Ecol 47:463–468

Ravindran KC, Venkatesan K, Balakrishnan V, Chellappan KP, Balasubramanian T (2007) Restoration of saline land by halophytes for Indian soils. Soil Biol Biochem 39:2661–2664
Rhoades JD, Loveday J (1990) Irrigation of agricultural crops. In: Stewart BA, Nielsen DR (eds) Salinity of irrigated agriculture. American Society of Agronomy, Madison, pp 1089–1142
Rogers ME, Craig AD, Munns RE, Colmer TD, Nichols PGH, Malcolm CV, Barrett-Lennard EG, Brown AJ, Semple WS, Evans PM, Cowley K, Hughes SJ, Snowball R, Bennett SJ, Sweeney GC, Dear BS, Ewing MA (2005) The potential for developing fodder plants for the salt-affected areas of southern and eastern Australia: an overview. Aust J Exp Agric 45:301–329
Rozema J, Flowers T (2008) Crops for a salinized world. Science 322:1478–1480
Sabir M, Waraich EA, Hakeem KR, Ozturk M, Ahmad HR, Shahid M (2015) Phytoremediation: mechanisms and adaptations. In: Hakeem S, Ozturk M (eds) Soil remediation and plants: prospects and challenges. Academic/Elsevier, New York, pp 85–106
Sakai Y, Ma Y, Xu C, Wu H, Zhu W, Yang J (2012) Phytodesalination of a salt-affected soil with four halophytes in China. J Arid Land Stud 22:17–20
Schwitzguébel JP, Kumpiene J, Comino E, Vanek T (2009) From green to clean: a promising and sustainable approach towards environmental remediation and human health for the 21st century. Agrochimica 53:209–237
Seaman J (2005) Mechanisms of salt tolerance in halophytes: can crop plants resistance to salinity be improved? Ph.D. dissertation, University of Sheffield
Seemann JR, Critchley C (1985) Effects of salt stress on the growth, ion content, stomatal behavior and photosynthetic capacity of a salt-sensitive species, *Phaseolus vulgaris* L. Planta 164:151–162
Shafi M, Zhang J, Bakht J, Khan MA, Islam EU, Khan MD, Raziuddin GZ (2010) Effect of cadmium and salinity stresses on root morphology of wheat. Pak J Bot 42:2747–2754
Shamsutdinov ZS, Savchenko IV, Shamsutdinov NZ (2000) Halophytes of Russia and their ecological characteristics and use. Academy of Agricultural Sciences. Russian Fodder Institute, Moscow, 399p
Shamsutdinov ZS, Shamsutdinov NZ (2007) Biogeocenotechnology of remediation of disturbed arid pasturable ecosistems. Messenger Russ Acad Agric Sci 3:37–38 (in Russian)
Shamsutdinov ZS, Shamsutdinov NZ (2010) Doctrine of Nechaeva NT about deserted pastures. Arid Ecosyst 16:11–29 (in Russian)
Sharma KK, Gupta JK (1986) Optimum reliability of gain scores. J Exp Educ 54:105–108
Shevyakova NI, Netronina IA, Aronova EE, Kuznetsov VV (2003) Compartmentation of cadmium and iron in *Mesembryanthemum crystallinum* plants during the adaptation to cadmium stress. Russ J Plant Physiol 50:678–685
Shukurov E, Askerov F, Zaytsev Yu (2008) Biodiversity of desert and semi-desert plants of Azerbaijan. BP Caspian Sea Ltd, Baku, 143 p
Soil Atlas of Azerbaijan Republic (2007) Mammadov GSh et al. (eds) The State Committee of Land and Cartography of Azerbaijan Republic and Institute of Soil and Agrochemistry of Azerbaijan National Academy of Sciences. Cartography Factory, Baku
Squires VR, Glenn EP (2009) Salination, desertification and soil erosion. In: Squires VR (ed) The role of food, agriculture, forestry and fisheries in human nutrition, vol 3. Encyclopedia of Life Support Systems (EOLSS), pp 102-123
Sun J, Chen SL, Dai SX, Wang RG, Li NY, Zhou XY, Lu CF, Zheng XJ, Hu ZM, Zhang ZK, Song J, Xu Y (2009) Ion flux profiles and plant ion homeostasis control under salt stress. Plant Signal Behav 4:261–264
Swingle RS, Glenn EP, Squires V (1996) Growth performance of lambs fed mixed diets containing halophyte ingredients. Anim Feed Sci Technol 63:137–148
Takeda R, Yoshimura N, Matsumoto S, Komemushi S (2005) Accumulation of heavy metals by Japanese weeds and their seasonal movement. Contamin Soils 9:349–359
Thomas JC, Malick FK, Endreszl C, Davies EC, Murray KS (1998) Distinct responses to copper stress in the halophyte *Mesembryanthemum crystallinum*. Physiol Plant 102:360–368
Toderich KN, Tsukatani T (2007) New approaches for biosaline agriculture development, Management and conservation of Central Asian degraded drylands. Discussion paper no 638, Kyoto Institute of Economic Research, Kyoto, Japan 19 p

Toderich KN, Black CC, Juylova E, Kozan O, Mukimov T (2007) C3/C4 plants in the vegetation of Central Asia, geographical distribution and environmental adaptation in relation to climate, In: Lal R, Steward BA, Suleimenov M (eds) Carbon management and sequestration in Drylands of Central Asia. Balkema Publishers, Netherlands, pp 33–63

Toderich K, Tsukatani T, Ismail Sh, Massino I, Wilhelm M, Yusupov S, Kuliev T, Ruziev S (2008) Extent of salt-affected land in Central Asia: biosaline agriculture and utilization of the salt affected resources. Discussion paper № 648. Kyoto University, Kyoto, Japan, pp 1–36

Toderich KN, Tsukatani T, Petukhov OF, Gruthinov VA, Khujanazarov T, Juylova EA (2004) Risk assessment of environmental contaminants of Asiatic deserts ecosystems in relation to plant distribution and structure. J Arid Land Stud 14S:33–36

Toderich KN, Shuyskaya EV, Ismail S, Gismatullina L, Radjabov T, Bekhchanov BB, Aralova D (2009) Phytogenic resources of halophytes of Central Asia and their role for rehabilitation of sandy desert degraded rangelands. J Land Degrad Dev 20(4):386–396

Toderich KN, Shuyskaya EV, Ozturk M, Juylova A, Gismatulina L (2010a) Pollen morphology of some Asiatic species of genus *Salsola* (Chenopodiaceae) and its taxonomic relationships. Pak J Bot Spec Iss (S.I. Ali Festschrift) 42:155–174

Toderich KN, Shuyskaya EV, Khujanazarov TM, Ismail Sh, Kawabata Y (2010b) The structural and functional characteristics of Asiatic desert halophytes for phyto-stabilization of polluted sites. In: Ashraf M, Ozturk M, Ahmad MSA (eds) Plant adaptation and phytoremediation. Springer, New York, pp 245–274

Toderich KN, Shuyskaya EV, Faisal T, Ismail S, Gismatullina LG, Li EV (2012) Adaptive fruit structural mechanisms of Asiatic *Salsola* species and its germplasm conservation and utilization. J Arid Land Stud 22:73–76

Toderich KN, Shuyskaya EV, Rajabov TF, Ismail S, Shaumarov M, Yoshiko K, Li EV (2013a) Chapter 13: Uzbekistan: rehabilitation of desert rangelands affected by salinity, to improve food security, combat desertification and maintain the natural resource base. In: Hesmati GA, Squires V (eds) Combating desertification in Asia, Africa and the middle east: proven practices. Springer, Dordrecht, pp 249–278

Toderich KN, Shuyskaya EV, Taha FK, Matsuo N, Ismail S, Aralova DB, Radjabov TF (2013b) Integrating agroforestry and pastures for soil salinity management in dryland ecosystems in Aral sea basin. Chapter 39, In: Shahid SA, Abdelfattah MA, Taha FK (eds) Developments in soil salinity assessment and reclamation-innovative thinking and use of marginal soil and water resources in irrigated agriculture Springer, New York, pp 579–602

Tsukatani T, Katayama Y (2001) Water quality of Zarafshan River Basins. Discussion paper no 527. Kyoto Institute of Economic Research, Kyoto University, Japan, p 28

Waskiewicz A, Beszterda M, Golinski P (2013) ABA: role in plant signaling under salt stress. In: Ahmad P et al (eds) Salt stress in plants: omics, signaling and adaptation. Springer, New York, pp 175–196

Wen X, Qiu N, Lu Q, Lu C (2005) Enhanced thermotolerance of photosystem II in salt-adapted plants of the halophyte *Artemisia anethifolia*. Planta 220(3):486–497

Wu YV, Sessa DJ (2004) Protein fractiontion and properties of *Salicornia* meal. J Am Oil Chem Soc 81:173–176

Yokoi S, Quintero FJ, Cubero B, Ruiz MT, Bressan RA, Hasegawa PM, Pardo JM (2002) Differential expression and function of *Arabidopsis thaliana* NHX Na^+/H^+ antiporters in the salt stress response. Plant J 30:765–768

Zhao K-F (1991) Desalinization of saline soils by *Suaeda salsa*. Plant Soil 135:303–305

Zhao K-F, Hai F, Ungar IA (2002) Survey of halophyte species in China. Plant Sci 163:491–498

Zhu JK (2000) Genetic analysis of plant salt tolerance using *Arabidopsis*. Plant Physiol 124:941–948

Zhu JK (2001) Plant salt tolerance. Trends Plant Sci 6(2):66–71

Zhu Z, Wei G, Li J, Qian Q, Yu J (2004) Silicon alleviates salt stress and increases antioxidant enzymes activity in leaves of salt stressed cucumber (*Cucumis sativus* L). Plant Sci 167:527–533

Zorrig W, Rabhi M, Ferchichi S, Smaoui A, Abdelly C (2012) Phytodesalination: a solution for salt – affected soils in arid and semiarid regions. J Arid Land Stud 22-1:299–302

Nickel Metal Uptake and Metal-Specific Stress Alleviation in a Perennial Desert Grass *Cenchrus ciliaris*

Faiz-ul Hassan Nasim, Rabia Khalil, Ayesha Sumreen, Muhammad Shafiq Chaudhry, and Muhammad Ashraf

Contents

Abstract Associations of Vesicular Arbuscular Mycorrhizal Fungi (VAMF) with plant roots are known to function from Stress Alleviation to Bioremediation in metal polluted soils. We have studied uptake and accumulation of Nickel metal in a perennial grass, *Cenchrus ciliaris*, from Cholistan desert in the presence or the absence of mycorrhizal colonization of its roots by a fungus *Glomous mosseae*. Our results show that *Cenchrus ciliaris* has a tendency to absorb and tolerate the Nickel metal present in the contaminated soils and its shoots accumulate more Nickel than its roots. Introduction of Nickel in the plant rhizosphere generates stress that at some

---

The original version of this chapter was revised: The affiliation of the author M. Ashraf's are corrected. The erratum to this chapter is available at: http://dx.doi.org/10.1007/978-94-017-7194-8_21

F.-u.H. Nasim (✉) • R. Khalil • A. Sumreen
Chemistry Department, The Islamia University of Bahawalpur, Bahawalpur, Pakistan
e-mail: faiz.nasim@iub.edu.pk

M.S. Chaudhry
Department of Life Sciences, The Islamia University of Bahawalpur, Bahawalpur, Pakistan

M. Ashraf
Chemistry Department, The Islamia University of Bahawalpur, Bahawalpur, Pakistan

M. Öztürk et al. (eds.), *Plants, Pollutants and Remediation*,
DOI 10.1007/978-94-017-7194-8_5

concentrations creates anatomical changes both at macro and microscopic levels. Such concentrations also adversely affect fungal colonization. Nature and extent of the stress directly correlate with the concentration of the Nickel metal in the soil. Metal exposure alone or in combination with the mycorrhiza produces specific changes in some enzyme activities both in the root and the shoot tissues suggesting that these changes are involved in progression/regression of the metal-specific stress. While *Glomous mosseae* association is not required for Nickel uptake and accumulation by *Cenchrus ciliaris*, it appears to be helpful in alleviating the stress exerted by the presence of this metal in soil. We conclude that the perennial desert grass *Cenchrus ciliaris* is capable of mobilizing and up taking Nickel metal from soil, its transportation from roots to shoots and is equipped with the machinery that helps tolerate the metal-specific stress to some extent.

Keywords Nickel metal pollution • Mycorrhiza • *Glomous mosseae* • *Cenchrus ciliaris* • Cholistan

1 Introduction

Majority of the higher plants establish a mutual symbiotic relationship with Arbuscular Mycorrhizal Fungi (AMF). This interaction provides, in exchange for the photosynthetic material, mineral nutrients, moisture and increased resistance to the environmental stresses like temperature, drought and disturbance including those originating from soil contamination by heavy metals (Borges and Chaney 1989; van der Heijden et al. 1998; Quilambo 2000; Al-Karaki et al. 2004; Cruz et al. 2004; Martin and Stutz 2004; Barea et al. 2005; Citterio et al. 2005; Hakeem et al. 2015).

Plants have the ability to uptake a number of heavy metals such as Co, Cu, Fe, Mn, Mo, Ni, Zn and toxic organic compounds (Lambert et al. 1979; Gildon and Tinker 1983; Killham and Firestone 1983; Sabir et al. 2014). Plants that, in leaf dry matter, are able to accumulate greater than 0.1 % of elements such as Ni, Co & Pb, greater than 1 % of Zn or greater than 0.01 % of Cd are termed as hyper-accumulators (Verbruggen et al. 2009). This remarkable ability of the plants to absorb, concentrate and metabolize elements and compounds from the environment has lead to the development of a cost effective, non-intrusive and aesthetically pleasing way of eco-remediation termed as Phytoremediation (Hovsepyan and Greipsson 2004; Liu et al. 2004; Al Agely et al. 2005). This technology has shown its promise for removal of various pollutants, including toxic heavy metals, from large volumes of contaminated soil and water (Vassilev et al. 2004; Hildebrandt et al. 2007).

Plants respond to the heavy metal toxicity in a variety of ways including immobilization, exclusion, chelation and compartmentalization of metal ions, and the expression of more general stress mechanisms like ethylene and stress proteins

(Hakeem et al. 2015). Mycorrhizal infection has been reported to enhance plant uptake of certain metals (Galli et al. 1994; Salt et al. 1995), lower heavy metal content of some plants inoculated with specific mycorrhiza (Ouziad et al. 2005), and increase or decrease various enzymatic activities in plant tissues (Aggangan et al. 1998; Ajungla et al. 2003). In general, AMF have been reported to affect plant growth and health positively by increasing nutrient uptake (Cruz et al. 2004), nitrogen fixation (Ibekwe et al. 1995) and controlling harmful bugs (Diedhiou et al. 2003). Plant growth and health has also been reported to be associated with the diversity and abundance of the AMF population (Chaudhry et al. 2009). These AMF contributions are known to be adversely affected by the presence of heavy metal ions in the soil (Del Val et al. 1999).

How do plants deal with and detoxify heavy metal ions and whether AMF association with the metal tolerant plants, the metallophytes, help in the accumulation of heavy metals in plant organs in non-toxic form have not yet been clearly answered. Our group is involved in studying role of mycorrhiza in heavy metal uptake and accumulation by certain perennial grasses from a nearby desert, Cholistan, located in the south east area of Pakistan. The study being presented was conducted using *Cenchrus ciliaris* as a model system to investigate uptake and accumulation of Nickel metal, role of *Glomus mosseae* in the process of uptake and accumulation, and the variations in some of the associated stress related biochemical parameters. Commonly found and widespread in nature in harsh and hard conditions this grass can tolerate strong winds, low rainfall, acute erosion and nutrient adapted soil profile and is capable of accumulating various metals (Rao et al. 1989). To our knowledge this is the first report of such a study involving a desert grass from this part of the world.

2 Materials and Method

The experiment was planned to regenerate the plants from stubs in sterile soil followed by inoculation with mycorrhizal spores or/and metal treatment. Sandy soil collected from Cholistan desert was passed through a 200 size mesh in order to remove any particulate matter and impurities. The sieved soil was autoclaved at 120 °C for about 50 min, cooled and filled in labeled polythene bags to prepare the experimental pots. Stubs were obtained from whole plants that were carefully uprooted from the desert, taken to the laboratory and washed with potassium permanganate solution (70 %) which acts as disinfectant. Stubs of 2–3 in. size were sown in the experimental pots in the month of March. Analytical grade chemicals used during the experimental work were purchased from Merck/Fluka or BDH.

2.1 Inoculation

Sown stubs sprouted after 8–10 days of continuous watering. Regeneration percentage was 70. When the regenerated plants attained a length of about 4–8 cm, 2–3 cm long pieces of *Cymbopogan jawarancuse* roots infected with *Glomus mosseae* were buried around the base region of each plant with the help of cork borer. A total of six sets of plants were prepared that were distributed in to an "All Blank" control that was given no treatment, a Metal Blank control referred to as "Myco" that was treated with mycorrhiza only, two Mycorrhiza Blank controls referred to as "Ni 1" and "Ni 2" that were treated with two selected concentrations of the experimental metal only (see next) and the "Tests" that were treated both with mycorrhiza and the selected metal concentrations. Each experiment was conducted in triplicate, unless otherwise described.

2.2 Metal Treatment

Solutions of Nickle chloride hexahydrate, 100 ppm or 1000 ppm, were used for the study. The Mycorrhiza Blank and the Test sets of the growing plants were treated with a selected concentration of the Nickel metal by pouring 10 ml of 100 ppm or 1000 ppm solution in the pot soil near the plant root area. The 'All Blank' and the 'Metal Blank' controls were given 10 ml of distilled water instead.

2.3 Harvesting

After the mycorrhizal inoculation and metal treatment watering of plants was continued for 30 days. After 30 days plants were harvested and placed in labeled cellophane bags. Various physical parameters including health of the plant, shoot length, fresh weight of the sample and flowering were recorded during the growth period and at the time of harvest.

2.4 Mycorrhizal Root Analysis

The roots were taken out from the soil after harvesting and washed with water to remove sandy soil. A part of each root sample was fixed in FAA mixture (Formaline:Ethyl alcohol:Acetic acid::5:90:5) and stored at 4 °C for mycorrhizal analysis. Root samples were stained in a mixture of Lactophenole and blue stain and examined under microscope for AMF colonization. Mycorrhizal infection was determined by grid line method (Giovannetti and Mosse 1980).

2.5 Moisture Content Determination

AOAC official method (AOAC 1995) was used for moisture content determination. Briefly, accurately weighed crushed sample was kept in an oven for 5–6 h at 110 °C, cooled in a desiccators till a constant weight.

2.6 Guaiacol Peroxidase Activity

To assay Guaiacol peroxidase activity plant sample was homogenized in a pestle and mortar in 10 ml of 1 M phosphate buffer of pH 5.8. The homogenized sample was filtered through cheese cloth and centrifuged at 13000 rpm for 10 min. The supernatant was stored on ice and used as the enzyme source. For the assay H_2O, Guaiacol and 1 μl of the enzyme source were mixed in a total volume of 2 ml. The content were mixed and used to calibrate the spectrophotometer at 471 nm. The reaction was started by adding H_2O_2 and increase in absorbance was noted after a lag phase of 10–15 s for upto 2–3 min at regular intervals.

2.7 Amylase Activity

Amylase activity in the root and shoot samples was determined by extracting a known weight of the sample in a known volume of Phosphate buffer pH 6.8. Amylase activity was assayed by incubating the reaction mixture, a total volume of 3 ml, containing 100 μl of 0.03 % Starch solution, 2 ml of detection reagent (0.06 % solution of KI in 0.004 N HCl and 0.02 % Iodine), 880 μl of water and 20 μl Enzyme source at 37 °C for half an hour and measuring the OD at 565 nm. Standard curve was prepared by mixing different concentrations of starch with 2 ml of the detection reagent in a total volume of 3 ml as mentioned except the enzyme source followed by measuring the OD after incubation at 37 °C for half an hour.

2.8 Protein Estimation

Protein content of each extract was estimated by Biuret method (Switzer and Garrity 1999). Values obtained were used to calculate the Specific activities of the enzymes investigated.

2.9 Nickel Estimation

Plant samples were converted in to ash in a furnace at 600–700 °C followed by Diacid digestion where a known weight of the sample was boiled with a known volume of diacid mixture, HNO_3:$HClO_4$::1:4, for 15–20 min, diluted and filtered to get a clear solution. Atomic absorption Spectrophotometer (Perkin Elmer Analyst 100) was used for the estimation of Nickel in the sample solutions after calibration with Nickel chloride hexahydrate solutions (2, 4, 6 and 8 ppm) as the standard.

2.10 Translocation Factor (Tf%)

The ratio of metal concentration in the shoot to root was used to determine the Translocation Factor and expressed as TF% (Bose and Bhattacharyya 2008).

3 Results and Discussion

The study being presented was planned to determine uptake and accumulation of nickel metal by a Cholistan desert perennial grass, *Cenchrus ciliaris* in the presence or the absence of mycorrhizal colonization. The plants were grown and subjected to various treatments as described earlier. Variations in plant health and shoot length etc. were examined to ascertain nature of the host plant-fungus relationship. Since the regenerated plants were given different treatments after achieving certain height, plant health was determined in terms of gain in shoot length and the corresponding dry mass. While the metal challenge was expected to create stressful conditions for plant growth it was the 10 ppm metal treatment that proved to be the most detrimental resulting in significant decreases in plant height and total dry mass (Fig. 1a, b). The 1 ppm metal treatment on the other hand generated some morphological changes in the treated plants producing significantly taller plants with broad leaves (Photo 1). These changes can be considered an attempt to minimize the toxic effects of the metal through dilution. Our hypothesis of the dilution effect is further supported by the observation (Fig. 1c) that the plants treated with 10 ppm of nickel in the absence of mycorrhiza had the highest moisture content (70.8 %).

Compared with the non-mycorrhizal control, the *Glomus mosseae* treatment produced less increase in the shoot length and a somewhat lower dry mass, both in the aerial (shoot) and the non-aerial (root) parts of the plants, thereby affecting plant growth negatively (Fig. 1a, b). This observation is in line with the reports that plant growth may vary with the type of the mycorrhiza being used for colonization (Sailo and Bagyaraj 2005) or even depressed by mycorrhiza under certain conditions (Reynolds et al. 2005).

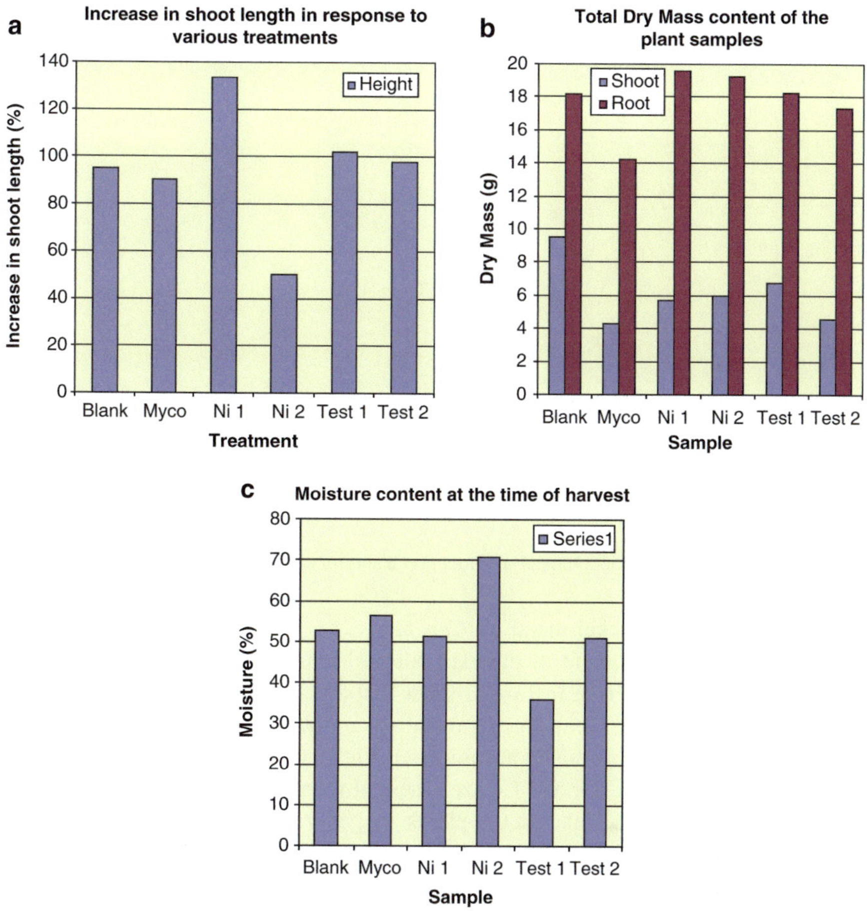

Fig. 1 Effect of mycorrhiza and Nickel metal treatments on the growth of *C. ciliaris* plants. (**a**) Increase in the shoot length, (**b**) variations in the total dry mass content and (**c**) moisture content of the experimental plants at the time of harvest

Effects of the mycorrhiza-metal combined treatments were however interesting. Mycorrhizal inoculation of the nickel treated plants, Test 1 & 2, alleviated adverse effects of the heavy metal stress on plant growth. These plants grew like the all blank control both in terms of shoot length and the total dry mass (Fig. 1a, b). Interestingly, the relief was mostly seen in the aerial parts (shoot) of the treated plants. The metal treatments also adversely affected mycorrhizal colonization of the *C. ciliaris* roots. While in the absence of any exogenous metal added to the soil 99.5 % of the roots were colonized by mycorrhiza, only about 57 % of the roots could be colonized by mycorrhiza in the presence of the added nickel metal (Table 1). Similar observations had earlier been reported by other workers (Chen et al. 2005a, b).

Photo 1 Effect of 1 ppm Nickel metal treatment on anatomy of the *C. ciliaris* plants

Accumulation of Nickel metal by *Cenchrus ciliaris* was investigated by determining Nickel metal content of the experimental plants using Atomic Absorption Spectrometry. Results given in Table 1 show that this perennial grass is itself capable of absorbing Nickel metal from the soil since both the shoots and the roots of even the all blank control contain accumulated Nickel metal. Since heavy metal ions such as Cu, Zn, Mn, Fe, Ni and Co are essential micronutrients required for plant metabolism (Williams et al. 2000), plants in general would be expected to be equipped with mechanisms for absorption and transport of these metals. *Cenchrus ciliaris* is therefore no surprise. It is however the amount of Nickel that matters. Hyperaccumulator plants can accumulate in their leaf dry matter greater than 0.1 % of Ni, Co & Pb, greater than 1 % of Zn or greater than 0.01 % Cd (Verbruggen et al. 2009). Results presented in the Table 1 show that when provided a chance, *Cenchrus ciliaris* can accumulate at least 0.02 % Nickel (186 μg/g) metal in its total dry mass out of which almost half is accumulated in the aerial part (90.5 μg/g shoot dry mass).

Roots are the first to come in contact with the metal ions present in the soil and so they absorb and accumulate the metal in question both in the stress and non-stress conditions. Our results indicate that the nickel uptake and accumulation behaviour of *Cenchrus ciliaris* varies in stress and non-stress conditions (Table 1). Under non-stress conditions *C. ciliaris* uptakes Nickel from soil both in the absence and the presence of the mycorrhiza and with a $TF > 1$ a large part of it is transported in to the shoot tissue. Heavy metal stress in the absence of mycorrhiza significantly increases nickel uptake by the *Cenchrus ciliaris* roots but it is not efficiently translocated to the shoot ($TF < 1$) and so the metal taken up accumulates in the roots. Under the same conditions of heavy metal stress the presence of mycorrhizal colonization, Test 1 and Test 2, though results in a reduced uptake of the nickel from the

Table 1 Effect of Myccorhizal colonization and metal treatment on Nickel uptake by *Cenchrus ciliaris*

	Controls				Test	
Mycorrhiza	–	+	–	–	+	+
Metal	–	–	1	10	1	10
Mycorrhiza colonization (%)	Nil	99.5	Nil	Nil	56.66	56.6
Shoot Ni content (μg/g)	81.8	89.2	84.5	90.5	77.30	93.8
Root Ni content (μg/g)	63.2	78.05	95.30	95.4	51.90	52.3
Total Ni content of the plant	145	167	180	186	129	146
Translocation factor (%)	1.29	1.14	0.89	0.95	1.49	1.79
Soil Ni content (μg/g)	50.00	40.50	45.5	56.5	45.50	63.3

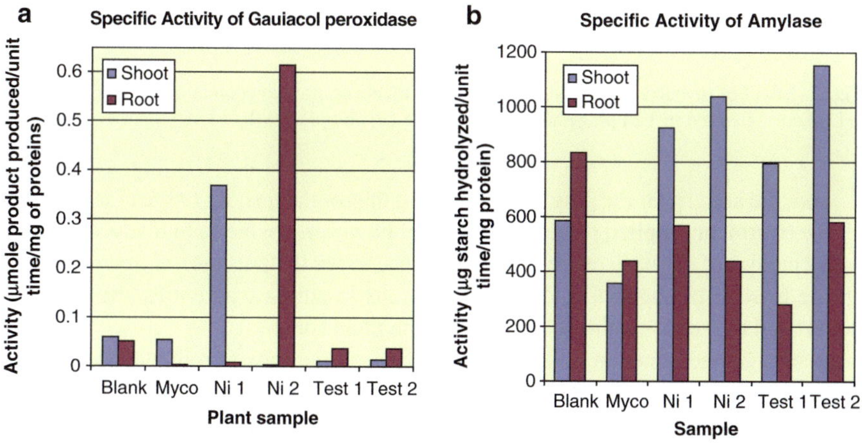

Fig. 2 Effect of mycorrhiza and Nickel metal treatments on specific activities of guaiacol peroxidase (**a**) and amylase (**b**) in the *C. ciliaris* plants

contaminated soil and with a TF > 1.5 it significantly increases translocation of the metal to the shoot (Table 1).

Since heavy metals are known to generate oxidative stress for the growing plants (Andrade et al. 2009), we investigated possible changes in the activities of two enzymes, guaiacol peroxidase and amylase along with soluble protein content both in the root and the shoot tissues. The results shown in Fig. 2a indicate that compared with non-stress conditions exposure of the plant root system to the heavy metal stress of 1 or 10 ppm in the absence of mycorrhiza elicits a very high level of specific activity of guaiacol peroxidase in the shoots or the roots, respectively. Exposure of the plant to the same level of heavy metal stress in the presence of mycorrhiza, however, brings down the activities back to the normal levels in the roots or even below normal levels in the shoots. Since guaiacol peroxidase plays an important role in stress tolerance, these results suggest that the mycorrhiza is involved in alleviation of the stress induced by the heavy metal treatment through a mechanism that parallels the anti-oxidant activities of peroxidase like enzymes.

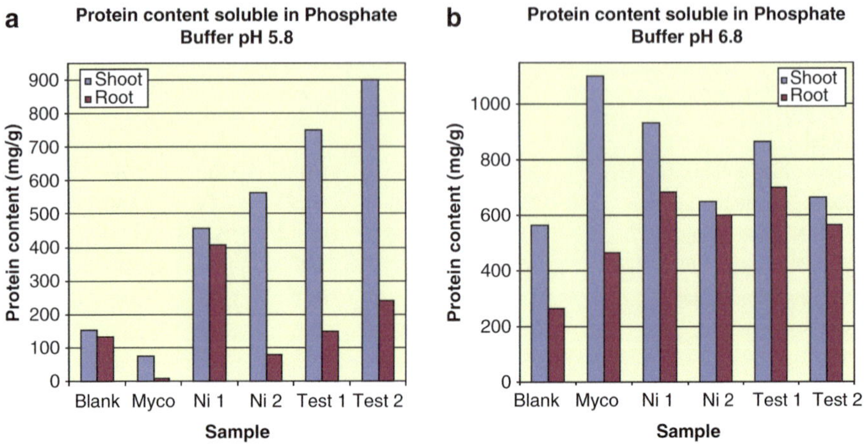

Fig. 3 Effect of mycorrhiza and Nickel metal treatments on protein content of the *Cenchrus ciliaris* plants extracted in 1 M phosphate buffer pH 5.8 (**a**) and pH 6.8 (**b**)

Specific activity of the amylase enzyme in the root and shoot tissues increases in all the treatments applied (Fig. 2b). We were, however, unable to find any variations in the pattern of amylase expression in the absence or the presence of mycorrhiza in all the treatments suggesting that the variations in amylase activities are probably not related to the mycorrhizal infection of the plant roots.

A strange phenomenon was observed during estimation of protein content of the different plant samples. The plants given no treatment or mycorrhizal treatment alone, the no-stress group, contained proteins that were extracted the most in Phosphate buffer pH 6.8 (Fig. 3b). Expression of these proteins increased in the roots but decreased in the shoots during metal alone or metal-mycorrhiza combined treatment. These proteins generally did not elute in the Phosphate buffer pH 5.8 (Fig. 3a). The plants treated with the nickel metal alone or a combination of nickel and mycorrhiza, the stress group, exhibited a different pattern of protein expression in the roots and shoot systems. While the roots of the plants facing metal stress expressed proteins that were mostly extracted in the 1 M phosphate buffer pH 6.8, the aerial parts, the shoots, expressed proteins that were mostly extracted in the 1 M phosphate buffer of pH 5.8. These observations suggest that the heavy metal stress generated during nickel treatment affects gene expression. We hypothesize that these proteins may be related to the stress alleviation; however, detailed investigations are required before making any final conclusion.

Thus, it is concluded that the *Cenchrus ciliariss* plants are capable of mobilizing and uptaking nickel metal from soil through their roots from where the accumulated metal is transported up into the shoot. Although, higher concentrations of nickel in the soil exert a lot of stress on the plant tissue, the plant appears to be equipped with the machinery that can help tolerate this stress upto a reasonable extent. Guaiacol peroxidase and amylase activities appear to be related to the stress however, the former is related to the stress alleviation by the mycorrhiza *Glomus mosseae*. The mechanism and role of these components in stress alleviation need to be investigated in more detail.

References

Aggangan NS, Dell B, Malajczuk N (1998) Effects of chromium and nickel on growth of the ectomycorrhizal fungus Pisolithus and formation of ectomycorrhizas on Eucalyptus urophylla S.T. Blake. Geoderma 84(1–3):15–27

Ajungla T, Sharma GD, Dkhar MS (2003) Heavy metal toxicity on dehydrogenase activity on rhizospheric soil of ectomycorrhizal pine seedlings in field condition. J Environ Biol 24(4):461–463

Al Agely A, Sylvia DM, Ma LQ (2005) Mycorrhizae increase arsenic uptake by the hyperaccumulator Chinese brake fern (Pteris vittata L.). J Environ Qual 34(6):2181–2186

Al-Karaki G, McMichael B, Zak J (2004) Field response of wheat to arbuscular mycorrhizal fungi and droughtstress. Mycorrhiza 14:263–269

Andrade SAL, Gratão PL, Schiavinato MA, Silveira APD, Azevedo RA, Mazzafera P (2009) Zn uptake, physiological response and stress attenuation in mycorrhizal jack bean growing in soil with increasing Zn concentrations. Chemosphere 75(10):1363–1370

AOAC (1995) AOAC official method 985.26: solids (Total) in processed tomato products. In: Official methods of analysis of AOAC International. AOAC International, Arlington

Barea JM, Pozo MJ, Azcon R, Azcon-Aguilar C (2005) Microbial co-operation in the rhizosphere. J Exp Bot 56:1761–1778

Borges R, Chaney W (1989) Root temperature affects mycorrhizal efficacy in Fraxinus pennsylvanica Marsh. New Phytol 112:411–417

Bose S, Bhattacharyya AK (2008) Heavy metal accumulation in wheat plant grown in soil amended with industrial sludge. Chemosphere 70(7):1264–1272

Chaudhry MS, Rahman SU, Ismaiel MS, Sarwar G, Saeed B, Nasim FH (2009) Coexistence of arbuscular mycorrhizae and dark septate endophytic fungi in an undisturbed and a disturbed site of an arid ecosystem. Symbiosis 49:19–28

Chen B, Tang X, Zhu Y, Christie P (2005a) Metal concentrations and mycorrhizal status of plants colonizing copper mine tailings: potential for revegetation. Sci China C Life Sci 48(Suppl 1):156–164

Chen X, Wu C, Tang J, Hu S (2005b) Arbuscular mycorrhizae enhance metal lead uptake and growth of host plants under a sand culture experiment. Chemosphere 60(5):665–671

Citterio S, Prato N, Fumagalli P, Aina R, Massa N, Santagostino A, Sgorbati S, Berta G (2005) The arbuscular mycorrhizal fungus Glomus mosseae induces growth and metal accumulation changes in Cannabis sativa L. Chemosphere 59:21–29

Cruz C, Green JJ, Watson CA, Wilson F, Martins-Loução MA (2004) Functional aspects of root architecture and mycorrhizal inoculation with respect to nutrient uptake capacity. Mycorrhiza 14(3):177–184

Del Val C, Barea JM, Azcon-Aguilar C (1999) Diversity of arbuscular mycorrhizal fungus populations in heavy-metal-contaminated soils. Appl Environ Microbiol 65(2):718–723

Diedhiou PM, Hallmann J, Oerke EC, Dehne HW (2003) Effects of arbuscular mycorrhizal fungi and a non-pathogenic Fusarium oxysporum on Meloidogyne incognita infestation of tomato. Mycorrhiza 13(4):199–204

Galli U, Schuepp H, Brunold C (1994) Heavy metal binding by mycorrhizal fungi. Physiol Plant 92(2):364–368. doi:10.1111/j.1399-3054.1994.tb05349.x

Gildon A, Tinker PB (1983) Interactions of vesicular-arbuscular mycorrhizal infection and heavy metals in plants. I. The effects of heavy metals on the development of vesicular-arbuscular mycorrhizas. New Phytol 95(2):247–261. doi:10.1111/j.1469-8137.1983.tb03491.x

Giovannetti M, Mosse B (1980) An evaluation of techniques for measuring vesicular arbuscular mycorrhizal infection in roots. New Phytol 84(3):489–500. doi:10.1111/j.1469-8137.1980.tb04556.x

Hakeem KR, Sabir M, Ozturk M, Mermut A (2015) Soil remediation and plants: prospects and challenges. Academic Press, Elsevier, New York, p 724

Hildebrandt U, Regvar M, Bothe H (2007) Arbuscular mycorrhiza and heavy metal tolerance. Phytochemistry 68(1):139–146

Hovsepyan A, Greipsson S (2004) Effect of arbuscular mycorrhizal fungi on phytoextraction by corn (Zea mays) of lead-contaminated soil. Int J Phytoremediation 6(4):305–321

Ibekwe AM, Angle JS, Chaney RL, van Berkum P (1995) Sewage sludge and heavy metal effects on nodulation and nitrogen fixation of legumes. J Environ Qual 24(6):1199–1204

Killham K, Firestone M (1983) Vesicular arbuscular mycorrhizal mediation of grass response to acidic and heavy metal depositions. Plant and Soil 72(1):39–48

Lambert DH, Baker DE, Cole H Jr (1979) The role of mycorrhizae in the interactions of phosphorus with zinc, copper, and other elements. Soil Sci Soc Am J 43(5):976–980

Liu SL, Luo YM, Cao ZH, Wu LH, Ding KQ, Christie P (2004) Degradation of benzo[a]pyrene in soil with arbuscular mycorrhizal alfalfa. Environ Geochem Health 26(2–3):285–293

Martin CA, Stutz JC (2004) Interactive effects of temperature and arbuscular mycorrhizal fungi on growth, Puptake and root respiration of Capsicum annuum L. Mycorrhiza 14:241–244

Ouziad F, Hildebrandt U, Schmelzer E, Bothe H (2005) Differential gene expressions in arbuscular mycorrhizal-colonized tomato grown under heavy metal stress. J Plant Physiol 162(6): 634–649

Quilambo O (2000) Functioning of peanut (Arachis hypogeae L.) under nutrient deficiency and drought stress in relation of symbiotic associations. University of Groningen, the Netherlands., Van Denderen, B.V.Groningen

Rao A, Arshad M, Shafiq M (1989) Perennial grass germplasm of Cholistan desert and its phytosociology. Cholistan Institute of Desert Studies, The Islamia University of Bahawalpur, Bahawalpur

Reynolds HL, Hartley AE, Vogelsang KM, Bever JD, Schultz PA (2005) Arbuscular mycorrhizal fungi do not enhance nitrogen acquisition and growth of old-field perennials under low nitrogen supply in glasshouse culture. New Phytol 167(3):869–880

Sabir M, Hakeem KR, Aziz T, Zia-ur-Rehman M, Rashid I, Ozturk M (2014) High Ni levels in soil can modify growth performance and mineral status of wheat cultivars. Clean Soil Air Water 42:1263–1271

Sailo GL, Bagyaraj DJ (2005) Influence of different AM-fungi on the growth, nutrition and forskolin content of Coleus forskohlii. Mycol Res 109(Pt 7):795–798

Salt DE, Blaylock M, Kumar NPBA, Dushenkov V, Ensley BD, Chet I, Raskin I (1995) Phytoremediation: a novel strategy for the removal of toxic metals from the environment using plants. Nat Biotechnol 13(5):468–474

Switzer R, Garrity L (1999) Experimental biochemistry, 3rd edn. W. H. Freeman and Company, New York

van der Heijden MGA, Klironomos JN, Ursic M, Moutoglis P, Streitwolf-Engel R, Boller T, Wiemken A, Sanders JR (1998) Mycorrhizal fungal diversity determines plant biodiversity, ecosystem variability and productivity. Nature 396:69–72

Vassilev A, Schwitzguebel JP, Thewys T, Van Der Lelie D, Vangronsveld J (2004) The use of plants for remediation of metal-contaminated soils. ScientificWorldJournal 4:9–34

Verbruggen N, Hermans C, Schat H (2009) Molecular mechanisms of metal hyperaccumulation in plants. New Phytol 181:759–776

Williams LE, Pittman JK, Hall JL (2000) Emerging mechanisms for heavy metal transport in plants. Biochim Biophys Acta 1465:104–126

Taming Food Security Through Wastewater Irrigation Practices

Zeshan Ali, Riffat Naseem Malik, Alvina Gul, and A. Mujeeb-Kazi

Contents

Abstract Water scarcity, deteriorating water quality and wastewater irrigation is a regional as well as global problem affecting human livelihood and economic prosperity in developing countries. Pakistan's economy is based on agriculture, which supports livelihood of more than 50 % people relying heavily on available fresh water supplies. Around 90 % of the food and fiber requirements of the country are met from irrigated areas (86.25 % of the total cultivated area) and remaining 10 % from rain-fed areas (13.75 % of the area under cultivation). Water availability is maximum in the Kharif (summer) season and lowest in the Rabi (winter) season. Since Pakistan's creation water availability has reduced from 5600 m^3 to <1000 m^3

Z. Ali
Environmental Biology and Ecotoxicology Laboratory, Department of Plant Sciences, Faculty of Biological Sciences, Quaid-i-Azam University, Islamabad PO 45320, Pakistan

R.N. Malik (✉)
Environmental Biology and Ecotoxicology Laboratory, Department of Environmental Sciences, Faculty of Biological Sciences, Quaid-i-Azam University, Islamabad PO 45320, Pakistan
e-mail: r_n_malik2000@yahoo.co.uk; eco4nd@yahoo.com

A. Gul
Atta-ur-Rahman School of Applied Biosciences, National University of Sciences and Technology, Islamabad, Pakistan

A. Mujeeb-Kazi
Wheat Wide Crosses, National Institute of Biotechnology & Genetic Engineering (NIBGE), Faisalabad, Pakistan

M. Öztürk et al. (eds.), *Plants, Pollutants and Remediation*,
DOI 10.1007/978-94-017-7194-8_6

placing immense pressure on the domestic, industrial and agricultural sectors. Reduction in available fresh water supplies has attracted farmers in rural, urban and peri-urban areas to harvest untreated wastewater in agriculture. Wastewater is a complicated mixture of chemical pollutants in which heavy metals are of significant concern. Heavy metals laden wastewaters have adverse impacts on the plants, soils, humans and livestock health. Grave food safety issues arise from the consumption of metal polluted agricultural produce leading to serious clinical conditions. Around 4369 million m^3 of wastewater is generated per annum in Pakistan, which is roughly equivalent to standing wheat crop water requirements. Wheat is grown in the Rabi season (November to May), the season, which is already facing serious water deficit. Voluminous amounts of wastewater generated from municipal and industrial sectors can be efficiently utilized in safer crop (wheat) production on treatment. As raw wastewater irrigation can result in serious toxicological conditions in the wheat crop therefore diverse wheat genetic makeup (different wheat varieties/advanced lines) may be used to determine the extent of adverse impacts of heavy metal containing municipal/industrial wastewater on growth/metabolism of different Pakistani wheat cultivars. Identified tolerant wheat varieties then can be recommended for areas where wastewater irrigation is common or fresh water is scarce that will also minimize contaminants (metals) exposure to human and livestock through wheat consumption. Related information on wastewater antagonistic impacts on different wheat varieties is not available from Pakistan. Food safety issues arising from consumption of metal contaminated wheat grains/flour irrigated with municipal/industrial wastewater have only been explored on one of the most commonly growing wheat varieties in various studies. Wheat is the conduit to food security, so harvesting precious wastewater with its potential reuse in wheat irrigation can be of great competitive advantage in achieving food security in rapidly populating country i.e. Pakistan, where growing population needs continuous food supplies with an ever growing water needs. Future research endeavors if focused on gauging impact of wastewater irrigation practices on different wheat cultivars, information generated would be helpful in combating water scarcity, food safety and food security related issues in Pakistan.

Keywords Water scarcity • Wastewater irrigation • Food security and Food safety • Agriculture • Wheat • Heavy metals

1 Introduction

Growing water scarcity places immense pressure on human livelihood and economic development in arid/semi-arid regions globally specially in developing countries (Ensink et al. 2004). General water scarcity and its un-availability at optimum irrigation times are contributing significantly to wastewater exploitation in current agricultural practices (Drechsel and Evans 2010). The practice of wastewater irrigation in agriculture is a century old phenomenon. Farmers have preferred raw wastewater as compared to clean water/treated wastewater due to its high nutritive value, continuous availability and elevated crop yields (Darvishi and Farahani 2010). Wastewater being a low cost alternative to fresh water is currently being used in peri-urban agriculture at an ever-increasing rate. Increased industrial activities and urban expansion include the sectors where fresh water consumption has tremendously increased depriving the agriculture sector from the fair water share it enjoyed earlier (Ensink et al. 2002). Hence it is acceptable to state that water scarcity coupled with wastewater irrigation practices are together compromising agricultural productivity and produce quality in developing countries (Qadir et al. 2010; Ozturk et al. 2011).

Decline in fresh water availability with the development of dams in our neighboring territory represents a key issue in Pakistan's crop production and food safety/security problems (Shakir et al. 2011). A shortfall of 48.6 MAF (million acre feet) is expected by 2024 due to increasing water demands in different sectors. All water-utilizing sectors (i.e. domestic, agricultural and industrial) are influenced by increasing population, which is expected to grow to 260 million by the year 2025 with the global estimate being 8.2 billion (Kahlown and Majeed 2001). Pakistan so far has exploited to a maximum its available fresh water resources and is on its way to swiftly become a water deficit country in near future if alternates are not harnessed. Agriculture is the largest user of the available fresh water supplies with flood irrigation as the dominant surface irrigation technique. Efficiency of our conventional irrigation systems is lowest and leads to maximum water loss at the farm level including evapotranspiration etc. Total available fresh water/annum ranges between 180,000–185,000 million m^3, out of which 96 % is utilized in agriculture, 2 % in the industrial sector and remaining 2 % for domestic use (Pakistan Economic survey 2009–2010). Corresponding to Pakistan's total fresh water available/annum 4369 million m^3 of wastewater is generated/annum including 1309 million m^3 from industrial and 3060 million m^3 from the municipal sector (Pakistan water sector strategy 2002). Voluminous amounts of wastewater generated have considerable impacts on the hydrology and quality of surface and ground aquifers. Increased industrial activities and population boom have not only deteriorated the wastewater quality further but also increased wastewater volumes in the past decades (Qadir et al. 2008). Adverse quality of irrigation wastewater is seriously threatening the environment, food safety and human/livestock health (Ali et al. 2013a, b; Ali et al. 2015). Food safety and food quality are related terms. Food safety specifies hazards to consumer health whereas food quality concerns contamination, spoilage, bad odors, processing origin etc. Food safety hazards mostly include contamination via microbial, metal, pesticide and food additives. Wastewaterirrigation is practically concerned with all these food safety hazards (Mutengu et al. 2007; Habbari et al. 2000). Gastrointestinal disease, cholera, typhoid, diarrhea etc. outbreaks in the urban populations via consumption of wastewater irrigated plants are some of the negative outcomes of wastewater irrigation (Shuval et al. 1986). Wastewater irrigation of vegetables, fodder and cereals serves as a potential source for transmitting microbial and chemical pollutants to the consumers through the food chain contamination. Heavy metal contamination of milk through wastewater irrigation of fodder and its subsequent usage in cattle feed has been reported in the South Asia establishing a connection with the food chain contamination through wastewater irrigation (Delgado et al. 1999). Similarly cereal and vegetable contamination with heavy metals and their worse impacts on the human and livestock health has been widely reported in literature (Rattan et al. 2005; Muchuweti et al. 2006; Abbas et al. 2007; Chandra et al. 2009; Ashraf et al. 2015). Santos et al. (2004) determined heavy metal intake by the local population of Rio de Janeiro (Brazil) through vegetables, cereals, their derivatives, animal products and reported highest heavy metal intake from the cereal (rice and wheat flour) consumption.

In Pakistan like many other developing countries farmers irrigate crops (vegetables, fodder and cereals) with municipal and industrial effluents having high levels of heavy metals (Co, Cr, Cu, Fe, Ni, Pb, Mn, Cd and Zn etc.) due to water scarcity or its un-availability (Ensink et al. 2004). Approximately 32,500 ha of land in Pakistan is presently irrigated with wastewater including 26 % of the country's vegetable production acreage (Ensink et al. 2004). Wheat (*Triticum aestivum* L.) is the conduit to food security and the most important staple food of people in Pakistan contributing 3.1 % to agriculture GDP. It is cropped over an area of 8.6 Mha (million hectares) which is 36.78 % of the total area under cultivation (23.38 Mha; GoP 2008). Wheat production target and cultivation area for the year 2009–2010 was set at 25 million tons and 9.045 Mha respectively. However due to low rainfall and water shortage it was cultivated over an area of 9.042 Mha, approximately 0.04 % less than the 2008–2009 wheat acreage of 9.046 Mha. Thus the production target of 25 million tons was not achieved due to water shortage and low acreage of cultivation. These figures represent wheat production targets hindered by water scarcity that was maximized in the Rabi season (wheat growing period). Around 95 % of the total national wheat production is in the irrigated areas whereas the remaining 5 % is in the rain-fed areas (Agricultural Statistics of Pakistan 2011). Self-sufficiency in production of this staple food relies upon ample water supply and efficient management of water resources. Achievements in conventional development of disease resistant and high yielding varieties will make less impact under deficit irrigation waters in different agro-ecological systems of Pakistan (Farooq and Iqbal 2000). Besides implementation of good agronomic practices which will help to bridge the gap between actual (2.5–3.0 t/ha) and potential yield (6–8 t/ha) of existing wheat varieties, water availability/scarcity in irrigated and rain-fed areas will continue to be pivotal in wheat production. The production target is 33 million tons for 2030 which was 23 million tons in 2010 (Rajaram et al. 1998; Agricultural Statistics of Pakistan 2011). With expanding population, increasing industrial activities, silting of the canal system, low rainfall, climate change and water scarcity the wheat production target for 2030 seems challenging to achieve, raising serious questions with respect to food security.

Only one province (i.e. Punjab) produces surplus wheat due to ample water supplies and appropriate agro-climatic conditions. Other provinces depend on Punjab for their wheat supplies. Food security is directly linked with water availability which is of course less than optimum (Shakir et al. 2011). To ensure national food security efficient utilization of water and alternatives of fresh water irrigation need to be explored and implemented. Irrigationwater demands are not only on the rise for wheat but for other crops as well i.e. sugar cane, oil seed and horticultural crops. Competitive water demands for these crops and other water utilization sectors will deprive wheat crop from its water share, intensifying food security problems (Qadir et al. 2003). It has been estimated that the total wastewater generated in Pakistan is roughly equivalent to water supplies of the winter grown spring wheat crop. This scenario is not without interest which reveals immense wastewater importance for reuse in agriculture as fresh water resources decline. Such a strategy will not only reduce pressure on fresh water resources but will help in ensuring wheat based food security.

Incidentally wheat (*Triticum aestivum* L.) is the staple crop of Pakistan with the per capita wheat consumption of 125 kg/year (among the highest in the world) and sustaining its output potential is vital. It is also the most important crop worldwide grown over large areas. Wastewaterirrigation can help achieving the esteemed goals of food security to combat rapid population increase but surely not at the cost of food safety and environmental health. Municipal and industrial wastewaters are laden with varying amounts of hazardous heavy metals. The fear of production of heavy metal contaminated wheat with different wastewater irrigation practices will not only pose potential human health risks but will indefinitely harbor serious environmental concerns. Thus wheat varietal performance in terms of food safety, growth and yield in response to wastewater irrigation are some key issues which need to be addressed. Usually the varietal performance differs with respect to abiotic stresses. Similarly the sensitivity of wheat varieties to different wastewater irrigation practices also differs. The challenge is to identify wheat varieties least affected by heavy metals from different wastewater irrigation practices helping to achieve the goal of food security/food safety in limited water regimes. Since wheat is the main food for the inhabitants of Pakistan, metal accumulation behavior in different wheat varieties and those currently evolving through unique genomic diversity usage can help in ensuring human, livestock and environmental health.

Baseline for this chapter has thus been laid above and the coming sections will deal with overview of the water resources of Pakistan, national water scarcity scenario, national water quality in context of irrigation, wheat response to wastewater irrigation practices, wastewater re-use in irrigation, concluding with recommendations to target our 2050 vision of ample wheat for a 9.2 billion global populace.

2 Overview of Water Resources of Pakistan

Water is a basic necessity and an important abiotic factor in all ecological systems. Fresh water resources support the agrarian economy of Pakistan contributing 25 % of the GDP. This economy in turn supports the livelihood of more than 50 % of the labor force. Approximately 22.05 Mha is cultivated land out of 79.61 Mha of the total land of the country (Agricultural Statistics of Pakistan 2005–2006). In 1947 only 14.7 Mha was cultivated which increased up to 22.05 Mha at present times due to the development of the extensive Indus river supply system. Using available water resources around 86.25 % of the cultivated area is irrigated which fulfills almost 90 % of the food and fiber requirements of the country. Water resources of Pakistan are categorized into three main sources i.e. surface water, groundwater and rainfall.

Pakistan's surface water main source comprises of the Indus river system having important eastern (Siran, Chenab, Ravi, Sutlej, Haro, Soan, Beas and Jhelum) and western (Kora, Kabul, Kurram, Gomal, Punj, Kunar and Tai) tributaries with the catchment area of 944,569 km^2 (Pakistan water sector strategy 2002). Lion's share to the Indus inflow is from rainfall and glaciers melt. Some of the above mentioned rivers flow only in the rainy season and are trivial contributors to Pakistan's surface water resources. Pakistan is signatory of Indus Basin Treaty (1960) and is allocated only the western rivers i.e. Indus, Jhelum, Chenab whereas the eastern rivers i.e. Beas, Sutlej and Ravi have been allocated exclusively to India. India after the Indus treaty has constructed Ranjit Sagar Dam also called Thein dam (Ravi), Pong dam (Beas) and Bhakra dam (Sutlej) on respective rivers to efficiently utilize their water. Average inflow of eastern rivers in Pakistan is 8.40 MAF fluctuating from 1.56 MAF in Rabi to 6.85 MAF in Kharif seasons. Kharif and Rabi are two main agricultural seasons in Pakistan. To meet the deficiency of the irrigation water in the eastern rivers; dams, barrages and inter-river link canals were built (Kamal 2008). Annual average inflow of the western rivers is around 143.2 MAF fluctuating from 25.73 MAF in Rabi to 117.5 MAF in Kharif seasons. This average inflow in Indus river greatly depends on the associated tributaries and varies yearly. Around 104.7 MAF is used for irrigation from the annually available water. On an average Indus river along with its western and eastern tributaries bring 151.6 MAF water/annum.

Pakistan's ground water adds 40 % of total water supplies/annum to meet the food and fiber requirements. Around 40 MAF is pumped per annum and main recharge sources include precipitation and infiltration from surface waters from the last 90 years (Agricultural Statistics of Pakistan 2011). Its quality varies under the earth crust and useable portion is restricted to 10 Mha. Unfortunately no estimates are available regarding quantity of underground water in Pakistan. The amount of ground water available globally is 500,000 MAF, found up to 750 m from the surface. This volume is many times greater than the surface water (rivers, streams, lakes etc.; Ahmad 1993). According to Kahlown et al. (2004) ground water resource has a potential of 55 MAF and exploited to maximum for household, industrial and agricultural purposes. Punjab (42.7 MAF) is the highest user of the ground water following Sindh (3.5 MAF), Khyber-Pakhtunkhwa (KPK; 2 MAF), Balochistan (0.5 MAF) and Azad Jammu Kashmir (AJK, 4300 AF i.e. acre feet). Ground water resources have been exhausted in the heavily populated areas of Pakistan (Pakistan Water Partnership 1999). In Balochistan the ground water situation is worse where this precious resource has been utilized near to exhaustion with little or no alternatives available. Underground water table is falling at an average of 30–40 cm/year in populated areas of the country (Shakir et al. 2011). Besides development of ground water resources much attention is currently needed for its sustainable development and long term availability. Brutal exploitation through tube wells without any monitoring system and continuous lowering of the underground water table has aggravated the situation. Installation of new tube wells without knowing the rate of underground aquifer recharge will leave no water for future generations. The equilibrium between the water recharge and water withdrawal is a must and need of the hour (PCWRAS 2005). With the declining water table, cost involved in pumping ground water will increase further hampering appropriate irrigation needs of field crops especially wheat. Ground water resource has utmost importance in the barani areas of Pakistan where except for rain it is the sole irrigation source.

Pakistan's major rainfall sources include western disturbances and monsoons. Around 70 % of the monsoon rains are witnessed from July to September and are of substantial importance in the arid and semi-arid agro-climatic regions of the country (Shaheen and Baig 2011). Average rainfall/annum is around 250–300 mm (Ahmad et al. 2011). Indus river system receives 40 MAF of rainfall/annum fluctuating from 53 mm in Rabi to 212 mm in Kharif seasons (Agricultural Statistics of Pakistan 2011). Rainfall pattern varies across the country and northern areas receive more rains generally (Agricultural Statistics of Pakistan 2011). Rainfall harvesting potential except irrigated areas is as high as 20 MAF, major portion of which is lost through evaporation. Rain-fed (non-irrigated) area accounts to 4 Mha approximately and relies on rain and ground water sources.

In spite of prevailing water scarcity 38.1 MAF water/annum from Indus river system drains into the Arabian Sea. Around 35.61 MAF flows during Kharif and 2.40 MAF flows during the Rabi season. In some winter months there is no flow to the Arabian Sea. Again this flow depends on the tributaries feeding Indus, rain fall patterns and cropping season involved, therefore flow as high as 91.8 MAF to as low as 0.74 MAF escapages have been recorded from Indus river to the sea. It has been calculated that in the last two decades almost 988.3 MAF water has flowed to the sea. By conservation this water could be efficiently utilized in power generation, irrigation demands and saving saline water inland intrusion from sea by keeping to a minimum flow from Indus system to the sea. Just water distribution among all the provinces is urged especially in Sindh and Punjab provinces where water allocation controversy remains unresolved. Even within provinces irrigation water distribution between farmers is uneven and remains a bone of contention. Consensus between the provinces for construction of large reservoirs and water distribution plans within provinces for peaceful settlement of water issues are need of the time. Farmers near the canal systems or water distribution points pump most of the irrigation water as no policy exists for its regular distribution and the tail end farmers are frequently deprived of their water share (Agricultural Statistics of Pakistan 2011). Conveyance losses (up to 55 %) are mostly responsible for water unavailability at downstream of the prevalent canal system (Shakir and Qureshi 2007). These losses are due to the inefficient conventional irrigation system. Maintenance and operational measures for the existing irrigation system are not optimized which leads to its inefficiency and water losses (Saeed et al. 2002). Water use efficiency cannot be achieved without water resources development/management and via varietal research and development (Hsiao et al. 2007).

3 National Water Scarcity Scenario

Scarcity and contamination of existing water resources are on rise corresponding to their demands in the domestic, agricultural, industrial and recreational purposes. Agricultural productivity is seriously compromised as water availability is receding and conventional flood irrigation methods remain popular with farmers in this era of high efficiency irrigation systems (HEIS). Dropping per capita water availability is a regional as well as a global problem (Ahmad and Yasin 2006; Kahlown et al. 2007). Pakistan's per capita water availability has reduced drastically from 5600 m^3 since independence to <1000 m^3 at present (Martin et al. 2006). Due to declining fresh water resources our country has already become one of the most water-stressed countries globally and prevailing conditions will turn the situation into one of outright water scarcity (PCWRAS 2005).

Pakistan witnessed the "Green revolution" in the last century based on water resources development and related agricultural product expansion through adoption of dwarf wheat/rice varieties. With time water storage of the reservoirs is lost due to sedimentation and overall water availability has been reduced owing to low riverine flows. Less water availability has impacted crop yields in the recent years as water and yield are directly related. Water consumption in the urban centers has further pressurized fresh water resources and population increase (50 %) by 2025 is an added concern. Industrial and rural water supplies similarly have also increased. In the perspectives of water availability and consumption Pakistan now needs to develop a "Blue revolution" to suffice for the industrial, agricultural, urban, rural and recreational water needs. Reservoir development, increased water usage efficiency in all sectors, combating conveyance losses, de-siltation/de-sedimentation of canal systems, equal distribution and wise use of raw/treated wastewater should be the part and parcel of this Blue revolution. Thus water would play the same role as other identical inputs played in the Green revolution of the last century. Every drop of water should therefore be conserved and harvested for poverty alleviation, self-reliance in food, food safety, combating drought and preventing environmental degradation. Any delay in the development of supplementary water resources is detrimental and a common policy is required for the initiation of all pending projects that are either controversial or non-controversial i.e. Kalabagh dam.

Over exploitation of water resources in the past raised questions about astute water resource management. To obtain more crop per drop a realistic National Water Policy (NWP) is required which can sustainably optimize surface and ground water resources of the country. NWP can probably contribute to food safety, food security, poverty alleviation, support livelihood of small farmers by sustainable and ample supply of water to all provinces with wise and smart management. Prime utilization of fresh water resources through NWP is of extreme importance as water scarcity is a global issue. Existence of an ineffective NWP, its in-elaborate coverage of domestic water problems, pathetic implementation at the institutional and grass root level reflect grave issues associated with its poor structure. Water deficit is growing not only in Pakistan but also in other developing (India, Morocco, Algeria, Libya, Egypt, Somalia, South Africa, Saudi Arabia, Yemen, UAE etc.) and developed (Australia, USA) countries. Most of the countries suffering from the water scarcity and drought conditions are managing the water shortage problems wisely with their policy set up. Some countries do not have renewable fresh water resource and rely only on the alternate options i.e. importing food and water goods. It is estimated that production of one kg beef and wheat require 15,000 and 1500 l of water respectively (UNCCD 2013). Countries sanctified with plenty of fresh water resources have competitive economic advantage over less blessed ones.

Pakistan's agricultural productivity depends on ample water supplies in Kharif and Rabi seasons. Average water availability for both seasonal crops as an example have remained from 2.5 to 20.6 % less in the years 2005–2006 and 2004–2005 respectively. Usually the water flow in the Indus river system is low in winter which hampers the Rabi crops especially wheat. In the year 2009–2010 available water for the Kharif crops (cotton, sugar cane and rice) was 0.3 % higher than the average resulting in increased yield. Irrigation system's efficiency can help to achieve higher agricultural production, however poor efficiency of Pakistan's irrigation system leads to precious water wastage in voluminous amounts despite good networking (Saeed et al. 2002). Major Kharif (cotton, sugar cane and rice) and Rabi (wheat) crops contribute 7 % to country's GDP and 33 % value addition in overall agriculture. This economy besides farm management practices, policy framework and varietal performance depends on the sufficient water supplies in the Kharif and Rabi seasons. During the low flow regimes storage reservoirs i.e. Mangla, Tarbela and Chashma recharge the natural flows to serve irrigation purposes. Around 27 % (4.18 MAF) of the live storage of these reservoirs is lost due to sedimentation whereas the original storage capacity was 18.37 MAF. Indus River System Authority (IRSA) states that this shortage has increased up to 30 % therefore it is becoming increasingly difficult for IRSA to provide requisite water to provinces in the Rabi season when water in the system is already low. From this discussion it is clear that water availability in the two seasons markedly varies; around 16 % flows in Rabi and 84 % flows during Kharif season of the available annual water supplies. Also around 80 % of the average river flows and 70 % of the monsoon rains occur in the Kharif season (Shaheen and Baig 2011). View point of Bastiaanssen et al. (2002) is worth mentioning here that there could be drastic differences between actual irrigated areas of Pakistan to official figures by the Government. Irrigation expansion strategies to the remote areas low in underground water and precipitation are demanding as river basins have already been exploited to their full capacity.

From the discussion so far irrigated agriculture is the major consumer of surface and underground water resources of Pakistan. Irrigated agriculture is similarly the largest consumer of fresh water supplies worldwide and in Asia (>80 %). Largest irrigated land areas lie in Asia (370 Mha or 79 %) followed by North America (7 %) and Europe (7 %). Around 50 % of the total irrigated area on earth comprises of 6

Asian states i.e. Thailand (1.7 %), Indonesia (1.8 %), Iran (2.8 %), Pakistan (6.6 %), China (19.4 %) and India (21.7 %; Bhatti et al. 2009). Pakistan's irrigated agriculture requires around 117 MAF for irrigation purposes (IUCN 2010). Besides irrigation Pakistan needs 8 MAF for municipal, rural, industrial, recreational and environmental purposes to meet national water demands (NWP 2003). Majority of the irrigated area lies in Punjab (77.4 %) following Sindh/Balochistan (19.8 %) and KPK (2.8 %). Total irrigated area of Punjab (13.84 Mha), Sindh (2.52 Mha), KPK (0.89 Mha) and Balochistan (0.81 Mha) is mentioned respectively which reflects their irrigation water necessities. AJK and Gilgit-Baltistan (GB) consume insignificant irrigation waters as compared to other provinces as there exists no substantial infrastructure for irrigated agriculture. Average surface irrigation water consumed by Punjab (Kharif 34.3, Rabi 19.87), KPK (Kharif 2.35, Rabi 1.46) and Sindh/Balochistan (Kharif 31.4, Rabi 16.06) in MAF is given respectively. Another 41.6 MAF is obtained annually from underground aquifers to fulfill the irrigation demands, very little portion of which is replenished by rainfall and surface water flows causing decline in this resource. Drought 2000–2001 reduced irrigation water supplies by 18 % due to which agriculture showed negative growth of 2.5 % adversely impacting the country's GDP. To meet the irrigation water requirements with the growing population a 10 year plan of amplifying water supplies by 12 MAF is envisioned at national level. This plan included construction of new storage facilities, de-sedimentation of canals and improved management practices. According to some estimates both surface and ground water contributed 135.7 MAF to meet irrigation requirements by 2011; surely the demand is on increase. The future requirements cannot be met even with the development of supplementary storages. Future requirements need to be focused on careful selection of varieties, increased water usage efficiency, re-use of municipal/industrial effluents through treatment and safe use of brackish/saline waters.

Water availability and scarcity affect provincial crop yields. Punjab and Sindh lie in the Indus catchment areas and their agriculture is profoundly irrigated, therefore crop yields are relatively higher in presence of sufficient water resources. Most of the area of KPK is rain-fed and cropped with wheat and gram. Their yields are considerably lower than the irrigated areas of Punjab and Sindh. Balochistan is the most water scarce province of Pakistan with alarmingly receding water table which is pumped for domestic and agricultural needs. Acute water shortage is expected if underground aquifers continue to be exploited at current rates. All provinces vary in the available fresh water supplies therefore production of agricultural commodities also differs. Wise and enhanced management of irrigation systems in irrigated areas are mostly neglected and draw attention towards better water resource management and related policy framework development. Improved yields and water use efficiency go hand in hand by following modern agricultural practices i.e. best choice of varieties, timely sowing, laser leveling, balanced use of fertilizer and organic amendments, optimally scheduled irrigation, weed control etc. In irrigated areas of Pakistan average wheat yield (approximately 2.8 t/ha) is recorded less than most other irrigated areas of the world mainly due to farmer unawareness of modern agricultural practices. High crop yields in irrigated areas cannot be expected due to the rapidly declining water resources of the country. The low yield scenario in irrigated regimes urges policy makers and researchers to disseminate latest knowledge to farmers to fully reap the blessings of available fresh water. If the wheat yield increases up to 5 t/ha by better farm practices and water use efficiency then national wheat requirements can be met by growing wheat on around 6 Mha rather than 8.6 Mha. The yield increase will not only suffice for the expanding population, but will also spare 2.6 Mha of land and water required to irrigate this land. Thus precious water resources conserved will be available for other arid areas having limited or no water availability. If wheat average yield remains the same (2.8 t/ha), then to meet the population demands in 2025 wheat needs to be grown over an area of 12.5 Mha or a drastic increase in the yield will be required.

Domestic urban and industrial sectors are also important consumers of available fresh water supplies. Domestic and industrial sectors consume 3.2 and 1.18 MAF of water/annum. Due to expanding population the domestic urban and industrial water demand will increase up to 12.1 and 1.84 MAF by the year 2025 (Pakistan Statistical Yearbook 2001). Many industries are based on agriculture that also runs rural economies. Rural domestic needs at present are estimated at 0.8 MAF, which will increase up to 3.2 MAF by year 2025. At present urban and rural domestic needs are mostly met by the underground water resources. If domestic sector needs are not met by the underground resources then surface water will be utilized depriving the irrigation sector. State of the Environment Report (SOE 2005) describes 29 % and 33 % irrigation water shortage for the years 2010 and 2025 respectively. To bridge the gap between irrigation water supplies and crop water requirements by year 2025, an additional 18 MAF water storage will be required besides execution of up-to-the-minute agricultural innovations. Thus water availability will continue to play a critical role in sustainable economic development.

Pakistan's stressed water resources, crop varieties and cropping patterns are influenced by the climate change/global warming. Their impact on rainfall patterns and temperature is erratic and adverse to the agricultural production of the country. Climate change is altering the water cycle in unpredictable ways and disturbing water availability. Pakistan is considered as the most susceptible country due to change in rainfall patterns, rising temperatures, inadequate water supply for crops, droughts, flash floods and rapid glacier melting etc. Most significant impacts are seen on the water availability and low crop productivity (caused by water unavailability and temperature shift). Rising temperature can be beneficial in the northern mountain regions of the country where it favorably shrinks the growing period of winter crops (Hussain and Mudasser 2004). However this rise in temperature has adverse effects on crops in southern parts of the country where it increases crop water requirements. Dire need is to adapt with the changing climatic conditions and develop strategies to combat its negative effects.

4 National Water Quality in Context of Irrigation

Competitive rising demands of water in different sectors have given birth to water quality problems regionally and globally. Water availability and or scarcity have been prioritized from the past decades whereas deteriorating water quality has been less emphasized which exhibited serious ecological implications (Ensink et al. 2002). Water shortage and pollution have given rise to multiple problems i.e. water borne diseases, reduced biodiversity, decline in agricultural production, food safety and food security issues (SOE 2005). Thus poor water quality plays a significant role in controlling the socio-economic scenario of developing countries i.e. Pakistan. Due to increasing fresh water degradation, many species of plants and animals are being eliminated from riverine and the associated terrestrial ecosystem of Pakistan (SOE 2005). Domestic, irrigation, drinking and industrial water supplies are declining in quality whilst the contamination sources remain active to pollute them. Safe municipal and industrial effluent discharge regulations exist but are not practiced. Safe environmental water use should be encouraged to overcome adverse water pollution effects downstream. Significant efforts in this regard are required to strengthen related institutions which in turn educate general public and farmers. Pakistan for sure will not have adequate water resources in the coming future; with growing population in urban centers surface water quality will further decline requiring every individual to be environmentally educated. Anthropogenic influences on quality/contamination of agricultural water supplies are briefly described hereunder. Regionally contamination from one water source is not restricted; instead pollutants find a way to other clean water sources and contaminate them too. Water once polluted moves through the ecosystem leaving contaminant traces everywhere.

Agricultural consumption of water i.e. irrigation is met from three sources i.e. surface water, rainfall and ground water. Rainfall is erratic and plays a significant role in only rain-fed areas of Pakistan. Surface and ground water are contaminated to varying degrees by various municipal and industrial activities. Pollution levels are high around urban centers and their municipal and industrial discharge contaminates irrigation water supplies with municipal wastes, heavy metals and pathogens etc. Due to high wastewater volumes from urban centers and low dilution factors, irrigation sources are anticipated to be more metal contaminated than earlier and resultant food produce is less safe. There is unrestricted use of this wastewater in irrigation. Around 60 % of the total wastewater generated by this sector in Pakistan comes from 10 large urban centers i.e. Lahore, Faisalabad, Gujranwala, Rawalpindi, Sheikhopura, Multan, Sialkot, Karachi, Hyderabad and Peshawar (PCWRAS 2005). Nature of domestic effluents that arise from mentioned urban centers are highly toxic as they are mixed with industrial effluents leading to higher heavy metal loads. Mostly there is no separation between the domestic/municipal and industrial effluents. These effluents find an easy way to the surface water channels (streams and rivers) through nullahs/drains which carry discharge from the urban centers. Surface water pollution in Pakistan is largely associated with these effluents and around 6140 AF of sewage is discharged to surface water bodies/day (Martin et al. 2006).

Industrial sectors generate voluminous amounts of wastewater which are discharged untreated to the surrounding ecosystems. Different districts of Punjab (Lahore, Sialkot, Faisalabad, Gujranwala) contain significant number of industries without any serious approach towards environmental protection. Surface water channels (streams, nullahs and rivers) in the vicinity of these industries show alarmingly high levels of heavy metals, Chemical Oxygen Demand (COD) and Biological Oxygen Demand (BOD). Ravi river receives municipal and industrial loads from Lahore and its water is unfit for irrigation from Lahore to Balloki. After Balloki, river water quality improves due to dilution received from Qadirabad-Balloki link canal. Sialkot and Gujranwala are the hub of tannery, cutlery, electronics, ceramics, metallurgy, surgical and sports goods with inadequate or no wastewater treatment facilities (Qadir et al. 2008). Chenab river receives industrial effluents from Sialkot (through nullah Aik and nullah Palkhu) and Gujranwala and its water is utilized for irrigation in Punjab. Soan river receives Islamabad and Rawalpindi municipal effluents through nullah Lei and drains in to Indus river. Likewise in Multan and Faisalabad industrial wastewater treatment facilities do not exist and their metal laden effluents drain to nearby rivers. Jhelum and Indus rivers are comparatively less impacted in terms of water quality degradation, which is attributable to high flows in these rivers. In Sindh most of the industry is located at Karachi industrial zones i.e. Korangi Industrial and Trading Estate (KITE), Sindh Industrial Trading Estate (SITE), Karachi Export Processing Zone and Northern Bypass Industrial Zone etc. KITE and SITE are the largest industrial hubs with no industrial wastewater treatment facility. Untreated heavy metal bearing effluents are discharged to surface water bodies (Lyari and Malir rivers) which take the discharge to the coastal areas threatening Mangrove and coastal ecosystems.

In KPK alone 80,000 m^3 of wastewater having very high metal load is discharged to River Kabul/day producing frequent public health problems. Agricultural production is facing low water quality problems (SOE 2005). Major contributors to industrial wastewater generation include sugar, tanneries, paper and pulp, textile, refineries, food processing, cement, petrochemicals, polyester yarn and fertilizer industries. These industries are estimated to generate almost 80 % of the total industrial wastewaters in KPK (PCWRAS 2005). Textile and its associated industries is the largest producer of industrial wastewaters (SOE 2005). After textile, sugarcane is the second largest industry and a major producer of polluted wastewaters in KPK and Pakistan. Around 76 sugarcane industries having installed capacity of 0.36 million tons sugar/day generate 0.5 million m^3 of wastewater/day. This wastewater is also discharged untreated to the surface water channels which are used for irrigation.

Like surface water, groundwater quality has also declined with the passage of time owing to various anthropogenic activities. Around 52 % of the underground water supplies are reported to be of marginal or low quality (Zuberi 1999). Aboveground anthropogenic activities have a significant role in polluting underground water resources. Toxic metals i.e. arsenic, lead, chromium, fluoride and zinc have been found in the underground water supplies where aboveground industrial activities are prevalent or industrial discharge carrying drains flow (SOE 2005). Tanneries in Kasur and Sialkot have greatly degraded underground aquifers with heavy metals and dissolved solids (TDS; Qadir et al. 2008; Ullah et al. 2009; Ali et al. 2013a). Metals long term persistence in aquifers might take hundreds of years to reduce or flush out and can produce serious toxicological conditions on irrigation.

Underground water salinity is another major problem seen in vast areas of Pakistan. Near major water ways the underground water is less saline (<1000 mg/l) however salinity increases as distance increases from the water bodies. Salinity as high as >3000 mg/l is recorded at considerable distance from the major water channels. Saline waters are unfit for irrigation usages. National statistics regarding salinity affected areas are: 4.28 Mha of land with >3000 mg/l, 1.84 Mha of land with 1000–3000 mg/l and 5.75 Mha of land with <1000 mg/l salinity. Provincially Punjab (79 %) and Sindh (28 %) have access to corresponding underground safe freshwater supplies. Within Punjab salinity affected areas cover; 3.96 Mha of land with <1000 mg/l, 1.22 Mha of land with 1000–3000 mg/l and 1.32 Mha of land with >3000 mg/l salinity. Cholistan and central Doab areas are particularly famous for high underground salinity levels. Besides high salinity levels, Kasur, Bahawalpur, Mianwali, Salt Range underground water supplies have high fluoride contents and Gujrat, Rahim Yar Khan, Jhelum, Sargodha have high arsenic contents beyond the WHO guidelines. In absence of surface water for irrigation over exploitation of the groundwater resource has resulted in pumping of sodic-saline water in Sindh and Balochistan which is highly injurious to soil and plant health (SOE 2005). Increased soil sodicity can reduce water infiltration degrees in heavy textured soils consequently affecting plant health. In Sindh only around 28 % of the area contains fresh groundwater supplies suited for irrigation at a depth of 20–25 m. Remaining areas i.e. Thar, Kohistan, Nara and Umarkot have brackish/saline underground water quality and unfit for drinking or irrigation. Irrigation water quality of Indus river and associated tributaries is generally safe from salinity point of view. Salinity levels normally vary between 60 and 375 mg/l. Indiscriminate disposal of saline drainage from irrigated lands have resulted in significant increases in TDS levels in the lower river reaches of the Punjab. The deterioration advances downstream and the highest salinity levels (207–907 mg/l) are recorded at river Chenab and Ravi convergence. Any improvement downstream comes only by dilution from river Sutlej at Panjnad and Indus at Guddu. Without this dilution, water is not considered fit for irrigation due to high salinity levels. Irrigation drainage system in Punjab also carries sodic-saline waters with high TDS, COD, BOD, sodium absorption ration (SAR) and residual sodium carbonate (RSC) levels. Silt loads and suspended matter also add to the problem. Environmentally safe disposal of saline wastewater is a substantial issue in the irrigation sector of the country. Agricultural drainage also contributes to the general contamination of the water resources but less than municipal and industrial sources. An example of Sindh can be quoted where irrigation contributes only 3.21 % of the total pollution (SOE 2005).

5 Wastewater Treatment Systems; A National Perspective

Pakistan's Environmental Protection Agency (EPA) has legislated national environmental quality standards (NEQS) for safe effluent discharge to inland surface waters, wastewater treatment facilities and in to sea, which are seldom observed by almost all municipal and industrial sectors across the country. Wastewater/effluent treatment facilities are originally not constructed within urban/industrial premises and wastewaters having high pollution loads on discharge not only effects riverine ecosystem but also leads to agricultural land salinization. Therefore contaminated water channels threat associated biodiversity and agricultural produce. Industrial hubs are located in or around big cities and only a few national industries (5 %) have installed effluent treatment systems (SOE 2005). Only 1 % of the total volume of wastewater produced is treated by industries and rest is destined to spread contamination. In Lahore alone industries are discharging hazardous metal pollutants and only 3 have wastewater treatment facilities. Also infra-structure for the municipal/domestic wastewater treatment is inefficient and insufficient across the urban hubs of the country, according to some estimates only 8 % of the total wastewater generated from this sector is treated. Sewage treatment plants exist in the main cities where majority are either over loaded or abandoned. Associated sewerage system is also constructed with little care and not functioning properly. Nationally a fraction of sewage is collected in the sewage treatment systems and only around 10 % is treated (Martin et al. 2006).

Karachi and Islamabad treat a small proportion of their municipal wastewater before disposal through biological/mechanical treatment procedures. Treatment of this type is absent in other cities of Pakistan. In Islamabad three municipal sewage treatment plants are constructed but only one is operational that partially treats the capital's effluents. In Karachi, effluents are treated only by screening and sedimentation by trickling filters. Lahore is the second most populated urban center in Pakistan and performs only screening and grit removal from municipal effluents at WASA (water and sanitation agency) collection stations. Non-functioning sewage treatment plants, unavailability of technical/trained staff, continued discharge of untreated municipal/domestic effluents to open fields and surface water channels, persistence of related environmental health problems are indicative of lack of Government interest in this regard.

Regular water quality monitoring programs are needed to address health repercussions related to irrigation water. Pakistan has multiple pressures on degrading water quality and no Government interventions are visible in this sector. Primary challenge in this regard is financing the establishment and operational cost of the wastewater treatment plants. Strict enforcement of the existing environmental laws and revisions in the loopholes will help in managing the municipal and industrial wastewater problems wisely. Federal and provincial departments are required to play their part and Government should support them with consistent policies, financial inputs and trained human resource. Establishment of environmental tribunals besides environmental protection agencies/departments can also help caring for our agriculture and environment.

6 Wastewater Re-use in Irrigation

Production of markedly higher agricultural supplies to meet growing demands of population with an ever increasing water scarcity seems an impossible target. Latest agricultural innovations coupled with water use efficiency will help in sufficing food demands for next one or two decades but water supply will be a limiting factor. As irrigation is the largest consumer of available water, any water conservation strategy will ensure sustainable economic livelihood of the country (Ensink et al. 2002). Modern innovations in wastewater treatment reveal potential of its possible re-use in irrigation. Effluent treatment to safer degrees is required prior to irrigation which otherwise will compromise plant and environment health. Effluents/wastewaters alone are toxic and have negative impacts on the environment. Municipal and industrial effluents on treatment cannot only be utilized in irrigation but will also reduce pressure on fresh water supplies of the country along with reversing ill effects related with their untreated discharge. On recycling, the purpose of water conservation will be well served with achievement of environmental safety. Recycling of wastewater or wastewater treatment has been considered in the past with less gratitude which should now be prioritized in the water sector strategies. Utilization of reclaimed/recycled wastewater has an edge over freshwater for having comparatively higher nutrient value, continuous supplies and negligible production costs (Ali et al. 2014; Farid et al. 2014). Continuous supplies of treated wastewater will also serve to increase food production ensuring food security issues in our country. Treated wastewater can be also be efficiently utilized in the rehabilitation of natural ecosystems. Globally 90 % of the wastewater generated remains untreated, similarly negligible proportions of municipal and industrial wastewaters are treated in Pakistan causing water pollution problems. Israel employs around 70 % of its efficiently treated wastewater in irrigation due to the water scarcity problems.

In general there is no appreciation for municipal and industrial wastewater treatment systems throughout the country. Surface or groundwater is acquired at cheaper rates in KPK and Punjab where supply is copious. After use the wastewater is inevitably discharged into open lands or water channels without any consideration. Water scarcity in cities alike Karachi and Quetta has made people to economically use water in domestic needs. Similarly in the arid areas water utilization is done with extreme caution. Still majority takes this renewable resource for granted either in domestic or irrigation needs. Disposal of recycled wastewater in the environment or its re-use in irrigation depends on the treatment efficiency. Adequate treatment is a must to prevent pollution of recipient environments. Since large volumes are subjected to wastewater treatment therefore appropriate physical, biological and chemical treatment processes are usually carried out in either batch operations or on the continuous flow. All treatment processes (i.e. physical, chemical and biological) have explicit significance in wastewater treatment systems and a normal municipal/industrial wastewater treatment plant usually comprises of any combination of the abovementioned processes depending upon the contamination levels. Physical processes include sedimentation, equalization, degasification, filtration, aeration and skimming etc. Chemical (ion exchange, adsorption, ozonation, coagulation, chlorination and neutralization) and biological treatment processes (trickling filters, lagoons, oxidation ponds, anaerobic digestion and septic tanks etc.) are given in parenthesis. These technologies are mostly not practiced due to unavailability of technical and financial inputs.

Treated wastewater re-use is of immense importance in irrigation due to declining water resources (Ali et al. 2013a, b). Low rainfall results in reduced riverine water flows and low water storages in the reservoirs leaving less water for the Rabi crops (especially wheat). Using HEIS technologies less available water can be evenly utilized. Water available from the conventional irrigation systems is very cheaply/readily available to most farmers and its over utilization generates economic benefits to the farmer communities residing near water heads, farmers at the supplies tail end are usually deprived creating economic inequality. Land holdings of more than 80 % of Pakistani farmers is <5 ha, with continuous availability of cheap water supplies economic benefits achieved by integrative crop/livestock farming approaches will be much more higher than assessed. Installation and management of wastewater treatment plants can generate cheap water supplies for the small farmers as abundant fresh water supplies with changing climate/global warming are not going to last long. Many developed countries have recycled/treated their domestic/industrial wastewaters to safer levels and currently utilizing in industries, domestic needs and public parks. Anyhow underdeveloped and developing countries in spite of facing acute water shortage are directly exploiting wastewater in agricultural activities with long term deleterious effects on their abiotic/biotic environments (Ensink et al. 2002). Strict monitoring of treated wastewater and vigilant comparison with safety levels are needed for sustainable utilization of this resource for irrigation purposes.

To meet the food and fiber requirements of the country by 2025 additional water resources (around 28–37 MAF) and increased crop yields will be required. Available water resources cannot be increased except by constructing large reservoirs or wastewater re-use in agriculture. Similarly crop yields cannot be increased by simply increasing the area under cultivation which will essentially affect other important crops. Yield maximization can only be done within existing resources with effective utilization of agricultural innovations and efficient water use either fresh or treated wastewater. By the year 2025 availability of 28–37 MAF added water seems difficult; to overcome this tremendous water shortage in the near future we will have to adopt to efficient irrigation systems, drastically improve the existing irrigation system, set in place excellent crop planting field facilities (station and farm management), rain water harvesting and re-utilization of treated wastewater in agriculture. Except for these we are left with few other options. Surface and ground water resources are already exploited to a maximum. Another strategy for modest water budgeting can be the reduction in the cultivation area of water intensive crops i.e. sugarcane and rice and their replacement with water efficient crops. All agro-ecological zones need to be focused with respect to these water scarcity combating strategies and re-use of treated wastewater for irrigation in particular. Food safety and food security are at stake in Pakistan. Wheat is the recognized conduit for food security and staple food of the people of Pakistan. It is a victim of serious water shortage in rain-fed areas and of moderate water shortage in irrigated areas (Ali et al. 2013b). Wastewater irrigation in cereal (wheat) production contains less deleterious effects as compared to vegetables, pulses and fodder crops. Therefore raw/treated wastewater re-use in wheat production may be explored to reduce futuristic pressure on fresh water supplies and aid to tame food security in times of increasing water scarcity keeping food/environmental safety a priority.

7 Wheat Response to Wastewater Irrigation Practices

Application of municipal and industrial effluents in cropland irrigation is a custom which affects plant growth and metabolic machinery (Barman et al. 2000). Accumulation of contaminants (heavy metals) in crops from the irrigation water is important for food safety and general public health (Nan et al. 2002). Diluted municipal and industrial effluents besides affecting plant health have been reported to be beneficial in growth or productivity (Bose and Bhattacharyya 2008). Metal uptake, accumulation, toxicity and translocation are strongly impacted by agronomic practices, climatic conditions, plant genotypes, soil pH, electrical conductivity (EC), metal concentration/speciation, bioavailability, organic matter % (OM), cation exchange capacity (CEC) and texture (Kabata-Pendias 2001; Gupta and Sinha 2007). Mode of action for the metal contaminants is largely unknown; however plant responses may include changes in proteins especially enzymes conformations, production of reactive oxygen species, inhibition of transpiration, photosynthesis and respiration etc. (Halliwell et al. 1987; Assche et al. 1988).

Chandra et al. (2009) have reported reduction in overall chlorophyll contents, chlorophyll a/b ratio, increased protein contents in leaves/roots, high malondialdehyde (MDA) production, increased levels of ascorbic acid, non-protein thiols/cysteine in wheat plants when grown with distillery and tannery effluents. Similarly Singh et al. (2010) have reported higher metal contents in wheat plants when grown over continuous wastewater irrigation as compared to clean water irrigation which can disturb the metabolic machinery. Same authors calculated higher human health risks associated with cereal's (wheat) consumption as compared to vegetable/dairy products consumption owing to comparatively high per capita wheat consumption. Bermudez et al. (2011) studied accumulation of heavy and trace metals in the wheat plants (Table 1). They described relatively lower mean levels of Pb, As, Ni, Cd in the wheat grains however higher levels of Fe, Cr, Zn, Cu and Mn then prescribed WHO/FAO limits for human consumption limits. Based on the elevated metal levels they described significant non-carcinogenic health risks associated with the contaminated wheat consumption. Lamhamdi et al. (2011) have reported that increased Pb contents have the ability to inhibit wheat seed germination and seedling growth. Deleterious effects of Pb are more pronounced on the root growth. Higher Pb concentration also resulted in elevated antioxidant enzymes in wheat plants experiencing Pb stress from the wastewater. Zheng et al. (2007) have reported mercury (Hg) in wheat flour from wheat growing areas of Huludao, PR China due to significant industrial activities nearby. Hassan et al. (2013) have reported higher Cu, Zn and Cd in the water used to irrigate wheat in KPK, Pakistan. The reported levels were beyond US irrigation water quality standards resulting in their higher accumulation in the growing wheat plants coupled with toxicological concerns. Table 1 summarizes recent studies that show high metal accumulation in wheat grains and flour owing to different municipal and industrial contamination sources. Food chain (wheat) contamination followed by ingestion is the primary route contaminants take to enter in human or livestock body (Wang et al. 2003). Acute intoxication in consumers can result when the contaminants pass the threshold concentrations, therefore the clinical symptoms appear after prolonged exposures of continuous consumption (UNEP/FAO/WHO 1988). Lacatusu et al. (1996) has shown that deleterious health effects on contaminant exposures are most conspicuous in children. As wheat is the most consumed crop in Pakistan therefore health implication from wastewater borne contaminants in wheat can be severe. Wheat straw is utilized as a major livestock feed in dairy farming communities. Nolan et al. (2005) with wheat showed that metal concentration in different plant parts increased as plants aged. Higher accumulation of contaminants/metals in the grain or straw and consumption may result in carcinogenic, neurotoxic, teratogenic or mutagenic affects.

Researchers have investigated metal accumulation, translocation and toxicity in most commonly growing wheat cultivars when irrigated with municipal/industrial wastewaters. However as wheat varieties differ in their growth performance and yield outputs, future research should focus on the varietal performance under different wastewater irrigation practices. In this regard very little or no scientific information is available on local wheat varieties or advanced lines from Pakistan. To bridge the demand (population expansion) and supply (wheat production) gap under limited water supply conditions, wastewater irrigation impacts on different wheat varieties must be determined. This will help in identification of wheat varieties/lines with ability to perform well under metal stress arising from different wastewater irrigation practices. It will also serve as a way to minimize the heavy metals exposure to human and livestock through wheat consumption by adopting resistant varieties/lines for areas where wastewater irrigation is common or in rain-fed areas with very limited fresh water supplies and wastewater is a potential irrigation source.

8 Conclusions and Recommendations

An overview is provided of water resources with a Pakistan's focus relative to scarcity, available water quality, impact of wastewater irrigation on wheat's growth/metabolism and reuse of wastewater in irrigation. Agricultural economy of the country heavily relies on the water availability in the rain-fed and irrigated areas. Available fresh water supplies are continuously declining and all water consuming sectors are facing water shortage; agriculture in particular. Growing water needs in the industrial sectors and urban centers due to population increase have diversified

Table 1 Metal concentration in the wheat grains and flour in response to municipal and industrial contamination sources

References	Study area	Contamination source	Wheat part analyzed	Cr	Ni	Mn	Na	K
1. Hassan et al. 2013	KPK (Pakistan)	Municipal + industrial	Wheat grain	0.05	0.2	4.9	–	–
2. Zhu et al. (2011)	Beijing (China)	Municipal	Wheat grain	4.62	–	–	–	–
3. Bermudez et al. (2011)	Córdoba (Argentina)	Municipal + industrial	Wheat grain	0.54	0.237	49.8	20	–
4. Hussain et al. (2011)	Peshawar (Pakistan)	Municipal + industrial	Wheat grain	0.03	0.053	3.06	–	–
5. Ahmed and Bouhadjera (2010)	Hammam Boughrara (Algeria)	Municipal + industrial	Wheat grain	–	6.09	–	–	–
6. Jamali et al. (2009)	Jamshoro (Pakistan)	Sewage sludge	Wheat grain	0.17	6.2	–	–	–
7. Wang et al. (2009)	Jinchang (China)	Mining/smelting	Wheat grain	–	3.75	–	–	–
8. Chandra et al. (2009)	Unnao-Uttarpardesh (India)	Industrial	Wheat grain	8.16	4.12	18.4	–	–
9. Karami et al. (2009)	Isfahan (Iran)	Sewage sludge	Wheat grain	–	–	–	–	–
10. Huang et al. (2008)	Kunshan (China)	Industrial	Wheat grain	0.1	0.148	–	–	–
11. Zheng et al. (2007)	Huludao (China)	Industrial	Wheat flour	–	–	–	–	–
12. Karatas et al. (2006)	Konya (Turkey)	Municipal + industrial	Wheat grain	5.47	9.92	38.4	–	–
13. Mantovi et al. (2005)	Po Valley (Italy)	Sewage sludge	Wheat grain	1.06	1.03	–	–	–
14. Rattan et al. (2005)	Western Delhi (India)	Municipal	Wheat grain	–	20	15.3	–	–
15. Santos et al. (2004)	Rio de Janeiro (Brazil)	Municipal + industrial	Wheat flour	0.022	0.062	15	–	–
16. Ensink et al. (2004)	Faisalabad (Pakistan)	Municipal + industrial	Wheat grain	–	–	–	–	–
17. Nan et al. (2002)	Baiyin (China)	Industrial	Wheat grain	–	–	–	–	–
18. Yadav et al. (2002)	Haryana (India)	Municipal	Wheat (boot leaf stage)	–	17	23.3	–	1.91
19. Cuadrado et al. (2000)	Madrid (Spain)	Different anthropogenic	Wheat flour	–	0.065	8.18	–	–
20. Frost and Ketchum (2000)	Paris (France)	Sewage sludge	Wheat grain	1.2	–	–	–	–

[a]Not detected

Fe	Co	Cu	Zn	Cd	Pb	Al	Se	Mo	U	Ba	Br	Cs	Hg
28.5	–	4.1	35.3	0.28	0.4	–	–	–	–	–	–	–	–
–	–	6.09	52.38	0.04	0.2	–	–	–	–	–	–	–	–
47.7	0.05	3.79	29.2	0.017	0.1	–	0.3	–	–	7.3	9.3	0.02	–
4.38	–	0.42	2.73	0.01	0.1	–	–	–	–	–	–	–	–
–	ND[a]	57.12	165.7	3.1	17	–	–	–	–	–	–	–	–
–	–	11.8	133	0.89	23	–	–	–	–	–	–	–	–
–	–	9.69	–	–	–	–	–	–	–	–	–	–	–
90.2	-	5.06	28.26	1.06	8.1	–	–	–	–	–	–	–	–
–	–	4.9	38.5	0.8	–	–	–	–	–	–	–	–	–
–	–	5.229	27.78	0.055	0.2	–	–	–	–	–	–	–	–
–	–	3.117	6.45	0.021	0	–	–	–	–	–	–	–	0.002
–	–	4.21	22.31	–	1.6	–	–	–	–	–	–	–	–
–	–	7.72	54.6	0.16	ND	–	–	–	–	–	–	–	–
404	–	9.39	65.3	–	–	–	–	–	–	–	–	–	–
–	–	0.9	5	0.006	0	6.1	–	–	0	–	–	–	–
–	–	–	–	0.1	–	–	–	–	–	–	–	–	–
–	–	4.72	38.02	0.23	1	–	–	–	–	–	–	–	–
85.7	–	9.5	263.8	1	19	–	–	–	–	–	–	–	–
–	–	1.88	8.63	0.033	0	–	0.04	0.21	–	–	–	–	–
–	–	8.42	132.3	0.05	12	–	–	–	–	–	–	–	–

the issues of continuous water supplies for agriculture. Water use efficiency in agriculture with a wise and integrative approach seems a logical and sustainable solution in the prevailing scenario. Besides educating farmers and construction of reservoirs for water storage, high efficiency irrigation methods can help in meeting agricultural water demands, which is the largest single user of available fresh water supplies of the country. Reduced water supplies motivate farmers to use wastewater alternate in crop production at the cost of plant and environmental health. Reuse of wastewater is practiced with slight consideration and little or no dilution of municipal/industrial effluents is done prior to irrigation. Recycling and reuse of wastewater in agriculture is a common practice in developed countries and needs to be focused in Pakistan's future water policies or revising the existing policy set up. Water policy implementation at the grass root level is urged to have efficient, judicious and scientific use of water in agriculture sector. Treatment of municipal and industrial effluents prior to discharge or irrigation should be given extreme attention in this policy to avoid heavy metal contamination of related ecological compartments, which will help in achieving food safety. Wheat alone supplies 20 % of the nutrients globally and is the major nutrient source to the people of Pakistan. Impact of untreated and treated wastewater on growth, physiology and productivity of different wheat cultivars may be carried out to determine their performance under wastewater borne metal stresses. This marginal harvesting of wastewater in wheat irrigation will help in identification of local varieties with ability to withstand wastewater related stresses i.e. particularly heavy metals. Related research on wastewater quality, treatment and optimum utilization in wheat production can help meeting growing population demands and ensuring food security issues inspite of growing water scarcity. For a look to the future apart from identifying conventional wheat varieties with diversity performance profiles it behooves us to probe the unexploited or underexploited genetic diversity that exists in the wheat family (Triticeae) close and distant gene pools exhibiting heavy metal tolerance to ensure human, livestock and environmental safety. Such avenues would unravel for addressing water related issues profiles and broaden the spectrum of beneficial wheat productivity. Allelic richness would be harnessed that can compliment existent genetic structure via new diversity that is currently at the forefront of genomic research (Mujeeb-Kazi 2006; Mujeeb-Kazi et al. 2008, 2013; Trethowan and Mujeeb-Kazi 2008). Unison of such water focused issues and genetic diversity to alleviate heavy metal constraints will allow in the decades ahead to add greater precision through integration of outsourcing high tech research facets to groups involved in genotypic and phenotypic platforms. Refinement will emerge and global wheat research/production will acquire added value to serve as a more potent conduit to the futuristic 2050 vision of food security that keeps the 9.2 billion populace projection focus in front in times of increasing water scarcity. Cross match of physico-chemical characteristics of raw/untreated wastewater and performance of tolerant/susceptible cultivars in terms of growth, physiology and yield traits will provide valuable insights to benefits of wastewater irrigation practices overcoming its adverse effects. Pursuit of heavy metal tolerance in released varieties is possibly the key to combat water scarcity and food insecurity through wastewater irrigation practices without compromising food and environmental safety. Both (food safety and food security) can operate in tandem through wise wastewater irrigation to tolerant wheat varieties that are the nucleus of this discourse.

References

Abbas ST, Sarfras M, Mehdi SM, Hassan G, Rehman OU (2007) Trace element accumulation in soil and rice plants irrigated with the contaminated water. Soil Tillage Res 94:503–509

Agricultural Statistics of Pakistan (2005–2006) different issues, Ministry of Food, Agriculture & Livestock, Government of Pakistan

Agricultural Statistics of Pakistan (2011) 43rd edition, Ministry of Food and Agriculture (economic wing). Government of Pakistan, Islamabad, Pakistan, p 273

Ahmad N (1993) Water resources of Pakistan and their utilization. Shahid Nazir publications, Lahore, pp 20–100

Ahmad M, Yasin M (2006) Low quality water management for sustainable irrigation and crop production. Pak J Water Res 10:23–31

Ahmad Z, Ashraf A, Fryar A, Akhter G (2011) Composite use of numerical groundwater flow modeling and geo-informatics techniques for monitoring Indus basin aquifer, Pakistan. Environ Monit Asses 173:447–457

Ahmed AB, Bouhadjera K (2010) Assessment of metals accumulated in Durum wheat (*Triticum durum* desf), pepper (*Capsicum annuum*) and agricultural soils. Afr J Agric Res 5:2795–2800

Ali Z, Malik RN, Shinwari ZK, Qadir A (2015) Enrichment, risk assessment and statistical apportionment of heavy metals in tannery-affected areas. Int J Environ Sci Technol 12(2):537–550

Ali A, Ali Z, Qureshi UM, Kazi AG, Malik RN, Sher H, Kazi AM (2014) Integrating physiological and genetic approaches for improving drought tolerance in crops. In: Ahmed P (ed) Emerging technologies and management of crop stress tolerance: vol. 2 a sustainable approach. Elsevier, Burlington, pp 315–345

Ali Z, Malik RN, Qadir A (2013a) Heavy metals distribution and risk assessment in soils affected by tannery effluents. Chem Ecol 29:676–692

Ali Z, Ali B, Mohammad A, Ahmad M, Ahmad I, Napar AA, Kazi AG, Ali A, Shah SS, Mujeeb-Kazi A (2013b) Combating water scarcity for global food security. In: Raza A (ed) Agricultural systems in the 21st century. Nova Science Publishers, Inc, Hauppauge, New York, USA

Ashraf S, Abbas F, Ibrahim M, Rashid U, Khalid S, Raza HA, Hakeem KR (2015) Application of Geographical Information System (GIS) for the identification and demarcation of selective heavy metal concentrations in the groundwater of a metropolitan city. J Geogr Sci 25(2):225–235

Assche FV, Cardinaels C, Clijsters H (1988) Induction of enzyme capacity in plants as a result of heavy metal toxicity: dose-response relation in *Phaseolus vulgaris* L, treated with zinc and cadmium. Environ Pollut 52:103–115

Barman SC, Sahu RK, Bhargava SK, Chaterjee C (2000) Distribution of heavy metals in wheat, mustard, and weed grown in field irrigated with industrial effluents. Bull Environ Contam Toxicol 64:489–496

Bastiaanssen WGM, Ahmad MD, Chemin Y (2002) Satellite surveillance of evaporative depletion across the Indus Basin. Water Resour Res 38:1273–1282

Bermudez GMA, Jasan R, Pla R, Pignata ML (2011) Heavy metal and trace element concentrations in wheat grains: assessment of potential non-carcinogenic health hazard through their consumption. J Hazard Mater 193:264–271

Bhatti AM, Suttinon P, Nasu S (2009) Agriculture water demand management in Pakistan: a review and perspective. Society for Social Management Systems (SSMS), Kochi, Japan

Bose S, Bhattacharyya AK (2008) Heavy metal accumulation in wheat plant grown in soil amended with industrial sludge. Chemosphere 70:1264–1272

Chandra R, Bharagava RN, Yadav S, Mohan D (2009) Accumulation and distribution of toxic metals in wheat (*Triticum aestivum* L) and Indian mustard (*Brassica campestris* L) irrigated with distillery and tannery effluents. J Hazard Mater 162:1514–1521

Cuadrado C, Kumpulainens J, Carbajal A, Moreiras O (2000) Cereals contribution to the total dietary intake of heavy metals in Madrid, Spain. J Food Compos Anal 13:495–503

Darvishi H, Farahani A (2010) Economical value of irrigation with wastewater instead of agronomical water at Iran. Curr Res J Econ Theory 2:8–10

Delgado C, Rosegrant M, Steinfeld H, Ethui S, Courbois C (1999) Livestock to 2020: the next food revolution. J Food Agric Environ discussion paper 228. IFPRI, FAO and ILRI IFPRI, Washington, DC

Drechsel P, Evans AEV (2010) Wastewater use in irrigated agriculture. Irrig Drain Syst 24:1–3

Ensink JHJ, van der Hoek W, Matsuno Y, Munir S, Aslam MR (2002) Use of untreated wastewater in peri-urban agriculture in Pakistan: risks and opportunities. IWMI research report no 64. International Water Management Institute (IWMI), Colombo, Sri Lanka

Ensink JHJ, Mahmood T, van der Hoek W, Raschid-Sally L, Amerasinghe FP (2004) A nationwide assessment of wastewater use in Pakistan: an obscure activity or a vitally important one? Water Policy 6:197–206

Farid M, Irshad M, Fawad M, Ali Z, Eneji AE, Aurangzeb N, Mohammad A, Ali B (2014) Effect of cyclic phytoremediation with different wetland plants on municipal wastewater. Int J Phytorem 16(6):572–581

Farooq U, Iqbal M (2000) Attaining and maintaining self-sufficiency in wheat production: institutional efforts and farmers limitations. Pak Dev Rev 39:487–514

Frost HL, Ketchum H (2000) Trace metal concentration in durum wheat from application of sewage sludge and commercial fertilizer. Adv Environ Res 4:347–355

GoP (2008) Economic survey of Pakistan (2007–08). Ministry of Finance, Government of Pakistan, Islamabad, Pakistan

Gupta AK, Sinha S (2007) Phytoextraction capacity of the plants growing on tannery sludge dumping sites. Bioresour Technol 98:1788–1794

Habbari K, Tifnouti A, Bitton G, Mandil A (2000) Geohelminthic infections associated with raw wastewater use for agricultural purposes in beni-mellal. Prasitol Int 48:249–254

Halliwell B (1987) Oxidative damage, lipid peroxidation and antioxidant protection in chloroplasts. Chem Phys Lipids 44:327–340

Hassan NU, Mahmood Q, Waseem A, Irshad M, Faridullah PA (2013) Assessment of heavy metals in wheat plants irrigated with contaminated wastewater. Pol J Environ Stud 22:115–123

Hsiao TC, Steduto P, Fereres E (2007) A systematic and quantitative approach to improve water use efficiency in agriculture. Irrig Sci 25:209–231

Huang M, Zhou S, Sun B, Zhao Q (2008) Heavy metal in wheat grain: assessment of potential health risk for inhabitants in Kunshan, China. Sci Total Environ 405:54–61

Hussain SS, Mudasser M (2004) Prospects for wheat production under changing climate in mountain areas of Pakistan-an econometric analysis. Water Int 29:189–200

Hussain I, Khan MA, Ali J (2011) Comparative study of heavy metals in of wheat crops growing in different environmental conditions. J Chem Soc Pak 33:499

IUCN (2010) Pakistan water apportionment accord for resolving inter-provincial water conflicts–policy issues and options. IUCN Pakistan, Karachi, pp 1–11

Jamali MK, Kazi TG, Arain MB, Afridi HI, Jalbani NJ, Kandhro GA, Shah AQ, Baig JA (2009) Heavy metal accumulation in different varieties of wheat (*Triticum aestivum* L) grown in soil amended with domestic sewage sludge. J Hazard Mater 164:1368–1391

Kabata-Pendias A (2001) Trace elements in soils and plants, 3rd edn. CRC Press, Boca Raton

Kahlown MA, Majeed A (2001) Presentation during COMSATS 1st meeting on water resources in the south: present scenario and future proposals, 1–2 November 2001, Islamabad, Pakistan

Kahlown MA, Tahir MA, Sheikh AA (2004) Water quality status in Pakistan second report 2001–2003, Pakistan Council of Research in Water Resources, Ministry of Science & Technology, Government of Pakistan, Islamabad, Pakistan

Kahlown MA, Raoof A, Zubair M, Kemper WD (2007) Water use efficiency and economic feasibility of growing rice and wheat with sprinkler irrigation in the Indus basin of Pakistan. Agric Water Manage 87:292–298

Kamal A (2008) Environmental flows Indus river system in Pakistan In: Proceedings of 3rd international conference on water resources and arid environments and the 1st Arab Water Forum Riyadh, Saudi Arbia, 16–19 November

Karami M, Afyuni M, Rezainejad Y, Schulin R (2009) Heavy metal uptake by wheat from a sewage sludge-amended calcareous soil. Nutr Cycl Agroecosyst 83:51–61

Karatas M, Dursun S, Guler E, Ozdemir C, Argun E (2006) Heavy metal accumulation in wheat plants irrigated by waste water. Cellul Chem Technol 40:575–579

Lacatusu R, Răuţă C, Cârstea S, Ghelase I (1996) Soil–plant–man relationships in heavy metal polluted areas in Romania. Appl Geochem 11:105–107

Lamhamdi M, Bakrim A, Aarab A, Lafont R, Sayah F (2011) Lead phytotoxicity on wheat (*Triticum aestivum* L) seed germination and seedlings growth. C R Biol 334:118–126

Mantovi P, Baldoni G, Toderi G (2005) Reuse of liquid, dewatered, and composted sewage sludge on agricultural land: effects of long-term application on soil and crop. Water Res 39:289–296

Martin P, Nishida J, Afzal J, Akbar S, Damania R, Hanrahan D (2006) Pakistan strategic country environmental assessment, South Asia Region, vol I, 36946-PK. World Bank, Washington D.C., United States

Muchuweti M, Birkett JW, Chinyanga E, Zvauya R, Scrimshaw MD, Lester JN (2006) Heavy metal content of vegetables irrigated with mixture of wastewater and sewage sludge in Zimbabwe: implications for human health. Agric Ecosyst Environ 112:41–48

Mujeeb-Kazi A (2006) Utilization of genetic resources for bread wheat improvement. In: Singh RJ, Jauhar PP (eds) Genetic resources, chromosome engineering, and crop improvement cereals, vol 2. CRC Press, Boca Raton, pp 61–97

Mujeeb-Kazi A, Gul A, Farooq M, Rizwan S, Ahmad I (2008) Rebirth of synthetic hexaploids with global implications for wheat improvement. Aust J Agric Res 59:391–398

Mujeeb-Kazi A, Kazi AG, Dundas I, Rasheed A, Ogbonnaya F, Kishi M, Bonnett D, Wang RRC, Xu S, Chen P, Mahmood T, Bux H, Farrakh S (2013) Genetic diversity for wheat improvement as conduit to food security. Adv Agron 122:179–258

Mutengu S, Hoko Z, Makoni FS (2007) An assessment of the public health hazard potential of wastewater reuse for crop production; A case of Bulawayo city, Zimbabwe. Phys Chem Earth 32:1195–1203

Nan Z, Zhao C, Li J, Chen F, Sun W (2002) Relations between soil properties and selected heavy metal concentrations in spring wheat (*Triticum aestivum* L) grown in contaminated soils. Water Air Soil Pollut 133:205–213

National Water Policy (NWP-2003), Government of Pakistan, Islamabad, Pakistan

Nolan AL, Zhang H, Mclaughlin MJ (2005) Prediction of zinc, cadmium, lead, and copper availability to wheat in contaminated soils using chemical speciation, diffusive gradients in thin films, extraction, and isotopic dilution techniques. J Environ Qual 34:496–507

Ozturk M, Gucel S, Sakcali S, Guvensen A (2011) An overview of the possiblities for wastewater utilization for agriculture in Turkey. Isr J Plant Sci 59:223–234

Pakistan Country Water Resources Assistance Strategy, Water Economy: Running Dry (PCWRAS 2005) South Asia region agriculture and rural development unit, South Asia region, report no: 34081-PK

Pakistan Economic Survey (2010) 2009–10; finance division, EA Wing, Government of Pakistan, Printing Corporation, Pakistan Press, Islamabad, Pakistan

Pakistan Statistical Yearbook (2001) Federal Bureau of Statistics, Statistics Division, Government of Pakistan

Pakistan Water Partnership (1999) Pakistan country report: vision for the 21st century, first draft (Global Water Partnership South Asia Technical Advisory Committee), Islamabad, Pakistan

Pakistan Water Sector Strategy (2002) National water sector profile, Ministry of water and power, national water sector profile, vol-5, Government of Pakistan, Islamabad, Pakistan

Qadir M, Boers TM, Schubert S, Ghafoor A, Murtaza G (2003) Agricultural water management in water-starved countries: challenges and opportunities. Agric Water Manage 62:165–185

Qadir A, Malik RN, Husain SZ (2008) Spatio-temporal variations in water quality of Nullah Aik-tributary of the river Chenab Pakistan. Environ Monit Assess 140:43–59
Qadir M, Wichelns D, Raschid-Sally L, McCornick PG, Drechsel P, Bahri A, Minhas PS (2010) The challenges of wastewater irrigation in developing countries. Agric Water Manage 97:561–568
Rajaram S, Hobbs PR, Heisey PW (1998) Review of Pakistan's wheat and maize research systems. Report submitted to PARC by a multidisciplinary team of CIMMYT Scientists visited Pakistan
Rattan RK, Datta SP, Chhonkar PK, Suribabu K, Singh AK (2005) Long-term impact of irrigation with sewage effluents on heavy metal content in soils, crops and groundwater-a case study. Agric Ecosyst Environ 109:310–322
Saeed M, Ashrif M, Bruen M (2002) Diagnostic analysis of framers skimming well technologies in the Indus basin of Pakistan. Irrig Drain Syst 16:139–160
Santos EE, Lauria DC, Porto da Silveira CL (2004) Assessment of daily intake of trace elements due to consumption of foodstuffs by adult inhabitants of Rio de Janeiro city. Sci Total Environ 327:69–79
Shaheen A, Baig MA (2011) Drought severity assessment in arid area of Thal doab using remote sensing and GIS. Int J Water Res Arid Environ 1:92–101
Shakir AS, Qureshi M (2007) Irrigation water management on canal command level: a case study of lower Bari doab canal system. In: Proceedings of international conference on water and flood management. Institute of Water and Flood Management, Dhaka, pp 215–222
Shakir AS, Rehman H, Khan NM, Qazi AU (2011) Impact of canal water shortages on groundwater in the lower Bari doab canal system in Pakistan. Pak J Eng Appl Sci 9:87–97
Shuval HI, Adir A, Fattal B, Rawitz E, Yekutel P (1986) Wastewater irrigation in developing countries: health effects and technical solutions, Technical paper no: 51. The World Bank, Washington, DC
Singh A, Sharma RK, Agrawal M, Marshall FM (2010) Health risk assessment of heavy metals via dietary intake of foodstuffs from the wastewater irrigated site of a dry tropical area of India. Food Chem Toxicol 48:611–619
State of the Environment Report (2005) Ministry of Environment, Government of Pakistan
Trethowan RM, Mujeeb-Kazi A (2008) Novel germplasm resources for improving environmental stress tolerance of hexaploid wheat. Crop Sci 48:1255–1265
Ullah R, Malik RN, Qadir A (2009) Assessment of groundwater contamination in an industrial city, Sialkot, Pakistan. Afr J Environ Sci Technol 3(12):429–446
UNCCD (2013) Information accessed from the URL: http://www.unccd.int/en/programmes/Thematic-Priorities/water/Pages/defaultaspx. 30 Sept 2013
UNEP, FAO, WHO (1988) Assessment of chemical contaminants in food Report on the results of the UNEP/FAO/WHO programme on health-related environmental monitoring
Wang QR, Cui YS, Liu XM (2003) Soil contamination and plant uptake of heavy metals at polluted sites in China. J Environ Sci Health Part A-Tox/Hazard Subst Environ Eng 38:823–838
Wang S, Nan Z, Liu X, Li Y, Qin S, Ding H (2009) Accumulation and bioavailability of copper and nickel in wheat plants grown in contaminated soils from the oasis, northwest China. Geoderma 152:290–295
Yadav RK, Goyal B, Sharma RK, Dubey SK, Minhas PS (2002) Post-irrigation impact of domestic sewage effluent on composition of soils, crops and ground water-a case study. Environ Int 28:481–486
Zheng N, Wang Q, Zhang X, Zheng D, Zhang Z, Zhang S (2007) Population health risk due to dietary intake of heavy metals in the industrial area of Huludao city, China. Sci Total Environ 387:96–104
Zhu YE, Zhao Y, Sun K, Chen ZF, Qiao JJ, Yan-Qin JI (2011) Heavy metals in wheat grain and soil; assessment of the potential health risk for inhabitants in sewage irrigated area of Beijing, China. Fresen Environ Bull 20:1109
Zuberi FA (1999) Integrated surface and groundwater management programme for Pakistan: groundwater resource study, IWASRI international report no 99/7, pp 6–12

The Potentiality of Wastewater Use for Irrigation in Turkey

İsmail TAŞ, Yusuf Ersoy Yildirim, and Fatma Özkay

Contents

Abstract Water is one of the most essential elements of life which makes up 3/4th of our planet. However, only 2.6 % of this water is available for usage. More water is needed for the world with increasing industrialization and population. Yet, both in Turkey and the world in general, the current use of water lacks a sustainability mentality. This, coupled with the increased need for more water, leads to a search for new sources of water. Thus, efforts are geared toward finding ways of utilizing the previously unused sources of water and/or recycling waste water. Even if these projects succeed, changing climates and global warming might have considerable negative effects on water resources, then sustainability and effective use become vital issues for these new sources of water as well the existing ones. This study examined possible use of treated wastewater over agricultural lands for irrigation purposes. It focused on effective ways in which treated water can be used for agricultural purposes. Sustainability and effective use are the vital issues for these new sources of water as well as the existing ones. Hence sustainable wastewater management was also discussed in this chapter.

İ. TAŞ (✉)
Canakkale Onsekiz Mart University, Faculty of Agriculture, Agricultural Structures and Irrigation Department, Canakkale, Turkey
e-mail: tas_ismail@yahoo.com

Y.E. Yildirim
Ankara University, Faculty of Agriculture, Agricultural Structures and Irrigation Department, Ankara, Turkey

F. Özkay
Soil, Fertilizer and Water Resource Central Research Institute, Ankara, Turkey

M. Öztürk et al. (eds.), *Plants, Pollutants and Remediation*,
DOI 10.1007/978-94-017-7194-8_7

Keywords Wastewater • Irrigation • Sustainability • Wastewater management

1 Introduction

Scarcity of water is important in worldwide. In particular this is more emphasis in arid and semi arid region like Turkey. As taken of the fact that this long process has been completed, several countries today regularly face imbalances of water demand and water supply, especially in the summer period, due to simultaneous occurrence of low precipitation, high evaporation and increased demands for irrigation. Not surprisingly, the decrease in resources in natural waters brought about by drought and population growth is inciting authorities to establish and to encourage the reuse of wastewater.

Major portion of irrigated agriculture was supported by fresh irrigation water resources which is surface and groundwater. Excessive withdrawal of ground water is declining water table in the most of groundwater irrigated areas. Groundwater is important resource in arid end semi arid region. So groundwater management is essential to combat the emerging problems of water scarcity. There is requiring for scientific planning and development of groundwater under different hydrogeological and agro-climatcal situations and to evolve suitable management practices. In view of the reducing availability of fresh water resources and increasing demands, it is essential to have emphasis on development of poor quality groundwater resources. On the other hand conservation of fresh groundwater resources is so important. In particularly, polluters for water has to be blocked in water logged areas. In this reason, use of saline groundwater with wastewater must be promoting.

Wastewater may be used directly or after mixing with sewage channeled into natural drainage systems, from where the polluted water is used for farming (Qadir et al. 2010). Most commonly, a year-round vegetable production is practiced, for which farmers have a good market. In many places in the world, this form of production has great importance as a source of income and livelihood for many people. Huibers and Raschid-Sally (2005) observed that farmers usually have no land rights and make use of available urban land belonging to property owners or the state, until they are thrown out.

Water needed for irrigation in some places accounts for 3/4 of total demand, regulations and guidelines vary widely about reuse in the world. Benefits of agricultural reuse could be suggested as high concentrations of nutrients may reduce the demand for fertilizers, long-term soil enrichment, decreases demand of potable water supply, additional treatment in soil, and water being not discharged to receiving waters. Thus, water pollution would be prevented. There are also some disadvantages of agricultural reuse such as health risk from associated pathogens, health risk from other contaminants (e.g., heavy metals, chemicals, and pharmaceuticals), decrease in soil quality from accumulation of salts and soil acidification, and infiltration of pollutants into groundwater (Lund 1980; Pair et al. 1983; Ashraf et al. 2015).

Water amount in Turkey is declining since 1927 considering the population growth. It is estimated 1000 m^3 water potential per person in 2030, This amount puts Turkey in water poor countries class border. Moreover, this water potential is not distributed evenly. Larger part of Turkey is in a semi-arid climatic area. To overcome this water shortage various strategies could be suggested such as development of fossil water stores, draining or redirection of surface waters, more efficient use of the available water resources, water saving measures, desalination of seawater, and reusing wastewater after treatment for irrigation in agriculture (Aydin and Ozcan 2008).

Agriculture is the main user of water in Turkey, as in most regions of the world. Particularly in arid and semi arid region, where it is more difficult to meet the agricultural water demand with conventional resources, wastewater reuse represents a viable option. It would be an important step forward if waste water could be recycled, especially in view of the expectation that fresh water will become ever scarcer in the next few decades – considering both climate change and increasing demand. Application of treated wastewater for irrigation of plants and crops is gradually becoming a common practice worldwide. Wastewater reclamation and reuse is of great interest and a viable option for many industrial sectors and countries which suffer from water scarcity problems. Although the amount of reusable domestic and industrial wastewater is much lower than the wastewater generated, many countries show an increasing interest in wastewater reclamation and reuse. In many parts of the world, such as India, Pakistan, Tunisia, South Africa, Jordan, Mexico, Spain, Australia, Italy, Japan and United States treated wastewater has been successfully used for irrigation and many researchers have recognized its benefits.

In other words whenever good quality water is scarce, water of marginal quality will have to be considered for use in agriculture. Although there is no universal definition of marginal quality water, for all practical purposes it can be defined as water that possesses certain characteristics which have the potential to cause problems when it is used for an intended purpose. For example, brackish water is a marginal quality water for agricultural use because of its high dissolved salt content, and municipal wastewater is a marginal quality water because of the associated health hazards. From the viewpoint of irrigation, use of a 'marginal' quality water requires more complex management practices and more stringent monitoring procedures than when good quality water is used. This publication deals with agricultural use of municipal wastewater, which is primarily domestic sewage but possibly contains a proportion of industrial effluents discharged to public sewers (Pescod 1992).

Irrigationwater quality is one of the main factors limiting plant growth. Recycled wastewater effluent is an important source of irrigation water in arid and semiarid regions. Wastewater effluents generally contain high concentrations of suspended and dissolved solids, both organic and inorganic. Conventional sewage treatment (seconder or tertiary) can removed the effluent from inorganic dissolved solids in wastewater. Actually most of the salts added during domestic and industrial usage remain in the irrigation water and may eventually reach the soil. This salt with water reach the soil by irrigation. A number of researchers have reported reduced hydraulic conductivity for soils to which treated wastewater has been applied.

Some positive effects of reusing wastewaters for irrigation could be suggested as following. Reuse of wastewater for irrigation increase yield in agriculture and serves for the reforestation and, therefore, prevents erosion, Reuse of reclaimed wastewater could encourage a sustained environmental awareness. Types of reuse applications are urban reuse irrigation, of restricted-access-area, agricultural reuse, recreational impoundments, golf links, fire extinguishing, cooling water, conserves water, low-cost method for sanitary disposal of municipal wastewater, reduces of water resources pollution, reducing the need for artificial fertilizer, landscape impoundments, contraction uses, industrial reuse, groundwater recharge, and indirect potable reuse. Disinfected, tertiary treated effluent can be used in all of these applications (Anonymous 1992; Aydın and Gür 2002; Yurseven et al. 2010; Fatta et al. 2005).

2 Literature Review

There is at least 3.5 million ha are irrigated globally with untreated, partly treated, diluted, or treated wastewater (Jimenez and Asano 2004 and Anonymous 2006). In Tula Valley, Mexico, almost half of the untreated wastewater infiltrates through soil, which acts as a filter and removes pollutants. However, salinity and nitrate levels in groundwater are increasing. Continuous monitoring of the aquifer is needed to identify emerging health problems (Jimenez and Chávez 2004).

Cities in developing countries are experiencing unparalleled growth and rapidly increasing water supply and sanitation coverage that will continue to release growing volumes of wastewater. In many developing countries, untreated or partially treated wastewater is used to irrigate the cities' own food, fodder, and green spaces. Comprehensive management approaches in longer term will need to encompass treatment, regulation, farmer user groups, forward market linkages that ensure food and consumer safety, and effective public awareness campaigns (Scott et al. 2004).

Irrigated agriculture is the biggest consumer of water in the world. In areas with dry climates, crop irrigation requires from 50 to 85 % of total water use (Hamdy 2001; Ragab 2001). In most countries, increasing urbanisation is producing large volumes of wastewater that have become a serious environmental problem. An effective solution for both needs is the reuse of municipal effluents for irrigation. Wastewater has been applied to crops, rangelands, forests, parks and golf courses in many parts of the world (Steward et al. 1986; Angelakis et al. 1999; Al-Shreideh, 2001; Al-Jamal et al. 2002): in Israel wastewater irrigation uses more than 65 % of the total municipal sewage production of the country (Haruvy 1996; Friedler 2001).

Comparative studies between wastewater and non-wastewater farmers have shown that the former make more income not only from savings in fertilizer but additionally the reliable wastewater supply allows them to grow short-cycle cash crops (van der Hoek 2004; van der Hoek et al. 2002; Ensink et al. 2004; Karanja et al. 2010).

Long-term use of industrial and/or municipal wastewater in irrigation is known to make significant contribution to the trace elements load (Cd, Cu, Zn, Cr, Ni, Pb, and Mn) in surface soils (Mapanda et al. 2005). Excessive accumulation of trace elements in agricultural soils through wastewater irrigation may not only result in soil contamination but also affect food quality and safety (Muchuweti et al. 2006; Sharma et al. 2007).

Accumulation of metals in root from soil and subsequent translocation to other parts of plant like stem, leaves and fruits is important for the selection of plant specially crops and vegetables. Plant accumulating least quantity of metals in the edible parts, with the concentration within the permissible limit than the other varieties or species can be selected for the cultivation on the field having high level of metal contamination (Barman and Bhargava 1997). In contrast, plants accumulating high concentration of heavy metals from contaminated soil can be used for detoxification/ phytoremediation of metals from soil or growing medium (Sarma 2011).

Despite farmers good reasoning and the advantages accruing from wastewater use, this practice can severely harm human health and the environment (Qadir et al. 2007). This is mainly from not only the associated pathogens, but also heavy metals and other undesirable constituents depending on the source. Furthermore, farmers, consumers, and some government agencies in many countries are not fully aware of the potential impacts of irrigation with wastewater (Qadir et al. 2010).

The major challenge is to optimize the benefits of wastewater as a source of both water and nutrients it contains, and to minimize the negative impacts on human health. From the environmental point of view, there are potential positive and negative impacts that should be considered. There are existing international guidelines for waste water quality and use in agriculture. These standards can only be met by proper treatment processes. Because of high treatment costs, most cities in low-income developing countries will not have wastewater treatment facilities in the foreseeable future. However, while the use of untreated wastewater has become a routine practice in most developing-countries, policies on utilization of waste water have not taken this reality into consideration. Such policies range from active enforcement of legislation that totally prohibits the use of untreated wastewater, to turning a blind eye. Clearly, there is a need for better-informed decision-making (Hoek 2004).

Among the inorganic contaminants, heavy metals are important due to their non-degradable nature leading to bioaccumulation through tropic level which may have deleterious biological effects. Even at low concentrations, elements such as nickel (Ni), cadmium (Cd), chromium (Cr) and lead (Pb) are harmful to plants and humans (Emongor 2007). Compared to heavy metals in soil, various crop parts accumulated more heavy metal loads. The potential metal load for uptake depends on soil conditions and concentrations.

Heavy metal contamination caused by natural processes or by human activities is one of the most serious ecotoxicological problems. Long term irrigation can induce changes in the quality of soil as trace element inputs are sustained over long periods. When wastewater is used for the irrigation of edible plants for prolonged period, soil health is affected (Barman et al. 2000; Singh et al. 2004). Heavy metal ions like Fe, Cu, Zn, Mn etc. at appropriate concentrations are required for structural and catalytic components of proteins and enzymes as cofactors, essential for normal growth and development of plants. However, supra-optimal concentrations of these micronutrients and other heavy metals in plants operate as stress factors (Sinha et al. 2002; Singh et al. 2004).

Strategies for managing health risks to achieve the health targets include wastewater treatment to achieve appropriate microbiological quality guidelines, crop restriction, waste water application methods, control of human exposure, chemotherapy, and vaccination. Phased implementation of the WHO microbial water quality standards may be necessary as treatment is gradually introduced and improved over a period of time, e.g. 1–15 years. For optimal public health effect, the guidelines should be co-implemented with such other health interventions as hygiene promotion, provision of adequate drinking water and sanitation, and other health-care measures (Carr et al. 2004).

The use of urban wastewater in agriculture is a centuries-old practice that is receiving renewed attention with the increasing scarcity of freshwater resources in many arid and semi- arid regions. Driven by rapid urbanization and growing wastewater volumes, wastewater is widely used as a low-cost alternative to conventional irrigation water; it supports livelihoods and generates considerable value in urban and peri-urban agriculture despite the health and environmental risks associated with this practice. Though pervasive, this practice is largely unregulated in low-income countries, and the costs and benefits are poorly understood (Scott et al. 2004).

Farmers prefer using wastewater to freshwater for irrigation, as they immediately see higher profits. However, few take precautions to protect themselves, and as a result, 60 % of them are plagued with intestinal parasites. Additionally, the practice poses a significant public health risk because some vegetables irrigated with waste water are eaten raw. Urban agriculture itself is constrained by the insecurity of land tenure, as the constant threat of losing their land makes farmers unwilling to commit to major investments. Thus the potential for safer and more convenient irrigation methods, such as hoses fitted with drip irrigators, is limited (Faruquie et al. 2004a).

In the foreseeable future, many towns in developing countries will continue or expand the direct or indirect irrigation of crops with untreated wastewater. Current government policies focus on regulation of wastewater use and wastewater treatment and are unable to offer practical solutions to the users. An important input into more realistic policies on wastewater use is information on the area irrigated with urban wastewater at national and global levels. Such macro-level estimates can only be obtained when there is a common understanding of the different types of wastewater use (Hoek 2004).

Planned reuse that seeks to maintain the benefits and minimize the risks will require an integrated approach. Key to the success of endeavours to make the transition to planned strategic reuse programs are a coherent legal and institutional framework with formal mechanisms to coordinate the actions of multiple government authorities, sound application of the 'polluter pays' principle, conversion of farmers towards more appropriate practices for wastewater use, public awareness campaigns to establish social acceptability for reuse, and consistent government and civil society commitment over the long term with the realization that there are no immediate solutions (Scott et al. 2004).

3 Materials and Methods

Turkey's surface area is namely 78 million ha. Almost one third of this, 28 million ha, can be classified as cultivable land. Recent studies indicate that an area of about 8.5 million ha is economically irrigable under the available technology. Until now, 5.6 million ha has been under irrigation.

Classified as salty and containing sodium of land in Turkey is about 1.5 million ha. Problematic basin in terms of salt and sodium are widely seen especially in first irrigation areas such as Gediz, Buyuk Menderes and Konya basin. Salinity and drainage problems are seen in parallel to the irrigation in a lot of irrigated areas.

Mean precipitation in Turkey is 643 mm/year amounting to 501 billion m^3/year. A volume of 274 billion m^3/year water evaporates from water bodies and soils to atmosphere. 69 billion m^3/year water leaks into groundwater, whereas 28 billion m^3/year is retrieved by springs from groundwater contributing to surface water. Also, 7 billion m^3/year water comes from neighboring countries. Thus, total annual surface runoff amounts to a volume of 193 billion m^3/year of water. Including 41 billion m^3/year net discharging into groundwater (covering safe yield extraction, unregistered extraction, emptying into the seas, and transboundary), the gross (surface and groundwater) renewable water potential of Turkey is estimated as 234 billion m^3/year. However, under current technical and economic constraints, annual exploitable potential has been calculated as 112 billion m^3/year of net water volume, as 95 billion m^3/year from surface water resources, as 3 billion m^3/year from neighboring countries, as 14 billion m^3/year from groundwater safe yield (Anonymous 2012).

According to municipality sewage statistics of Turkey for the year 2010, 2235 of 2950 municipalities serve with sewage system. In 2010, 48.6 % of 3.58 billion m^3 wastewater was discharged 48.6 % into rivers and streams, 41.8 % into sea, 3.6 % into dams and reservoirs, 2.1 % into lakes and ponds, 1 % into land surfaces and 2.8 % into other receiving bodies. About 2.72 billion m^3 of 3.58 billion m^3 (76 %) discharged water was treated at treatment facilities. By the year 2010, there were 326 wastewater treatment facilities serving for 438 municipalities in Turkey. Wastewater was treated biologically by 61 %, physically by 12 %, advanced by 16 and %16 by natural treatment systems (Anonymous 2011).

4 Result and Discussion

4.1 *State of Wastewater in Turkey*

Figure 1 shows that wastewater and treatment rates produced by municipalities in Turkey from 1994 to 2010. In 1994, approximately 90 % of the wastewater coming from municipal sewerage system had been discharged without treatment and this rate decreased to the level of 25 % in 2010. As it is clear that, the wastewater

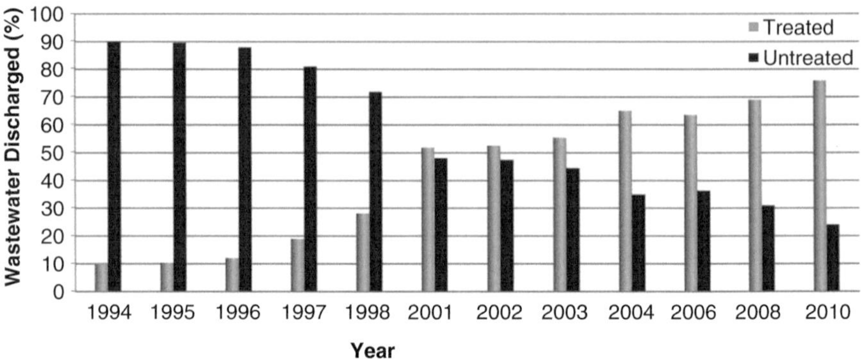

Fig. 1 Changing of the wastewater treatment in Turkey from 1994 to 2010

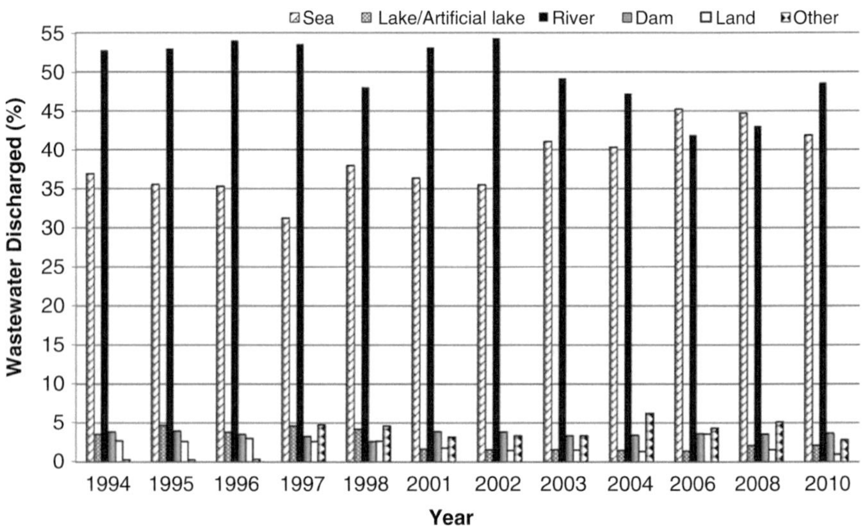

Fig. 2 Changing of the wastewater in receiving environments in Turkey from 1994 to 2010

treatment in Turkey has increased rapidly. As of 2010, approximately 75 % of the wastewater was discharged into receiving environment after being treated.

Figure 2 illustrates that the wastewater collected by municipal sewerage in Turkey from 1994 to 2010, was discharged into receiving environment. Also it can be inferred from this figure that, the largest receiving environment is rivers. While sea comes in the second place of this rank, dam is the third. Approximately half of the produced wastewater was discharged into rivers whereas nearly 40 % was discharged into the seas. In 1994, the discharge rate into rivers was 52.8 % but increased to 54.3 % in 2002, and eventually this rate decreased to 48.6 % in 2010. But the case

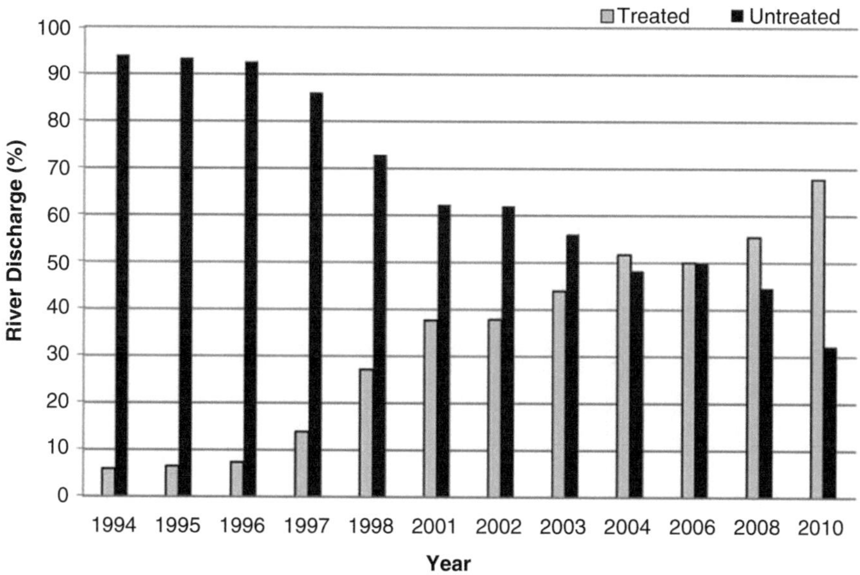

Fig. 3 State of wastewater discharged into river

is considerably different in discharging into sea. In other words, in 1994, this rate was 36.9 % and illustrated fluctuations in time period and it slightly increased to 41.8 % in 2010. Discharge into Lake and Artificial lake showed decreases in same time period. In 1994, rate of discharge was 3.5 %. This rate has slightly fluctuated from 1994 to 2010 and it decreased slightly to 2.1 % in 2010. Similar case was observed in dams. While discharge rate into dams was 3.8 % in 1994 it slightly increased. In 2010, the decrease was very little and the discharge rate decreased to 3.6 %. Discharge into lands showed a slight fluctuation as occurred in discharging into dams. In 1994, discharge into land was 2.7 %. This rate declined 1 % in 2010. Discharge into other receiving environments was 0.3 % in 1994. Then it was dramatically increased to 6.2 % in 2004, and decreased to 2.8 % in 2010.

As illustrated on Fig. 3, the rate of discharge wastewater in rivers was 90 % which was directly discharged without treatment, in 1994. This rate was decreased to 32 % in 2010. In other words, the rate of discharging treated wastewater in rivers has increased from 6 to 68 % in the same period. Considering the use of treated wastewater, instead of discharging into river, can be more useful for irrigation. Thereby it will be possible to fresh water resources.

Figure 4 shows that the rate of discharge wastewater in sea. In 1994, discharging of wastewater without treatment into sea was approximately 85 %. This rate was decreased by 10 % in 2010. In other words, the rate of discharging treated wastewater was 15 % in 1994 and has increased to 90 % in 2010. As can be understood from this figure discharged of wastewater into sea is large part treated in Turkey.

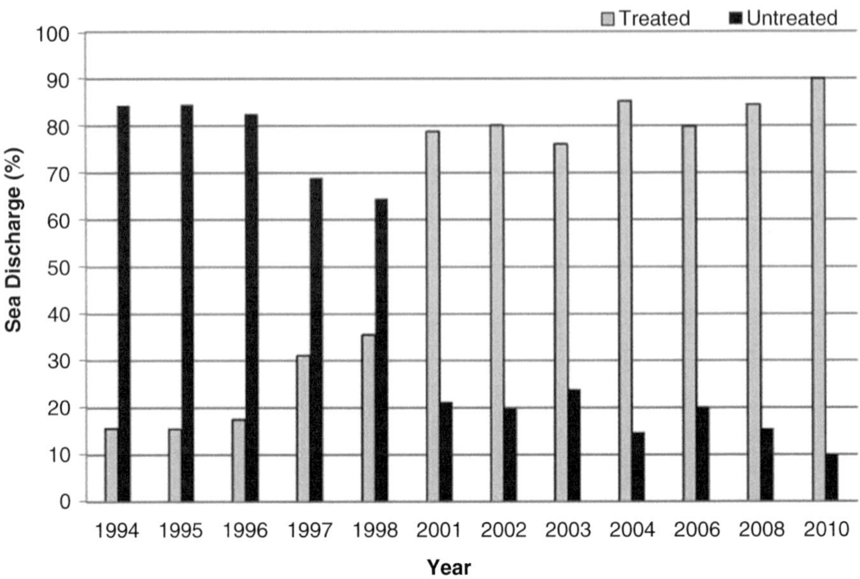

Fig. 4 State of the wastewater discharged into the sea

5 Discussion

About 75 % of the wastewater produced by the municipalities in Turkey is given to discharge points after treatment. In view of Turkey, The two most important receiving environments are river and sea. As of 2010, 90 % of treated wastewater produced by municipality is discharged into rivers and the sea. Depending on the quality of treated wastewater, it can be used by suitable methods for irrigation. Thereby, it can both protect fresh water resources and the pollution can be remained in region.

Irrigationwater quality is one of the main factors limiting plant growth. Recycled wastewater effluent is an important source of irrigation water in arid and semiarid regions. Wastewater effluents generally contain high concentrations of suspended and dissolved solids, both organic and inorganic. Conventional sewage treatment (seconder or tertiary) can removed the effluent from inorganic dissolved solids in wastewater. Actually most of the salts added during domestic and industrial usage remain in the irrigation water. This salt in water reach the soil by irrigation. A number of researchers have reported because of solved salts reduced hydraulic conductivity of soils to which treated wastewater has been applied.

The use of wastewater in irrigation depends on a series of factors, such as community size, fresh water resources, socio-economic aspects, soil and plant requirements relative location to other communities, education of framer and land availability for effluent reuse. Taking into account the requirements of different growth stages of the plant, for example, in the form of seedlings during the nitrogen, potassium and potassium (N:P:K) ratio 2:1:2, form of the pre-harvest during 1:1:1 and form of harvest during should be 1:1:1. Higher Ca and Mg in the recycled wastewater combined with gypsum application helped prevent a greater degree of Na build up in the soil.

Examples of the use of treated wastewater in irrigation water, Ankara (capital of Turkey) Wastewater Treatment Plant are examined and give to Tables 1 and 2. The facility is located along the flow direction of Ankara Stream in Tatlar village of Sincan, Ankara and located 45 km west of the city. The topography of the city and system do not require a pump station to convey the wastewater into the facility. Whole wastewater collected in the sewage system arrives to plant with gravity (Anonymous 2013).

Treated wastewater is discharged into Ankara Stream. Ankara Stream is one of the most important branches of Sakarya River. The stream is named as "Ankara Stream" after the intersection point of Çubuk and Hatip Streams within the city borders of Ankara. There are a few other streams out of those in the border of Ankara. These other ones are incesu, Dikmen and Balgat Streams. The biggest branch of Ankara stream is Ova Stream. The total length of Ankara Stream is about 140 km from Ankara to Sakarya River (Gokalp et al. 2011).

Numerous irrigation water quality guidelines have been proposed. The guidelines presented were developed by university of California Committee of Consultants and were expanded subsequently by Ayers and Westcot (1985). The long term influence of water quality on crop production, soil conditions, and farm management are emphasized, and the guidelines are applicable to both freshwater and reclaimed water. Four categories of potential management problems associated with water quality in irrigation are (a) salinity, (b) water infiltration rate, (c) specific ion toxicity and (d) other problems (Ayers and Westcot 1994).

Not all trace elements are toxic and in small quantities many are essential for plant growth (Fe, Mn, Mo, Zn). However, excessive quantities will cause undesirable accumulations in plant tissue and growth reductions. There have been few field experiments from which toxic limits could be established, especially for irrigation water. However, research dealing with disposal of wastewater has gained sufficient experience to prove useful in defining limitations. It is now recognized that most trace elements are readily fixed and accumulate in soils, and because this process is largely irreversible, repeated applications of amounts in excess of plant needs eventually contaminate a soil and may either render it non-productive or the product unusable. Recent surveys of wastewater use have shown that more than 85 % of the applied trace element accumulates in the soil and most accumulates in the surface few centimetres. Although plants do take up the trace elements, the uptake is normally so small that this alone cannot be expected to reduce appreciably the trace element in the soil in any reasonable period of time (Ayers and Westcot 1994).

According to this classification, long term Water quality of Ankara Wastewater Treatment Plant can be used attentively for salinity, alkalinity, phosphate-phosphorus and potassium. Particularly salinity and alkalinity are very important for infiltration. Phosphate-phosphorus and potassium are necessary fertilizer for plant growing. In this respect, this wastewater can be useful for saving of fertilizer.

Table 1 Water quality parameters of wastewater used for irrigation in Ankara

Water parameter	Symbol	Unit[a]	Usual range in irrigation water	AWTP[d]
Salinity				
Salt content				
Electrical conductivity	EC_w	dS/m	0–3	1.13
Total dissolved solids	TDS	mg/l	0–2000	815
Cations and anions				
Calcium	Ca^{++}	mg/l	0–400	54.9
Magnesium	Mg^{++}	mg/l	0–61	20.1
Sodium	Na^{+}	mg/l	0–920	112.7
Carbonate	$CO^{--}{}_3$	mg/l	0–3	–
Bicarbonate	HCO_3^-	mg/l	0–610	383.4
Chloride	Cl^-	mg/l	0–1065	89.2
Sulphate	SO_4^{--}	mg/l	0–960	96.5
Nutrients[b]				
Nitrate-nitrogen	NO_3-N	mg/l	0–10	0.4
Ammonium-nitrogen	NH_4-N	mg/l	0–5	29.2
Phosphate-phosphorus	PO_4-P	mg/l	0–2	**5.4**
Potassium	K^+	mg/l	0–2	**17.7**
Miscellaneous				
Boron	B	mg/l	0–2	0.7
Acid/basicity	pH	1–14	6.0–8.5	7.7
Sodium adsorption ratio[c]	SAR	(me/l)[a,b]	0–15	**18.4**

[a]dS/m = deciSiemen/metre in S.I. units (equivalent to 1 mmho/cm = 1 millimmho/centi-metre)
mg/l = milligram per litre ≃ parts per million (ppm)
me/l = milliequivalent per litre (mg/l ÷ equivalent weight = me/l); in SI units, 1 me/l = 1 millimol/litre adjusted for electron charge
[b]NO_3 -N means the laboratory will analyse for NO_3 but will report the NO_3 in terms of chemically equivalent nitrogen. Similarly, for NH_4-N, the laboratory will analyse for NH_4 but report in terms of chemically equivalent elemental nitrogen. The total nitrogen available to the plant will be the sum of the equivalent elemental nitrogen. The same reporting method is used for phosphorus
[c]SAR is calculated from the Na, Ca and Mg reported in me/l
[d]Long Term Water quality of Ankara Wastewater Treatment Plant (2004–2007)

For sustainability, entire plant, soil and water factors should be taken into consideration. Discharge water of AWTP can readily be used for irrigation by selecting proper cropping pattern, proper irrigation methods and monitoring tools. In this way, significant water saving can be achieved and farmer incomes can be raised (Gokalp et al. 2011).

While the current N:P:K rates of effluent water from Ankara Water and Sewage Administration (AWTP) are 1:0.2:0, the required rates should be as 1:0.8:1.5. It is considered that 200 kg N/da was rather high for plant demand. Soil salinity and heavy metal pollution were assumed as the possible risks. These two problems can better be evaluated by taking sustainable agriculture and environmental risks into consideration. In a long-term project, the factors causing salinity, heavy metal

Table 2 Recommended maximum concentrations of trace elements in irrigation water[a]

Element	Recommended maximum concentration[b] (mg/L)	AWTP[c]
Aluminium	5	0.12
Arsenic	0.1	0.03
Beryllium	0.1	–
Cadmium	0.01	0.01
Cobalt	0.05	–
Chromium	0.1	0.03
Copper	0.2	<0.01
Fluoride	1	–
Iron	5	0.21
Lithium	2.5	
Manganese	0.2	0.05
Molybdenum	0.01	–
Nickel	0.2	0.02
Lead	5	0.02
Selenium	0.02	0.01
Tin	–	–
Titanium	–	–
Tungsten	–	–
Vanadium	0.1	–
Zinc	2	0.19

[a]Adapted from National Academy of Sciences (1972) and Pratt (1972)
[b]The maximum concentration is based on a water application rate which is consistent with good irrigation practices (10,000 m^3 per hectare per year). If the water application rate greatly exceeds this, the maximum concentrations should be adjusted downward accordingly. No adjustment should be made for application rates less than 10,000 m^3 per hectare per year. The values given are for water used on a continuous basis at one site
[c]Long Term Water quality of Ankara Wastewater Treatment Plant (2004–2007)

pollution and endangering human health should be prevented. For example, proper irrigation systems should be implemented; the species should be selected carefully and the plants with edible leaves that have a direct contact with the irrigation water should be avoided. Untreated wastewater irrigation poses serious health risks that cannot be ignored. Heavy metal accumulation from wastewater in the crops is important as this would affect human and animal health directly through the food chain.

The challenge of water reuse is to maintain long-term sustainability. Two main concerns over the use of recycled wastewater for irrigation are (1) potential problems caused by excessive sodium and salinity, and (2) excessive nutrients or nutrient imbalance. Soil salinity is a function of soil type, management, salinity of water used for irrigation, and the depth of water table. Clay soil is more prone to salt accumulation and sodium deterioration. A shallow water table can reduce leaching and introduce salts to the root zone. Therefore, the most salinity susceptible sites are sites with shallow water table, high clay content, poor drainage, and great soil compaction. Management practices that reduce water table, cap the topsoil with sand (especially for sports fields), improve drainage, and reduce compaction would reduce the potential sodium problems (Qian 2006).

Long-term waste water irrigation did not negatively affect the studied soil processes. The predominant effect was an increase in microbial biomass and its activities due to the addition of easily decomposable organic material and nutrients and a more humid water regime. On the other hand, this study also shows that the increasing salinization, and especially Na saturation (Siebe 1998) exerts a negative effect on the basal respiration in Leptosols. AEC ratios were decreased probably due to salt accumulation in the irrigated soils, while the denitrification capacity increased. Both effects in combination indicate a shift in soil functional diversity. Negative effects on soil microbial biomass, its activities and functional diversity have been described in the literature at sites both with similar and with slightly higher total metal contents. Alkaline pH values, as well as large organic matter and clay contents in the soils of Irrigation district have kept metal solubility low. Available metals, however, are increasing with irrigation duration (Siebe and Cifuentes 1995) and this will presumably lead to Cd contents in alfalfa and maize reaching unacceptable levels after 120 years of irrigation. Thus, further applications of untreated waste water to the fields will probably evoke more severe detrimental effects on soil microbial communities. Additionally, changes in soil management or wastewater quality leading to increased mineralization rates may increase the availability and mobility of pollutants adsorbed to the soil organic matter (Friedel et al. 2000).

Faruquie et al. (2004b) stated that complex challenges of managing wastewater require a pragmatic, proactive and forward-looking perspective. The lessons learned from past experience with wastewater use and management suggest that:

- Comprehensive realization of the importance of wastewater use in agriculture is still on the peripheral edges of public awareness, and is not always clear to many policy-makers and donors;
- There is insufficient understanding of the social and economic factors that drive farmers to use wastewater, and thus inadequate consideration of these in policy formulation;
- The protection of public health and the alleviation of poverty are not mutually exclusive outcomes when it comes to wastewater use, however, one may have to be given greater emphasis than the other in different contexts;
- Effective measures do exist to protect health and environmental quality, particularly when these are included in integrated, multi-barrier approaches to wastewater management;
- Rigid wastewater use guidelines tend to become targets rather than norms;
- Effective, lower-cost, decentralized treatment systems exist; conventional, northern treatment technologies tend to be unsustainable, in part because of high capital and recurring costs;
- Many forms of wastewater use are practiced in various contexts for different reasons, and individual socioeconomic contexts contribute to varying levels of acceptability of wastewater use;

- Increasing year-round demand for fresh fruits and vegetables in developed countries, and increasing tourism in a globalize world, make wastewater use an issue for more than just developing countries;
- Sound legal and regulatory frameworks require sustained application and enforcement;
- Insecure land tenure mitigates against farmer investment in safer and more efficient wastewater irrigation technologies;
- The informal nature of wastewater irrigation tends to leave it in institutional no-man's land; and
- A lack of coordination among institutions within and outside of government, and the tendency towards isolated, uni-disciplinary research on wastewater, has inhibited the testing and design of integrated, workable solutions.

It is known that wastewater has the potential to increase yields while saving on fertilizer and irrigation costs. However, under these conditions, it can be possible that is important to choose groves that are well designed without low spots to facilitate maximum drainage. The potential negative effects of wastewater irrigation are considered to be the following: (i) health risks; (ii) contamination of groundwater (iii) Soil pollution (iv) creation of habitats for disease vectors (v) excessive growth of algae and vegetation in canals carrying wastewater (eutrophication) (vi) some problems on pressurized irrigation system.

In general, the main problem that can create significant obstacles in the safe reuse of the treated wastewater in agriculture is the lack of information of all the involved actors, namely (Fatta et al. 2005):

Governmental authorities: lack of legislation and guidelines on the reuse of treated wastewater

Local authorities and authorities responsible in wastewater treatment: (i) lack of information on innovative cost effective technologies for wastewater treatment, (ii) difficulties in the development of technical specifications for the construction and operation of appropriate wastewater treatment systems (in terms of technology, size, quality of the outflow), (iii) difficulties in the development of specifications for the proper use of the final outflow, (iv) difficulties in finding the appropriate funds for the improvement of the wastewater treatment system

Operators: lack of knowledge for the efficient operation, control and monitoring of the wastewater treatment system

Farmers: lack of information on the health risks related to the use of treated wastewater and the appropriate management procedures

6 Conclusion and Recommendations

Both problems and opportunities exist in using wastewater for irrigation. Water reuse in irrigation is a powerful means of water conservation and nutrient recycling, thereby reducing the demands of freshwater and mitigating pollution of surface and ground water. Wastewater treatment facilities may realize cost savings due to disposal costs and the sale of the recycled water. Communities can benefit from reuse by eliminating or delaying the cost associated with obtaining additional sources and facilities for freshwater. Due to these reasons, currently there are a lot of successful water reclamation and reuse operations in the world.

In Turkey, water resource is scarce and agriculture is the biggest user. It would be an important step forward if wastewater could be recycled, especially in view of the expectation that fresh water will become ever scarcer in the next few decades – considering both climate change and increasing demand.

From the foregoing, it is clear that there is varied and overwhelming information in respect to wastewater use. The use of wastewater has great potential of transforming poor urban and peri-urban agricultural activities and livelihood of the informal irrigators. However, before such potential is realized, there is urgent need of various players to work together. These actors include agronomists, plant nutritionists, policy makers, irrigation engineers, vegetable produce regulators, health workers, environmentalists, farmers and consumers among others. There is need for research in terms of crop choice, soil interventions, diseases and pathogen to be conducted within agreeable policy framework (Gweyi-Onyango and Osei-Kwarteng 2011).

Application of treated wastewater for irrigation of plants and crops is gradually becoming a common practice worldwide. Wastewater reclamation and reuse is of great interest and a viable option for many industrial sectors and countries which suffer from water scarcity problems. Although the amount of reusable domestic and industrial wastewater is much lower than the wastewater generated, many countries show an increasing interest in wastewater reclamation and reuse.

The major challenge is to optimize the benefits of wastewater as a resource of both the water and the nutrients it contains, and to minimize the negative impacts of its use on human health. From the environmental aspect, there are potentially positive and negative impacts that should be considered. There are international guidelines for reuse and quality standards of wastewater in agriculture and these standards can only be achieved through proper wastewater treatment practices. Because of high treatment costs, most cities in low-income developing countries will not have wastewater treatment facilities in the foreseeable future. However, while the use of untreated wastewater has become a routine practice in most developing countries, policies on its usage have not taken this reality into consideration. Such policies range from active enforcement of legislation that totally prohibits the use of untreated wastewater to turning a blind eye. Clearly, there is a need for better-informed decision-making (Hoek 2004).

It is currently used widely in the world which is Water policies such as price reforms from local or central government to specific consumers. Reuse practices and stimulate of wastewater can be contribution on relative costs. Suitable planning of regional or national policies can be provide new opportunities of wastewater treatment and reuse infrastructure construction in future. Also it can be supported regional or national economic developing.

Provided that the wastewater is used for irrigation, features of wastewater should be considered necessarily as well as the conditions of soils and plants being applied.

An acceptable level for soil parameter can be a toxic level for the plant. In addition to this, the careful examination of possible pollution is necessary when it is used for the production of nutritious crops. In the event that continuous installation is made to a live structure of an ion having high concentration, vital risks may occur. These two components must be carefully examined in the future studies and recommendations. Sustainable wastewater management can be ensured solely with this method.

References

Al-Jamal MS, Sammis TW, Mexal JG, Picchioni GA, Zachritz WH (2002) Agrowth-irrigation scheduling model for wastewater use in forest production. Agric Water Manag 56:57–79

Al-Shreideh B (2001) Reuse of treated wastewater in irrigation and agriculture as a non-conventional resource. In: Advanced short course on water saving and increasing water productivity: challenges and options. Faculty of Agriculture, University of Jordan, Amman, Jordan, March 2001, pp b 18.1–b 18.30

Angelakis AN, Marecos Do Monte MHF, Bontoux L, Asano T (1999) The status of wastewater reuse practice in the Mediterranean basin: need for guidelines. Water Res 10:2201–2217

Anonymous (1992) Manual guidelines for water reuse. US EPA/625/R-92/004

Anonymous (2006) Recycling realities: managing health risks to make wastewater an asset. Water policy briefing 17, IWMI, Colombo

Anonymous (2011) News bulletin. Municipal wastewater statistic. www.tuik.gov.tr. Accessed 15 Aug 2013

Anonymous (2012) Water and DSI. Ministry of forestry and water affaires general directory of state hydraulic works, 1954–2012. 58 years full of realized projects. Ankara

Anonymous (2013) Web page. www.aski.gov.tr. Accessed 01 Sept 2013

Ashraf S, Abbas F, Ibrahim M, Rashid U, Khalid S, Raza HA, Hakeem KR (2015) Application of Geographical Information System (GIS) for the identification and demarcation of selective heavy metal concentrations in the groundwater of a metropolitan city. J Geogr Sci 25(2):225–235

Aydın ME, Gür K (2002) Using reclaimed municipal wastewater for irrigation. In: International conference on the environmental problems of the Mediterranean region, 12. 15 April 2002, Near East University, Nicosia, TRNC

Aydin ME, Ozcan S (2008) Sustainable adopted wastewater treatment and reuse in agriculture. In: Proceedings of the NATO advanced research workshop on the role of ecological chemistry in pollution research and sustainable development Chisinau, Moldova 8–11 October 2008. Library of congress control number: 2009928842. ISBN 9 78–9 0–481 -2902 -7 (PB) ISBN 973-90-461-2901-0 (HB) ISBN 9 7 8–90 -481 -2 90 3–4 (e- book)

Ayers RS, Westcot DW (1994) Water quality for agriculture. F.A.O. irrigation and drainage paper 29, Rev. 1, 1994, FAO, Rome

Barman SC, Bhargava SK (1997) Accumulation of heavy metals in soil and plants in industrially polluted fields. In: Cheremisinoff PN (ed) Ecological issues and environmental impact assessment. Gulf Publishing Company, Houston, pp 289–314

Barman SC, Sahu RK, Bhargava SK, Chatterjee C (2000) Distribution of heavy metals in wheat, mustard and weed grains irrigated with industrial effluents. Bull Environ Contam Toxicol 64:489–496

Carr RM, Blumenthal UJ, Mara DD (2004) Health guidelines for the use of wastewater in agriculture: developing realistic guidelines. In: Scott CA, Faruquie NI, Reschid-Sally L (eds) Waste water use in irrigated agriculture: confronting the livelihood and environmental realities. CABI Publishing, Wallingford

Emongor V (2007) Biosorption of lead from aqueous solutions of varied pH by kale plants (Brasicca oleraceae var. acephala). J Agric Food Environ Sci 1:90–91
Ensink JHJ, Mahmood T, Van der Hoek W, Raschid-Sally L, Amerasinghe FP (2004) A nationwide assessment of wastewater use in Pakistan: an obscure activity or a vitally important one? Water Policy 6:1–10
Faruquie NI, Scott CA, Reschid-Sally L (2004a) Confronting the realities of wastewater use in agriculture: lesson learned and recommendations. In: Scott CA, Faruquie NI, Reschid-Sally L (eds) Waste water use in irrigated agriculture: confronting the livelihood and environmental realities. CABI Publishing, Wallingford Oxfordshire OX10 8DE. UK
Faruquie NI, Scott CA, Reschid-Sally L (2004b) Confronting the realities of wastewater use in agriculture: lesson learned and recommendations. In: Scott CA, Faruquie NI, Reschid-Sally L (eds) Waste water use in irrigated agriculture: confronting the livelihood and environmental realities. CABI Publishing, Wallingford
Fatta D, Alaton IA, Gokcay C, Rusan MM, Assobhei O, Mountadar M, Papadopoulos A (2005) Wastewater reuse: problems and challenges in cyprus, Turkey, Jordan and Morocco. Eur Water 11(12):63–69
Friedel JK, Langer T, Siebe C, Stahr K (2000) Effects of long-term waste water irrigation on soil organic matter, soil microbial biomass and its activities in central Mexico. Biol Fertil Soils 31:414–421
Friedler E (2001) Water reuse—an integral part of water resources management: Israel as a case study. Water Policy 3:29–39
Gökalp Z, Tas I, Ozkay F, Akgül S (2011) Wastewater use in irrigation – Ankara case study. Afr J Agric Res 6(7):1807–1812
Gweyi-Onyango JP, Osei-Kwarteng M (2011) Safe vegetable production with wastewater in developing countries: demystifying the negative notions. Afr J Hortic Sci 5:70–83
Hamdy A (2001) Agricultural water demand management: a must for water saving. In: Advanced short course on water saving and increasing water productivity: challenges and options. Faculty of Agriculture, University of Jordan, Amman, Jordan, March 2001, pp b 18.1–b 18.30
Haruvy N (1996) Wastewater for agriculture economic and environmental perspectives. In: Proceedings of the seventh international conference on water and irrigation, Tel Aviv, Israel, 1996, pp 64–72
Hoek WVD (2004) A framework for a global assessment of the extent of wastewater irrigation: the need for a common wastewater typology. In: Scott CA, Faruquie NI, Reschid-Sally L (eds) Waste water use in irrigated agriculture: confronting the livelihood and environmental realities. CABI Publishing, Wallingford
Huibers FP, Raschid-Sally L (2005) Design in domestic wastewater irrigation. Irrig Drain 54:113–118
Jimenez B, Asano T (2004) Acknowledge all approaches: the global outlook on reuse. Water 21:32–37
Jimenez B, Chávez A (2004) Quality assessment of an aquifer recharged with wastewater for its potential use as drinking source: El-Mezquital Valley case. Water Sci Technol 50:269–276
Karanja N, Njenga M, Prain G, Kang'ethe E, Kironchi G, Gichuki C, Kinyari P, Mutua GK (2010) Assessment of environmental and public health hazards in wastewater used for urban agriculture in Nairobi, Kenya. Trop Subtrop Agroecosyst 12:85–97
Lund E (1980) Health problems associated with the reuse of sewerage. WHO seminar on health aspects of treated sewage reuse, Regional Office for Europe. World Health Organization, Copenhagen. ISBN 92 890 120S 0
Mapanda F, Mangwayana EN, Nyamangara J, Giller KE (2005) The effect of long-term irrigation using wastewater on heavy metal contents of soils under vegetables in Harare, Zimbabwe. Agric Ecosyst Environ 107:151–165
Muchuweti M, Birkett JW, Chinyanga E, Zvauya R, Scrimshaw MD, Lester J (2006) Heavy metal content of vegetables irrigated with mixture of wastewater and sewage sludge in Zimbabwe: implications for human health. Agric Ecosyst Environ 112:41–48
Pair CH, Hinz WH, Frost KR, Sneed RE, Schiltz TJ (1983) Irrigation, 5th edn. The Irrigation Association, Silver Spring

Pescod BM (1992) Wastewater treatment and use in agriculture. FAO irrigation and drainage paper 47. FAO, Roma

Qadir M, Sharma BR, Bruggeman A, Choukr-Allah R, Karajeh F (2007) Non-conventional water resources and opportunities for water augmentation to achieve food security in water scarce countries. Agric Water Manag 87:2–22

Qadir M, Wichlns D, Raschid-Sally L, McCornick PG, Dreschsel P, Bahri A, Minhas PS (2010) The challenges of wastewater irrigation in developing countries. Agric Water Manag 97:561–568

Qian Y (2006) Urban landscape irrigation with recycled wastewater. Colorado Water Resource Research Institute, Completion Report Number, Funding provided by: United States Geological Survey: Grant Number 01HQGR0077; Project Number 2003CO71B; Colorado Agricultural Experiment Station Project Number 658 USA

Ragab R (2001) Policies and strategies on water resources in the European Mediterranean region. In: Proceedings of the water and irrigation development international conference, Mantova, Italy, 2001. Editoriale Sometti, Mantova, Italy, pp 37–55

Sarma H (2011) Metal hyperaccumulation in plants. A review focusing on phytoremediation technology. J Environ Sci Technol 4(2):118–138

Scott CA, Faruquie NI, Reschid-Sally L (2004) Wastewater use in irrigated agriculture: management challenges in developing countries. In: Scott CA, Faruquie NI, Reschid-Sally L (ed) Waste water use in irrigated agriculture: confronting the livelihood and environmental realities CABI Publishing, Wallingford Oxfordshire OX10 8DE. UK

Sharma RK, Agrawal M, Marshall F (2007) Heavy metal contamination of soil and vegetables in suburban areas of Varanasi, India. Ecotoxicol Environ Saf 66:258–266

Siebe C (1998) Nutrient inputs to soils and their uptake by alfalfa through long-term irrigation with untreated sewage effluent in Mexico. Soil Use Manag 13:1–5

Siebe C, Cifuentes E (1995) Environmental impact of wastewater irrigation in central Mexico: an overview. Int J Environ Health Res 5:161–173

Singh KP, Mohan D, Sinha S, Dalwani R (2004) Impact assessment of treated/untreated wastewater toxicants discharged by sewage treatment plants on health, agricultural, and environmental quality in the wastewater disposal area. Chemosphere 55:227–255

Sinha S, Saxena R, Singh S (2002) Comparative studies on accumulation of Cr from metal solution and tannery under repeated metal exposure by aquatic plants: its toxic effects'. Environ Monit Assess 80:17–31

Steward HTL, Allender E, Sandell P, Kube P (1986) Irrigation of tree plantations with recycled water in Australia: research developments and care studies. In: Cole DH, Henry CL, Nutter WL (eds) The forest alternative for treatment and utilization of municipal and industrial wastes. University of Washington Press, Seattle, pp 431–441

Van der Hoek W (2004) A framework for a global assessment of the extent of wastewater irrigation: the need for a common wastewater typology. In: Scott CA, Faruqi NI, Raschid-Sally L (eds) Wastewater use in irrigated agriculture. Confronting the livelihood and environmental realities. CABI Publishing, Wallingford, pp 11–24

Van der Hoek W, Mehmood ul H, Ensink JHJ, Feenstra S, Raschid-Sally L, Safraz Munir S, Aslam R, Hussain Rand Matsuno Y (2002) Urban wastewater: a valuable resource for agriculture, a case study from Haroonabad, Pakistan. Research report 63. IWMI, Colombo, Sri Lanka

Yurtseven E, Çakmak B, Kesmez D, Polat HE (2010) Tarımsal Atık Suların Sulamada Yeniden Kullanılması. TMMOB Ziraat Mühendisleri Odası Türkiye Ziraat Mühendisliği VII.Teknik Kongresi 11–15 Ocak 2010, Cilt:1, s. 135–154, Ankara

Plant Secondary metabolites: Deleterious Effects, Remediation

(Special Reference to Forage)

Salah A. Attia-Ismail

Contents

Abstract A major constraint to the use of some of the livestock feeds is the presence of toxic and plant secondary metabolite (PSM's) constituents. These constituents have different but adverse effects on animal performance including loss of appetite and reductions of dry matter intake and nutrients digestibility. They are produced in plants for protective purposes for the plants itself and to adapt to environmental stresses. Some of them are deleterious and some are beneficial, some of which may be nutritionally valuable but many have no nutritional value or nutritionally detrimental effects. It is estimated that almost 80,000 PSM compounds are found in plants and occur naturally. Toxicity of most poisonous plants is associated with stage of growth, temperature, site, rainfall precipitation, light, soils, weather conditions and kind of animal. Toxicity is rather affected by numerous factors such as rate of ingestion, types and rates of microbial transformations in the rumen, rates of gastro-intestinal absorption, rates and pathways of biotransformation in gut tissues, liver and kidney and effect of enzyme induction or inhibition. Plant secondary metabolites can be divided into five major groups: the phenolic compounds, Glycosides, Alkaloids, Nitrates and Others like Oxalate and Lectins (Haemagglutinins). Tannins are a group of poly phenolic compounds that are produced naturally in some plants. They are distinguished from other polyphenolic

S.A. Attia-Ismail (✉)
Department of Animal and Poultry Nutrition, Desert Research Center, Cairo, Egypt
e-mail: saai54@hotmail.com

M. Öztürk et al. (eds.), *Plants, Pollutants and Remediation*,
DOI 10.1007/978-94-017-7194-8_8

compounds by their ability to precipitate proteins. They form complexes also with carbohydrates in the feeds, and with digestive enzymes. Tannins could be beneficial to ruminants. They may be involved in bloat prevention. Monogastric animals are highly susceptible to tannins than ruminant animals. Saponins can also exhibit a range of biological properties, both beneficial and deleterious. Hydrocyanic acid poisoning of sorghum is the most known problem of cyanogens. Drying considerably reduces the HCN level and sun drying has been shown to be more effective than oven drying. Many of the alkaloids are toxic to humans and animals. Nitrate itself is relatively nontoxic, but its metabolites, i.e. nitrite, nitric oxide, etc. are poisonous. Oxalates are poisonous when animals consume large quantities of oxalate containing plants without adaptation. If oxalates combine with calcium, they become insoluble. Lectins are glycoproteins that have the ability to bind to carbohydrate-containing molecules which cause the agglutination of red blood cells as well as reduced growth, diarrhea, and interference with nutrient absorption. Several methods were used to overcome the adverse effects of tannins such as, alkali treatments including ferrous sulphate and calcium hydroxide, Polyethylene glycol (PEG). Physical methods like soaking and drying and heat treatment before feeding of forage may reduce the toxic level of tannin. Potential methods for reducing the effects of tannin include drying, chemical agents like urea, wilting and wetting with chemical agents, as well as gelatin high in proline content and ensiling.

Keywords Plant secondary metabolites • Phytoremediation • Forage

1 Introduction

The recent shifts in both technological and practical potentialities in animal feeding have been strongly reinforced by the economic pressure. The ever increasing gap between developed countries and those striving for more development has impacts on human welfare. In the less developed countries, there are huge amounts of different natural resources that are not utilized or at least inefficiently utilized. However, there is a continuing doubt that those who need food will be able to afford buying it. Herein, there should be more emphasis on alternative concepts of productivity and more efficient utilization of local resources. The achievements, hence, were assessed as much on reduced or lower cost inputs as on greater output. Optimizing the input/output relations is of great importance. This may suggest that more attention is given to most economical feeding while maintaining animal production at optimal levels. It is, then, that the use of non conventional feed resources is of more importance. The greater demand for feed in most developing countries imposes certain pressures.

Ruminant production in most of the world is constrained by inadequate supply of feeds in terms of quality and quantity. Natural vegetations in arid, semi arid and coastal areas constitute the main feed resources for the indigenous herbivores. The native pastures and crop residues are the major feed sources available in these areas for the ruminants (Osuga et al. 2005) Utilization of the marginal resources such as saline soils and underground water for producing feed for animals (Fodder crops) becomes necessary in order to improve nutritional status of these livestock. Some of these fodder crops are toxic. Toxicity may result from number of secondary metabolites or anti- nutritional factors.

A major constraint to the use of some of the livestock feeds is the presence of toxic and plant secondary metabolite (PSM's) constituents. These constituents have different but adverse effects on animal performance including loss of appetite and reductions in dry matter intake and protein digestibility.

The plant secondary metabolite is a term that is used widely to refer to a broad spectrum of plant metabolically produced compounds. They are produced in plants for protective purposes for the plants and to adapt to environmental stresses. Some of them are deleterious and some are beneficial, some of which may be nutritionally valuable but many have no nutritional value or nutritionally detrimental (Bento et al. 2005). PSM's are an extremely large group of compounds referred to also as phytochemicals (Acamovic and Brooke 2005). They are of small molecular weights and they are largely distributed in plants (Edreva et al. 2008). Wink (1999a, b) estimated that almost 80,000 PSM compounds are found in plants and occur naturally.

Toxicity of most poisonous plants is associated with stage of growth, temperature, site, rainfall precipitation, light, soils, weather conditions and kind of animal. For instance sheep are more resistance to certain type of fodder plants than are cattle.

PSM's are substances that when present in animal feed reduce the availability of one or more nutrients. They interfere with the intake, availability, or metabolism of nutrients in the animals (Attia-Ismail 2005). Their effects can also range from a mild reduction in animal performance to death, even at relatively small intakes. Harmful effects of plant secondary metabolites cause great economic losses to livestock producers. However, PSM's can inhibit the growth of microbes and fungi and they are often referred to as defences (Forbey et al. 2009).

Toxicity is rather affected by numerous factors such as rate of ingestion, types and rates of microbial transformations in the rumen, rates of gastro-intestinal absorption, rates and pathways of biotransformation in gut tissues, liver and kidney and effect of enzyme induction or inhibition. The PSM's may cause liver damage, renal failure, Anoxia, pancreatic hypertrophy, hypoglycemia, death and other pathological conditions for animals.

2 Divisions of Plant Secondary Metabolites

Most of secondary metabolites are classified based on their biosynthetic origin. Although this classification does not take into consideration all the groups of plant secondary metabolites, it gives a list of the frequently found in animal feeds. Plant secondary metabolites can, therefore, be divided into five major groups:

A. ***The phenolic compounds***:

Phenolic acids
Tannins
Gossypols

B. ***Glycosides***:

Saponins
Cyanogens

C. ***Alkaloids***:
D. ***Nitrates***:
E. ***Others***:

Oxalate
Lectins (Haemagglutinins)

Table 1 shows Endogenous plant secondary metabolites present in some halophytic fodder crops. Table 2 shows the PSM's detected in some *Acacia spp*.

Table 1 Endogenous plant secondary metabolites present in halophytic fodder crops

Fodder crops	Anti-nutritional factors
Atriplex nummularia	Saponin, Alkaloids, Tanins, Nitrate
Atriplex leucoclade	Saponin, Alkaloids, Tannins
Atriplex halimus	Saponin, flavonoids, Alkaloids, Tannins, nitrate
Diplache fusca	Flavonoids, Alkaloids
Halocnemum strobilecum	Saponin, flavonoids, alkaloids, tannins, nitrate
Haloxylon salicornicum	Saponin, flavonoids, alkaloids, tannins
Kochia eriophora	Alkaloids, Tannins
Juncus acutus	Flavonoids, Alkaloids, Tannins, nitrate
J. arabicus	Alkaloids, Tannins
J. subulatus	Alkaloids, Tannins, flavonoids
Limonium pruinosum	Saponin, alkaloids, Tannins
Nitraria retusa	Saponin, Tannins
Salsola glauco	Saponin, flavonoids, alkaloids
Suaeda fruticosa	Alkaloids, Tanins, nitrate
Tamarix aphylla	Saponin, Tanins
Salsola tetrandra	Nitrate
Tamarix mannifera	Saponin, tannins
Zygophyllum album	Saponin, flavonoids, alkaloids, tannins, nitrate
Sesbania sesban	Saponin, alkaloids

Adapted from Afify AF (2004) Anti-nutritional factors in fodder crops: constraints and solutions (unpublished data)

Table 2 Toxic compounds in some Acacia species for ruminants

Species	Part of plant	Toxin	References
A. aneura	Phyllode	Oxalate	Gartner and Hurwood (1976)
A. aneura	Phyllode	Tannin	
A. burrowii	Flowers	Hydrogen cyanide	Cunningham et al. (1981)
A. cambagei	Phyllode	Hydrogen cyanide	
A. cambagei	Timber, bark	Oxalate	
A. cana	Browse	Selenium	
A. deanei	Browse	Hydrogen cyanide	
A. decora	Browse	Abortive agent	
A. doratoxylon	Browse	Cyanogenic glycoside	
A. georgina	Browse	Hyrolytic enzyme only	Hall (1972)
A. georgina	Seeds/pods	Fluoroacetate	Everist (1969)
A. longifolia	Browse	Hydrogen cyanide	Cunningham et al. (1981)

Adapted from Dynes and Schlink (2002)

2.1 The Phenolic Compounds

Phenolic compounds vary in their structure from simple phenols such as the derivatives of hydrobenzoic acid (Hydrolysable tannins) to condensed tannins (Tania et al. 2012). The two types differ in their effects on animals both on nutritional and toxic levels. The condensed tannins may affect animals more than hydrolysable tannins with respect to nutrient digestibility. Hydrolysable tannins, however, may cause varied toxic manifestations due to hydrolysis in rumen (Smitha Patel et al. 2013).

Among the most important are flavonoids, phenolic acids, coumarins and isoflavones which are widespread in vegetable crops such as fruits, vegetables, herbs, grains and seeds (Miniati 2007). A phenol is a phenyl ($-C_6H_5$) bonded to a hydroxyl (–OH) group. Tannins are responsible for the astringent taste of some plants.

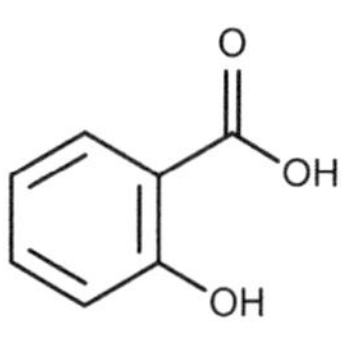

Hydrobenzoic acid (example of phenolic acids)

2.1.1 Effect of Phenolic Compounds on Animal

Tannins are a group of poly phenolic compounds that are produced naturally in some plants. They are distinguished from other polyphenolic compounds by their ability to precipitate proteins (Haslam 1989; Silanikove et al. 2001). The net tannin percent showed in the Sorghum tannins may bind and precipitate at least 12 times their weight of protein (Jansman 1993). They also have the ability to complex with

Table 3 Tannin contents of some tree leaves (% DM)

Tree	Total phenols	Net tannin	Condensed tannins
Acacia nilotica	16.2	14.6	1.1
Toona cililate	3.8	2.3	0.9
Bauhinia variegate	4.8	3.7	3.4
Phoenix acaulis	5.8	4.8	4.3
Anogeissus latifolia	17.4	15.9	0.4
Carrisa spinarum	6.6	4.5	4.6
Ougenia oojeiuealis	4.2	2.9	2.6
Leucaena leucocephala	4.9	2.1	0.8

Adapted from Rana et al. (2006)

minerals (Reed 1995). Tannins are widely found in most plants, especially in trees, shrubs and herbaceous leguminous plants. Tannin contents differ from one plant to another. Table 3 shows the variations in tannin percent on some tree leaves (Rana et al. 2006). They also differ within the same plant from season to season. The concentration of tannins was almost two-fold higher in the dry compared to the wet season of *Albizia procera* (Alama et al. 2005). There are reports of low digestibility for some accessions of *Calliandra calothyrsus*, a shrub legume that occurs throughout the tropics and sub-tropics, and this has been related to the variable concentrations of tannins (Salawu et al. 1999; Mupangwa et al. 2000).

2.1.2 Effects of Tannins on Animals

Tannins are the second most abundant group of plant phenolics after lignin. Tannins are classified into two main groups (hydrolysable and condensed tannins). The most important aspect of tannins in their nutritional and toxicological effects is forming strong complexes with proteins (Hagerman and Butler 1981). They form complexes also with carbohydrates in the feeds, and with digestive enzymes. Tannins could be beneficial or harmful to ruminant. It is said that tannin concentration from 2 to 4 % in rations help protein to escape rumen degradation and increases the absorption of essential amino acids whereas 4–10 % depresses feed intake.

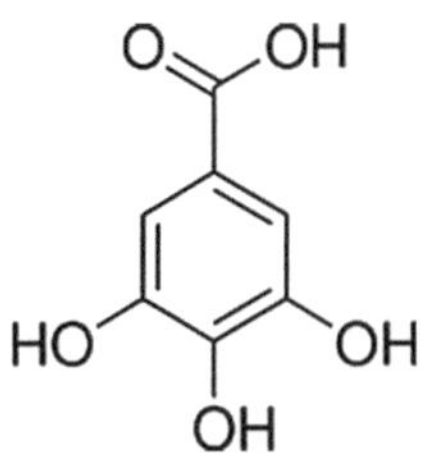

Simple tannins

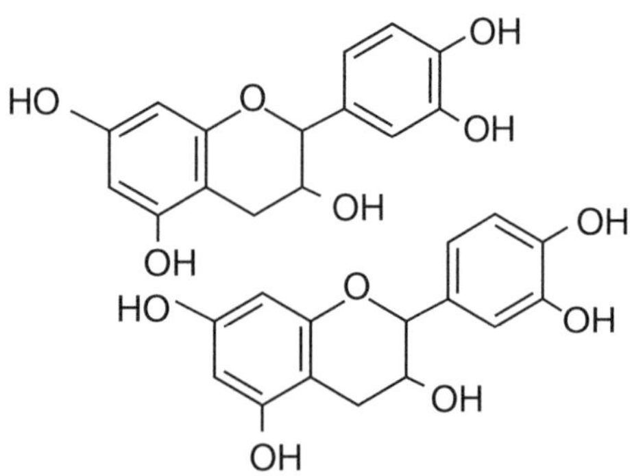

condensed tannins

Benefits reported include increases in wool production, milk protein secretion, ovulation rate and the development of more nutritionally based and ecologically sustainable systems for disease control in grazing animals (Ben Salem et al. 2005). Soluble dietary proteins stabilize foam (Nguyena et al. 2005) in the rumen that can entrap gas bubbles and cause bloat (McLeod 1974). Therefore, tannins may be involved in bloat prevention (Waghorn 1990). Ram'ırez-Restrepo et al. (2005) fed diet containing 18–29 g condensed tannins/kg DM for *Lotus corniculatus*. They found that under commercial dryland farming conditions, the use of *L. corniculatus* during the mating season in late summer/autumn can be used to increase reproductive efficiency and wool production. They attributed these findings to the higher digestibility and metabolizable energy of *L. corniculatus* than pasture, and to the condensed tannins in *L. corniculatus* improving both protein digestion and absorption.

Effects of Tannins on Poultry

Monogastric animals are highly susceptible to tannins than ruminant animals. Tannins cause leg abnormalities in chick which is due to defective formation of bone matrix (Armstrong et al. 1974). Tannins may cause depression in feed intake. The reduced feed intake may affect growth as a result. The complexity of tannins with proteins reduces not only intake but also the protein digestibility and, therefore, increased the fecal output of nitrogen. Alledredge (1994) has suggested that there is a considerable evidence to suggest that enzymatic proteins, as well as other endogenous proteins, comprise a considerable portion of excreted nitrogen. This also may result in a deficiency of one or more of the essential amino acids leading to reduced growth. A strong relationship between concentration of tannins in feeds and the palatability has been found (Mangan 1988). Adverse effects of tannins on food palatability and consumption have been repeatedly reported (Makkar 2003; Hassan et al. 2003; Kim and Miller 2005). Tannins can increase the size of the parotid glands and damage the mucosal lining of the gastro intestinal tract of chickens, but to a lesser extent in the laboratory rat (Oritz et al. 1994). Chicks fed diets high in condensed tannins (faba beans hulls) had poor digestibility of amino acids (Longstaff and McNab 1991). They explained that the low digestibility may be due to an increased excretion of inactivated enzymes. Studies on the effects of condensed tannins have given similar results. Negative effect of tannins on starch digestibility in 3 week old chickens was detected (Flores et al. 1994). The extent of the depression depended on the quantity of tannins ingested.

Sorghum is a feed that is commonly used as poultry feed. It has high content of tannin. All sorghum varieties contain phenolic compounds, which can influence the color, appearance and nutritive value of the grain. Tannin content in sorghum grains can vary considerably among different varieties (Gu et al. 2004). In the US the grain breeders have developed a new sorghum variety that has no tannins. This has its implications on the feeding value of sorghum. When new grain sorghum varieties were compared to other cereal grains using broilers, layers, and mature leghorn roosters, Huang et al. (2006) found that crude protein digestibility of sorghum versus corn in all three classes of birds was similar between the grain sources. Similar work using broilers by Ravindran et al. (2005) found that the digestibility of crude protein was higher for sorghum compared to corn (99 vs. 81 %).

Effects of Tannins on Ruminants

Ruminant animals are more tolerant to tannins than monogastric animals. Tannins have negative effects (Barry and Manley 1986) on protein metabolism and decrease palatability of feeds (because of the astringency through tannin–salivary protein complex formation in the mouth) at very high levels (>60 g/kg DM) and high levels (>50 g/kg DM) but at low (10–30 g/kg DM) or trace levels (<10 g/kg DM) tannins are beneficial (Balogun and Holmes 1998). Therefore, it is safe to say that condensed tannins can have both beneficial and detrimental effects on ruminants. Positive effects on ruminants include preventing bloating and anathematic effects. Among some of the beneficial effects, condensed tannins complex with soluble proteins in the rumen, thus, slowing the rapid microbial degradation, and, therefore, increasing ruminal escape protein (Waghorn et al. 1999), leading to increased amino acids absorption in the lower gut (Barry and Manley 1986). Condensed tannins may also contribute to animal health by reducing the detrimental effects of internal parasites in sheep and the risk of bloat in cattle (Niezen et al. 1998). They also have beneficial antibiotic effect on animals (Aengwanich et al. 2009). The degradation products of hydrolysable tannins within the gastro intestinal tract can be absorbed and cause toxicity (Acamovic and Brooker 2005). The tannins, however, have variable effects on rumen microorganisms. The results of Bento et al. (2005) indicated that there is considerable interaction between tannins, microbes and non-starch-polysaccharides (NSP) in animal feeds and that these interactions may influence the functional ability of microbes in the gastrointestinal tract of animals.

In some ruminants, particularly goats and camels, tannin-resistant rumen microbial populations have been described (Brooker et al. 1994). Tannin protein complexes in the rumen are considered stable (Andrabi et al. 2005). These complexes can dissociate in the region of post-rumen in response to the low pH that occur there (McNabb et al. 1996). The low pH in the abomasum as well as the high pH in the small intestine can stimulate dissociation. However, Rakhmani et al, (2005) investigated this relation where they fractionated condensed tannins into monomeric, oligomeric and polymeric components. A negative correlation between oligomers, flavonols and flavonol glycosides and DM digestibility in vitro was detected whereas a positive correlation between the polymeric proanthocyanidins and DM digestibility in vitro was observed. However, some rumen bacteria species that tolerate or degrade tannins have now been identified (McSweeney et al. 2001) in animals that used to consume tannin containing diets. Browse species with high tannin content had inhibitory effects on rumen microbial fermentation (Osuga et al. 2005). Tannins also inhibit abomasal and intestinal structure and function (Robins and Brooker 2005).

WocBawek-Potocka et al. (2013) mentioned that "Diphenolic compounds are a sort of phytochemicals that resemble estradiol. They belong to a heterogenous group of herbal substances and similar to estradiol-17β (E2). They are called estrogen-like molecules or nonsteroidal estrogens structurally similar to E2". In a research paper to study diverse effects of phytoestrogens on the reproductive performance of cattle, these authors concluded that there is evidence that phytoestrogen exposure (when feeding soy bean to cow) may significantly affect reproductive function of cattle. The decrease of fertility might be attributed to the direct effect of phytoestrogens on reproductive tract. Phytoestrogens can inhibit endogenous estrogen production in the ovary leading to disturbances in immune system regulation as well as in follicle development and lack of estrous (Rosselli et al. 2000). Scanlan and Skinner (2002) suggested that the plant-derived isoflavone, as 17β-estradiol, can be a stimulator to growth hormone secretion in ewes and may exert its effect at the level of the central nervous system.

Methods to Overcome Tannins Effects

Several methods were used to overcome the adverse effects of tannins such as, alkali treatments including ferrous sulphate (Smitha Patel et al. 2013) and calcium hydroxide (Alama et al. 2005), Polyethylene glycol (PEG) (Barry et al. 2001). Physical methods like soaking and drying (Reddy 2001) and heat treatment before feeding (Vitti et al. (2005) of forage may reduce the toxic level of tannin (Nuttaporn and Naiyatat 2009). Potential methods for reducing the effects of tannin include drying, chemical agents like urea, wilting and wetting with chemical agents, as well as gelatin high in proline content (Rusdi 2004) and ensiling (Attia-Ismail 2005).

Polyethylene glycol is commonly used as an additive to improve intake, digestibility, and live weight gain and wool growth in sheep and goats (Palmer and Jones 2000; Barry et al. 2001). Polyethylene glycol (molecular weight 4000 or 6000) suppresses negative effects of tannins. Polyethylene glycol-4000 prevents formation of complexes between tannic acid and protein and helps in the breakdown of already formed complexes thus liberating protein (Reddy 2001). PEG preferentially binds with tannins (Yildiz et al. 2005) who found that the addition PEG seems unnecessary as it did not improve crude protein digestibility, N retention and body weight. The amount of PEG used has a large impact on the utilization of tannin-rich shrub foliage (Makkar 2003). Chopping, water sprinkling, storage under aerobic and anaerobic conditions, urea, wood ash, activated charcoal and PEG 4000 treatments were evaluated for their efficiency in deactivating tannins in shrub (*Acacia cyanophylla* Lindl.), foliage (Ben Salem et al. 2005). They found that chopping, water soaking or storage under anaerobic conditions are efficient techniques and further improvement could be achieved when two or more of these techniques are combined. Economic considerations were an important barrier to the use of PEG and activated charcoal that proved effective in tannin deactivation in contrast to wood ash. Vitti et al. (2005) studied the effects of oven-, sun- and shade-drying and of urea treatment. A 30-d treatment with urea reduced extractable tannins. The urea treatment was also most effective at reducing the in vitro effects of tannins compared to the other drying treatments. Drying has no negative effect on the biological activity of the tannins examined (Muetzel and Becker 2005).

Hydrochloric acidic and calcium hydroxide solutions were used among attempts to deactivate tannins (Wina et al. 2005a). Soaking in calcium hydroxide solution, hydrochloric acid or water removed 41–76 % of tannin and total phenolics removed from the recovered leaves. Soaking of the leaves of acacia also removed fermentable materials.

Alkali treatment (either calcium hydroxide or potassium carbonate) reduced the concentrations of extractable tannin by as much as 92 % of *Albizia procera*. Calcium hydroxide alone did not improve the feeding value of Albizia. It was concluded that calcium hydroxide does not deactivate the tannins in Albizia (Alama et al. 2005).

Rusdi (2004) explained that gelatin high in proline content forms an imide instead of amide bond with tannins, which cannot be hydrolysed by endogenous enzymes in mammals. They concluded that gelatine may improve digestibility of nutrients in livestock given a tannin containing diet.

Abd El Halim (2003) and Abd El Rahman (1996) found that ensiling mixture of halophyte plants increased animal acceptability and feed intake by sheep and goats. The ensiling process sharply depressed the presence of some plant secondary metabolites (Saponin, Alkaloids, Coumarin, Volatile oils, flavonoids and tannins).

Another route of ameliorating the nutritive value of halophytes is the use of feed additives like monensin (Jones and Hegarty 1984). Fahmy et al. (2003) added monensin to a mixture of halophytes and obtained 14 % increment in body weight gain of growing lambs. Supplementation with energy and protein supplements plays an important role supporting animal performance. Shawkat et al. (2001) concluded that fresh saltbushes supplemented with 50 % ground barley and 50 % ground date seeds may be a good source of energy for growing goat kids.

2.1.3 Gossypols

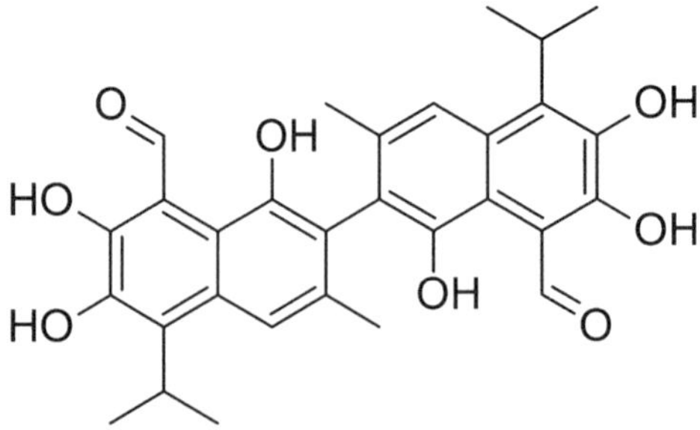

Gossypol is a toxic pigment ($C_{30}H_{30}O_8$) found in the glands of cottonseed from the genus Gossypium (http://www.thefreedictionary.com/gossypol). It is a natural toxin present in the cotton plant that protects it from insect damage. This compound is concentrated in the cottonseed but can also be found in other parts of the plant such as hulls, leaves and stems. Gossypol exists in a free form and a bound form. The free form is toxic, while the bound form (binds to proteins) is not toxic (EFSA 2008). Gossypol affects mainly the heart and liver. The reproductive tract (inhibition of sperm production), abomasum, and kidneys are also affected. Monogastric animals, such as poultry, are highly susceptible to gossypol toxicity.

(http://pearlsnapblog.com/2011/08/26/forage-facts-gossypol-toxicity-in-livestock/). Poultry are affected by dietary gossypol concentrations above 200 ppm. Ruminants can tolerate higher levels of free gossypol. Cattle have the ability to detoxify gossypol because free gossypol binds to proteins in the rumen. This ability can be overcome at very high levels of cottonseed feeding (Velasquez-Pereira et al. 1999). Feeding cottonseed meal caused death of some calves.

No treatment exists to cure gossypol toxicity. Vitamin E supplementation may present some plausible protection against gossypol toxicity (Velasquez-Pereira et al. 1999). High intakes of protein, calcium hydroxide, or iron salts appear to be protective in cattle (http://www.merckmanuals.com/vet/toxicology/gossypol_poisoning/overviewof_gossypol_poisoning.html#top). Mature cattle should also be given ≥40 % of dry matter intake from a forage source, and dietary gossypol concentrations should be limited to ≤1,000 ppm. Young ruminants are affected by dietary gossypol concentrations >100 ppm. . In normal feeding situations, the amount of cottonseed that can be fed is about 0.5 % of body weight for mature cows and .33 % of body weight for weaned calves (http://osufacts.okstate.edu). Added iron of up to 600 ppm in poultry diets may be effective in preventing symptoms of toxicity. Free gossypols are readily extractable with solvents (EFSA 2008). Storage, steam and heat, and extrusion of oil, reduce free gossypol concentrations.

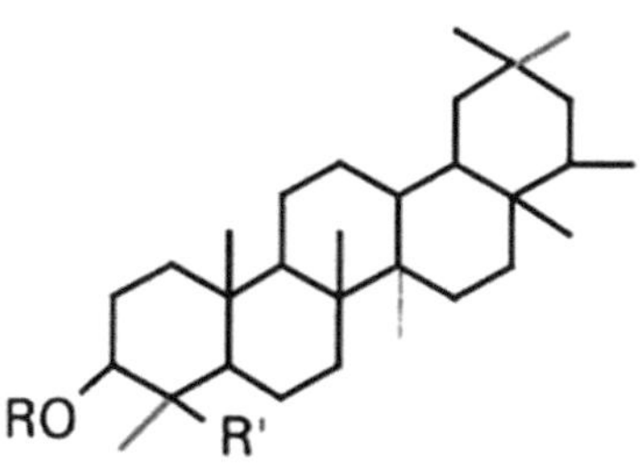

Triterpenoid

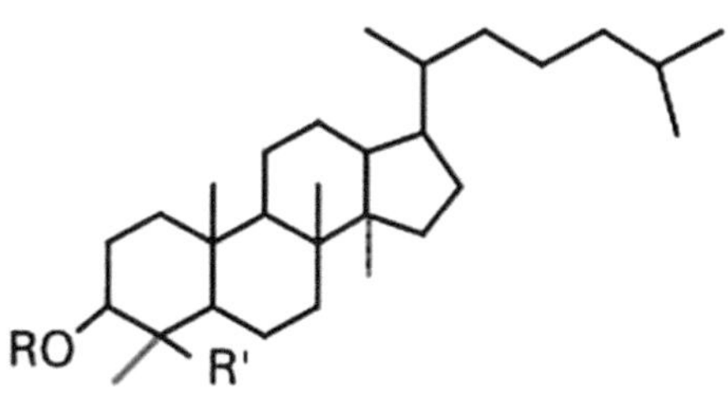

Steroid

2.2 *Glycosides*

2.2.1 Saponins

Saponins are plant glycosides that form soapy lathers when mixed and agitated with water, used in detergents, foaming agents, and emulsifiers according to on line free dictionary (http://www.thefreedictionary.com/saponin). Saponins have extremely diverse biological activities mostly related to their structure which is affected by their source and they also exhibit a range of biological properties, both beneficial and deleterious. Saponins act as membrane-permeabilizing, have immunostimulant properties and enhance antibody production (Wina et al. 2005b; Rajput et al. 2007) and found to have significantly affect growth and feed intake in animals (Das et al. 2012). Saponins contain a steroidal or triterpenoid aglycone (Oda et al. 2003). Triterpenoid saponins have been detected in many legumes such as soybeans, beans, peas, alfalfa (*Medicago sativa*), Berseem and sunflower.

Saponins in poultry rations depressed growth, feed consumption and egg production (Jenkins and Atwal 1994). Also, the formation of saponin-protein complexes reduces protein digestibility (Potter et al. 1993). Saponins can affect the count of ruminal protozoa. Saponins react with cholesterol in the protozoal cell membrane, causing the cell to rupture (Wina et al. 2005b). Some studies reported adaptation of the mixed microbial population of the rumen to saponins or saponin-containing plants (Das et al. 2012).

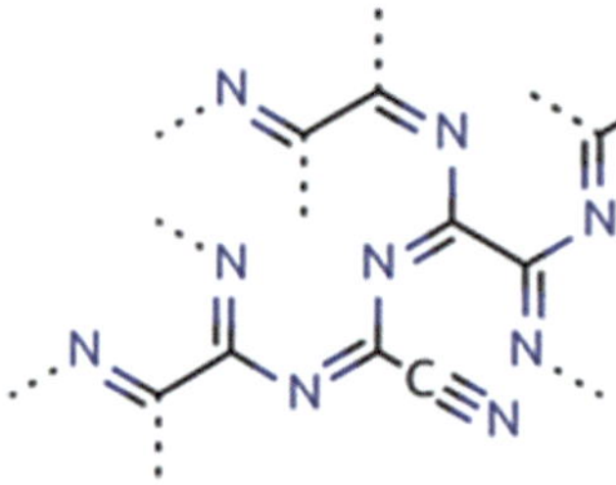

Paracyanogen

2.2.2 Cyanogens

Cyanogens are glycosides which belong to the phytotoxins. Cassava root, flax (linseed), white clover and the shoots of sorghum contain cyanogenic glycosides. Cyanogenic glycosides can convert to prussic acid. Microorganisms in the digestive tract of ruminants hydrolyze cyanogens to release hydrogen cyanide (HCN) and further (depending on rumen pH) to cynide anion (CN-). This makes ruminants be more susceptible to cyanogens than monogastrics (EFSA 2007).

Table 4 Recommended inclusion rates of sorghum (% dry matter) in diets of livestock and poultry based on nutritional values

	Ruminants					Poultry			
	Calf	Dairy	Beef	Lamb	Ewe	Chick	Broiler	Breeder	Layer
Sorghum	5	10	10	5	10	0	0	5	5
linseed meal	7.5	20	20	7.5	20	0	0	2.5	2.5
Cassava meal	5	30	30	5	30	5	10	10	15

Ewing (1998) as cited by EFSA (2007)

However, hydrocyanic acid poisoning of sorghum is rarely a serious problem if the crop is cut and wilted prior to feeding (EFSA 2007). The toxicity of cyanide from flax can largely be eliminated by soaking the meal in water for 24 h.

Drying considerably reduces the HCN level and sun drying has been shown to be more effective than oven drying (Tewe et al. 1980). In the recent decades, efforts focussed on the selection of cassava cultivars low in cyanogenic glycosides (EFSA 2007). The recommended inclusion rate of sorghum, linseed meal and cassava meal are present in Table 4.

2.3 *Alkaloids*

Alkaloids are a group of various organic compounds normally with basic chemical properties and usually containing at least one nitrogen atom in a heterocyclic ring (Bush and Fannin 2009). They are one of the most diverse groups of plant secondary metabolites. Many are toxic to humans and animals. Alkaloids are Ergot, Pyrrolizidine, or Tropane (Robert and Wink 1998). Ergot alkaloids are produced by several members within the fungal orders of Hypocreales and Eurotiales (EFSA 2012). Ergot alkaloids are fungal structures (called Ergots) from Claviceps species replace kernels on grain ears or seeds on grass heads three or four weeks following the invasion of the fungus and contain different classes of alkaloids. The most infected crop is rye in Europe, fiscue in the USA (Hannaway et al. 2009) and sorghum (Sorghum bicolor) in Africa, Central America, and South Asia (Bandyopadhyay et al. 1998). Continuous consumption of small quantities of the fungus on grass may lead to ergot toxicosis in ruminants (Strickland et al. 2011).

The pyrrolizidine alkaloids (PA) are a complex group of structurally related compounds that are composed of a necine base and one or two ester groups or a macrocyclic diester as stated by (Mulder et al. 2009). The PA's are toxins produced by plants as secondary metabolite to act as a defensive mechanism against herbivores. EFSA (2011) estimated that almost 6000 plant species, representing 3 % of all flowering plants, may contain pyrrolizidine alkaloids and that their contents vary from trace amounts up to 19 % based on dry weight. Ensiling plants PA's reduced to 4.5 % of the initial concentration while drying did not affect PA-containing plants (Gardner et al. 2006). Ruminant animals appear to be more resistant and tolerate to higher PA dosages (Wiedenfeld and Edgar 2011) while poultry are considered to be more sensitive (WHO-IPCS 1988).

Tropane alkaloids (TA's) are the third type of alkaloids and a kind of secondary metabolites which exist in several plant families. The most famous and important member of this type of alkaloids is cocaine and many other medicinal products (EFSA 2013). However, none of the TA's producing plants is cultivated for use as livestock feeds ((Panter 2005). The toxicosis associated with animal feeds is the result of contamination with foliage or seeds of the alkaloid-containing plants. However, TA-containing plant products appear to be unpalatable. Poultry are considerably less sensitive to TAs than other livestock due to the expression of specific hydrolysing enzymes that inactivate the alkaloids (EFSA 2013). Visual inspection of unground/uncrushed fruits for the presence of weed seeds is an accepted method and they can be removed by mechanical separation (List and Spencer 1976).

2.4 Nitrates

Nitrates are compounds that exist naturally in plants. Nitrates are used as fertilizers. Although nitrates poisoning in cattle occurred long before the use of nitrogen fertilizers, they can be fed if properly managed. Nitrate concentrations in feeds for livestock depend on plant species and environmental conditions. Nitrate itself is relatively nontoxic, but its metabolites, nitrite, nitric oxide, etc are poisonous. Rumen microorganisms can incorporate nitrates into microbial protein by converting the nitrates to ammonia. If the amount ingested is large enough, nitrates can not be converted completely to ammonia and toxic levels of nitrite are absorbed. The levels of nitrate nitrogen (feed total dry matter basis) of animal feeds as stated by Andrae (2008).

0–1000 ppm (level is safe to feed under all conditions),

1000–1500 ppm (level is safe for non-pregnant animals under all conditions. It may be best to limit its use to pregnant animals to 50 % of the total ration on a dry basis),

1500–2000 ppm (level is safe if limited to 50 % of ration's total dry matter),

2000–3500 ppm (feeds should be limited to 35–40 % of total dry matter in the ration. Over 2000 ppm nitrate nitrogen should not be used for pregnant animals), 3500–4000 ppm (feeds should be limited to 25 % of total dry matter in ration. Do not use for pregnant animals), and

>4000 ppm (Feeds containing over 4000 ppm nitrate nitrogen are potentially toxic. Do not feed)

2.5 *Others*

2.5.1 Oxalate

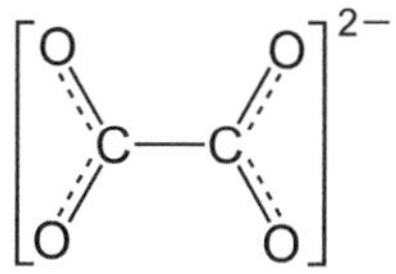

Oxalate is a common constituent of plants. Oxalates are poisoning when animals consume large quantities of oxalate containing plants without adaptation. If oxalates combine with calcium, they become insoluble (Gartner and Hurwood 1976). If this type of combination occurs in the rumen, insoluble oxalate may be excreted in the feces. If it occurs in the blood, the crystals may precipitate in the kidneys and can cause kidney failure (Lincoln and Black 1980). Soluble oxalate may be degraded by rumen microflora (Allison et al. 1977). The occurrence of insoluble oxalates leads to problems in calcium and phosphorus metabolism. The content of oxalate in forage can be controlled by certain agricultural practices (fertilizer application). For example nitrate application and increased rate of potassium application resulted in higher contents of oxalates while increased rate of calcium application soluble oxalate content showed a decreasing trend and insoluble oxalate content showed a reverse trend (Rahman and Kawamura 2011).

2.5.2 Lectins (Haemagglutinins)

Lectins are glycoproteins that have the ability to bind to carbohydrate-containing molecules which cause the agglutination of red blood cells as well as reduced growth, diarrhea, and interference with nutrient absorption. Lectin content in beans ranges from 1 to 3 %. In general, the lectins in common beans are highly toxic, while lectins in peas and faba beans appear to be the least toxic. Moist heat treatment will destroy much of the lectin present in grain legumes while dry heat treatment does not affect lectins that much (Anderson et al. 1992).

2.6 *Summary of Methods of Alleviating the Effects of PSM*

The greater demand for feed in most developing countries imposes certain pressures. Utilization of the marginal resources such as saline soils and underground water for producing feed for animals (Fodder crops) becomes necessary in order to improve nutritional status of these livestock. Some of these fodder crops are toxic. Toxicity may result from number of secondary metabolites. The consequences of toxic compounds in plants used as feedstuffs are not their direct toxicity to animals, but also the economic loss resulting from poisoning of domestic animals and the cost of preventing or reducing such compounds. The point sometimes is that ruminants, for instance, may convert a toxic compound to another toxic (cyanide to thiocyanate, which is goitrogenic) (Jones et al. 1997). The later is less toxic, yet still toxic. A single plant may contain two or more toxic compounds which add to the difficulties of detoxification (Soetan and Oyewole 2009). Some detoxification methods may result in losses of nutrients which in turn lowers the nutritive or feeding value of feeds. Cooking and/or germination treatments caused significant decreases in fat, total ash, carbohydrate fractions, anti-nutritional factors, minerals and B-vitamins of chickpeas (*Cicer arietinum* L.), (El-Adawy 2002). Yet, the germination treatment was less effective than cooking treatment in reducing trypsin inhibitor, hemagglutinin activity, tannins and saponins; it was more effective in reducing phytic acid, stachyose and raffinose. On the other hand, some detoxification treatments may improve the accessibility of nutrients by animals. Steam treatment, caused the PSM's to at least partially break down and some nutrients such as fats become better and, therefore, the nutritional value of the final animal feed increases (Van Bruggen et al. 1993).

Table 5 shows a summary of the most common plant secondary metabolites, their effect on animals and methods recognized so far to relief their effect on animals. Methods of lessening the effects of these compounds on animal performance include ensiling of the materials, yet it is not known yet whether the process of ensiling halophytes or the microbial community itself in the silage exert the effect of decreasing or demolishing the plant secondary metabolites.

Table 5 Plant secondary metabolites and their impact on animals and how to lessen them

Plant secondary metabolite	Impact on animal	Methods to relief
Phenolic compounds	Affect rumen fermentation	1- PEG
		2- Physical treatment
		3- Silage
Glycosides :	1- bloat	1- Rrepeated washing with water 2- Ensiling or wilting in the field
1. Saponins	2- Inhibit microbial fermentation	
	3- Formation of calcium salt	
	4- Decrease growth rate	
2. Cyanogens	Animal death due to its harmful on haemoglobin	1- Add methionine to animal diet (sulfur combines with cyanide to form thiocyanate (non toxic).
		2- sun drying
3. Goitergens	1- Enlargement of thyroid gland	Broken down in the rumen by rumen bacteria
	2- Rapid decline in serum thyroxine,	
	3- Decreased intake	
	4- Prolonged feeding has produced hair loss, excessive salivation and oesophageal lesions	

(continued)

Table 5 (continued)

Plant secondary metabolite	Impact on animal	Methods to relief
Alkaloids	1- ataxia	1- Air drying
	2- diarrhea	2- Ensiling
	3- decrease animal performance	
Nitrates	1- Inhıbıtıon of cellulose digestion	1- Add grains and vitamin A to the diet
	2- Combines with hemoglobin, thus reducing the oxygen	2- Mechanıcal treatment
	3- Hıgh nıtrates cause abortion in livestock	3- Add more soluble CHO to increase microbial nitrogen requirements
Oxalate	1- Excess oxalate may result in fatal intoxication with hypocalcaemia, metabolic disturbances and kidney failure	Anımal adadptaıon because rumen bacterıa can degrade ıt
	2- May result in fatal intoxication with hypocalcaemia, metabolic disturbances and kidney failure	
	3- Kıdney falıure due to the accumulatıon of oxlaıate crystals	
Phytates	Hypomagnesima (low WBC)	1- Mıneral balance
	Milk fever (decreased Ca & P)	2- Vıtamıne D ınjectıon
Tannins	1- Reduced voluntary feed intake	Add PEG
	2- Reduced digestibility of protein and carbohydrate through the inhibition of digestive enzymes	
	3- May reduce bacterial enzymes	
	4- Tannins/protein complex that survives in the ruminal environmental may not be digested in the lower tract	

References

Abd El Rahman HH (1996) Constraints and possibilities for their alleviation to improve utilization of desert natural range plants for grazing ruminants. Ph.D. thesis, Faculty of Agriculture Cairo University, Egypt

Abd El-Halim AM (2003) Studies of some anti-nutritional factors affecting forage utilization by ruminants. Ph.D. thesis, Faculty of Science Ain Shams University, Egypt

Acamovic T (2005) Biochemistry of plant secondary metabolites and their effects in animals. Symposium on 'plants as animal foods: a case of catch 2?' Symposium on 'plants as animal foods: a case of catch'. Proc Nutr Soc 64:403–412

Aengwanich W, Suttajit M, Srikhun T, Boonsorn T (2009) Antibiotic effect of polyphenolic compound extracted from Tamarind (Tamarindus indica L.) seed coat on productive performance of broilers. Int J Poult Sci 8(8):749–751

Agostini-Costa Tânia da S, Vieira RF, Bizzo HR, Silveira D, Gimenes MA (2012) Secondary metabolites, chromatography and its applications. In: Dr. Sasikumar Dhanarasu (eds) ISBN: 978-953-51-0357-8,InTech.Availablefrom:http://www.intechopen.com/books/chromatography-and-itsapplications/secondary-metabolites

Alama MR, Kabir AKMA, Amin MR, McNeill DM (2005) The effect of calcium hydroxide treatment on the nutritive and feeding value of *Albizia procera* for growing goats. Anim Feed Sci Technol 122:135–148

Alledrege J (1994) Effects of condensed tannins on browsers and grazers: qualitative and quantitative defense. Colorado State University, Fortcollins, pp 805–823

Allison MJ, Littledike ET, James LF (1977) Changes in ruminal oxalate degradation rates associated with adaptation of oxalate ingestion. J Anim Sci 45:1173–1179

Anderson-Hafermann JC, Zhang Y, Parsons CM (1992) Effect of heating on nutritional quality of conventional and Kunitz trypsin inhibitor-free soybeans. Poult Sci 71:1700–1709

Andrabi SM, Ritchie MM, Stimson C, Horadagoda A, Hyde M, McNeill DM (2005) In vivo assessment of the ability of condensed tannins to interfere with the digestibility of plant protein in sheep. Anim Feed Sci Technol 122:13–27

Andrae J (2008) Preventing prussic acid poisoning. Forage Leaflet, vol 13. Clemson University

Armstrong WD, Featherston WR, Rogler JC (1974) Effects of bird resistant sorghum grain and various commercial tannins on chick performance. Poult Sci 53:2137–2142

Attia-Ismail SA (2005) Factors limiting and methods of improving nutritive and feeding values of halophytes in arid, semi arid and coastal areas. In: International conference on biosaline agriculture & high salinity tolerance, Mugla, Turkey, 9–14 January 2005, pp 91–99

Balogun RO, Jones JJ, Holmes JH (1998) Digestibility of some tropical browse species varying in tannin content. Anim Feed Sci Technol 76:77–88

Bandyopadhyay R, Frederickson DE, McLaren NW, Odvody GN, Ryley MJ (1998) Ergot: a new disease threat to sorghum in the Americas and Australia. Plant Dis 82:356–367

Barry TN, Manley TR (1986) Interrelationships between the concentrations of total condensed tannins, free condensed tannins and lignin in Lotus sp. and their possible consequences in ruminant nutrition. J Sci Food Agric 37:248–254

Barry TN, McNeill DM, McNabb WC (2001) Plant secondary compounds; their impact forage nutritive value and upon animal production. In: Gomide JA, Mattos WRS, da Silva SC (eds) Proceedings of the XIX international grassland congress, Sao Paulo, Brazil, 2001/2002, pp 445–452

Ben Salem H, Saghrouni L, Nefzaoui A (2005) Attempts to deactivate tannins in fodder shrubs with physical and chemical treatments. Anim Feed Sci Technol 122:109–121

Bento MHL, Acamovic T, Makkar HPS (2005) The influence of tannin, pectin and polyethylene glycol on attachment of 15N-labelled rumen microorganisms to cellulose. Anim Feed Sci Technol 122(1–2):41–57

Brooker JD, Skene I, Clarke K, Blackall L, Muslera P, O'Donovan L (1994) Streptococcus caprinussp. nov. A tannin resistant ruminal bacterium from feral goats. Lett Appl Microbiol 18:313–318

Bush L, Fannin FF (2009) Alkaloids. In: Fribourg HA, Hannaway DB, West C. P (eds) Tall fescue for the twenty-first century. ASA, CSSA and SSSA, Madison, pp 229–249

Chanchay N, Poosaran N (2009) The reduction of mimosine and tannin contents in leaves of Leucaena leucocephala. Asian J Food Agric Ind 2009(Special Issue):S137–S144

Cunningham GM, Mulham WE, Milthorpe PL, Leigh JH (1981) Plants of Western New South Wales. Soil Conservation Service of N.S.W. and N.S.W. Government Printing Office, [no loc.]

Das TK, Banerjee1 D, Chakraborty D, Pakhira MC, Shrivastava B, Kuhad RC (2012) Saponin: role in animal system. Vet World 5(4):248–254

Dynes RA, Schlink AC (2002) Livestock potential of Australian species of Acacia. Conserv Sci W Aust 4(3):117–124

Edreva A, Velikova V, Tsonev T, Dagnon S, Gürel A, Aktaş L, Gesheva E (2008) Stress-protective role of secondary metabolites: diversity of functions and mechanisms. Gen Appl Plant Physiol Special Issue 34(1–2):67–78

EFSA (2007) Cyanogenic compounds as undesirable substance in animal feed scientific opinion of the panel on contaminants in the food chain. EFSA J 434:1–67

EFSA (2008) Gossypol as undesirable substance in animal feed scientific opinion of the panel on contaminants in the food chain. EFSA J 908:1–55

EFSA (2011) Scientific opinion on pyrrolizidine alkaloids in food and feed. EFSA J 9(11):2406

EFSA (2012) Scientific opinion on Ergot alkaloids in food and feed. EFSA J 10(7):2798

EFSA (2013) Scientific opinion on tropane alkaloids in food and feed scientific opinion on tropane alkaloids in food and feed. EFSA J 11(10):3386

Everist SL (1969) Use of fodder trees and shrubs. Queensland Department of Primary Industries Advisory Leaflet no. 1024

El-Adawy TA (2002) Nutritional composition and anti-nutritional factors of chickpeas (Cicer arietinum L.) undergoing different cooking methods and germination. Plant Foods Hum Nutr (formerly Qualitas Plantarum) 57(1):83–87

Ewing WN (1998) The feeds directory commodity products guide. Context, Ashby de la Zouch

Fahmy AA, Attia-Ismail SA, Afaf MF (2003) Effect of monensin on salt plant utilization and sheep performance. Egypt J Feeds Nutr 4:581

Flores MP, Castanan JI, Mcnabb JM (1994) The effect of tannin on starch digestibility and TMEN of triticale and semi purified starches form triticale, and field beans. Br Poultry Sci 34:281–286

Forbey JS, Harvey AL, Huffman MA, Provenza FD, Sullivan R, Tasdemir D (2009) Exploitation of secondary metabolites by animals: a response to homeostatic challenges. Integr Comp Biol 49:314–328

Gardner DR, Thorne MS, Molyneux RJ, Pfister JA, Seawright AA (2006) Pyrrolizidine alkaloids in Senecio madagascariensis from Australia and Hawaii and assessment of possible livestock poisoning. Biochem Syst Ecol 34:736–744

Gartner RJW, Hurwood IS (1976) The tannin and oxalic acid content of Acacia aneura (mulga) and their possible effects on suphur and calcium availability. Aust Vet J 52:194–195

Gu L, Kelm MA, Hammerstone JF, Beecher G, Holden J, Haytowitz D, Gebhardt S, Prior R (2004) Concentrations of proanthocyanidins in common foods and estimations of normal consumption. J Nutr 134:613–617

Hagerman AE, Butler LG (1981) The specificity of proanthocyanidin-protein interactions. J Biol Chem 256:4494

Hall N (1972) The use of trees and shrubs in the dry country of Australia. Australian Government Printer, Canberra

Hannaway DB, Daly C, Halbleib MD, James D, West CP, Volenec JJ, Chapman D, Li X, Cao W, Shen J, Shi X, Johnson S (2009) Development of suitability maps with examples for the United States and China. In: Fribourg HA, Hannaway DB, West CP (eds) Tall fescue for the twenty-first century.. ASA, CSSA, and SSSA, Madison, pp 33–47

Haslam E (1989) Plant polyphenols-vegetable tannins revisited. Cambridge University Press, Cambridge

Hassan IA, Elzuber EA, Tinay HA (2003) Growth and apparent absorption of minerals in broiler chicks fed diets with low or high tannin contents. Trop Anim Heal Prod 35:189–196

Huang KH, Li X, Ravindran V, Bryden WL (2006) Comparison of apparent ileal amino acid digestibility of feed ingredients measured with broilers, layers, and roosters. Poult Sci 85:625–634

Jansman AJM (1993) Tannins in feedstuffs for simple stomached animals. Nutr Res Rev 6:209–236

Jenkins KJ, Atwal AS (1994) Effects dietary saponins on fecal bile acids and neutral sterols, and availability of vitamins A and E in the chick. J Nutri Biochem 5:134–138

Jones RJ, Hegarty MP (1984) The effect of different proportions of Leucaena Leucocephala in the diet of cattle on growth, feed intake, thyroid function and urinary excretion of 3-hydroxy-4 LH pyridone. Aust J Agric Res 35:317

Jones TC, Hunt RC, King NW (1997) Veterinary pathology, 6th edn. Lippincott Williams and Wilkins, Baltimore

Kim HS, Miller DD (2005) Proline rich in proteins moderate the inhibitory effect of tea on iron absorption in rats. J Nutr 135:532–537

Lincoln SD, Black B (1980) Halogeton poisoning in range cattle. J Am Vet Med Assoc 176:717–718

List GR, Spencer GF (1976) Fate of Jimson weed seed alkaloids in soybean processing. J Am Oil Chem Soc 53:535–536

Longstaff MA, Mcnab JM (1991) The inhibitory effects of hull polysaccharides and tannins of filed beans (Vicia faba L.) on the digestion of amino acids, starch and lipid and on digestive enzyme activities in young chicks. Br J Nutr 65:199–216

Makkar HPS (2003) Effects and fate of tannins in ruminant animals, adaptation to tannins and strategies to overcome detrimental effects of feeding tannin rich feeds. Small Rumin Res 49:241–256

Mangan JL (1988) Nutritional effects of tannins in animal feeds. Nutr Res Rev 1:209–231

McLeod MN (1974) Plant tannins, their role in forage quality. Nutr Abstr Rev 44:803–815

McNabb WC, Waghorn GC, Peters JS, Barry TN (1996) The effect of condensed tannins in Lotus pedunculatus on the solubilization and degradation of ribulose-1 5-bisphosphate carboxylase (EC 4. 1. 1. 39; Rubisco) protein in the rumen and the sites of Rubisco digestion. Br J Nutr 76:535–549

McSweeney CS, Palmer B, McNeill DM, Krause DO (2001) Microbial interactions with tannins: nutritional consequences for ruminants. Anim Feed Sci Technol 91:83–93

Miniati E (2007) Assessment of phenolic compounds in biological samples. Ann Ist Super Sanita 43(4):362–368

Muetzel S, Becker K (2005) Extractability and biological activity of tannins from various tree leaves determined by chemical and biological assays as affected by drying procedure. Anim Feed Sci Technol 125:139–149

Mulder PPJ, Beumer B, Oosterink E, de Jong J (2009) Dutch survey pyrrolizidine alkaloids in animal forage. Analytical Services & Development, Veterinary Drugs, Institute of Food Safety Wageningen University & Research Centre. Wageningen University

Mupangwa JF, Acamovic T, Topps JH, Ngongoni NT, Hamudikuwanda H (2000) Content of soluble and bound condensed tannins of three tropical herbaceous forage legumes. Anim Feed Sci Technol 83:139–144

Nguyena TM, Binh DV, Ørskov ER (2005) Effect of foliages containing condensed tannins and on gastrointestinal parasites. Anim Feed Sci Technol 121:77–87

Niezen JH, Waghorn GC, Charleston WAG (1998) Establishment and fecundity of Ostertagia circumcincta and Trichostrongylus colubriformis in lambs fed lotus (Lotus pedunculatus) or perennial ryegrass (Lolium perenne). Vet Parasitol 78:13–21

Oda K, Matsuda H, Murakami T, Katayama S, Ohgitani T, Yoshikawa M (2003) Relationship between adjuvant activity and amphipathic structure of soyasaponins. Vaccine 21(17–18):2145–2151. doi:10.1016/S0264-410X (02)00739-9. Plohmann B, Bader

Oritz LT, Aizueta C, Tervino J, Castano M (1994) Effect of faba bean tannins on the growth and histological structure of the intestinal tract and liver of chicks and rats. Br Poultry Sci 35:743–754

Osuga IM, Abdulrazak1 SA, Ichinohe T, Fujihara T (2005) Chemical composition, degradation characteristics and effect of tannin on digestibility of some browse species from Kenya harvested during the wet season. Asian-Aust J Anim Sci 18(1):54–60

Palmer B, Jones RJ (2000) In vitro digestion studies using 14C-labelled polyethylene glycol (PEG): the effect of sample pretreatment on dry matter and nitrogen digestibility as well as PEG binding of Calliandra calothyrsus. Anim Feed Sci Technol 86:149–155

Panter KE (2005) Natural toxins of plant origin. In: Dabrowski WM, Sikorski ZE (eds) Toxins in food. CRC Press-Taylor/Francis Group, Boca Raton, pp 11–63

Potter SM, Jimenez-Flores R, Pollack J, Lone TA, Berber-Jimenez MD (1993) Protein 45. Saponin interaction and its influence on blood lipids. J Agric Food Chem 41:1287–1291

Rahman MM, Kawamura O (2011) Oxalate accumulation in forage plants: some agronomic, climatic and genetic aspects. Asian-Aust J Anim Sci 24(3):439–448

Rajput IZ, Song-hua HU, Chen-wen XIAO, Abdullah ARIJOG (2007) Adjuvant effects of saponins on animal immune responses. J Zhejiang Univ Sci 8(3):153–161
Rakhmani S, Brooker JD, Jones GP, Palmer B (2005) Composition of condensed tannins from Calliandra calothyrsus and correlation with in sacco digestibility. Anim Feed Sci Technol 121(1–2):109–124
Ram'ırez-Restrepo CA, Barry TN, L'opez-Villalobos N, Kempb PD, Harvey TG (2005) Use of lotus corniculatus containing condensed tannins to increase reproductive efficiency in ewes under commercial dryland farming conditions. Anim Feed Sci Technol 121:23–43
Rana KK, Wadhwa M, Bakshi MPS (2006) Seasonal variations in tannin profile of tree leaves. Asian-Aust J Anim Sci 19(8):1134–1138
Ravindran V, Hew LI, Ravindran G, Bryden WL (2005) Apparent ileal digestibility of amino acids in dietary ingredients for broiler chickens. Anim Sci 81:85–87
Reddy DV (2001) Principles of animal nutrition and feed technology. Oxford IBM Publishing Company Pvt. Ltd, New Delhi
Reed JD (1995) Nutritional toxicology of tannins and related polyphenols in forage legumes. J Anim Sci 73:1516–1528
Robert MF, Wink M (1998) Alkaloids: biochemistry, ecology and medicinal applications. Plenum Press, New York
Robins C, Brooker JD (2005) The effects of Acacia aneura feeding on abomasal and intestinal structure and function in sheep. Anim Feed Sci Technol 121:205–215
Rosselli M, Reinhart K, Imthurn B, Keller PJ, Dubey RK (2000) Cellular and biochemical mechanisms by which environmental oestrogens influence reproductive function. Hum Reprod Update 6(4):332–350
Rusdi (2004) In vitro studies of the interaction between selected proteins and condesed tannins. J Agroland 11(1):90–96
Salawu M, Acamovic T, Stewart CS, Hvelplund T, Weisbjerg MR (1999) The disappearance of dry matter, amino acids and proanthocyanidins in the gastrointestinal tract from different fractions of calliandra leaves. Anim Feed Sci Technol 79:289–300
Scanlan N, Skinner DC (2002) Estradiol modulation of growth hormone secretion in the ewe: no growth hormone-releasing hormone neurons and few somatotropes express estradiol receptor α. Biol Reprod 66(5):1267–1273
Shawkat SM, Khatab IM, Borhami BE, El-Shazly KA (2001) Performance of growing goats fed halophytic pasture with different energy sources. Egypt J Nutr Feeds 4:251–26
Silanikove N, Perevolotsky A, Provenza FD (2001) Use of tannin binding chemicals to assays for tannins and their negative postingestive effect in ruminants. Anim Feed Sci Technol 91:69–81
Smitha Patel PA, Alagundagi SC, Salakinkop SR (2013) The anti-nutritional factors in forages – a review. Curr Biotica 6(4):516–526
Soetan KO, Oyewole OE (2009) The need for adequate processing to reduce the antinutritional factors in plants used as human foods and animal feeds: a review. Af J Food Sci 3(9):223–232
Strickland JR, Looper ML, Matthews JC, Rosenkrans CFJ, Flythe MD, Brown KR (2011) Board invited review: St. Anthony's fire in livestock: causes, mechanisms, and potential solutions. J Anim Sci 89:1603–1626
Tewe OO, Gomez G, Maner JH (1980) Effect of extraneous linamarase on the hydrocyanic acid content of some tropical cassava varieties. Nigerian J Nutr Sci 1:27–32
Van Bruggen J, Veth P, Sebastiaan NL (1993) Method and device for reducing the amount of antinutritional factors in a mixture of raw material for animal feed. World Intellectual Property Organization (WO/1993/005664)
Velasquez-Pereira J, Risco CA, McDowell LR, Staples CR, Prichard D, Chenoweth PJ, Martin FG, Williams SN, Rojas LX, Calhoun MC, Wilkinson NS (1999) Long-term effects of feeding gossypol and vitamin E to dairy calves. J Dairy Sci 82(6):1240–1251
Vitti DMSS, Nozella EF, Abdalla AL, Bueno ICS, Silva Filho JC, Costa C, Buenod MS, Longo C, Vieira MEQ, Cabral Filho SLS, Godoy PB, Mueller-Harvey I (2005) The effect of drying and

urea treatment on nutritional and anti-nutritional components of browses collected during wet and dry seasons. Anim Feed Sci Technol 122:123–133

Waghorn GC (1990) Effects of condensed tannins on protein digestion and nutritive value of fresh herbage. Proc Aust Soc Anim Prod 18:412–415

Waghorn GC, Reed JD, Ndlovu L R (1999) Condensed tannins and herbivore nutrition. In: Proceedings of the 18th international grass congress, vol 111. International Grassland Congress

WHO-IPCS (World Health Organisation-International Programme on Chemical Safety) (1988) Pyrrolizidine alkaloids. Environmental health criteria, vol 80. WHO, Geneva, pp 1–345. Available from http://www.inchem.org/documents/ehc/ehc/ehc080.htm

Wiedenfeld H, Edgar J (2011) Toxicity of pyrrolizidine alkaloids to humans and ruminants. Phytochem Rev 10:137–151

Wina E, Tangendjaja B, Susana IWR (2005a) Effects of chopping, and soaking in water, hydrochloric acidic and calcium hydroxide solutions on the nutritional value of Acacia villosa for goats. Anim Feed Sci Technol 122:79–92

Wina E, Muetzel S, Becker K (2005b) The impact of saponins or saponin containing plant materials on ruminant production – a review. J Agric Food Chem 53:8093–8105

Wink M (ed) (1999a) Introduction: biochemistry, role and biotechnology of secondary metabolites. In Biochemistry of plant secondary metabolism annual plant reviews, vol. 2. CRC Press, Boca Raton, pp 1–15

Wink M (ed) (1999b) Introduction: biochemistry, role and biotechnology of secondary metabolites. In Biochemistry of plant secondary metabolism annual plant reviews, vol 3. CRC Press, Boca Raton, pp 1–16

Woclawek-Potocka I, Mannelli C, Boruszewska D, Kowalczyk-Zieba I, WaVniewski T, SkarHyNski DJ (2013) Diverse effects of phytoestrogens on the reproductive performance: cow as a model. Int J Endocrinol 2013. Article ID 650984, 15 pages. http://dx.doi.org/10.1155/2013/650984

Yildiz SI, Kaya Y, Unal DA, Elmali S, Kaya M, Cenesiz M, Kaya AO (2005) Digestion and body weight change in Tuj lambs receiving oak (Quercus hartwissiana) leaves with and without PEG. Anim Feed Sci Technol 122:159–172

Genotoxic Effects of Boron on Chickpea (*Cicer arietinum* L.) and Tomato (*Solanum lycopersicum* L.)

Guzin Tombuloglu, Huseyin Tombuloglu, and M. Serdal Sakcali

Contents

Abstract This study elucidates the genotoxic effect of boron (B) on chickpea and tomato. Experimental results have revealed a sharp growth rate inhibitions on plants (23 % chickpea; 31 % tomato), starting from 5 ppm. B-induced growth inhibition was confirmed by DNA alterations detected by RAPD profiles changes. DNA alteration was clear at the beginning from 10 ppm B in chickpea. Tomato, as a tolerant species, shows high genomic stability against to high B. These preliminary findings support the effective usage of RAPD-PCR in investigations of genotoxic effects of B for these crops and then for others.

Keywords RAPD-PCR • Genotoxicity • Boron • *Cicer arietinum* L • *Solanum lycopersicum* L • Growth rate

---

Dedicated to Dr. Irem Uzonur on her sad demise in 2013.

G. Tombuloglu (✉)
Vocational School of Medical Sciences, Fatih University, Istanbul 34500, Turkey
e-mail: gkekec@fatih.edu.tr

H. Tombuloglu (✉)
Department of Biology, Fatih University, Istanbul 34500, Turkey
e-mail: htombuloglu@fatih.edu.tr

M.S. Sakcali
Department of Biology, T. C Suleyman Demirel Universitesi, Isparta, Turkey

M. Öztürk et al. (eds.), *Plants, Pollutants and Remediation*,
DOI 10.1007/978-94-017-7194-8_9

1 Introduction

Boron (B) is an essential micronutrient for plants. The amount of B useful for plant growth varies between 0, 5–2 mg L^{-1} (Taiz and Zeiger 1991). Plant response to B is variable through the species even ecotypes (Ozturk et al. 2010). Either deficiencies or excessiveness causes many disorders. High level of B in the agricultural soils inhibits the crop productivity (Hale and Orcutt 1987; Nable et al. 1997; Cervilla et al. 2007), reduce vigor, slow development and growth (Cervilla et al. 2007). Reduced shoot and root elongation, chlorosis, and necrosis in leaves are characteristics of B toxicity (Ochiai et al. 2008). Necrosis of the leaf results in a loss of photosynthetic capacity, which reduces plant productivity (Lovatt and Dugger 1984). Pollen germination and pollen tube growth may also be inhibited (Versar, Inc. 1975).

Today, broad regions in the world contain high B where they reside particularly around B mining areas. South west of Anatolia (40,000 ha) and the western part of Australia (30 % of agricultural soil – 5 million ha) are the two remarkable examples for B contaminations (Kekec et al. 2010; Miwa et al. 2007). The level of B is much more than the threshold of normal plant growth (15 mg kg^{-1}) (Miwa et al. 2007). Accumulation of B in arid and semi-arid regions threats agricultural soils through the desalinized sea water, B carried by the winds (Versar, Inc 1975; Larsen 1988). Thus, B toxicity can easily be found with naturally in soils by exposing to the surface mining, fly ash or sewage sludge, irrigation water, and also industrial application (Nable et al. 1997). B-induced toxicity symptoms were extensively reported in these kinds of regions (Babaoglu et al. 2004; Türe and Bell 2004; Oz et al. 2009) that plants are suffering from high B and shows toxicity symptoms. High level of B is also affecting the antioxidant pathway (Keles et al. 2004) and chromosomal aberrations (Sakcali et al. 2015). In higher plants, the antioxidant reactions are critical defense systems against oxidative damage of reactive oxygen species.

In the last decade, some methods were developed to assess genotoxic effects of pollutants on plants, such as; comet, micronucleus, chromosome aberration assays, mitotic activity tests, RAPD (random amplified polymorphic DNA) – PCR (Steinkellner et al. 1999; Angelis et al. 2000; Reinecke and Reinecke 2004; Liu et al. 2005, 2007; Osman et al. 2008; Sakcali et al. 2015). Using sensitive but non-specific assays are suitable for the indication of DNA damage types related to monitoring of genotoxicity. RAPD-PCR is a widely applicable technique developed by Williams et al. (1990) and Welsh and McClelland (1990), that is capable of detecting variations in intensity as well as gain or loss of DNA bands following toxicant exposures that might be an indicator for DNA changes (Uzonur et al. 2004; Rong and Yin 2004; Liu et al. 2005; Atienzar and Jha 2006; Khan et al. 2013; Kekec and Cosgun 2015; Sakcali et al. 2015).

In the present study, we screened genome-wide DNA alterations in roots of chickpea (*Cicer arietinum* L. cv. Izmir-92) and tomato (*Solanum lycopersicum* L. cv. Menemen) plants exposed to various concentrations of B by using the RAPD-PCR method. It was aimed to investigate (1) Boron-induced genomic instability and (2) the correlation of RAPD profile change and the root growth inhibition on plants.

Our results suggest that B interfere the genomic instability under high level of B. And also, the growth rate inhibition is significantly correlated with the RAPD-PCR band profile that allowed us to use this method for genotoxicity monitoring for chickpea and tomato, may be for other crops for further studies.

2 Materials and Methods

The seeds of chickpea (*Cicer arietinum* L. cv. Izmir-92) and tomato (*Solanum lycopersicum* L. cv. Menemen) were surface-sterilized with 1 % sodium hypochlorite and placed in Petri dishes containing two layers of Whatman No 1 filter paper with test solutions and a control solution (distilled water). Boric acid (H_3BO_3) stock solution (150 ppm) was prepared and diluted to 5, 10, 25, 50, 100, 125 ppm concentrations with distilled water. Seeds were incubated in a climatic conditioner at 23 °C in the dark for 7 days. After 7 days of incubation, root and stem lengths of chickpea and tomato were measured and the inhibitory rates (%) of root growth were calculated using the following formula:

$$IR = \left(1 - \frac{x}{y}\right) x100$$

x: The average root lengths of control plants. y: The average root lengths of treated plants.

Experiments were followed by RAPD-PCR analysis of chickpea (*Cicer arietinum* L. cv. Izmir-92) and tomato (*Solanum lycopersicum* L. cv. Menemen). DNA isolation was performed by using the DNeasy Plant DNA Extraction Mini Kit (Qiagen) according to the supplier's instructions. DNA concentrations and sizes were measured by comparing with a standard sample (GeneRuler^{TM} 100 bp DNA Ladder, ready-to-use, MBI Fermentas) in a 2 % agarose gel. RAPD amplification was carried out in a 25 µl PCR mix, containing 1 x PCR buffer $(NH_4)_2SO_4$, 0.2 mM from each dNTP (2 mM dNTP mix), 25 pmol of primer OPA-08 5'CCACAGCAGT 3' (QIAGEN Operon RAPD® 10 mer Kits), 100 ng of genomic DNA, and 0.5 units of Taq DNA polymerase, filled up with sterile de-ionized water to the final volume. All PCR reagents were obtained from MBI Fermentas.

RAPD-PCR amplifications were performed by using Techne Endurance TC-512 Gradient Thermal Cycler, programmed for 3 min at 95 °C, followed by 45 cycles of 1 min at 94 °C, 1 min at 37 °C, 2 min at 72 °C, and 5 min at 72 °C as a final extension step.

RAPD-PCR amplification products were analyzed by 2 % agarose gel electrophoresis, stained with ethidium bromide and visualized under UV-light. Results were analyzed with GelDoc 2000 (Bio-RAD). Three replicates (i.e. nx = 3x) were prepared for each DNA sample for the confirmation of intra-individual variation in RAPD profiles. The prominent bands were detected for the monitoring of genetic profiles of each plant sample. The genomic instability and DNA variations were determined according to increase or decrease in band intensities, loss and gain of bands.

For each treatment and experiments three replicates were set up. Experimental results were statistically analyzed using *t*- tests.

3 Results and Discussion

Inhibitory rates of root growth were analyzed in chickpea and tomato in response to different B concentrations. In chickpea, inhibitory rate of root growth was decreased gradually from (–) 23 to (–) 60 % at 5, 10, 25 ppm B. Although it was calculated as 49 %, 20 %, 17 %, 42 % at 50, 100, 125, 150 ppm B exposures respectively ($p<0,05$) (Fig. 1). In tomato, inhibitory rate of root growth was decreased gradually at all concentrations from (–) 31 to (–) 92 % ($p<0.001$) (Fig. 2).

RAPD-PCR analysis was used to compare B exposed root DNA of chickpea and tomato. RAPD profile changes revealed that the DNA alterations; increase and decrease in band intensities, the loss and the gain of bands in B treated and the control plants. In chickpea, the molecular size of the bands obtained with OPA08 range from 193 to 976 bp and decreasing of band intensities were especially at 50 ppm B exposure (Table 1). Furthermore, the intensity of the RAPD band with 943 bp in molecular size substantially varied with increase of B concentration (Table 2). Whereas, intensity of RAPD band with approximately 516

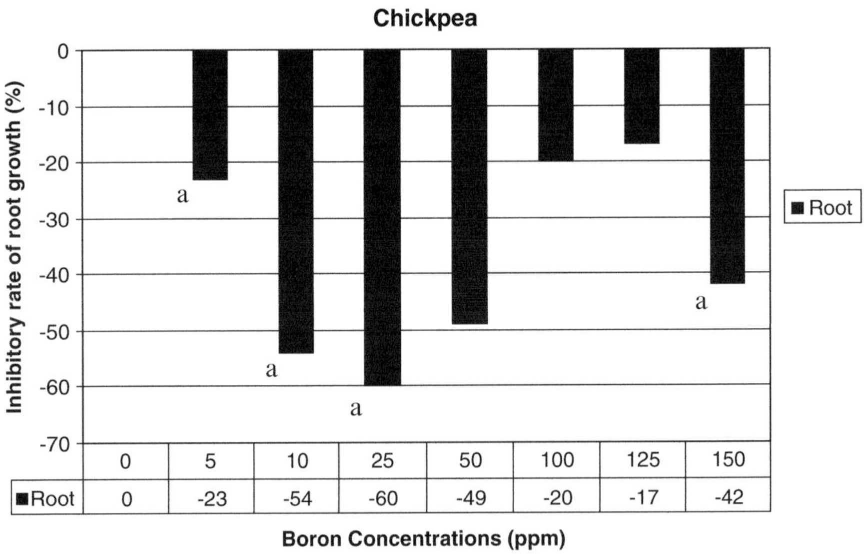

Boron (ppm)	0	5	10	25	50	100	125	150
■Root	0	-23	-54	-60	-49	-20	-17	-42

Fig. 1 Comparison of root growth inhibition in chickpea seedlings exposed to different B concentrations. [a]$p<0.05$, [b]$p<0.001$

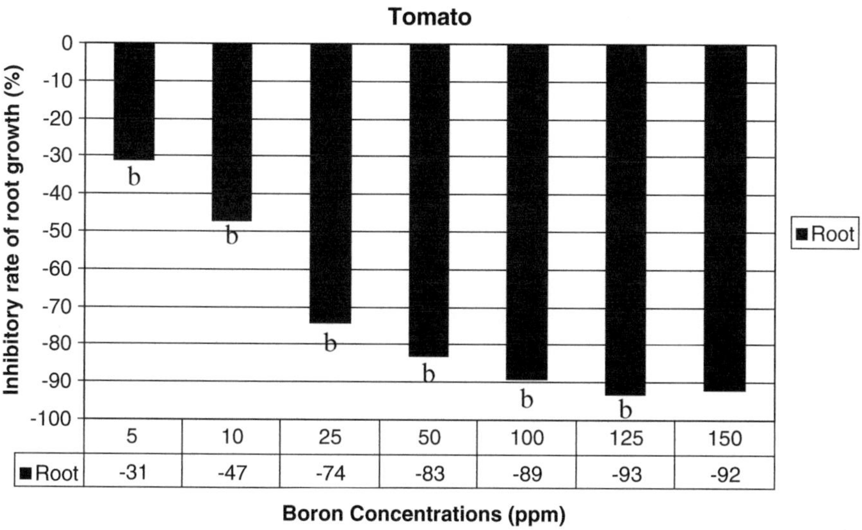

Fig. 2 Comparison of root growth inhibition in tomato seedlings exposed to different B concentrations ($^{a}p<0.05$, $^{b}p<0.001$)

Table 1 Changes of total bands, polymorphic bands and varied bands in B contaminated chickpea seedlings in comparison to the control

Primer: OPA08					
Boronconcentrations (ppm)		**Chickpea**			
	Total bands	**a**	**b**	**c**	**d**
Control	**8**				
5 ppm	**8**	–	–	–	**3**
10 ppm	**9**	**2**	**1**	**2**	**2**
25 ppm	**8**	–	–	**2**	–
50 ppm	**7**	–	**1**	**3**	–
100 ppm	**8**	**1**	**1**	**2**	**2**
125 ppm	**5**	–	**3**	–	–
150 ppm	**5**	–	**3**	**2**	–

a indicates appearance of new bands, **b** disappearance of normal bands, **c** decrease in band intensities and **d** increase in band intensities

bp substantially decreased with increase of B concentration. Extra bands of molecular sizes between approximately 210, 193 and 211 bp appeared at 10, 100 ppm B. At 10 and 50 ppm 643 bp; at 100 ppm 600 bp normal RAPD bands was disappeared. And the number of missing normal RAPD bands obviously increased with two highest concentration (125, 150 ppm); the bands of molecular sizes 361, 310, 236; 361, 310, 236 bp were disappeared for these concentrations respectively (Figs. 3a and 3b).

Table 2 Molecular sizes (bp) of appearing, disappearing bands, and bands of chickpea with changes in intensities

Primer: OPA08	**Chickpea**			
Boronconcentrations (ppm)				
	a	**b**	**c**	**d**
5 ppm	–	–	–	784, 680, 361
10 ppm	210, 193	643	700, 484	361, 236
25 ppm	–	–		–
50 ppm	–	643	516, 355, 283	–
100 ppm	211	600	921, 770	355, 223
125 ppm	–	361, 310, 236	–	–
150 ppm	–	361, 310, 236	784, 680	–

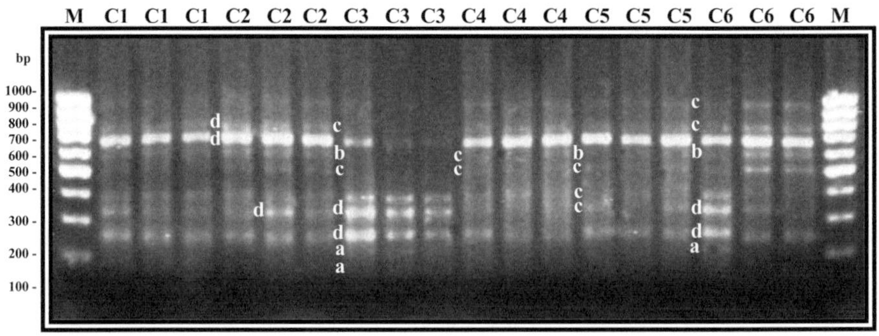

Fig. 3a RAPD profiles of genomic DNA from root-tips of chickpea seedlings exposed to 0, 5, 10, 25, 50, 100 ppm B

Fig. 3b RAPD profiles of genomic DNA from root-tips of chickpea seedlings exposed to 125 and 150 ppm B. *a*: indicates appearance of new bands, *b*: disappearance of normal bands, *c*: decrease in band intensities, *d*: increase in band intensities

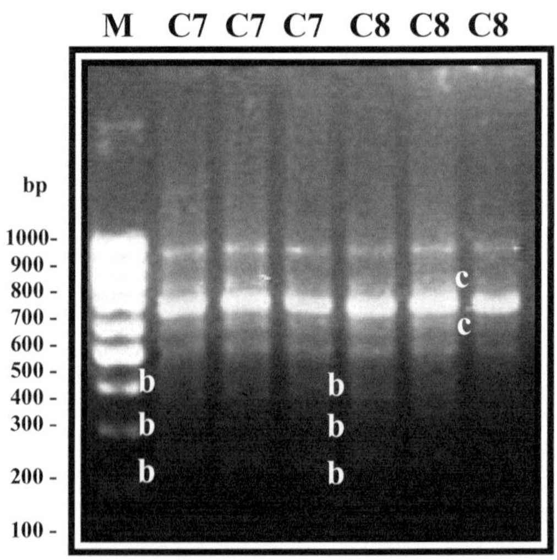

Table 3 Changes of total bands, polymorphic bands and varied bands in B contaminated tomato seedlings in comparison to the control

Primer: OPA08					
Boron concentrations (ppm)		**Tomato**			
	Total bands	a	b	c	d
Control	**8**				
5 ppm	**8**	–	–	**1**	–
10 ppm	**8**	–	–	**1**	–
25 ppm	**8**	–	–	**1**	–
50 ppm	**8**	–	–	**1**	–
100 ppm	**8**	–	–	**1**	–
125 ppm	**7**	–	–	**3**	**1**
150 ppm	**7**	–	–	**3**	**1**

a indicates appearance of new bands, **b** disappearance of normal bands, **c** decrease in band intensities and **d** increase in band intensities

Table 4 Molecular sizes (bp) of appearing, disappearing bands, and bands of tomato with changes in intensities

Primer: OPA08		**Tomato**		
Boronconcentrations (ppm)				
	a	b	c	d
5 ppm	–	–	359	–
10 ppm	–	–	346	–
25 ppm	–	–	440	–
50 ppm	–	–	346	–
100 ppm	–	–	340	–
125 ppm	–	–	428, 386, 332	254
150 ppm	–	–	428, 393, 337	254

In tomato, the molecular size of the bands obtained with OPA08 were ranged from 254 bp to 1216 bp and decreasing of band intensities were especially at 125 and 150 ppm B exposure (Table 3). Furthermore, intensity of the RAPD band with 643 bp in molecular size substantially varied with increase of B concentration (Table 4). Whereas, intensity of RAPD band with approximately 346 bp substantially decreased with increase of B concentration. In tomato there is no extra and disappearing band (Figs. 4a and 4b).

Boron toxicity is an important problem that can limit plant growth on soils of arid and semi arid environments throughout the world. Turkey has the 72 % of world B reserves in agronomical important areas. In this context, several studies have investigated the toxic effect of B with reference to the physiologic and metabolic defects, mitotic index, oxidative damage, transcript accumulation of stress-related genes, transcriptome analysis and DNA damage as well (Karabal et al. 2003; Papadakis et al. 2004; Molassiotis et al. 2006; Cervilla et al. 2007; Konuk et al. 2007; Kekec et al. 2010; Tombuloglu et al. 2012; Sakcali et al. 2015;

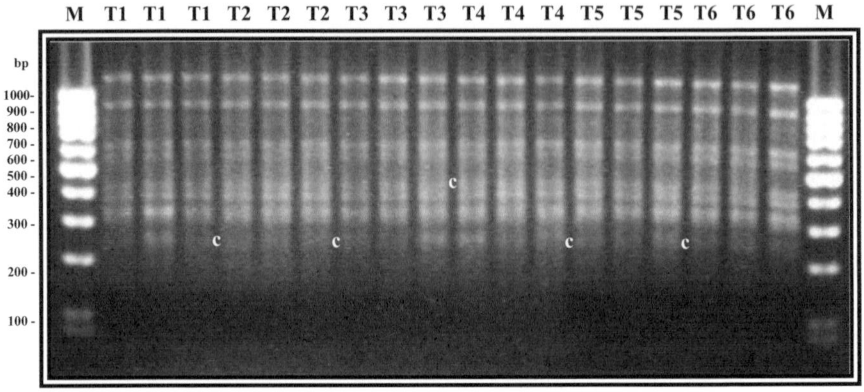

Fig. 4a RAPD profiles of genomic DNA from root-tips of tomato seedlings exposed to 0, 5, 10, 25, 50, 100 ppm B

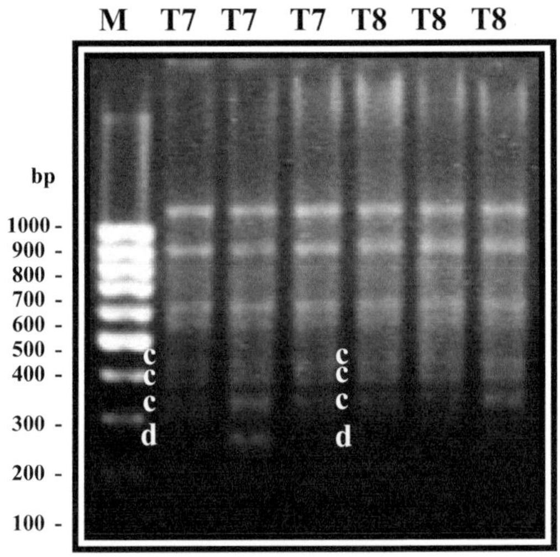

Fig. 4b RAPD profiles of genomic DNA from root-tips of tomato seedlings exposed to 125 and 150 ppm B. *a*: indicates appearance of new bands, *b*: disappearance of normal bands, *c*: decrease in band intensities, *d*: increase in band intensities

Tombuloglu et al. 2015). In addition to these, RAPD-PCR is a useful technique to detect genomic instability indicated by such as point mutations, genetic and chromosomal rearrangements, deletion and insertions. Toxicant-induced genotoxic effects, DNA variation, DNA damage, genetic instability and mutagenic effects have been evaluated successfully with RAPD analysis (Liu et al. 2007; Khan et al. 2013).

Recent studies have been reported that the antioxidant reactions are critical defense systems against oxidative damage of reactive oxygen species. There are strong reactive oxygen species, like superoxide ($O_2^{\cdot-}$) hydrogen peroxide (H_2O_2), hydroxyl radicals (OH^-) cause harmful effect on cell structure as lipid peroxidation, protein oxidation, and DNA damage (Gunes et al. 2006; Molassiotis et al. 2006; Gill and Tuteja 2010). So these findings were indicator of close relation between B toxicity induced oxidative damage and DNA damage. For this purpose, recent early studies demonstrated the genotoxic effect of excessive B on bean, wheat and maize (Cenkci et al. 2010; Kekec et al. 2010; Sakcali et al. 2015). In the present study, it was observed that RAPD band profile and intensities were altered conducted to B exposure. That would mainly cause gain or loss of bands. The appearances of new bands were detected at 10 and 100 ppm and disappearance of normal bands were detected at 10, 50, 100, 125 and 150 ppm B exposed chickpeas (Table 1). Appearance and disappearance of bands may be a result of the genomic instability related to DNA damage. However, there were no extra or missing bands in tomato, related to DNA damage (Table 3). According to physiological characterizations on tomato, it was considered as a B-tolerant plant (Maas 1990). RAPD profile changes confirmed B-tolerance with the occurrence of genomic stability on tomato. It revealed that no missing/extra band may infer to tolerance for plants. And also, the opposite is observable for chickpea that may imply to sensitiveness to B.

The current study has indicated the potential genotoxic effect of B in chickpea and tomato with considering the toxicity determination parameters; root growth inhibition, RAPD-PCR profile changes. These results may represent that short-term B exposure causes the specific alterations, as either increase or decrease of RAPD bands related to DNA damage. In conclusion, the present results recommend that RAPD-PCR method can be used as a useful marker for B-induced genomic alterations.

Acknowledgements This work was supported by the Scientific Research Fund of Fatih University under the project number P50030701.

References

Angelis KJ, McGuffie M, Menke M, Schubert I (2000) Adaptation to alkylation damage in DNA measured by the comet assay. Environ Mol Mutagen 36:146–150

Atienzar FA, Jha NA (2006) The random amplified polymorphic DNA (RAPD) assay and related techniques applied to genotoxicity and carcinogenesis studies: a critical review. Mutat Res 613:76–102

Babaoglu M, Gezgin S, Topal A, Sade B, Dural H (2004) *Gypsophila sphaerocephala* Fenzl ex Tchihat: a boron hyperaccumulator plant species that may phytoremediate soils with toxic B levels. Turk J Bot 28:273–278

Cenkci S, Cigerci İH, Yıldız M, Ozay C, Bozdag A, Terzi H (2010) Lead contamination reduces chlorophyll biosynthesis and genomic template stability in *Brassica rapa* L. Environ Exp Bot 67:467–473

Cervilla LM, Blasco B, Rios JJ, Romero L, Ruiz JM (2007) Oxidative stress and antioxidants in tomato (*Solanum lycopersicum*) plants subjected to boron toxicity. Ann Bot 100:747–756

Gill S, Tuteja N (2010) Reactive oxygen species and antioxidant machinery in abiotic stress tolerance in crop plants. Plant Physiol Biochem 48(12):909–930

Gunes A, Soylemezoglu G, Inal A, Bagcı EG, Coban S, Sahin O (2006) Antioxidant and stomatal responses of grapevine (Vitis vinifera L.) to boron toxicity. Sci Hortic 110:279–284

Hale MG, Orcutt DM (1987) The physiology of plants under stress. Wiley, New York

Karabal E, Yucel M, Oktem HA (2003) Antioxidant responses of tolerant and sensitive barley cultivars to boron toxicity. Plant Sci 164:925–933

Kekec G, Cosgun S (2015) Genotoxicity potentials of anionic and cationic amino acid-based surfactants. Toxicology and Industrial Health 31(4):377–385

Kekec G, Sakcali MS, Uzonur I (2010) Assesment of genotoxic effects of boron on wheat (*Triticum aestivum* L.) and bean (*Phaseolus vulgaris* L.) by using RAPD analysis. Bull Environ Contam Toxicol 84(6):759–764

Keles Y, Öncel I, Yenice N (2004) Relationship between boron content and antioxidant compounds in Citrus leaves taken from fields with different water source. Plant and Soil 265:345–353

Khan F, Hakeem KR, Sidiqqui TO, Ahmad A (2013) *RAPD* markers associated with salt tolerance in soybean genotypes under salt stress. Appl Biochem Biotechnol 170(2):257–272

Konuk M, Liman R, Cigerci İH (2007) Determination of genotoxic effect of boron on *Allium cepa* root meristematic cells. Pak J Bot 39:73–79

Larsen LA (1988) Boron. In: Seiler HG, Sigel H (eds) Handbook on toxicity of inorganic compounds. Marcel Dekker, New York/Basel, pp 129–141

Liu W, Li P, Qi X, Zhou Q, Sun T, Yang Y (2005) DNA changes in barley (*Hordeum vulgare*) seedlings induced by cadmium pollution using RAPD analysis. Chemosphere 61:158–167

Liu W, Yang Y, Zhou Q, Xie L, Li P, Sun T (2007) Impact assessment of cadmium contamination on rice (*Oryza sativa* L.) seedlings at molecular and population levels using multiple biomarkers. Chemosphere 67:1155–1163

Lovatt CJ, Dugger WM (1984) Boron. Biochemistry of the essential ultratrace elements. Plenum Press, New York, pp 389–421

Maas EV (1990) Crop salt tolerance. In: Tanji KK (ed) Agricultural salinity assessment and management. ASCE manual & reports on engineering practice, No. 71, New York, pp 262–304

Miwa K, Takano J, Omori H, Seki M, Shinozuka K, Fujiwara T (2007) Plants tolerant of high boron levels. Science 318:1417

Molassiotis A, Sotiropoulos T, Tanou G, Diamantidis G, Therios I (2006) Boron induced oxidative damage and antioxidant and nucleolytic responses in shoot tips culture of the apple rootstock EM9 (*Malus domestica* Borkh). Environ Exp Bot 56:54–62

Nable RO, Bañuelos GS, Paull JG (1997) Boron toxicity. Plant Soil 198:181–198

Ochiai K, Uemura S, Shimizu A, Okumoto Y, Matoh T (2008) Boron toxicity in rice (Oryza sativa L.). I. Quantitative trait locus (QTL) analysis of tolerance to boron toxicity. Theor Appl Genet 117:125–133

Osman GMA, Mekkawy AI, Verreth J, Wuertz S, Kloas W, Kirschbaum F (2008) Monitoring of DNA breakage in embryonic stages of the African Catfish *Clarias gariepinus* (Burchell, 1822) after exposure to lead nitrate using alkaline comet assay. Environ Toxicol 23:679–687

Oz MT, Yilmaz R, Eyidogan F, de Graff L, Yücel M, Oktem HA (2009) Microarray analysis of late response to boron toxicity in barley (*Hordeum vulgare* L.) leaves. Turk J Agric For 33:191–202

Ozturk M, Sakcali S, Gucel S, Tombuloğlu H (2010) Boron and plants. In: Ashraf et al. (eds) Plant adaptation & phytoremediation. Springer, The Netherlands, pp 275–311

Papadakis I, Dimassi A, Bosabalidis I, Therios A, Patakas, Giannakoula A (2004) Boron toxicity in 'Clementine' mandarin plants grafted on two rootstocks. Plant Sci 166:539–547

Reinecke SA, Reinecke AJ (2004) The comet assay as biomarker of heavy metal genotoxicity in earthworms. Arch Environ Contam Toxicol 46(2):208–215

Rong ZY, Yin HW (2004) A method for genotoxicity detection using random amplified polymorphism DNA with Danio rerio. Ecotoxicol Environ Saf 58:96–103

Sakcali MS, Kekec G, Uzonur I, Alpsoy L, Tombuloglu H (2015) Randomly amplified polymorphic-DNA analysis for detecting genotoxic effects of Boron on maize (Zea mays L.). Toxicology and Industrial Health 31(8):712–720

Steinkellner H, Kassie F, Knasmuller S (1999) Tradescantia-micronucleus assay for the assessment of the clastogenicity of Austrian water. Mutat Res 426:113–116

Taiz L, Zeiger E (1991) Plant physiol. The Benjamin/Cummings Publishing Company Inc., Redwood City

Tombuloglu H, Semizoglu N, Sakcali MS, Kekec G (2012) Boron induced expression of some stress-related genes in tomato. Chemosphere 86:433–438

Tombuloglu G, Tombuloglu H, Sakcali MS, Unver T (2015) High-throughput transcriptome analysis of barley (Hordeum vulgare) exposed to excessive boron. Gene 557:71–81

Türe C, Bell WR (2004) Plant distribution and its relationship to extractable boron in naturally-occurring high boron soils in Turkey. Israil J Plant Sci 52:125–132

Uzonur I, Abasiyanik MF, Bostanci B, Eyidemir M, Ocba N, Yanik C, Petek M (2004) Re-exploring planaria as a model organism for genotoxicity monitoring by an 'improved random amplified polymorphic DNA' approach. Fresenius Environ Bull 13:1420–1426

Versar Inc (1975) Preliminary investigation of effects on the environment of boron, indium, nickel, selenium, tin, vanadium and their compounds, volume 1: boron. Springfield, Virginia, Versar, Inc. (Prepared for the US Environmental Protection Agency, Washington, DC) (EPA 560/2-75-005a)

Welsh J, McClelland M (1990) Genomic fingerprinting using arbitrarily primed PCR and a matrix of pairwise combinations of primers. Nucleic Acid Res 18:7213–7218

Williams J, Kubelik AR, Livak KJ (1990) DNA polymorphisms amplified by arbitrary primers are useful as genetic markers. Nucleic Acid Res 18:6531–6535

The Influence of Some Particular Biotic and Abiotic Factors on Distribution of Metal Concentrations in the *Soil–Pine* System

Edita Baltrenaite, Arvydas Lietuvninkas, Pranas Baltrenas, Bal Ram Singh, Nina Moskvitina, and Olga Vaishlya

Contents

E. Baltrenaite (✉) • P. Baltrenas
Vilnius Gediminas Technical University, Lithuania, Saulėtekio Av. 11, LT-10223 Vilnius, Lithuania

Department of Environmental Protection, Vilnius Gediminas Technical University, Lithuania, Saulėtekio Av. 11, LT-10223 Vilnius, Lithuania
e-mail: edita.baltrenaite@vgtu.lt

A. Lietuvninkas
Department of Mineralogy and Geochemistry, Tomsk State University, Lenin Av. 36, Tomsk 634050, Tomsk, Russian Federation

B.R. Singh
Department of Environmental Sciences, Norwegian University of Life Sciences, P. O. Box 5003, N-1432 Ås, Norway

N. Moskvitina • O. Vaishlya
Department of Zoology of Vertebrates and Ecology, Institute of Biology, Soil Sciences, Ecology, Forestry and Agriculture, Tomsk State University, Lenin Av. 36, Tomsk 634050, Tomsk, Russian Federation

M. Öztürk et al. (eds.), *Plants, Pollutants and Remediation*,
DOI 10.1007/978-94-017-7194-8_10

Abstract All living systems, including the *soil–pine* system, are affected by some external and internal factors. The balance of a system depends on the nature of the influencing factors and the extent of their impact. When the balance is disturbed, the quantitative and qualitative changes in the chemical composition and the functions of a tree can be observed. Using the method of dynamic factors it has been determined that a biotic factor (a pathogen) stimulates biophilicity and bioaccumulation of macroelements in the pine tissue, whereas an abiotic factor (pollution) intensifies biophilicity and bioaccumulation of microelements. The external abiotic factor (pollution) has a stronger influence on the linear logarithmic distribution of metal concentrations in the soil–pine system than the external biotic factor (a pathogen). The strongest effect is produced on the balance between metal concentrations in pine tissue and the concentration of metals in the mobile form in soil.

Keywords Abiotic factor • Bioaccumulation • Biophilicity • Biotic factor • Dynamic factor of bioavailability • *Heterobasidion annosum* • Metals • *Pinus sylvestris* L • Uptake

1 Introduction

The condition of a tree as an organism, as well as its existence and development, directly depends on the surrounding environment. Depending on the type of the influencing factors and the extent of their impact, the balance between a tree and its environment, as well as the metabolism of materials and energy change vary, which, in turn, causes some relevant changes in the tree. The factors that cause the changes are commonly divided at least into two groups. The division into abiotic and biotic factors is based on the effect produced by the components of living and non-living nature. The division of factors into physiological and ecological ones is based on their general classification into internal and external factors, respectively (Lietuvninkas 2012). The insufficient or too strong impact of the factors of both types on trees disrupts physiological processes proceeding in them. The metals (Pb, Zn, Cu, Cd, Mg and K) actively participate in physiological processes of a tree. For example, potassium (K) regulates the opening of stomatal pores of the leaf, thereby ensuring photosynthesis. The opening of stomatal pores depends on bioavailable quantity of K, which accelerates the metabolism of gases (Noland and Kozlowski 1979). Due to air pollution, airborne particulates, which absorb metals, damage leaves and slow down photosynthesis (Reich 1987). The deficiency of metals, such as K, Mg and Ca, leads to the reduction of the amount of epicuticular wax, which limits the loss of transpirational water, controls the metabolism of gases, retains minerals in a plant, acts as a barrier to air pollutants and helps agricultural chemicals to get into leaves, fruit and stems (Kozlowski and Pallardy 1997). Most of the enzymes relevant for tree respiration, which are the catalysts of organic reactions in trees, can be active only in the presence of the ions of Mg, Mn, Cu, Ca, Zn and K (Markert et al. 2012, 2014; Youssef et al. 2014). Air pollutants can damage the wax layer of a leaf and increase the amount of the washed out materials.

It is interesting to note that the transformations caused by biotic and abiotic factors in the tree organism can vary, when the vector of an impact is changed. A question arises what changes take place in the *soil–tree* system under the influence of biotic and abiotic factors in terms of the composition of micro- and macroelements.

The Scots pines (*Pinus sylvestris* L.) were chosen for the investigation as the most widely spread tree species among their congeners (Navasaitis 2008). The environmental (external) factors causing a stress in a common pine, which are associated with a pathogen causing various diseases in a tree (a biotic factor) and aerogenic pollution with metals (an abiotic factor) were considered. According to Schulze et al. (2005), the first factor is referred to stress causing factors of living nature while the second one refers to the factors of non-living nature. Both factors produce a harmful effect (stress) on plants while this, which causes diseases can also cause their death (Kupčinskienė 2011).

The fungus *Heterobasidion annosum* that infects and damages the root system of pines is one of the main types of fungi damaging coniferous woods in Europe, Asia and North America. It grows both in coniferous and mixed forests, but can, usually, be found in wet and dark fir and pine forests. This fungus grows on the roots of coniferous trees, as well as at the base of the stem, on stumps, and sometimes, near the thick end of the lying stems. It is a very dangerous fungus, which first appears on the roots of living trees and later passes to their stems. The damaged trees perish and fall down (Gricius et al. 1999; Deacon 2005). The amount of wood lost in Europe each year due to the growth of this fungus is worth EUR 800 million (Woodward et al. 1998).

Pollution with heavy metals of aerogenic origin produced by anthropogenic source was selected for analysis, because technogenesis leads to the increase in technophilicity of metals (i.e. the increase in the use of metals with respect to metal concentrations in lithosphere), which has a negative effect on the forest ecosystem.

This work aims to assess the influence of biotic and abiotic factors (one of each type) on the concentration of microelements and macroelements in the *soil–pine* system and to determine the changes caused by them.

2 Research Methodology

2.1 The Investigated Territories

An experimental woodland (54°53′12″ N, 24°04′33″ S), which had previously been used for agricultural needs, has been selected as the territory, where pines affected by the biotic factor grow. The territory was forested in 1959–1963, when it had lost agricultural productivity. Scots pines (*Pinus sylvestris* L.) prevail on the investigated territory. Pines affected by the biotic factor (*Heterobasidion annosum*) were identified by some specific features. For example, the crowns of pines damaged by the fungus were particularly poor and uneven, while their upper branches were obviously dying off (Fig. 1a). The trees had yellow needles, and mycelia could be observed on their roots (Fig. 1b).

The territory in Panevėžys locates nearby the plant formerly producing TV sets has been selected as the territory covered with pines affected by the abiotic factor (55°44′04″N, 24°23′30″E). The company had been producing glass components and tubes for colour TV sets for almost 40 years. Since 1962, it had been the only plant producing TV sets in the Baltic States (with the territory of about 9000 m^2). The plant was closed in 2007. Emissions of various substances into the ambient air used to amount to 766 t per year, including 553 t of nitric oxides, 155 t of carbon monoxide, 46 t of volatile organic compounds and about 15 t of particulate matter. As for trace elements, Zn, Pb, Cr, Mn, Co, Sn, Mo, Ni, Ag, As and V are characteristic of the emissions in this plant. According to the data provided by the Environmental Protection Agency of Lithuania, in 2004, the company was among the largest polluters in Lithuania. Heavy aerogenic pollution with Pb, Cu, Cd and Zn negatively affected the geo-sanitary conditions of the environment (Kadūnas and Radzevičius 2001). The territory covered with pines, which was affected by the abiotic factor was located 200 m away from the anthropogenic source (a plant). The pines growing on the lee side of the contaminated territory with respect to the enterprise were examined. The southwest winds prevailed in the territory.

Fig. 1 Pines with root systems damaged by *Heterobasidion annosum*: (**a**) the dying off crown of a pine, (**b**, **c**) mycelia

The selected control territory (54°53′12″ N, 24°04′33″ S), where Scots pines prevail, was not far from the territory covered with pines affected by the biotic factor. The control territory was outside the range of influence of the sources of pollution because it was located on the outskirts of the city and was surrounded by a forest. The closest traffic road was more than 300 m away from it. Potential emitters of pollutants and shielding objects were more than 1 km away from the control territory and the territory affected by a biotic factor, and, therefore, the requirements to the selection of a sampling territory (Markert 1996; Youssef et al. 2014) were satisfied. The examined territories have similar microclimate and type of soil (gleyic, podsols) which allow us to state that, at the level of the ecosystem, the edaphic conditions associated with the 'soil–pine' system are mainly affected by the considered abiotic and biotic factors.

2.2 *Wood and Soil Sampling*

The sampling procedure was performed in May, at the beginning of the vegetation period of trees, when physiological processes in trees were most active (Hill 2002). Ten pines of similar age (about 30 years, 45–53 cm in diameter at the chest height), growing at a distance of more than 10 m from each other, were randomly selected for sampling in each investigated territory and for analysis. Only pines without noticeable signs of disease were used for sampling in the control territory and the territory affected by an abiotic factor (pollution). The pine age was preliminary assessed based on the number of annual rings and the available information about the investigated territory.

Wood samples were taken from 3 to 2 bores made in the trunk of each pine at a height of 1.5 m with the Pressler's increment borer (12 mm in diameter) (Baltrenaite et al. 2010). The external and internal bark layers were removed from the samples. Then, the samples were immediately placed into dry plastic bags, marked, sealed and left for the analysis in the laboratory.

Two composite samples of soil (2 kg of each) were taken near each pine at a depth of 0–40 cm, using a soil drill. The soil was sampled under the foliage of pines. Then, they were immediately placed into dry plastic bags, marked, sealed and left for the analysis in the laboratory.

2.3 *Physical and Chemical Preparation and Analysis of Samples*

The soil samples were dried at room temperature in the laboratory for 24 h and then passed through a sieve of 2-mm mesh. Each sample of 0.2 g was mineralized in the Milestone ETHOS mineralizer. The wood samples were dried at room temperature and then incinerated for 45 min in E5CK-T muffle furnace (450 °C) until they were ashed. The ash samples of 0.2 g were mineralized in the Milestone ETHOS mineralizer.

The concentrations of the mobile forms of metals were determined in the water extract of 20 g soil in 100 ml of deionised water. Soil-water samples were shaken in *Gerhardt*, *Rotoshake* RS 12 for an hour and left to settle. Then, the extracts were collected by filtration through filter paper and analyzed (Pundytė et al. 2011; Mancinelli et al. 2015).

The concentrations of metals in the samples were analyzed by the Buck Scientific 210 VGP atomic absorption spectrophotometer, using flame atomic absorption spectroscopy. When the metal concentrations were low, the graphite furnace atomic absorption spectrometry was used (Pundytė et al. 2011; Baltrenaite et al. 2014).

2.4 Dynamic Factors of Bioaccumulation, Biophilicity and Bioavailability

The comparative analysis of metal concentrations in plants can be performed at various levels: (a) direct comparison of their concentrations in biomass which is incorrect because the respective concentrations of metals in soil where the plants grow are not accounted for; (b) coefficients of bioconcentration (bioaccumulation) (in this case the effect produced by lower of higher concentration of metals in the soil on their uptake and accumulation in plants is not taken into account); (c) dynamic factors of bioaccumulation, taking into account the impact of metal concentration and various other factors (including biotic and abiotic as well as internal and external factors) on the soil of the control and affected territories and on uptake and accumulation of metals in the plants, growing on these territories. The concept of dynamic factors is wider discussed in the work of Baltrenaite et al. (2012) and the application of the dynamic factors is described in the work of Baltrenaite et al. (2015, 2016).

The methodology of calculating the *dynamic factors*,[1,2] was applied to determine the differences in bioaccumulation, biophilicity and bioavailability processes in the control territory and the territories which were affected by a biotic or an abiotic factor.

The dynamic factor referring to bioaccumulation (BA_{dyn}) is associated with the variation of bioaccumulation of a chemical element in the plants that grow in the soil affected by an abiotic or a biotic factor and in the soil in the control territory. The dynamic factor of biophilicity (BF_{dyn}) reflects the variation of the involvement of chemical elements in the metabolism reactions of the plants that grow in the soil affected by an abiotic or a biotic factor and in the soil in the control territory. In general, biophilicity of a chemical element is the ratio of accumulation of a chemical element in living biomass to its concentration in the Earth's crust.

---

[1] The word "factor", meaning a circumstance, fact or influence that contributes to a result or outcome, should not be confused with 'factor' as a mathematical term, meaning a coefficient (as used to define "dynamic factor").

[2] The factors are called dynamic because they reflect the dynamics of the processes taking place in trees and soil.

The dynamic factor associated with metal bioavailability (BIO_{dyn}) was calculated by Eq. (1). It is used to express the variation of the mobile portion of a chemical element in soil (thus, partially, it is the bioavailable portion of a chemical element). In the case of a dynamic factor, the variation of the mobile portion of a chemical element is expressed by the relationship between its concentration and the total concentration of a chemical element in the soil affected by an abiotic or a biotic factor and compared to the variation in the soil of the control territory:

$$BIO_{dyn} = \frac{C^{i}_{bio\_treated}}{C^{i}_{tot\_treated}} \times \frac{C^{i}_{tot\_control}}{C^{i}_{bio\_control}}, \quad (1)$$

where $C^{i}_{bio\_treated}$ is the concentration of chemical element i in the mobile form in the soil of the territory where a tree affected by a biotic or an abiotic factor grows, mg/kg DW; $C^{i}_{tot\_treated}$ is the total concentration of chemical element i in the soil of the territory, where a tree affected by a biotic or an abiotic factor grows, mg/kg DW; $C^{i}_{tot\_control}$ is the total concentration of chemical element i in the soil of the territory, where the control tree grows, mg/kg DW; $C^{i}_{bio\_control}$ is the concentration of the mobile form of chemical element i in the soil of the territory where the control tree grows, mg/kg DW.

2.5 *The Logarithmic Scale of Correlation*

The logarithmic scale was used to evaluate the correlation between metals in different components of the soil–pine system (Chvastov et al. 2011). The logarithmic scale was used in evaluating the correlation with the aim that the metals with high (e.g. K, Mg) and relatively low (e.g. Pb) concentrations could be compared at a similar level.

2.6 *The Coefficient of Concentration*

The coefficient of concentration is an ecogeochemical factor, allowing for quantitative evaluation of the technogenesis impact on a particular component of nature. It can be calculated for any dangerous material found in any landscape component (e.g. air, snow cover, soil, water, bottom sediment, plants, animals, etc.) by Eg. (2) (Lietuvninkas 2012; Baltrenaite et al. 2014):

$$K_{k,i} = \frac{C_i}{C_f}, \quad (2)$$

where $K_{k,i}$ is the coefficient of metal *i* concentration in the soil; C_i is metal *i* concentration in the considered soil, mg/kg; C_f is metal *i* concentration in the soil of the control territory, mg/kg.

2.7 Processing the Statistical Data and Ensuring the Research Quality

The samples were prepared in duplicates. Five blank samples were prepared for each test. Certified reference materials, BAM-U110 (contaminated soil-trace elements) and ERM®-CD100 (trace elements and pentachlorophenol (PCP)) were used to ensure the accuracy of analysis. To obtain the calibration curve, standard solutions were used. The statistical data analysis was performed by using *Excel* and *STATISTICA* 8.0 programs. The arrays of the initial data were evaluated using a 3D criterion. The tables present the average values of concentrations along with standard deviations. The values of the dynamic factors were calculated based on the mean values of the variables, which were also used for the logarithmic scale of correlation.

3 Results

3.1 Concentrations of Metals in the Investigated Soils

Concentrations of metals in the investigated soils, where pines affected by the biotic and abiotic factors and the control pines grew, are presented in Table 1.

Table 1 The total concentration of metals (the mean value ± SD) in various soils (mg/kg DW) and metal concentration coefficients, $K_{k,i}$ (in brackets)

Investigated territories	Zn	Cu	Mg	K	Pb	Cd
The control territory	30.2 ± 18.2	9.58 ± 0.97	497 ± 89	1282 ± 486	3.82 ± 0.63	0.25 ± 0.01
The territory covered with infected pines	15.8 ± 5.6 (0.52)	5.82 ± 0.50 (0.61)	450 ± 75 (0.91)	826 ± 271 (0.64)	3.37 ± 1.65 (0.88)	0.04 ± 0.02 (0.16)
The contaminated territory	36.9 ± 13.6 (1.22)	23.2 ± 3.4 (2.42)	498 ± 389 (1.00)	1107 ± 320 (0.86)	25.6 ± 10.1 (6.70)	0.28 ± 0.04 (1.12)

Table 2 The ratio between the concentrations of metals in mobile forms to their total concentrations, % (the largest values for each metal are italicized)

Investigated territories	Zn	Cu	Mg	K	Pb	Cd
The control territory	0.73	*0.80*	0.48	0.61	0.20	1.60
The territory covered with infected pines	*2.92*	0.02	0.54	1.62	*0.59*	2.50
The contaminated territory	1.39	0.31	*1.70*	*2.21*	0.46	*3.27*

As shown by the data in Table 1, the total concentration of metals in the soil of the contaminated territory was much higher than that in the control territory. Thus, the concentration coefficients of all microelements were considerably larger than 1.0. The increase in the concentrations of Cu and Pb on the contaminated territory was particularly high. Thus, their values were 2.42 and 6.70 times higher than the control values. It is clear that the increase in metal concentrations in the soil was caused by aerogenic emissions of metals from industrial enterprises located nearby. The concentrations of metals in the soil of the territory where infected pines grew were lower or close to the concentrations of metals in the soil of the control territory, with all concentration coefficients being less than 1.0 (Table 1).

Table 2 presents the concentration of metals in the mobile form versus their total concentration. The concentration values of such metals in the mobile form as Mg, K and Cd were larger, when the pH values of the soil were higher. The mobility of most of the analysed metals (except for Cu) in the soil of the polluted territory was higher than their mobility in the soil of the control territory. An increase in the amounts of all metals in the mobile form (except for Cu) in the soil of the territory covered with the infected pines could be observed compared to that in the control territory. The comparison of the mobility of Zn and Cd, as geochemically similar chemical elements, has confirmed (Table 2) that as shown by Alloway (1995), in the acid environment, Cd is more mobile that Zn. The soil in the territory covered with the infected pines is an exception because, in this case, the mobile Zn portion exceeds the mobile Cd portion by 17 %.

3.2 *The Variation of Metal Bioavailability in Different Types of Soil*

In the soil in the territories covered with pines affected by the abiotic and biotic factors, the values of the dynamic bioavailability factor for the considered metals (except for Cu) were higher than those found in the control territory (Fig. 2). Based on this characteristic, it is possible to forecast the change (increase) in the amount of metals bioavailable in trees due to the influence of the external environmental factors.

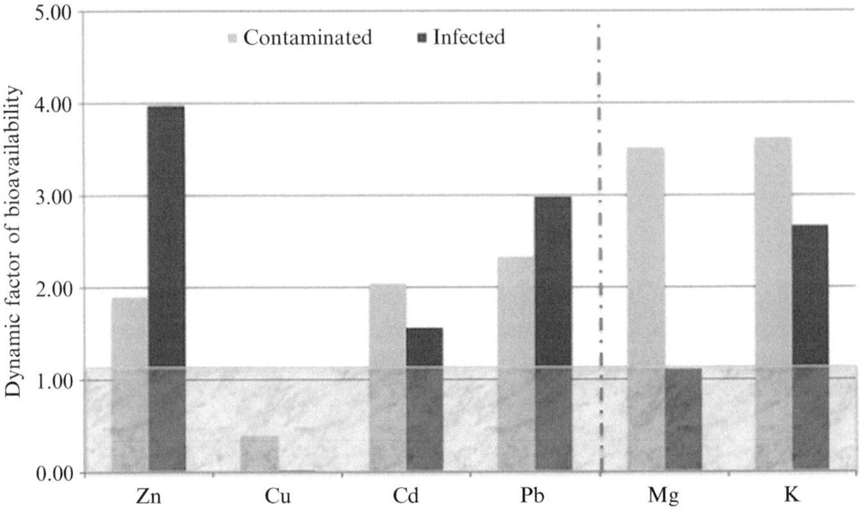

Fig. 2 Dynamic factors of metal bioavailability (BIO_{dyn}) in soil of the territory where infected pines grow and the territory affected by aerogenic pollution. The marked area indicates that factors of metal bioavailability in soil of the territories affected by a biotic or an abiotic factor, are not higher than in the control site, $BIO_{dyn} \leq 1$

Unlike other metals, the increase in the concentrations of Pb and Zn in mobile forms, compared to that in the control territory, was higher for the soil in the territory covered with the infected pines than in the territory with contaminated soil. In the case of macroelements K, Mg and microelement Cd, the opposite trend could be observed. In other words, it can be stated that the biotic factor affecting pines can cause a more prominent increase in the mobility of Pb and Zn, whereas the abiotic factor causes the increase in the mobility of K, Mg and Cd in the soil of the territories covered with growing pines.

3.3 *The Variation of Metal Bioaccumulation in Pine Tissue Affected by Different Factors*

Several main tendencies can be observed in studying the dynamic factors of bioaccumulation of metals in the pine tissue (Fig. 3). Bioaccumulation of Mg and K both in the infected pines and pines growing in the contaminated territory increased compared to the control pines. The increase in bioaccumulation of Zn and Cu was more prominent in the infected pines, whereas its decrease could be more clearly observed in pines growing in the contaminated territory. In the case of Pb, there was an opposite tendency. Thus, bioaccumulation of Cd in the infected pines increased, whereas it changed only insignificantly in the pines growing in the contaminated territory.

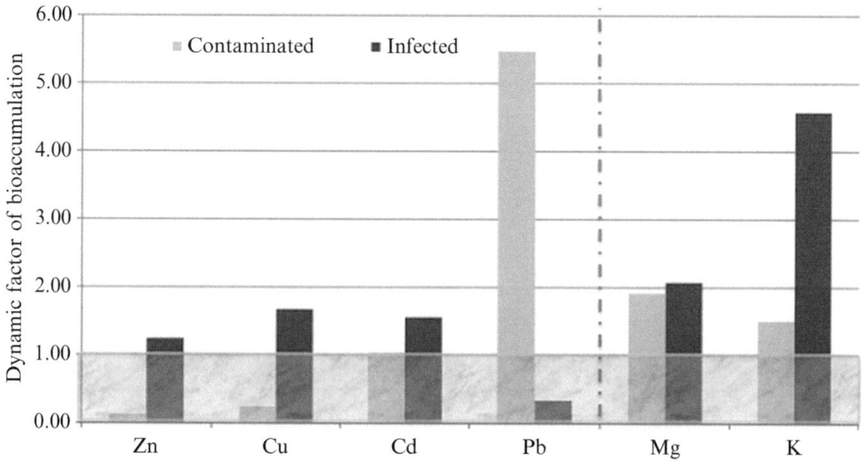

Fig. 3 Dynamic factors of metal bioaccumulation (BA_{dyn}) in the infected pines and the pines growing in the territory affected by the aerogenic pollution. The marked area indicates that factors of metal bioaccumulation in pines of the territories affected by a biotic or an abiotic factor, are not higher than in the control site, $BA_{dyn} \leq 1$

The increase in K was the highest in the tissue of the damaged pines (it was about 4.5 times that of the control pine tissue), whereas the increase in Pb was the highest in the tissue of the pines growing in the contaminated territory (about 5.5 times as large as the increase in this metal observed in the tissue of the control pines). Bioaccumulation of Mg influenced by both factors remained actually the same, though it was almost twice as large as that observed in the control pines (Fig. 3).

3.4 *The Variation of Metal Biophilicity*

Unlike the dynamic factor of metal bioaccumulation, the dynamic factor associated with metal biophilicity indicates the variation of metal concentration in pine tissue comparable to the average metal concentration in the Earth's crust. Therefore, the changes in the amount of metals transferred to trees can be evaluated from a biogeochemical perspective.

The analysis of the dynamic factors of biophilicity showed more active transfer of all the considered metals in the infected pines and pines growing in the contaminated territory, but the most prominent trends were as follows: in the case of pines growing in the contaminated territory, the biophilicity of Pb used to be higher, whereas in the case of the infected pines – the biophilicity of K was mostly prominent (Fig. 4). It can be stated that the biotic factors stimulated biophilicity and bioaccumulation of macroelements in pines, whereas pollution intensified biophilicity and bioaccumulation of microelements in pines.

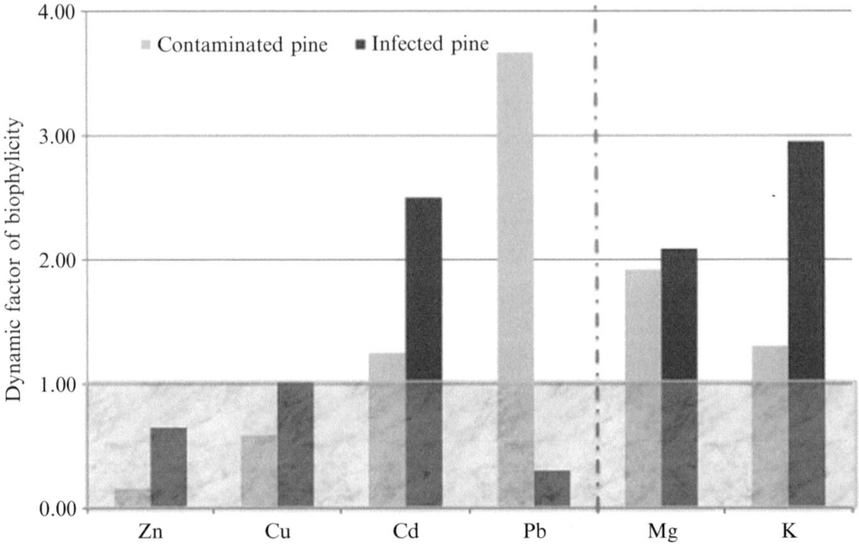

Fig. 4 Dynamic factors of metal biophilicity (BF_{dyn}) in the infected pines and the pines growing in the territory affected by the aerogenic pollution. The marked area indicates that factors of metal biophilicity in the territories affected by a biotic or an abiotic factor, are not higher than in the control site, $BF_{dyn} \leq 1$

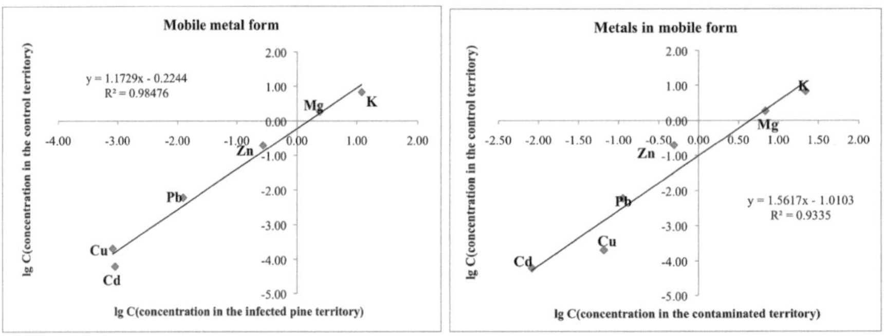

Fig. 5 The relationship between the concentrations of metals in the mobile form in soil of the territories affected by the abiotic and biotic factors and in the control territory (a logarithmic scale)

3.5 *The Variation of Balance of Metals in the Soil–Pine System*

Figures 5, 6, and 7 show the relationships between *soil–pine* systems affected by the abiotic and biotic factors and the control *soil–pine* system in terms of the concentration of metals in the mobile form and the total concentrations of metals in soil and

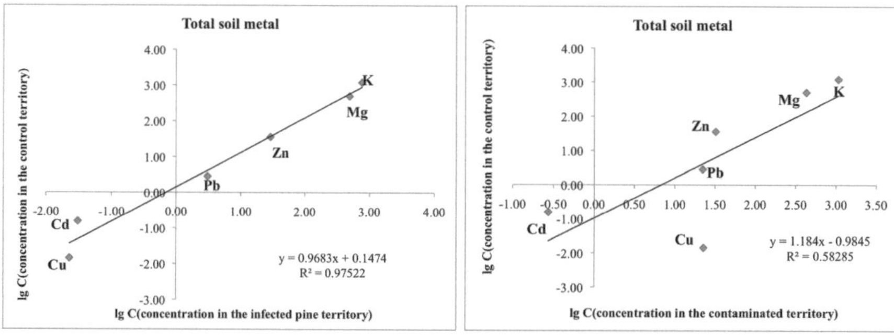

Fig. 6 The relationship between the total metal concentrations in soil of the territories affected by the abiotic and biotic factors and in the control territory (a logarithmic scale)

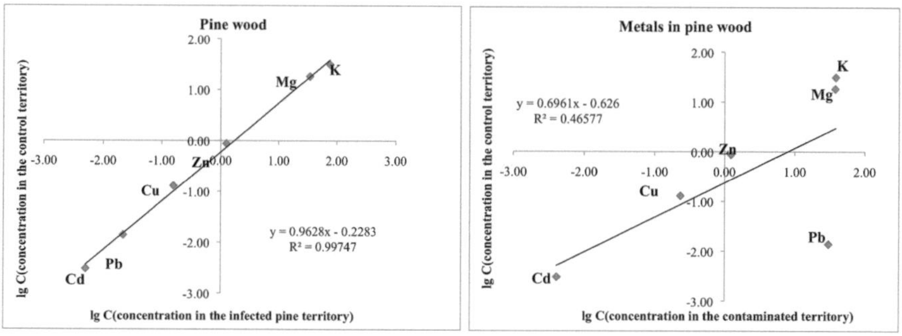

Fig. 7 The relationship between the total metal concentrations in pine tissue in the territories affected by the abiotic and biotic factors and in the control territory (a logarithmic scale)

pine tissue. It can be emphasized that the concentrations of metals (both in the mobile form and in their total concentrations in soil and in pine tissue) tend to be distributed on a logarithmic scale according to a linear relationship and only factors having the strongest influence can cause their deviations from a linear relationship.

It can be observed that a rather strong logarithmic linear relationship ($R^2 > 0.90$) is found between the concentrations of metals in the mobile form in various types of soil, where differently affected pines grow. In other words, the concentration of metals in the mobile form does not strongly depend on abiotic and biotic factors. Meanwhile, the observed general tendency of the total concentration of metals in soil and in pine tissue was somewhat different. The total concentrations of metals in soil demonstrated a less uniform logarithmic linear relationship, which could be explained by the influence of Cu, whereas in pine tissue it can be attributed to the influence of Pb. The lower uniformity of this relationship can be observed for the contaminated territory (affected by the abiotic factor) and probably depends on the type of contamination, as the concentrations of Cu and Pb in the soil of the contaminated territory are 2.42 and 6.70 times higher than those found in the control territory. Though the source of both metals' emissions (an industrial enterprise) and the

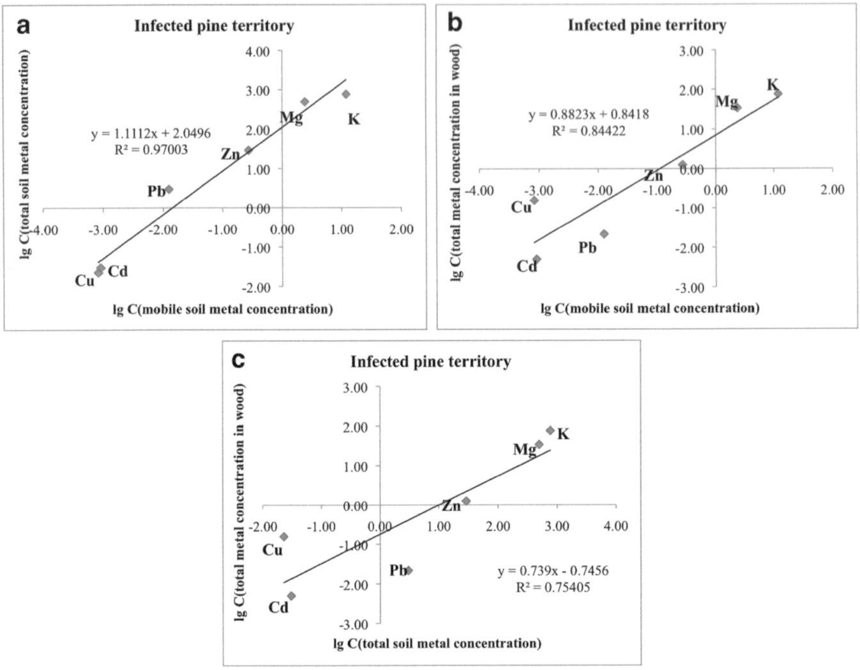

Fig. 8 Interdependence between metal concentrations in different elements of the *soil–pine* system affected by the biotic (a logarithmic scale)

character of transfer (which is aerogenic) are the same, however, they produce different effects on the logarithmic linear relationship in various media. This might have resulted from the contamination level, i.e. the concentration of Pb that was almost three times as high as that in the control territory.

The logarithmic linear relationship for the territory affected by the biotic factor is rather uniform.

Figures 8, 9, and 10 present the diagrams of the interdependence between the concentrations of the metals in the considered forms in different elements of the *soil–pine* system on the biotic and abiotic factors affecting pines. All the situations demonstrated in Figs. 8, 9, and 10 were similar because a logarithmic linear relationship was found between the concentrations of metals in the mobile forms and the total concentration, as well as between the concentrations of metals in the mobile form in soil and in pine tissue and their total concentration in soil and in pine tissue for all the analysed cases. The regression coefficient (R^2) in these cases ranged from 0.66 to 0.99.

The strongest positive relation persisted between the concentration of the mobile portion and the total concentration of metals in soil affected by all considered factors. This trend observed in all cases can be explained by the fact that the regression coefficient was always more than 0.90. The relation between the concentrations of metals in the mobile form in soil and the total concentrations in pine tissue was

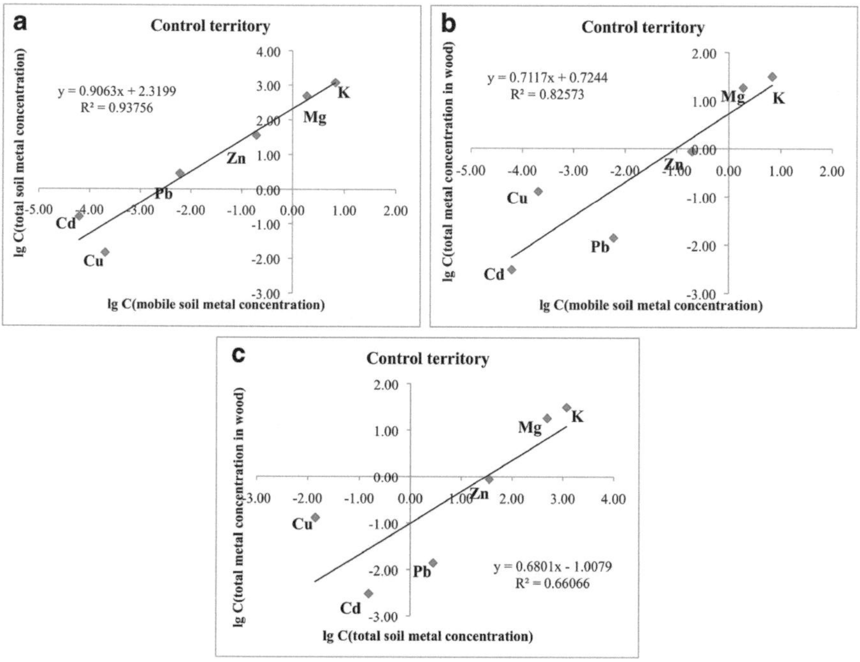

Fig. 9 Interdependence between metal concentrations in different elements of the *soil–pine* system in the control territory (a logarithmic scale)

weaker. It was even weaker between the total metal concentrations in soil and in pine tissue. The weakness of the relation may be explained by the influence of Pb and Cu, especially, taking into account the case of the abiotic factor influence.

The weakest positive relation was found between the concentrations of metals in the mobile form in soil and their total concentration in pine tissue in the contaminated territory. It can be stated that, in this case, the abiotic factor can have a stronger influence on the logarithmic linear dependence between the concentrations of metals in the mobile form in soil and in pine tissue (in this case, heavier pollution with Pb and Cu can be observed). A more stable relation was found, when the territory influenced by an abiotic factor and the control territory were considered.

4 Discussion

In the present study, the *soil–pine* systems found in the territories affected by the abiotic and biotic factors and the control territory were considered. The trends observed in variation of the concentrations of the selected microelements (Pb, Cu, Cd, Zn) and macroelements (K, Mg) found in different elements of the system (i.e.,

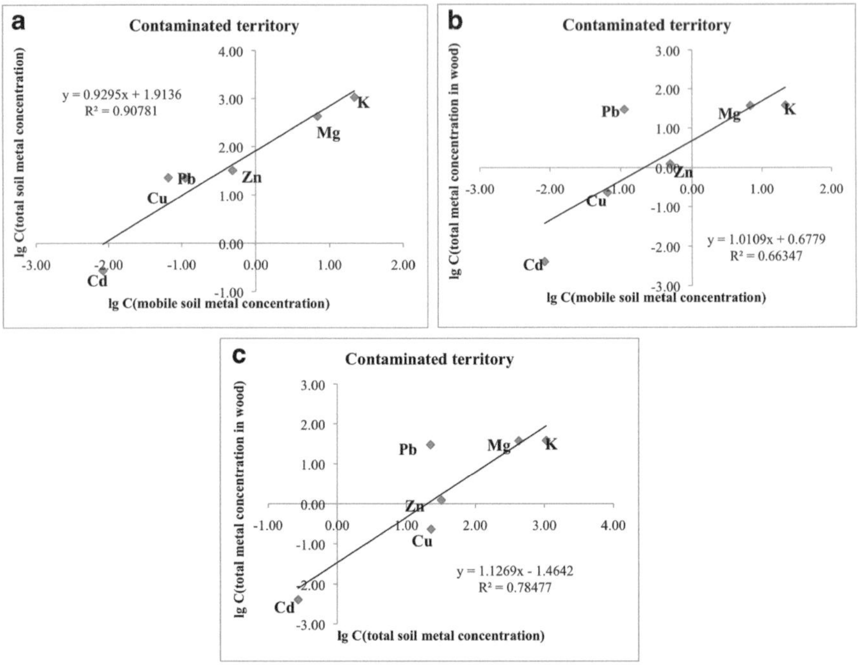

Fig. 10 Interdependence between metal concentrations in different elements of the *soil–pine* system influenced by the abiotic factor (a logarithmic scale)

the soil of the territories where pines grow, in pine tissue and soil solution) were analysed. In order to quantitatively define the variation of such processes as metal bioaccumulation, biophilicity and bioavailability, the methodology of dynamic factors was applied, while in order to analyse the interdependence between metal concentrations, a logarithmic correlation was used.

The concentrations of microelements and macroelements (both the concentration of metals in the mobile form and their total concentration in soil and in pine tissue) clearly distributed according to the linear dependence on a logarithmic scale, but it was assumed that certain factors caused deviations from the above linear dependence. The correlation between metal concentrations, expressed in the form of a logarithmic linear dependence, confirmed an important fact that in the absence of a source of anthropogenic metal pollution, the distribution of concentrations of the considered metals in a Scots pine expressed in a logarithmic form, followed a straight line. In the presence of an anthropogenic metal pollution source (for example, in the contaminated territory), the logarithmic linear dependence was violated, and the highest deviations of concentrations from the logarithmic line could be observed in metals of anthropogenic origin. In those cases, the greatest changes in metal concentrations could be observed in the pine tissue, while slightly smaller changes (of the total concentration of metals) were found in soil, and the smallest changes could be observed in the concentrations of metals in the mobile form in soil. In fact, this correlation can be used to determine some new elements in the *soil–pine* system, to identify the type of the anthropogenic source of metal pollution or even to determine the character of pollution. In the case of natural (not contaminated) soil or under the influence of an external but natural factor (for example, a biotic factor), the logarithmic linear dependence among the considered metals remained strong, especially, between the total concentration and the concentration of metals in the mobile form in soil, as well as, the concentration of mobile metal forms in soil and the total metal concentration in wood.

It was determined that both the abiotic and biotic factors caused the activation of the protective functions in trees (Poschenrieder et al. 2006), which were related to chemical mechanisms of protection against stress. Further analysis of the factors which had an influence on the *soil–pine* system showed that the biotic and abiotic factors determined the variation of metal concentration not only in pines, but also in the soil in the area where they grew. However, it should be noted that anthropogenic pollution (an abiotic factor) caused the increase in the total concentration of microelements, while the biotic factor caused the increase in the total concentration of macroelements in soil. It was determined that the abiotic factor (anthropogenic pollution) stimulated the increase in the concentration of microelements, particularly, typomorphic chalcophiles (Pb, Cu) (chemical elements characteristic of the research object), whereas the biotic factor stimulated the increase in the concentration of macroelement lithophile K in soil. It can be assumed that potassium actively suppresses pathogenic organisms which can cause a disease in a pine. Meanwhile, the concentration of the biophilic Zn in the soil in the area, where pines grew was lower than its concentration in the control soil. The concentration of such lithophile element as Mg in the examined soil compared to the soil in the control area differed insignificantly (Table 1).

The investigation results helped us to define the main trends in the variation of metal concentration in the contaminated territory. The concentrations of toxic elements Pb and Cd (especially, Pb) in the soil of the contaminated territory, which were larger than the respective concentrations in the control territory, as well as larger bioaccumulation values of K, Mg and Cd and biophilicity, allowed us to state that a disruption of photosynthesis in pines took place. Cu and Zn are known to be the essential elements for plants, which are required for photosynthesis reactions to proceed. Lower bioaccumulation of these elements observed in the present investigation can be evidence of a disruption of photosynthesis (Ayeni et al. 2010). When concentrations of Pb and Cd increase in the environment, a disruption of photosynthesis in trees can take place for several reasons. On the one hand, the increased concentrations of metals prevent from the fixation of carbon dioxide, which has a negative effect on gas exchange in the membranes of plant tissues (Baryla et al. 2001). Slow photosynthesis disturbs the synthesis of starch (Ericson 1979), thereby slowing down the growth of a tree. Besides, metal pollution stimulates the decay of a tree, which is caused by slow photosynthesis due to the decrease in chlorophyll amount, loss of deoxyribonucleic acid (DNA), as well as ribonucleic acid (RNA) and protein mass (Jana and Choudhuri 1982; Juneau et al. 2002).

According to the results obtained in the present study, the variation of metal concentrations can be also observed in pine tissue infected by a pathogenic fungus and in the soil of the area where pine grow. Metal concentrations in the soil in the areas where the infected pines grew were smaller than those found in the control soil, and, in some cases, this difference was considerable. For example, a decrease in Pb and Mg content was small, whereas the concentrations of Zn, K and Cu in the soil in the areas where the infected pines grew were approximately 1.5–2 times lower, while the concentration of Cd was even 6 times lower than the concentration of this metal in the control soil (Table 1). Therefore, it showed that the total concentrations of metals differed in the soil of the areas, where the infected pines grew, because the forms of metals changed and metals were uptaken by trees in different amounts. For example, the mobile portion of all examined metals (except for Cu) increased in the soil where the infected pines grew compared to the soil in the control territory. This effect is possible as it is known that the pathogenic fungus increases a bioavailable portion of metal in soil by secreting carboxylic and amino acids (Gramss 2010). In the present research, it could be observed, when the behaviour of Cu and Zn was studied.

During the investigation, the authors noticed that changes also occurred in the pine tissue infected by a pathogenic fungus, which, in turn, caused changes in metal concentrations. It is well known that when a pathogenic microorganism gets into a tree, it destroys lignin, which is a component part of its tissue. Then, it reaches cellulose in the cell walls, which is a very important source of energy for a tree. At this time, the stress signals, indicating that some dangerous processes are developing in a tree, activate its protective functions.

Lignin is one of the most abundant organic polymers and is thought to be responsible for the resistance of coniferous trees to potential pathogens (Vance et al. 1980). It is referred to quantitatively important secondary compounds of a tree, which, in the case of infection, reduce the decomposition of nutrients, which are vitally important for pathogens in a tree. Tissues of lignified plants, where lignin is abundant, contain a great number of phenolic compounds (e.g. ferulic acid or hydroxycinnamic acid), which participate in the formation of compounds with cellulose and hemicellulose (Hartley and Ford 1989). Peroxidase is one of the most important enzymes, which, by utilising phenolic precursors, stimulates biosynthesis of lignin. It is a protein, the formation of which is stimulated by infection or wounds (Asiegbu et al. 1994). The investigation results allow the authors to think that more intense lignification in the affected pine tissues intensifies bioaccumulation of microelements and macroelements in a tree (Fig. 3). Thus, bioaccumulation of the investigated metals (Mg, K, Zn, Cu and Cd), except for Pb, increased compared to this process in the control pines. This can be explained by the involvement of these metals in the process of lignification by their role in increasing the resistance of trees to infection. Copper (Cu) is a constituent part of the enzyme phenol oxidase. This enzyme is involved in the production of lignin, i.e. in lignification processes (Turvey et al. 1992; Kozlowski and Pallardy 1997). Many enzymes are active only when there is a sufficient concentration of the ions of such metals as Mg, K, Mn, Ca. The metals Cu, Zn (and Fe) are known as co-enzymes of enzyme systems. Zinc (Zn) is a co-factor of ribonucleic acid (*RNA)* polymerase and has an influence on the synthesis of enzymes. In the case of Zn deficiency, the synthesis of proteins is disturbed, while amino acids and amides are accumulated (Faust 1989). Magnesium (together with Ca^{2+}, Sr^{2+}, Ba^{2+}, NH_4^+) determines the peroxidase amount in the cell walls and, in this way, regulates the accumulation of lignin on the cell walls (Lipetz and Garro 1965). Johansson and Theander (1974) have found a significant increase in the amount of K^+ and Mg^{2+} ions in the infected wood tissues and explained it as a result of a decreased activity of pathogens. In this study, bioaccumulation of potassium was found to be most prominent and its bioaccumulation factor was the highest (4.58) compared to the respective factors of other investigated metals. Assessing the concentrations of all the above-mentioned metals, we determined that the factor of K biophilicity (2.95) was also the highest (Fig. 4). The regulation of Zn homeostasis is important for virulence of pathogens causing plant diseases (Tang et al. 2005). Finger-Teixeira et al. (2010) explain the impact of Cd on the lignification processes by the fact that this metal stimulate the production of monolignols which form lignin.

To conclude, the study of the variation of biophilicity and bioaccumulation of the considered metals in pines demonstrated that a biotic factor mainly stimulated biophilicity and bioaccumulation of macroelements in pines, whereas pollution intensified biophilicity and bioaccumulation of microelements characteristic of anthropogenic pollution.

5 Conclusions

1. The investigation made by the authors allowed them to establish that, in natural conditions, the concentrations of metals (K, Mg, Zn, Pb, Cd and Cu) found in the elements of the *soil–pine* system (such as wood, soil and metals in the mobile form) were distributed according to a logarithmic linear dependence. Thus, $R^2>0.90$ showed a **strong** linear dependence between the total concentrations of metals and the concentrations of metals in the mobile form in soil; $R^2>0.80$ indicated **a sufficiently strong** dependence between the concentrations of the metals in the mobile form in soil and the total metal concentration in pine tissue, while $R^2>0.60$ demonstrated **a higher than average** dependence between the total concentrations of metals in soil and pine tissue.
2. The external abiotic factor (associated with anthropogenic metals of aerogenic origin) had a stronger influence on the linear logarithmic distribution of metal concentrations in the soil–pine system than the external biotic factor (a pathogen). The strongest effect was produced on the balance between metal concentrations in pine tissue and the concentrations of metals in the mobile form found in soil while the balance between the total concentrations of metals and the concentrations of metals in the mobile form in soil was not strongly influenced.
3. The use of the method of dynamic factors of biophilicity and bioaccumulation allowed to the authors to find that a biotic factors stimulated the processes asso-

ciated with biophilicity and bioaccumulation of macroelement K in the pine tissue, whereas an abiotic factor (associated with anthropogenic pollution) intensified biophilicity and bioaccumulation of typomorphic microelements characteristic of anthropogenic pollution (Pb and Cu).

References

Alloway BJ (ed) (1995) Heavy metals in soils, 2nd edn. Chapman & Hall, London

Asiegbu FO, Daniel G, Johansson M (1994) Defence related reactions of seedling roots of Norway spruce to infection by Heterobasidion annosum (Fr.) Bref. Physiol Mol Plant Pathol 45:1–19

Ayeni OO, Ndakidemi PA, Snyman RG, Odendaal JP (2010) Chemical, biological and physiological indicators of metal pollution in wetlands. Sci Res Essays 5(15):1938–1949

Baltrenaite E, Butkus D, Booth CA (2010) Comparison of three tree-ring sampling methods for trace metal analysis. J Environ Eng Landsc Manag 18(3):170–178

Baltrenaite E, Lietuvninkas A, Baltrenas P (2012) Use of dynamic factors to assess metal uptake and transfer in plants – example of trees, water. Air Soil Pollut 223(7):4297–4306

Baltrenaite E, Baltrenas P, Lietuvninkas A, Ševičienė V, Zuokaitė E (2014) Integrated evaluation of aerogenic pollution by air-transported heavy metals (Pb, Cd, Ni, Zn, Mn and Cu) in the analysis of the main deposit media. Environ Sci Pollut Res 21:299–313

Baltrenaite E, Baltrenas P, Butkus D, Lietuvninkas A (2015) Using the dynamic factors method in bioindication and phytoremediation. In: Ansari AA, Gill SS, Gill R, Lanza GR, Newman L (eds) Phytoremediation: management of environmental contaminants, vol 1. Springer, Cham, 792 p, pp 21–30. http://www.springer.com/life+sciences/ecology/book/978-3-319-10394-5

Baltrenaite E, Baltrenas P, Lietuvninkas A (2016). The Sustainable Role of the Tree in Environmental Protection Technologies, Springer, Cham.

Baryla A, Carrier P, Franck F, Coulomb C, Sahut C, Havaux M (2001) Leaf chlorosis in oilseed rape plants (Brassica napus) grown on cadmium-polluted soil: causes and consequences for photosynthesis and growth. Planta 212:696–709

Chvastov IV, Kovalskaya GA, Pavlov VE (2011) Composition of elements in the annual rings of Scots pine in the regions of Chernonyl and river Podkamennaia Tunguska. Chem Veg Mater 153–158 (in Russian)

Deacon J (2005) Fungal biology. Blackwell Publishing, Oxford

Ericson A (1979) Effects of fertilization and irrigation on the seasonal changes of carbohydrate reserves in different age-classes of needle on 20-year-old pine trees (Pinus sylvestris). Physiol Plant 45:270–280

Faust M (1989) Physiology of temperate zone fruit trees. Wiley, New York

Finger-Teixeira A, Ferrarese ML, Soares AR, da Silva D, Ferrarese-Filho O (2010) Cadmium-induced lignification restricts soybean root growth. Ecotoxicol Environ Saf 73(8):1959–1964

Gramss G (2010) The universe of basidiomycetous ground fungi. In: Mendez-Vilas A (ed) Current research, technology and education topics in applied microbiology and microbial biotechnology. Formatex Research Center, Badajoz, pp 218–229

Gricius A, Urbonas V, Kutorga E, Matelis A (1999) Succession of fungi on dead timber of various trees. Botanica Lituanica 5(6):60–77

Hartley RD, Ford CW (1989) Phenolic constituents of plant cell walls and wall biodegradability. In: Lewis NG, Paice WG (eds) Plant cell wall polymers: biogenesis and biodegradation. American Chemical Society, Washington, DC, pp 138–145

Hill LJ (2002) Branching out into biogeochemical surveys: a guide to vegetation sampling. In: Roach IC (ed) Regolith and landscapes in Eastern Australia. CRC LEME, Perth, pp 50–53

Jana S, Choudhuri MA (1982) Senescence in submerged aquatic angiosperms: effects of heavy metals. New Phytol 90:477–484

Johansson M, Theander O (1974) Changes of sapwood of roots of Norway spruce, attacked by Fomes annosus. Part I Physiol Plant 30:218–225

Juneau P, Berdey A, Popovic P (2002) PAM fluorometry in the determination of the sensitivity of Chlorella vulgaris, Selenastrum capricornutum and Chlamydomonas reinhardtii to copper. Arch Environ Contam Toxicol 42:155–164

Kadūnas V, Radzevičius A (2001) Sunkiųjų metalų migracinės formos Panevėžio miesto įmonių technogeniškai užterštame grunte. Geologija 35:23–28

Kozlowski TT, Pallardy SG (1997) Physiology of woody plants. Academic, San Diego

Kupčinskienė E (2011) Phytoindication of Environment, Kaunas, p 752 (in Lithuanian)

Lietuvninkas A (2012) Environmental geochemistry [Aplinkos geochemija]. Technika, Vilnius

Lipetz J, Garro AJ (1965) Ionic effects on lignification and peroxidase in tissue cultures. J Cell Biol 25(1):109–116

Mancinelli E, Baltrėnaitė E, Baltrenas P, Paliulis D, Passerini G, Almas AR (2015) Trace metal concentration and speciation in storm water runoff on impervious surfaces. J Environ Eng Landsc Manag 23(1):15–27. doi:10.3846/16486897.2014.936441

Markert B (1996) Instrumental element and multi-element analysis of plant samples: methods and applications. Wiley, Chichester, p 312

Markert B, Wunschmann S, Baltrėnaitė E (2012) Innovative observation of the environment. Bioindicators and biomonitors: definitions, strategies and applications. J Environ Eng Landsc Manag 20(3):221–239 (in Lithuanian)

Markert B, Baltrėnaitė E, Chudzińska E, De Marco S, Diatta J, Ghaffari Z, Gorelova S, Marcovecchio J, Tabors G, Wang M, Yousef N, Fraenzle S, Wuenschmann S (2014) Multilingual education of students on a global scale and perspective – international networking on the example of bioindication and biomonitoring (B&B technologies). Environ Sci Pollut Res 21(8):5450–5456

Navasaitis M (2008) Dendrology. Margi raštai, Vilnius, 856 p [in Lithuanian]

Noland TL, Kozlowski TT (1979) Effect of SO_2 on stomatal aperture and sulfur uptake of woody angiosperm seedlings. Can J Forest Res 9:57–62

Poschenrieder C, Tolrà R, Barceló J (2006) Can metals defend plants against biotic stress? Trends Plant Sci 11(6):288–295

Pundytė N, Baltrėnaitė E, Pereira P, Paliulis D (2011) Anthropogenic effects on heavy metals and macronutrients accumulation in soil and wood of Pinus sylvestris L. J Environ Eng Landsc Manag 19(1):34–43

Reich PB (1987) Quantifying plant response to ozone: a unifying theory. Tree Physiol 3:63–91

Schulze ED, Beck E, Muller-Hohenstein K (2005) Plant ecology. Springer, Berlin/New York

Tang DJ, Li XJ, He YQ (2005) The zinc uptake regulator Zur is essential for the full virulence of Xanthomonas campestris pv. campestris. Mol Plant Microbe Interact 18:652–658

Turvey ND, Carlyle C, Downes GM (1992) Effects of micronutrients on the growth form of two families of Pinus radiata (D. Don) seedlings. Plant and Soil 139:59–65

Vance CP, Kirk TK, Sherwood RT (1980) Lignification as a mechanism of disease resistance. Annu Rev Phytopathol 18:259–288

Woodward J, Stenlid J, Karjalainen R, Hüttermann A (1998) Heterobasidion annosum. Biology, ecology, impact, and control. CAB International, Wallingford

Youssef N, Markert B, Gurbanov E, Sevnic H, Wünschmann S (2014) Bioindication of trace metal pollution in the atmosphere of Baku city using ligustrum japonicum, olea europea, and pyracantha coccinea leaves. J Environ Eng Landsc Manag 22(1):14–20

Plant-Pollutant Interaction

Rida Rehman and Alvina Gul

Contents

Abstract Plants are the primary reservoirs in food chain. They play a key role in the conversion of energy into food. The life on Earth is impossible without plants. With the increasing advances in different fields, the health of plants is deteriorating drastically. Heavy metals have become an environmental concern because of their adverse and detrimental effects on human life. Apart from that, one of the most important components of global change is increase in atmospheric nitrogen (N), which is threatening both the structure and function of the ecosystem. The N cycle is dominated by anthropogenic activity and the emission and deposition rates

R. Rehman • A. Gul (✉)
Atta-ur-Rahman School of Applied Biosciences,
National University of Sciences and Technology, Islamabad, Pakistan
e-mail: alvina_gul@yahoo.com

M. Öztürk et al. (eds.), *Plants, Pollutants and Remediation*,
DOI 10.1007/978-94-017-7194-8_11

of N are expected to be doubled by 2050, which will greatly increase the number of regions potentially receiving damaging levels of N inputs. Another harmful air pollutant is the tropospheric ozone (O_3) having negative impact on the growth and development of plants. Current concentration of O_3 is decreasing the productivity of forest and crop yields. Future O_3 concentration will increase if the current emission rate stays persistent. In addition, benzene; a volatile organic compound (VOC), is extracted from industries of petroleum and used widely as an additive or intermediate or solvent in many manufacturing industries. Ambient air pollution problems can be caused by the emission of benzene from many sources, even though several organizations and countries have standard guidelines on benzene use. A number of studies have been carried out to study the effect of petro-coke, cement, coal and dust, fly ash, automobile exhaust and other air-borne particulates on different physiological and morphological parameters in plants.

Keywords Phytoremediation • Chromated copper arsenate • Polyaromatic hydrocarbons • Benzene • Nanoparticles

1 Introduction

The evolution of Earth started nearly 4.5 billion years ago. Since that time, formation of earth has been all about changes in the balance and making life more adaptable. As the humans grew in number, evolution in humans also resulted as an integral element in the delicate balance of nature. The growth of humans increased at the expense of other living things as well as the ecosystem. (Purohit and Ranjan 2007). With development in the industrial sector, the numbers of pollutants affecting plant growth are increasing rapidly. Pollution harms the health or survival of all living organisms i.e. humans, animals and plants. Air is polluted significantly by the release of toxic materials and chemicals from different sources. Windcarries the emissions from the industries to hundreds of miles, where they come down from the sky in the form of acid rain, snow and dust rendering acidity in lakes and making them unsuitable for aquatic animals and plants. The humanrace is faced with the challenge of supporting enormous population growth. Basically, there are two types of pollutants, which are discussed below:

I. **Primary pollutants**
Primary pollutants are those particles or gases, which are pumped into the air to make it polluted. These include gases like carbon monoxide from different automobiles exhausts as well as sulfur dioxide from coal combustion.
II. **Secondary pollutants**
Pollutants have a tendency to mix up in the air and form even more complex compounds. When the primary pollutants mix up in the air in a chemical reaction, they result in the formation of even more dangerous chemical. Photochemical smog is an example of secondary pollutant which is formed in the atmosphere

by the reaction of two primary pollutants. Photochemical smog is the product of chemical reactions driven by sunlight and involving NO_x of urban and industrial origin and volatile organic compounds from either vegetation (*biogenic* hydrocarbons) or human activities (*anthropogenic* hydrocarbons).

Plants are affected by a number of pollutants whether it is soil or environmental pollution. Ozone depletion is one of the most alarming situation and is considered as a major concern of environment on Earth. Apart from that, the burning of hydrocarbons in engines of motor vehicles results in the formation of hazardous gases, which include CO, SO_2 (sulfur dioxide), NO_x (NO [nitrogen monoxide]) and NO_2^-—in varying proportions—and C_2H_4 (ethylene), as well as a variety of other hydrocarbons. Additionally SO_2 also originates from domestic and industrial burning of fossil fuels. Industrial plants, such as chemical works and metal-smelting plants, release SO_2, H_2S, NO_2, and HF (hydrogen fluoride) into the atmosphere. Tall chimney stacks may be used to carry gases and particles to a high altitude and thus avoid local pollution, but the pollutants return to Earth, sometimes hundreds of kilometers away from the original source. Ozone (O_3) and peroxyacetylnitrate (PAN) produced in these complex reactions can become injurious to plants and other life forms, depending on the concentration and duration of exposure. Hydrogen peroxide; another potentially injurious molecule, can form by the reaction between O_3 and naturally released volatiles (terpenes) from forest trees (Ozturk 1989; Zeiger 2006).

The concentration of polluting gases, or their solutions, to which plants are exposed are thus highly variable, depending on location, wind direction, rainfall, and sunlight. In urban areas, concentrations of SO_2 and NO_x in the air are typically 0.02–0.5 mL L^{-1}, the upper value being within the range that is inhibitory to plant growth. Relatively long-term experiments at appropriate concentrations of pollutants are necessary to establish the real impact of air pollution on vegetation. The reaction of plants to high concentration of pollutants in short-term experiments may overwhelm the plant's defense mechanisms (Zeiger 2006).

The responses of plants to polluting gases can also be affected by other ambient conditions, such as light, humidity, temperature, and the supply of water and minerals. Experiments aimed at determining the impact of chronic exposure to low concentrations of gases should allow plants to grow under near-natural conditions. One method is to grow the plants in open-top chambers into which gases are carefully metered, or where plants receiving ambient, polluted air are compared with controls receiving air that has been scrubbed of pollutants.

Unpolluted rain is slightly acidic, with a pH close to 5.6, because the CO_2 dissolved in it produces the weak acid, H_2CO_3. Dissolution of NO_x and SO_2 in water droplets in the atmosphere causes the pH of rain to decrease to 3–4, and in southern California polluted droplets in fog can be as acidic as pH 1.7. Dilute acidic solution can remove mineral nutrients from leaves, depending on the age of the leaf and the integrity of the cuticle and surface waxes. The total annual contributions to the soil of acid from acid rain (*wet deposition*) and from particulate matter falling on the soil plus direct absorption from the atmosphere (*dry deposition*) may reach 1.0–3.0 kg

H^+ per hectare in parts of Europe and the northeastern United States. In soils that lack free calcium carbonate, and therefore are not strongly buffered, such additions of acid can be harmful to plants. Furthermore, the added acid can result in the release of aluminum ions from soil minerals, causing aluminum toxicity. Air pollution is considered to be a major factor in the decline of forests in heavily polluted areas of Europe and North America. There are indications that fast-growing pioneer species are better able to tolerate an acidifying atmosphere than the climax forest trees, possibly because they have a greater potential for assimilation of dissolved NO_x, and more effective acid buffering of the leaf tissue cell sap (Zeiger 2006).

2 Effect of Pollutants on Pollinators

Wild bees play an essential role in pollination in temperate zones and are considered as major pollinators (Kevan 1999). The diversity of wild plants is maintained properly by bees (Ashmann et al. 2004; Aguilar et al. 2006; Brittain et al. 2010) as well as the agricultural productivity (Klein et al. 2007; Gallai et al.2009; Lenda et al. 2010). A large number of plant species that depend on insect pollination for seed and fruit production (Wilkaniec et al. 2004; Morandin and Winston 2005; Velthuis and Van 2006) may experience reduced pollination and production when the pollinator species are scarce (Ashmann et al. 2004). Therefore, the decrease in number of bees in the prominent regions of the globe (Steffan-Dewenter et al. 2005; Brittain et al. 2010; Biesmeijer et al. 2006) is pretty alarming.

A decrease in bee population has detrimental effects on both crops and wildflower pollination. It has been recognized that heavy metals are a problem that are affecting not only the large parts of the European Union but also have detrimental effects on bees. A study was carried out to investigate whether heavy metal pollution is a potential threat to communities of wild bees by carrying out the comparison in (i) number of species, (ii) their diversity and (iii) percentage abundance as well as natural deaths or mortality of the emerging bees along two gradients that are independent of pollution caused by heavy metals. These studies were carried out at Olkusz (OLK), Poland and Avonmouth (AVO), UK. The study was designed to measure the richness in bee species and their abundance. The concentration of heavy metal was also recorded in the pollen that was collected by a specific bee species in order to measure the pollution being caused by heavy metals. The heavy metals found were cadmium, lead and zinc in varying amounts. As a result of this study, it was found out that with an increase in the concentration of heavy metals, there is a decrease in the number, abundance as well as diversity of wild bees. Above all, an inverse trend was observed among wild bees and pollution caused by heavy metals. Applying protection plans to wild pollinating bee communities in heavy metal-contaminated areas will contribute to integrated land rehabilitation to minimize the impact of pollution on the environment. This will aid in not only the betterment of the environment but will also be helpful for plants and their pollination (Moron et al. 2012).

Table 1 Heavy metals affecting the type of pollinator population

S. No.	Pollinators	Heavy metals that effect them	Place reported
1.	Butterfly	Cd, Cu, Fe, Mn and Zn	Finland
2.	Butterfly	Cd, Cu and Zn	Netherlands
3.	Bees	Zn, Cd and Pb	Europe, Poland, UK

There are a number of factors, which decrease the communities of wild bees. One of the major factors includes agricultural intensification (Le Féon et al. 2010; Steffan-Dewenter 2003). Others include use of pesticides (Alston et al. 2007; Brittain et al. 2010), the effect of non-native invasive species (Moron et al. 2009), competition with other populations (Walther-Hellwig et al. 2006; Kenta et al. 2007, spread of pathogens (Colla et al. 2006) and genetic introgression (Kraus et al. 2011). Heavy metals have negative effects not only on one or two species but on all the invertebrate diversity (Syrek et al. 2006; Beyrem et al. 2007; Piola and Johnston 2008) Many studies have figured out the impact of heavy metals on pollinators (Nieminen et al. 2001; Mulder et al. 2005). The following table further explains the type of pollinator and heavy metals affecting its population (Table 1).

3 Soil Pollution

3.1 Chromated Copper Arsenate (CCA)

A large number of heavy metals are reported in air, water, soil and plants in the last few decades. Heavy metals have become an environmental concern because of their adverse and detrimental effects on human life (Yang et al. 2007; Ok et al. 2011a, b). Primary source of toxic metals in environment are anthropogenic systems (Ok et al.2007; Ahmad et al. 2012a, b). Soil is the major sink for inorganic pollutants. Many studies have been conducted to evaluate the extent of heavy metal contamination in soil from anthropogenic sources such as mining and smelting activities, steel and iron manufacturing, incineration of waste, production of cement, phosphate fertilizers and pesticides exhaust from automobiles (Chen et al. 2005; Moon et al. 2011; Ahmad et al. 2012a, b).

A preservative known as Chromated Copper Arsenate (CCA) has been used for the protection of wood products from fungal, bacterial and insect decay (Chirenje et al. 2003; Saxe et al. 2007). CCA treated wood serves as a material for construction of private/public housing, fences, playground equipment, picnic tables, walkways, sound barriers and other outdoors wood products (Kim et al. 2007). In CCA-treated wood, arsenic (As) and copper (Cu) act as an insecticide and a fungicide, respectively, and the chromium (Cr) fixes the As and Cu into the structures of cellulose or hemicellulose, as well as the lignin of the wood (Dawson et al.1991).

There are three forms of CCA; A, B and C which are commercially available. Among them the most popular is type C, which is composed of different constituents; namely CrO_3, CuO and As_2O_5 comprising of 47.5 %, 18.5 % and 35.0 % of C type, respectively (APWA 1991). However, in many countries, usage of woods treated with CCA has been banned. The wood treated with CCA leaches Cr, Cu and As into the surrounding environment, thereby increasing the levels of Cr, Cu and As in the soil thus causing negative effects on the environment and humans (Hingston et al. 2001; Solo-Gabriele et al. 2002; Gezer et al. 2005; Kumpiene et al. 2008). Hence, remediation of CCA contaminated soils is necessary to eliminate risks to human and the environment. Several remediation techniques are being used for soil contaminated with heavy metals (Isoyama and Wada 2007; Ok et al. 2011a). Due to several economic and logistic reasoning as well as possible detrimental effects on the ecological equilibrium and sustainable restoration, they have been restricted (Cao et al. 2002; Fries et al. 2003; Yang et al.2006).

Recently, Phytoremediation; a plant based technique for remediation is found out to be cost effective and an environment friendly method of removing heavy metals from the soil (Hakeem et al. 2015). Phytoremediation can be categorized into two types:

(i) Phytoextraction, which is the removal of heavy metals by aboveground parts of the plant,
(ii) Phytostabilization, which is the reduction of heavy metal mobility and availability around the rhizospheric soil.

Three categories of hyper accumulators, indicators and excluders are made depending upon the ability of plants with medium to absorb, accumulate and tolerate heavy metals (Usman et al. 2012). Specifically, plant species that aid in being hyper accumulators can accumulate extreme levels of heavy metals. Studies show that plant species that are native to a particular area can survive better under the stress of toxic metal relative to invasive plant species (Yoon et al. 2006). In the recent decades, contamination of heavy metals by Cr, Cu and As in soil that is adjacent to CCA treated wood has received great attention. A study was carried out to determine the level of pollution based on the concentration of aforementioned heavy metals to evaluate the remediation capacity of plant species native to that particular area grown in CCA contaminated site (Usman et al. 2012). It was noted that the concentrations of metal decrease as the distance between structure of wood treated with CCA and the point of sampling was increased. The results revealed that *Iris ensata* is a hyper accumulator with the highest Cr accumulation (Usman et al. 2012).

3.2 *Nitrogen; Its Role Towards Plant Growth*

One of the most important components of global change is increase in atmospheric nitrogen (N). This increase is threatening both the structure as well as the functioning of the ecosystem. The global N cycle is dominated by anthropogenic activity (Galloway et al. 2004) and it has been predicted that the emission and deposition rates of N will be doubled by 2050, greatly increasing the number of regions that are potentially receiving damaging levels of N inputs (Bobbink et al. 2010; Galloway et al. 2004). By understanding the responses of ecosystems and the mechanisms by which these mechanisms are driven, these responses continue to be of major importance for the conservation of ecosystems; both natural and semi-natural types, biodiversity preservation and ecosystem services sustainability. Over the least two decades, a number of studies are carried out to understand the N deposition through application of N to experimental plots. Along with these studies, certain survey studies evaluated the patterns of ecosystem response along gradients of N deposition in space and time (Dupre et al. 2010; Maskell et al. 2010) to reveal the diverse impacts on ecosystem structure and function. The major impacts are:

(i) Accumulation of nitrogen causing biodiversity declines through/by the expansion of nitrophilous species and competitive exclusion of others;
(ii) Accumulation of NH_4^+ ions that lead to toxic effects on sensitive species in ecosystems where NO_3 is usually the dominant N form (Stevens et al. 2011);
(iii) Acidification of soil, cation depletion of the base and enhanced toxic metals availability (e.g. Al_3^+, Fe_3^+) which can result in the reduction of health of plants and productivity, alteration in community composition, and cause declines in richness of; and
(iv) Increase in the susceptibility of plants to secondary stresses including increased herbivory, reduced resistance to attack by pathogen or increase in susceptibility to drought or freezing damage (Phoenix et al. 2012).

The role of deposition of N as of driver of biodiversity loss has recently been reviewed (Bobbink et al. 2010; Dise et al. 2011). However, there are certain limitations to these studies as well. Hence, the assessment of deposition of N threat can also be facilitated by the analysis of long-term experiments, across multiple numbers of ecosystems, application of techniques that adequately stimulate deposition of N and also employing realistic doses of N. Deposition of atmospheric nitrogen (N) is becoming global threat to biodiversity and ecosystem function. Much of the current understanding of deposition of N impact comes from manipulation of field studies, although interpretation may require caution where simulations of N deposition (in terms of dose, application rate and N form) have limited realism (Phoenix et al. 2012).

3.3 Stone Crushing

The stone crushing units and the traffic associated with it generates a large number of air pollutants. These pollutants degrade the quality of air in the specific area. The two main operations include:

- Mining or quarrying operations
- Crushing operations

These two operations are responsible for high load that causes pollution. But a lack in the environmental governance in both of the above-mentioned operations has resulted in degradation of environment to a considerable amount around the locations where the stone crushing industry is set up (ES 1998).

The particulates resealed by the above mentioned activities vary a lot in their sizes and weights. They remain as a part of air for varying time length and affect plants and human beings differently (Mishra 2013). Stone dust acts as 'primary aerosol' imparting a detrimental effect not only on the environment but also on people and other flora and fauna. The examples include change in the productivity and pH of soil, haze formation leading to reduced visibility in the surrounding areas, habitat destruction, damage to natural resources like important and valuable vegetation and wild lives and at last but not the least, promotion of spreading of many diseases (Semban and Chandrasekhar 2000; Das and Nandi 2002; Mishra 2004).

A number of studies have been carried out to study the effect of petro-coke, cement, coal and dust, fly ash, automobile exhaust and other air-borne particulates on many parameters that include physiological and morphological status in different plants (Singh and Rao 1980; Prasad and Rao 1981; Naidoo and Chirkoot 2004; Prajapati and Tripathi 2008).

3.4 Phthalic Acid Esters (PAEs)

Phthalic Acid Esters (PAEs) accumulation in the plants and soil in agricultural land near a recycling site of electronic waste in east China has become an increasingly evident threat to the quality of neighboring environment and human health. Collection of soil and plant samples from the land under different utilizations including fallow plots, plots for vegetable, alfalfa plots for green manure, fallow plot under long term flooding as well as under alternating dry and wet periods, along with the samples of plants for relative plots was done. Assessment for health risks was conducted on target PAEs, which are known as analogs of typical environmental estrogens based on their accumulation in edible vegetable parts. The potential damage that the target compounds of PAE may pose to human health should be taken into account in further comprehensive risk assessments of recycling sites areas of electronic waste. Moreover, alfalfa removed substantial amounts of PAEs from the soil, and its use can be considered a good strategy for *in situ* remediation of PAEs. Recycling of electronic wastes and dismantling has become an important activity of industrial sector to the local economy in many areas. However, improper and arbitrary methods of recycling have resulted in multiple contamination of environment proceeded by heavy metals, polychlorinated biphenyls (PCBs), polychlorinated dibenzo-p-dioxins/dibenzo-furans (PCDD/Fs), polycyclic aromatic hydrocarbons (PAHs), and polybrominateddiphenyl ethers (PBDEs). DnBP, DEHP and DnOP are supposed to be the three most frequently detected phthalic acid ester (PAE) compounds in most of the electrical wastes, which explains the discovery of elevated PAE compounds in agricultural soils in recent years (Ma et al. 2013).

3.5 Heavy Metals

Soil pollution with heavy metals is a worldwide environmental problem. Phytoremediation through phytoextraction and phtyostabilization appears to be promising technology for the remediation of polluted soil. It is important to emphasize strongly that the ultimate goal of remediation of heavy metals from the soil is to reduce their mobility and bioavailability. Quality of soil is defined as the capacity of a given soil to perform its ecological functions. Microbial properties of soil are increasingly being used as biological indicator of soil quality due to their quick response, increased sensitivity and capacity to provide information that integrates many factors of environment. Indeed, properties of microbes are among ecologically most relevant indicators of soil quality. Consequently, monitoring of microbial activity is often carried out during phytoremediation of heavy metal processes. However, microbial properties of soil are highly dependent on the context and difficult to interpret. For better interpretation of properties of microbial activity of soil, they may be grouped within categories of higher relevance of ecology such as functions of soil, health attributes of ecosystem and its services (Maria et al. 2012).

Pollution of soil by heavy metals is a worldwide important environmental problem. Anthropogenic activities result in pollution of soil with toxicity of heavy metals. Metallurgy, mining, agriculture, tanning, disposal of waste, fossil fuel combustion and other human activities are responsible for the enormous heavy metals amount and metalloids presently found in our soils. Among these, the most frequent heavy metals found in polluted soil are lead (Pb), cadmium (Cd), cobalt (Co), mercury (Hg), copper (Cu), selenium (Se), zinc (Zn) and nickel (Ni) (Kavamura and Esposito 2010; Hakeem et al. 2015).

The functionality of soil ecosystem is interfered by both inorganic and organic pollutants but special concern is the concentration of heavy metal due to their immutable nature i.e. they do not undergo any biologically or chemically induced degradation. They are highly persistent in the soil. Moreover, heavy metals are potentially harmful to all biota and tend to accumulate in the food chain (Maria et al. 2012).

Unfortunately, traditional way of management of polluted soil with heavy metals using a variety of physicochemical remediation methods (approaches to conventional remedial often involve excavation and washing, landfilling, replacement of soil with clean material, or capping the soil with an impermeable layer to reduce exposure to pollutants) has proven to be economically unattractive, particularly for sites that are largely polluted. Most importantly, some of the technologies that include physiochemical engineering result in deterioration to considerable level of the soil ecosystem. At times, they cause even more damage to soil ecosystem than the pollutants themselves and do not allow reshaping of soil ecosystem naturally. It has been noted that one of our most important resources is soil with a value of estimated $20 trillion (Maria et al. 2012).

Currently, development sectors are focusing on the environment friendly and cost effective methods for the remediation of soil that is polluted with heavy metals. The use of microorganisms for the detoxification or removal of pollutants, also known as, bioremediation is a well-known low-cost option for *in-situ* remediation of polluted soils (Maria et al. 2012).

Nonetheless, pertaining to the remediation of soils polluted with heavy metals, microorganisms have some important limitations:

- They have the ability of heavy metal toxification by valence transformation, precipitation of extracellular material or volatilization, but they cannot remove metals from sites that is polluted. Therefore, it is not a long term, large-scale solution to the highlighting problem of heavy metal pollution in the soil. On the contrary, plants have the ability to extract heavy metals from the soil theoretically rendering them clean (Maria et al. 2012).
- Furthermore, during remediation processes based on plants, plants have the ability to improve the chemical, physical and biological properties of the polluted soil by increasing the porosity, prevention of erosion, addition of nutrients and promising growth of microbes.

3.6 Polycyclic Aromatic Hydrocarbons (PAHs)

Polycyclic aromatic hydrocarbons (PAHs) are pollutants of ubiquitous nature with known carcinogenic and mutagenic effects. Several studies showed that PAHs are taken up by plants grown in contaminated soils. Many factors, like the initial concentration of soil PAH, soil physical and chemical properties, the microbial population, and physiological characteristics affect the contaminants uptake. In plants, it is generally agreed upon that sorption controls uptake of chemicals. Carbonaceous materials like unburnt coal, black carbon (soot and charcoal), and kerogen have a great capacity to adsorb hydrophobic pollutants like PAHs. Hence they can control the bioavailability of PAHs (Jakob et al. 2012).

Since a large amount of money is required in order to carry out the excavation of contaminated soil, the *in-situ* stabilization of organic pollutants in soils and sediments has attracted increasing attention in the recent years. The addition of small amount of AC (activated carbon) has been revealed to decrease the bioavailable concentrations of organic pollutants. AC is a processed form of charcoal with a high sorption capacity, likely because of its chemical structure, large surface area and high porosity. Strong sorption of pollutants to AC has been shown to decrease the accessibility of pollutants to microorganisms. In order to check the bioavailability of organic components for soil fauna, earthworms are often used. It is suggested that AC reduces the biota to sediment or soil accumulation factor (BSAF). Little attention has been directed towards the possible toxic effects of AC on sediment or soil dwelling organisms. In a study, Jonker et al. (2009)found ecological toxicity affects of AC in AC-water and AC-enriched sediment systems. However, in two other studies, changes were observed in the lipid concentration in different plants affected with AC.

In addition to that, two other studies revealed the possibility that AC results in the reduction of the bioavailable concentration of essential substances leading to the nutrient deficiency for soil and sediment organisms as well as for plants. The physical, microbial and chemical characteristics of the soil may also be affected by AC. Additionally, this may also have harmful or adverse effects on plant and soil organisms. A study was carried out which involved the investigation of AC performance (powder and granular) amendments as an *in-situ* technique for stabilization of soil that is contaminated. The study involved the elucidation of AC impact on the uptake of PAH by earthworms and plants in an urban soil contaminated with PAH. It was also noticed whether the addition of AC had toxic effects on plants and earthworms and whether or not the different kinds of AC affected the chemical and physical parameters of soil. This is, till now, the first field study for the investigation of *in-situ* sequestration capacity of AC in a soil contaminated with PAH as well as the effects of different AC amendments on plants and earthworms (Jakob et al. 2012).

3.7 Zinc; Afront Line Pollutant

Zinc (Zn) pollution is very prominent and plays an important role in plant growth. A study was conducted on the interaction between Zn pollution and rhizosphere microorganisms as well as plants. The influence of two strains of bacteria isolated from a Zn polluted soil was tested. Zn polluted soil affected the growth of plant and the efficiency of symbiotic native ArbuscularMycorrhizal Fungi (AMF). Zn tolerance was exhibited by two strains of bacteria cultivated under increasing levels of Zn in the medium. However, strain B-I showed higher Zn tolerance than strain B-II at the two highest levels of Zn in the medium. Results revealed that strain B-I consistently enhanced growth of plant, phosphorus and nitrogen accumulation and number of nodules and infection of mycorrhiza, which demonstrated activity of its plant growth promoting (PGP). It was recorded that the strain B-I produces IAA and accumulate 5.6 % of Zn from the growing medium. The increased nutrition and growth of plants dually inoculated with the AMF and B-I bacterium was observed at three levels of Zn. This effect can also be related to the stimulation of symbiotic structures and decreased concentration of Zn in tissues of plants. The amount of Zn acquired per unit weight of root was reduced by each of these strains of bacteria or AMF and particularly by the mixed AMF-bacterium inocula. This clearly explains the alleviation of toxicity of Zn by selected microorganisms and indicates that metal-adapted bacteria and AMF play a key role in enhancing growth of plant in soil contaminated with Zn (Vivas et al. 2006).

Zn is an essential metal for normal growth of plant and its development since it is a constituent of many enzymes as well as proteins. However, increased concentrations of this metalis toxic to many living organisms. Elevated Zn concentrations exist in many agricultural soils due to management practices that include sludge sewage application or usage of manure of animal and from activities that involve mining. Many evidences suggest that microorganisms are far more sensitive to stress caused by heavy metals than plants or animals (Vivas et al.2006).

In recent years, several studies have revealed the harmful effects caused by metals in increased concentrations on diversity of microbes and activities carried out by them in the soil. Zinc occurs only as the divalent cation Zn^{+2}, which does not undergo the redox changes under the biological conditions. It is a component in many enzymes and proteins that involve DNA-binding e.g. zinc-finger proteins, which exist in bacteria. As far as humans are concerned, toxicity of zinc may be based on zinc-induced deficiency of copper. However, the toxicity of zinc is much less as compared to copper. In *E.coli*, zinc toxicity is similar to that of nickel, cobalt and copper. It is observed that AMF are microorganisms of soil that establish mutual symbiotic relation with majority of the roots of higher plants, providing a direct physical link between plant and soil roots. They are found in usually all habitats and all climates. Thus, changes in population of AMF diversity produced by the presence of increased amounts of metals are expected to interfere with the possible beneficial effects of this association (Vivas et al. 2006).

The mycorrhizal symbiosis generally occurs in the presence of many microorganisms and this hypothesis is supported by abundant amount of literature. The literature further reveals that interaction of some of the microbes is specific and there are certain ways of influencing the mycorrhizal relationship and its effects on growth of plants. Thus, the microorganisms are associated with complement activity of mycorrhizae. One of these groups of bacteria, the so-called rhizobacteria which enhances the growth of plants known as plant-growth-promoting rhizobacteria (PGPR) has been reported by a number of others to have interacted with AMF. The final effect of microorganisms of soil that includes AMF, on the development of plants is the result of interactions among different microbial components that are involved. In contrast to that, only a few studies focus on the interactions between PGPR and AMF along with heavy metals as source of disturbance of soil. A significant increase in the growth of plant has been observed in the presence of heavy metals that include lead, zinc and nickel. However, the manipulation of combination of beneficial microorganisms is dependent upon the proper understanding of ecosystem in order to apply a suitable selection of microbes. In a study, the inoculation effect with two indigenous isolates of bacteria on a plantwas observed. Along with that, Zn tolerance on AMF was studied in terms of growth of plants, uptake of nutrients, acquisition of zinc and development of symbiosis. The strains of microbes used were isolated from areas contaminated with Zn for a long term. Microorganisms were assayed in single or dual coinoculation artificially in the soil contaminated with a range of Zn levels. Production of bacterial Indole Acetic Acid (IAA), ability of zinc biosorption and the number of viable bacterial cells at increasing levels of Zn were also determined (Vivas et al. 2006).

4 Air Pollution

4.1 Benzene

Benzene; a volatile organic compound (VOC), is extracted from industries of petroleum and used widely as an additive or intermediate or as a solvent in many manufacturing industries. Ambient air pollution problems can be caused by the emission of benzene from many sources, even though several organizations and countries have standard guidelines on benzene concentrations (PCD 2007). It is found that concentration of benzene in the atmosphere is higher than the local standard. In addition to this, 9 cohort and 13 case control studies confirmed that benzene can clearly induce many types of cancers including acute myelogenous leukemia. Many researchers have classified benzene in IA group, which is composed of carcinogens having high potential in human body by IARC. Other diseases that are likely to be caused by benzene include allergies, asthma, dizziness, tremors, eye irritation, restlessness and disorders of nervous system. Environment also has a tendency to accumulate and store benzene. Studies revealed that some plant species have the ability to take up gaseous benzene through the automata and wax on leaf surface (Treesubsuntorn et al. 2013).

A recent study showed that even when plants are grown under dark conditions, the plant could still take benzene up via cuticular wax because at night the stomata are closed. Many scientists found benzene accumulation in the cuticular wall. These days, for the treatment of benzene, activated carbon is widely used. However, there is a problem in the secondary waste disposal as well as high control cost. Use of plant leaf material for adsorption of benzene is beneficial in a way of a low cost adsorbent (Treesubsuntorn et al. 2013).

4.2 Ozone

Another harmful air pollutant is the tropospheric ozone (O_3) that has negative impact on the growth and development of plants. Current concentration of O_3 is decreasing the productivity of forest and crop yields. Future O_3 will increase if the current emission rate continues. When data from different studies were compared, it was noted that ambient O_3 decreases the seed number by nearly 16 % as well as fruit number and fruit weight by 9 % and 22 % respectively as compared to charcoal filtered air. Additionally, germination of pollen and tube growth were also decreased by elevated concentration of O_3 compared to charcoal filtered air. Relative to ambient air, fumigation with O_3 between 70 and 100 ppb decreased fruit yield by 27 % and individual seed weight by 18 %. Reproductive development of both C_3 and C_4 plants was found to be sensitive to elevated O_3 and lifecycle, flowering class and reproductive growth habit did not significantly affect a plant's response to elevated O_3 for many components of reproductive development. However, elevated O_3

decreased weight of fruit and its number significantly in indeterminate plants, and had no effect on these parameters in determinate plants. While gaps in knowledge remain about the effects of O_3 on plants with different growth habits, reproductive strategies and photosynthetic types, the evidence strongly suggests that detrimental effects of O_3 on reproductive growth and development are compromising current crop yields and the fitness of native plant species (Ainsworth and Leisner 2012).

Studies reveal that development of sexual reproduction is a critical stage in the plant's life cycle and many other stages of reproductive development are sensitive to this ozone. Ozone is basically a secondary pollutant that is formed by the oxidation via photochemical pathway of methane, carbon monoxide and other volatile compounds that are organic in nature (VOCs) in the presence of oxides of nitrogen (NOx). During warm and sunny weather, the formation of O_3 is the greatest, which coincides with times of maximum plant growth and reproductive development. Therefore it is considered as one of the most damaging tropospheric air pollutants. A considerable increase in the concentration of O_3 is observed since the industrial revolution in past 60 years. It is likely that the concentration will further increase by 10–30 parts per billion (ppb) by 2100 if the practices in Northern hemisphere are not put to an end (Ainsworth and Leisner 2012).

This increase is substantial considering the current tropospheric concentration less than 40 ppb in most parts of the world. These concentrations are believed to increase and exceed the international environmental criteria for protection of crop as well as natural vegetation and human health. Ozone enters the plants through their stomata, which is also the site of carbon dioxide (CO_2) uptake. Once taken up by the plant, O_3 rapidly reacts to form other reactive oxygen species (ROS), including hydrogen peroxide, singlet oxygen and hydroxyl radicals. Increased levels of ROS can lead to programmed cell death. ROS influx can also change the redox potential, levels of hormone and peroxidation of lipids in the apoplastic and intercellular spaces within plant cells. Reduction in photosynthesis and conductance by stomata in leaves are reduced by chronic exposure to O_3 that leads to reduced production in biomass and reproductive crop output. However, it still remains unclear that what proportion of decrease in the reproductive output is caused by direct damage to reproductive processes such as initiation of flower, development of ovary and pollen and abortion of seed as opposed to damage to the vegetation that subsequently reduces availability of assimilate to reproductive development. Ozone can have a direct effect on the reproductive structures of plant i.e., stylar and stigmatal surfaces, anthers, pollens, floral sites, seeds as well as fruits. The entry of O_3 via apoplastic space gives rise to the production of ROS, which eventually causes changes in surfaces of stigma and membranes of cells. The ability of pollen to germinate stigmatal surface is also affected by ROS. Moreover, a change is mediated by ROS in the cellular redox environment that can potentially lead to changes in the growth of pollen tubes, which is critical for embryo fertilization. Ozone (O_3) affects all the reproductive growth and development of plants. Nearly a decade ago, effect of O_3 on reproductive development of plant was reviewed, but recent quantitative assessments have focused on the influence of O_3 on growth and development of vegetative parts of plants or yield of crop. The complex nature of O_3 affects the vegetative and reproductive structures, a range of compensatory mechanisms available to plants with different habits of reproductive growth, the dependence of development stage of plants on sensitivity level and the consequences of environmental stresses, in addition, make it difficult to generalize O_3 effects on reproductive development (Black et al. 2010). Still, understanding O_3 effects on the reproductive development has significant ecological and agronomic consequences that include securing food resources for the future and ensuring the fecundity and composition of species of native flora. For this verification, meta-analysis was used to test the following:

1. Moderate increase in O_3 is enough to cause decrease in reproductive growth parameters in plants with increasing O_3 resulting in more severe effects on reproductive growth
2. Plants exposed to increased levels of O_3 with additional abiotic stresses will not experience a larger decrease in growth of reproductive parts and output when compared to plants that are exposed to elevated O_3 alone
3. Less sensitivity will be shown by C4 plants towards O_3 due to lower conductance in stomata and O_3 uptake
4. At last, determinate plants with decreased flexibility in flowering will have a limited compensatory response for loss of reproductive sites, rendering them more sensitive to increased O_3 concentration than indeterminate plants (Ainsworth and Leisner 2012).

4.2.1 Tropospheric Ozone; Deeper Roots

Tropospheric ozone is currently considered to be the most important air pollutant affecting vegetation and a further increase in background ozone concentration over the coming century throughout the northern hemisphere has been predicted as a result of large increases in emissions of precursor molecules from transport and industry. Summer mean ozone concentrations across Europe are expected to reach 40–60 ppb by 2030 and some models predict that annual mean ozone concentrations could exceed 75 ppb over much of the northern hemisphere by 2100. Several studies have shown that responses of plants are better related to accumulate stomatal fluxes of ozone than to the external ozone concentration, and the flux method is being used by the Convention on Long-range Transboundary Air Pollution (LRTAP Convention) to assess the risk of ozone damage to vegetation across Europe. Ozone flux models currently takes into account the effect of climatic conditions such as temperature, vapor pressure deficit (VPD) and photosynthetically active radiation (PAR), soil moisture content, and plant growth stage on stomatal opening, and thus are well suited to modeling ozone effects in a changing climate. Changes in precipitation patterns are likely to occur in the next few decades, and although there are predictions of increased humidity in some areas, it is likely that there will be reduced soil moisture across much of Europe. Predictions of future ozone impacts therefore need to take into account the modifying effects of prolonged ozone exposure on plant responses to soil moisture (Hayes et al. 2012).

A generalized flux model has been described by Emberson et al. (2000), which is a multiplicative model based on a Jarvis approach using temperature, VPD, PAR, phenology, ozone and soil moisture as model inputs. However, few studies have included consideration of soil moisture when measuring ozone effects even though soil moisture is a component of the stomatal flux model. In most cases, calculations of stomatal flux have been performed for experiments involving well-watered (WW) plants where restriction of stomatal conductance due to reduced soil moisture was not thought to occur. Until recently, it has been assumed that reduced soil moisture is associated with stomatal closure, leading to reduced ozone uptake, providing some protection to plants from the negative effects of ozone exposure. Such an effect was commonly reported in earlier studies for some species using relatively high ozone treatments, for example in wheat (*Triticum aestivum*; 80 ppb), tomato (*Lycopersicon esculentum*; 68 ppb) and common ash (*Fraxinus excelsior*; 150 ppb). Although the assumption that reduced soil moisture protects plants from ozone exposure is largely accepted, there are a growing number of reports where the reduction in stomatal conductance due to drought has not been as large as expected, particularly in the presence of ozone, and to date these effects have been largely overlooked. For example, in the field, with 8-h mean ozone concentrations of 51 ppb, ozone-induced visible injury in black cherry (*Prunus serotina*) was greater at drier sites than at wetter sites, with higher stomatal conductance measured in the drier than the wetter sites for both black cherry and white ash. Although no mechanism was presented, the authors hypothesized that either the trees had acclimatized to the drier conditions or that water availability was not low enough to induce stomatal closure. In a different study, severe drought stress protected birch (*Betula pendula*) from ozone injury as predicted, but under less severe drought-stress enhanced ozone damage was observed compared to that of WW plants using ozone concentrations 1.89 ambient. Again, the authors suggested no mechanism to explain this effect. The ability of drought to protect natural vegetation plants from ozone injury using ozone concentrations of up to ca. 200 ppb was found to be species-specific in a study by Bungener et al. (1998). While some species showed reduced ozone-induced injury symptoms due to drought-induced stomatal closure (e.g. white clover; *Trifolium repens*), for other species there was either no stomatal closure or an increase in stomatal opening in the presence of drought (e.g. false oat grass; *Arrhenaterum elatius*).

Some recent studies have shown that interactions between soil moisture and ozone can occur which could explain some of these published anomalies. For example, a decreased ability of stomata to close in response to drought in the presence of increasing ozone exposure has been demonstrated for ca. Ten grassland species including *Leontodon hispidus* and *Dactylis glomerata* and the widespread grasses *Anthoxanthu modoratum*, *Lolium perenne* and *Phleum pratense*. This has been attributed to a reduced responsiveness of stomata following elevated ozone exposure, and in particular an ozone-induced decrease in sensitivity to abscisic acid, which is produced in roots as a response to drought and transported to shoots in the xylem, where it induces stomatal closure via a network of chemical messengers.

The increasingly large number of 'anomalies' where drought-induced stomatal closure and protection from ozone-induced injuries does not happen suggests that this effect may possibly be fairly widespread among vegetation.

5 Nanoparticles; a New Threat

Among all the factors that influence the quality of soil, biological indicators are reported to be critically important because organisms of soil influence directly the processes in soil ecosystem, especially the decomposition of organic matter of the soil and nutrient cycling. Hence, any factor that affects the biomass of soil microbes, their activity and populations would necessarily affect the quality of soil and its sustainability. At present, an increasing number of engineered nanoparticles (ENPs), that are employed for environmental and industrial applications or are formed as by-products of activities by humans, are finding their way into soils. The antimicrobial activity of these ENPs has been studied extensively with human pathogenic bacteria. Similarly, studies also exist on the effect of ENPs on beneficial microbes *in vitro* under controlled conditions. But very little information is available on how these ENPs affect microbial communities in soil under field conditions. According to some of the published data, among ENPs, the fullerenes and their derivatives are less toxic whereas metal oxide ENPs and small sized metals pose detrimental effect to the microbial communities in the soil. However, under field conditions, organic matter of the soil and its related components, like fulvic and humic acids, could possibly negate the toxic effects of these ENPs through a number of mechanisms. Also, the resilience as well as resistance of microbial communities of soil to such perturbations cannot be discounted (Dinesh et al. 2012).

The quality of soil is usually defined as the capacity of a soil to function, within natural or managed ecosystem boundaries, to sustain plant and animal productivity, maintain or enhance water and air quality, and support human health and habitat. Among the factors that influence quality of soil, biological indicators are reported as critically most important because soil organisms have a direct influence on the ecosystem of soil processes, especially in soil decomposition of the organic matter and the nutrient cycling. Therefore, the protection of microbial biomass of the soil and its diversity is among the major challenges for use of sustainable resource. It is because the increased levels of microbial diversity and biomass means greater turnover in the nutrients and disease soil suppressiveness. The opposite is true for a sick soil with low nutrient and reserves of carbon and greater levels of contaminants that is caused by the presence of chemicals of xenobiotic source or other alteration in the environment of soil. As far as the xenobiotics are concerned, there are staggering number of new nanoparticles being engineered for application in the industrial and environmental sector or are formed as a by-product of human activity which are already finding their way into the soils. While the concentrations of most of the ENPs in the environments still remain unknown, exposure modeling suggests that the soil could act out as a major ENPs sink that is released into the environment.

5.1 Engineered Nanoparticles and Their Toxicity to Microorganisms

Engineered nanoparticles are artificially manufactured particles that are engineered by man because of their specific properties in the field of nanotechnology in terms of properties, size, behavior etc. The question to the answer that why ENPs built material have different electrical, optical, magnetic, mechanical and chemical properties from their bulk counterparts are that in this range of size, quantum effects start to reveal their predominate nature and there is an increased surface-area-to-volume ratio (sa/vol).

The sa/vol of most materials increases gradually as their particles become smaller in size, which results in the increased adsorption of the surrounding atoms and bring about the changes in their behavior and properties, subsequently. Once these particles are small enough, they work under the principles of quantum laws of mechanics. Suddenly, the particles that have been reduced to nano-scale display extremely different properties compared to what they exhibit on macro-scale. This enables unique applications e.g. opaque substances become transparent (copper), the materials that are stable becomes combustible (aluminum), materials that are inert start acting as catalysts (platinum), insulators act as conductors (silicon), some turn to liquids when kept at room temperatures (gold) (Hristozov and Malsch 2009).

ENPs can also be made up of single elements like carbon (C) or silver (Ag) or elements/molecules mixture. ENPs are often classified on the basis of their chemical composition, supplemented with size or morphological characteristics. Many ENPs are described viz. oxides (TiO_2, CuO, FeO_2, ZnO, Al_2O_3, CeO_2, SiO_2), Fullerenes (grouping of Buckminister fullerenes, carbon nano tubes (CNTs), nanocones etc.), ENPs metals (elemental Ag, Au, Fe etc.) and complex compounds like Co-Zn-Fe oxide and quantum dots are often coated with a polymer e.g. cadmium-selenide (CdSe) and polymers of organic compounds (dendrimers, polystyrene etc.). The increasing entry of the ENPs will lead to their accumulation in soil inevitably, which raise concerns about their potential adverse effects on the microbial activity of soil and its diversity. At present very little is known about the effect of these ENPs on the microbial community of the soil. They may have an impact on the microorganisms of soil via the following:

1. a direct effect (toxicity)
2. changes in the bioavailability of nutrients or toxins
3. indirect affects resulting from their interaction with natural compounds of organic nature
4. interaction with toxic organic compounds, which would alleviate or amplify their toxicity (Simonet and Valcarcel 2009).

Mechanisms of toxicity have not yet been completely elucidated for most ENPs. The possible mechanisms include protein oxidation, genotoxicity, energy transduction interruption, formation of reactive oxygen species (ROS) and the release of constituents that are toxic in nature. However, close contact is necessary for disruption of membrane to occur and it is likely that NPs cross into the cytoplasm although accumulation within the cytoplasm, probably after the disruption of membrane is often observed. It was also noted that the antibacterial activity possessed by NPs might involve ROS production as well as NPs accumulation in the cytoplasm or on the outer membranes. Structural changes may also be caused by ENPs on the microbial surface of cells that may eventually lead to cell death. It is, therefore, apparent that ENPs stimulate the production of ROS in organisms and cause damage in possibly every cell component (Bhatt and Tripathi 2011).

6 Role of Forests

European forests are facing significant changes in climate, which, in interaction with changes in the quality of air may affect significantly the productivity of forest, stand composition and sequestration of carbon in both soil and vegetation. Gaps in the identified knowledge and research needs include:

(i) Interactions between changes in quality of air (concentrations of trace gases), climate and other site factors in response to ecosystem of forest.
(ii) Significance of biotic processes in response to system
(iii) Up scaling and understanding diagnostic and mechanistic tools
(iv) Need for empirical research and unifying modeling for synthesis.

The forest ecosystems are crucial ecologically which cover 30 % of the land area of earth. Globally it has been recorded that forests store more than 805 of all the terrestrial carbon (C) aboveground and more than 70 % of all-organic soil carbon. Overall, currently global C sinks are considered to be forests whereas croplands are the sources because of more frequent disturbance in soil associated with practices involving agriculture. By the changes in photosynthesis, soil respiration and respiration, it will be determined whether forests will remain carbon sinks, affecting the net ecosystem of carbon flux. Such processes are strongly affected by the changes in:

(a) Quality of air comprising of deposition of nitrogen (N), atmospheric carbon dioxide (CO_2) concentration, ozone (O_3) exposure and fine particulates/aerosols
(b) Warming of climate with effects on water availability
(c) Soil acidity and availability of non-N nutrients (De Vries and Posch 2010)

These drivers affect sequestration of carbon in both above as well as below ground biomass in the forest ecosystem. Insights into the multi-factorial, influences on quality of air and changes in the climate, is crucial to provide a robust evidence base to policy makers. Such influences arise from the resources of fossil energy combustion and changes in land-use (i.e. clear cutting of forest and its burning), altogether releasing CO_2, oxides of nitrogen (NOx) and other climate related trace gases. Emissions from forests add to those from intense practices of agriculture, specifically ammonia (NH_3) and methane (CH_4) from livestock (Matyssek et al. 2012).

Additionally, natural emissions of organic compounds, which are volatile in nature (VOCs) from vegetation, increase the mixture of reactants in the atmosphere. With the accumulation of trace gases, there might be an increase in their reactivity, if there is an enhancement in the isolation as an effect of warming atmosphere. Under increased irradiance, some of the reactants become precursors to the formation of secondary pollutants such as O_3. With its concentration well above the levels of pre-industrial era and given the recent decrease in the emissions of sulfur in Europe and North America, O_3 is regarded as the pollutant of air potentially most detrimental to the vegetation (Matyssek et al. 2012).

On such grounds, currently, forests face significant pressures from changes in climate and the pollution in air. Until recently, global dynamic vegetation models predicted an increase in the global productivity and terrestrial sequestration of carbon in response to the climate shifts and concentration of CO_2. There have been many criticisms on these studies for over estimating the potential feedback of carbon climate because accumulation of carbon may be constrained by nutrients particularly by nitrogen (N) and by the negative impacts of increased exposure of O_3. Ozone may cause a limitation on substantial global scale by the end of this century with consequences of significant value for radiative forcing in the atmosphere by elevated levels of CO_2 (Matyssek et al. 2012).

7 Future Prospects

7.1 Synthetic Soil

In a study, the synthetic soil prepared from sand, organic material and topsoil was treated with magnetite powder for the stimulation of contaminated soil. Two soil types were chosen to prepare six soil treatments. Low contamination treatments, high contamination treatments and control. Above all, the contaminated soil had a greater decrease in the susceptibility of magnetism than the control and the difference in magnetic susceptibility (MS) decrease between the treatments was found to be statistically significant for both types of soil. Possible reasons for the overall decrease in MS were explored and among them trace uptake of elements by plants probably had a minor contribution as the differences in concentration of Fe and other trace elements (Mn, Ni) between treatments were not significant statistically. In soils, weakly or oxidized magnetic minerals, which include maghemite, hematite and goethite, were common after the growth of plants, when compared with the untreated soil. Overall decrease in MS can be contributed by transformation of minerals. The result reveals that exposure of contaminants of Fe can affect the growth of plants and suggest that growth of plant can measurably change the magnetic properties of their growth media. While thepotential variables affecting growth of plant were controlled as much as possible, but the possibility that biotic and abiotic chemical reactions could have affected the results, remains there. Thus, continuous monitoring of the changes in chemical and magnetic properties of soil in more complex soil–plant systems is required (Sapkota et al. 2012).

In soils, for the measurement of toxic metals, magnetic measurements have been used as a proxy. This may also be used to measure the sediments caused by emissions from industries. Magnetic studies of many pollutants have been studied by many researchers (Spassov et al. 2004) as well as toxic metals in polluted soils (Schmidt et al. 2005). In some of the studies, positive correlation between chemical analysis and magnetic measurement of soils and sediments have been presented, and they reveal that the concentration of magnetic minerals, that is present in the samples, exposed to toxic elements from an anthropogenic source is related to the concentration of these elements. In recent years, researchers have extended the application of methods using magnetic streaks in the study of environmental pollution by its integration with other science disciplines, which includes biological applications e.g. the bio-monitoring of air pollution traffic using the tree leaves magnetic properties, and also civil/environmental engineering applications, such as studies of building facades to measure the rates of erosion and monitoring of contamination of urban atmosphere. Microbe mediated magnetite formation was detected precisely by *in-situ* susceptibility of magnetite (Porsch et al. 2010). This study examined the potential of using magnetic methods to monitor the changes in magnetic properties of soil during the growth of plant in soil contaminated with metal and the possible use of these changes as a proxy for monitoring the uptake of toxic metals by plants. It has been found that plant species such as mustard, cucumber and dandelion are metal accumulating and they transport and concentrate metals from the polluted soil into their harvestable parts like above ground shoots and roots, in a process known as phytoextraction. Iron uptake mechanism is also observed in tomato plants. Tomato plants have also been used to understand iron deficiency stress and the metabolic response, the uptake of trace elements with time and the fly ash effect amendments in uptake of trace elements. It was noted by Jensen et al. (2004)that when the fly ash content of soil was increased, tomato plant increased the uptake of Cd, Co and Mo.

7.2 *Phytoremediation*

Phytoremediation is a cost effective strategy for remediation of soil polluted with heavy metals. Phytoremediation, or the use of green plants in order to clean up the polluted sites, is promising technology for cleaning up of the polluted sites with heavy metals. In some studies, phytoremediation from technical aspect together with its advantages and limitations has also been studied. Different strategies come under the term "phytoremediation" such as:

1. Phytoextraction
2. Phytostabilization
3. Phytodegradation/phytotransformation

4. Rhizofiltration/phytofiltration/blastofiltration
5. Phytovolatilization
6. Rhizoremediation/rhizodegradation/plant-assisted degradation/plant-aided *in-situ* biodegradation/phytostimulation
7. Phytosorption
8. Phtyocapping (Alkorta et al. 2010)

In case of organic pollutants, rhizoremediation appears to have increased potential because it combines plant's ability in detoxification of xenobiotics (the so-called "green liver") with the capacity of rhizosphere microorganisms in degradation of organic compounds. On the other hand, in case of heavy metals, phytostabilization and phytoextraction are the strategies of choice (Maria et al. 2012).

7.3 Activated Carbon

For improvement in the land for better production of plants, certain measures are under practice. The amendment of 2 % powder and granular activated carbon, also known as PAC and GAC, to a soil having moderate contamination of PAH had an impact on the PAH bioaccumulation of plants and earthworms, since AC is known to be a sorbent of strong nature for organic pollutants. Furthermore, AC secondary effects on the earthworms and plants were studied through the uptake of nutrient and growth as well as weight gain and survival. In addition, the effects of AC amendments on the characteristics of soil like capacity of holding water, pH and retention of water curve of soil were also studied. Results revealed that the amendment of 2 % PAC had a negative effect on the growth of plant while the increased GAC helped in the increase of plant growth rate. PAC had a toxic effect on earthworms that was demonstrated by weight loss to a significant level, while the results for GAC were less clear due to ambiguity in the result of a field and a parallel laboratory study. Both kinds of AC significantly reduced biota to soil accumulation factors (BSAFs) of PAHs in earthworms and plants. The GAC reduced the BSAFs of earthworms by an average of 47 ± 44 % and the PAC amendment reduced them by 72 ± 19 %. For the investigated plants, the BSAFs were reduced by 46 ± 36 % and 53 ± 22 % by the GAC and PAC, respectively.

8 Conclusions

Plants are the most important entities of food chain. Along with the above-mentioned pollutants, a number of other pollutants are also causing damage to the plant population either directly or indirectly. There are certain remedies to be done otherwise it will impose a negative effect on not only the plants but on humans and animals as well. The population of pollinators is also being effected. Certain measures are to be taken to decrease the concentration of heavy metals. Air and soil needs filtering agents and several check points are also required to control the immense chaos and loss being done to the plants. If this goes untreated or uncontrolled, the future of plants will be in grave danger.

References

Aguilar R, Ashworth L, Galetto L, Aizwen MA (2006) Plant reproductive susceptibility to habitat fragmentation: revive and synthesis through a meta-analysis. Ecol Lett 9:968–980

Ahmad M, Lee SS, Yang JE, Ro HM, Lee YH, Ok YS (2012a) Effects of soil dilution and amendments (mussel shell, cow bone, and biochar) on Pb availability and phytotoxicity in military shooting rnage soil. Ecotoxicol Environ Saf 79:225–231

Ahmad M, Usman AR, Lee SS, Kim SC, Joo JH, Yang JE et al (2012b) Eggshell and coral wastes as low cost sorbents for the removal of Pb2+, Cd2+ and Cu2+ from aqueous solutions. J Ind Eng Chem 18:198–204

Ainsworth EA, Leisner CP (2012) Quantifying the effects of ozone on plant reproductive growth and development. Glob Chang Biol 18:606–616

Alkorta I, Becerril JM, Garbisu C (2010) Phytostabilization of metal contaminated soils. Rev Environ Health 25:135–146

Alston DG, Tepedino VJ, Bradley BA, Toler TR, Griswold TL, Messinger SM (2007) Effects of the insecticide phosmet on solitary bee foraging and nesting in orchards of Capitol Reef National Park, Utah. Environ Entomol 36:811–816

APWA (1991) American wood-preservers' association book of standards. American Wood-Preservers Association, Grandbury

Ashmann TL, Knight TM, Streets JA, Amarasekare P, Burd M, Campbell DR et al (2004) Pollen limitation of plant reproduction: ecological and evolutionary causes and consequences. Ecology 85:2408–2421

Beyrem H, Mahmoundi E, Essidi N, Hedfi A, Boufahaja F, Aissa P (2007) Indiviual and combined effects of cadmium and diesel on a nematode community in a laboratory microcosm experiment. Ecotoxicol Environ Saf 68:412–418

Bhatt I, Tripathi BN (2011) Interaction of engineered nanoparticles with various components of the environment and possible strategies for their risk assessment. Chemosphere 82:308–317

Biesmeijer JC, Roberts SP, Reemer M, Ohlemuller R, Edwards M, Peeters T et al (2006) Parallel declines in pollinators and insect-pollinated plants in Britain and the Netherlands. Science 313:351–354

Black VJ, Stewart CA, Roberts JA, Black CR (2010) Direct effects of ozone on reproductive development in Plantago major L. populations differing in sensitivity. Environ Exp Bot 69:121–128

Bobbink R, Hicks K, Galloway J (2010) Global assesment of nitrogen deposition effects on terrestrial plant diversity: a synthesis. Ecol Appl 20:30–59

Brittain C, Bommarco R, Vighi M, Barmaaz S, Settele J, Potts SG (2010) The implication of an insecticide on insect flower visitation and pollination in an agricultural landscape. Agric For Entomol 12:259–266

Bungener P, Balls GR, Nussbaum S, Geissmann M, Grub A, Fuhrer J (1998) Leaf injury characteristics of grassland species exposed to ozone in relation to soil moisture condition and vapour pressure deficit. New Phytol 132:271–282

Cao X, Ma LQ, Chen M, Singh SP, Harris WG (2002) Impacts of phosphate amendments on lead biogeochemistry at a contaminated site. Environ Sci Technol 36:5296–5304

Chen TB, Zheng YM, Lei M, Huang ZC, Wu HT, Chen H et al (2005) Assessment of heavy metal pollution in surface soils of urban parks in Beijing, China. Chemosphere 60:542–551

Chirenje T, Ma LQ, Clark C, Reeves M (2003) Cu, Cr and As distribution in soils adjacent to pressure-treated decks, fences and poles. Environ Pollut 124:407–417
Colla SR, Otterstatter MC, Gegear RJ, Thomanson JD (2006) Plight of the bumble bee: pathogen spillover from commercial to wild population. Biol Conserv 129:416–467
Das DK, Nandi A (2002) Health hazard in stone quarries and crushers. NISM, Kolkata, Science and technolgy project
Dawson BS, Parker GF, Cowan FJ, Hong SO (1991) Interlaboratory determination of copper, chromium, and arsenic in timber treated with preservative. Analyst 116:339–346
De Vries W, Posch M (2011) Modelling the impact of nitrogen deposition, climate change and nutrient limitations on tree carbon sequestration in Europe for the period 1900–2050. Environ Pollut 159(10):2289–2299
Dinesh R, Anandaraj M, Srinivasan V, Hamza S (2012) Engineered nanoparticles in the soil and their potential implications to microbial activity. Geoderma 173:19–27
Dise NB, Ashmore M, Belyazid S (2011) Nitrogen deposition as a threat to European terrestrial biodiversity. In: The European nitrogen assessment: sources, effects and policy perspectives. Cambridge University Press, Cambridge, pp 463–494
Dupre C, Stevens CJ, Ranke T (2010) Changes in species richness and composition in european acidic grasslands over the past 70 years: the contribution of cumulative atmospheric nitrogen deposition. Glob Chang Biol 16:344–357
Emberson LD, Ashmore MR, Cambridge HM, Simpson D, Tuovinen JP (2000) Modelling stomatal ozone flux across Europe. Environ Pollut 109:403–413
ES (1998) 'Status of environment in West Bengal', part-II: vectors of environmental impacts, Pub. Department of Environment, West Bengal
Fries W, Lombi E, Horak O, Wenzel WW (2003) Immobilization of heavy metals in soils using inorganic amendments in a greenhouse study. J Plant Nutr Soil Sci 166:191–196
Gallai N, Salles JM, Settele J, Vaissiere BE (2009) Economic valuation of the vulnerability of world agriculture confronted with pollinator decline. Ecol Econ 68:810–821
Galloway JN, Dentener FJ, Capone DG (2004) Nitrogen cycles: past, present and future. Biogeochemistry 70:153–226
Gezer ED, Yildiz UC, Temiz A, Yildiz S, Dizman E (2005) Cu, Cr and As distribution in soils adjacent to CCA-treated utility poles in Eastern Blacksea Region of Turkey. Build Environ 40:1684–1688
Hakeem KR, Sabir M, Ozturk M, Mermut A (2015) Soil remediation and plants: prospects and challenges. Academic/Elsevier, New York, 724 pp
Hayes F, Wagg S, Mills G, Wilkinson S, Davies W (2012) Ozone effects in a drier climate: implications for stomatal fluxes of reduced stomatal sensitivity to soil drying in a typical grassland species. Glob Chang Biol 18:948–959
Hingston AJ, Collins CD, Murphy RJ, Lester JN (2001) Leaching of chromated copper arsenate wood preservative: a review. Environ Pollut 111:53–66
Hristozov D, Malsch I (2009) Hazards and risks of engineered nanoparticles for the environment and human health. Sustainability 1:1161–1194
Isoyama M, Wada S (2007) Remediation of Pb-contaminated soils by washing with hydrochloric acid and subsequent immobilization with calcite and allophonic soil. J Hazard Mater 413:636–642
Jakob L, Hartnik T, Henriksen T, Elmquist M, Brandli RC, Hale SE et al (2012) PAH-sequestration capacity of granular and powder activated carbon amendments in soil, and their effects on earthworms and plants. Chemosphere 88:699–705
Jensen RR, Brake SS, Mattox JM (2004) Trace element uptake in plants grown on fly ash amended soils. Toxicol Environ Chem 86:219–230
Jonker MT, Suijkerbuijk MP, Schmitt H, Sinnige TL (2009) Ecotoxicological effects of activated carbon addition to sediments. Environ Sci Technol 43:5959–5966
Kavamura VN, Esposito E (2010) Biotechnological strategies applied to the decontamination of soils polluted with heavy metals. Biotechnol Adv 28:61–69

Kenta T, Inari N, Nagamitsu T, Goka K, Hiura T (2007) Commercialized European bumblebee can cause pollination disturbance: an experiment on seven native plant species in Japan. Biol Conserv 134:298–309
Kevan PG (1999) Pollinators as bioindicator of the state of environment: species, activity and diversity. Agric Ecosyst Environ 74:373–393
Kim H, Kim DJ, Koo JH, Park JG, Jang YC (2007) Distribution and mobility of chromium, copper, and arsenic in soils collected near CCA-treated wood structure in Korea. Sci Total Environ 374:273–281
Klein AM, Vaissiere BE, Cane JH, Steffan-Dewenter I, Cunninhgham SA, Kremen C et al (2007) Importance of pollinators in changing landscapes for world crops. Biol Sci 274:303–313
Kraus FB, Szentgyorgyi H, Rozej E, Rhode M, Moron D, Woyciechowski M (2011) Greenhouse bumblebees (Bombus terrestris) spread their genes into the wild. Conserv Genet 12:187–192
Kumpiene J, Lagerkvist A, Maurice C (2008) Stabilization of As, Cr, Cu, Pb and Zn in soil using amendments – a review. Waste Manage 28:215–225
Le Féon V, Schermann-Legionnet A, Delettre Y, Aviron S, Billeter R, Bugter R, et al (2010) Intensification of agriculture, landscape composition and wild bee communities: a large scale study in four European countries. Agric Ecosyst Environ 137(1):143–150
Lenda M, Skórka P, Moroń D (2010) Invasive alien plant species a threat or a chance for pollinating insects in agricultural landscapes? In: Lee TH (ed) Agricultural economics: new research. Nova Science Publishers, New York, pp 67–87
Ma TT, Christie P, Luo YM, Teng Y (2013) Phthalate esters contamination in soil and plants on agricultural land near an electronic waste recycling site. Environ Geochem Health 35:465–476
Maria TG-S, Alkorta I, Becerril JM, Epelde L, Anza M et al (2012) Microbial monitoring of the recovery of soil quality during heavy metal phytoremediation. Water Air Soil Pollut 223:3249–3262
Maskell LC, Smart SM, Bullock JM, Thompson K, Stevens CJ (2010) Nitrogen deposition causes widespread loss of species richness in British habitats. Glob Chang Biol 16:671–679
Matyssek R, Wieser G, Calfapietra C, De Vries W, Dizengremel P, Ernst D, et al (2012) Forests under climate change and air pollution: gaps in understanding and future directions for research. Environ Pollut 160:57–65
Mishra M (2004) Project achievement reports: update, Nov. 2004, on demonstration of dust suppression technology to reduce occupational and environmental health hazards in stone crusher industries, a project study in Khurda, Orissa, India by JRP, VJSS (India) and OKI (USA)
Mishra PC (2013) Effects of stone crushers' pollution on plant morphology and biochemistry. Ecoscan 3:307–313
Moon DH, Kim KW, Yoon IH, Grubb DG, Shin DY, Cheong KH et al (2011) Stabilization of arsenic-contaminated mine tailings using natural and calcined oyster shells. Environ Earth Sci 64:597–605
Morandin LA, Winston ML (2005) Wild bee abundance and seed production in conventional, organic and genetically modified canola. Ecol Appl 15:871–881
Moron D, Szentgyorgyi H, Settele J, Woyciechowski M, Skorka P (2009) Wild pollinator communities are negatively affected by invasion of alien goldenrods in grassland landscapes. Biol Conserv 142:1322–1332
Moron D, Grzes IM, Skorka P, Szentgyorgyi H, Laskowski R, Potts SG et al (2012) Abundance and diversity of wild bees along gradients of heavy metal pollution. J Appl Ecol 49:118–125
Mulder C, Aldenberg T, De Zwart D, Van Wijnen HJ, Breure AM (2005) Evaluating the impact of pollution on plant- Lepidoptera relations. Environmetrics 16:357–373
Naidoo G, Chirkoot D (2004) The effects of coal dust on photosynthetic performance of the mangrove, Avicennia marina in Richards Bay, South Africa. Environ Pollut 127:359–366
Nieminen M, Nuorteva P, Tulisalo E (2001) The effect of metals on the morality of Parnassius apollo larvae (Lepidoptera: Papilionidae). J Insect Conserv 5:1–7
Ok YS, Yang JE, Zhang YS, Kim SJ, Chung DY (2007) Heavy metal adsorption by a formulated zeolite-Portland cement mixture. J Hazard Mater 147:91–96

Ok YS, Kim SC, Skousen JG, Lee JS, Cheong YW, Kim SJ et al (2011a) Ameliorants to immobilize Cd in rice paddy soils contaminated by abandoned metal mines in Korea. Environ Geochem Health 33:23–25

Ok YS, Usman AR, Lee SS, Abd El-Azeem SA, Choi BS, Hashimoto Y et al (2011b) Effects of rapeseed residue on lead and cadmium availability and uptake by rice plants in heavy metal contaminated paddy soil. Chemosphere 85:677–682

Ozturk M (ed) (1989) Plants and pollutants in developed and developing countries. Ege University Press, Izmir, 759 pp

Phoenix GK, Emmet BA, Britton AJ, Caporn SJ, Dise NB, Helliwell R et al (2012) Impacts of atmospheric nitrogen deposition: responses of multiple plant and soil parameters across contrasting ecosystems in long-term field experiments. Glob Chang Biol 18:1197–1215

Piola RF, Johnston EL (2008) Pollution reduces native diversity and increase invader dominance in marine hard-substrate communities. Divers Distrib 14:329–342

Pollution Control Department (PCD) (2007) Development of environmental and emission standards of volatile organic compounds (VOCs) in Thailand. Pollution Control Department (PCD), Bangkok

Porsch K, Dippo U, Rijal ML, Appel E, Kappler A (2010) In Situ magnetic susceptibility measurements as a tool to follow geomicrobiological transformation of Fe minerals. Environ Sci Technol 44:3846–3852

Prajapati SK, Tripathi BD (2008) Seasonal variation of leaf dust accumulation and pigment content in plant species exposed to urban particulates pollution. J Environ Qual 37(3):865–870

Prasad BJ, Rao DN (1981) Growth responses of phaseolus aureus plants to petro-coke pollution. J Exp Bot 32:1343–1350

Purohit SS, Ranjan R (2007) Ecology, environment and pollution. Agrobios, Jodhpur

Sapkota B, Cioppa MT, Gagnon JE (2012) Investigation of the changes in magnetic and chemical properties of soil during plant growth in a controlled environment. Environ Earth Sci 65:385–399

Saxe JK, Wannamaker EJ, Conklin SW, Shupe TF, Beck BD (2007) Evaluating landfill disposal of chromated copper arsenate (CCA) treated wood and potential effects on groundwater: evidence from Florida. Chemosphere 66:496–504

Schmidt A, Yarnold R, Hill M, Ashmore M (2005) Magnetic susceptibility as proxy for heavy metal pollution: a site study. J Chem Explor 85:109–117

Semban T, Chandrasekhar S (2000) Impact of crusher pollution on workers in Trichy. Environ People 7(7):29

Simonet BM, Valcarcel M (2009) Monitoring nanoparticles in the environment. Anal Bioanal Chem 393:17–21

Singh SN, Rao DN (1980) Growth on wheat plants exposed to cement dust pollution. Water Air Soil Pollut 14:241–249

Solo-Gabriele HM, Townsend TG, Messick B, Calitu V (2002) Characteristics of chromated copper arsenate-treated wood ash. J Hazard Mater 89:213–232

Spassov S, Egli R, Heller F, Nourgaliev DK, Hannam J (2004) Magnetic quantification of urban pollution sources in atmospheric particulate matter. Geophys J Int 159:555–564

Steffan-Dewenter I (2003) The importance of habitat area and landscape context for species richness of bees and wasps in fragmented orchard meadows. Conserv Biol 17:1036–1044

Steffan-Dewenter I, Potts SG, Packer L (2005) Pollinator diversity and crop pollination services are at risk. Trends Ecol Evol 20:651–652

Stevens CJ, Manning P, Van den Berg LJ (2011) Ecosystem responses to reduced and oxidised nitrogen inputs in european terrestrial habitats. Environ Pollut 159:665–676

Suzuki T, Akari S, Teranishi K, Fujii S (2014) Environmental monitoring system for map Ta Phut industrial estate in Thailand. FUJITSU Sci Tech J 50(3):16–23

Syrek D, Weiner WM, Wojtylak M, Olszowska GY, Kwapis Z (2006) Species abundance distribution of collembolan communities in forest soil polluted with heavy metals. Appl Soil Ecol 31:239–250

Treesubsuntorn C, Suksabye P, Weangjun S, Pawana F, Thiravetyan P (2013) Benzene adsorption by plant leaf material: effect of quantity and composition by wax. Water Air Soil Pollut 224:1736

Usman AR, Lee SS, Awad YM, Kyoung JL, Yang JE, Ok YS (2012) Soil pollution assessment and identification of hyperaccumulating plants in chromated copper arsenate (CCA) contaminated sites, Korea. Chemosphere 87:872–878

Velthuis HH, Van Doorn A (2006) A century of advances in bumblebee domestication and the economic and environmental aspects of its commercialization for pollination. Apidologie 37(4):421–451

Vivas A, Biro B, Ruiz-Lozano JM, Barea JM, Azcon R (2006) Two bacterial strains isolated from a Zn-polluted soil enhance plant growth and mycorrhizal efficiency under Zn-toxicity. Chemosphere 62:1523–1533

Walther-Hellwig K, Fokul G, Frankl R, Buchler R, Ekschmitt K, Wolters V (2006) Increased density of honeybee colonies affects foraging bumblebees. Apidologie 37:517–532

Wilkaniec Z, Giejdasz K, Prószyński G (2004) Effect of pollination of onion seeds under isolation by red mason bee (Osmia rufaL.) (Apoidea: Megachilidae) on the setting and quality of obtained seeds. J Apicult Sci 49:35–41

Yang JE, Skousen JG, Ok YS, Yoo KY, Kim HJ (2006) Reclamation of abonded coal mine waste in Korea using lime cake by-products. Mine Water Environ 25:227–232

Yang JE, Kim JS, Ok YS, Yoo KY, Kim HJ (2007) Mechanistic evidence and efficiency of Cr(VI) reduction in water by different sources of zerovalent irons. Water Sci Technol 55:197–202

Yoon J, Cao X, Zhou Q, Ma LQ (2006) Accumulation of Pb, Cu, and Zn in native plants growing on a contaminated Florida site. Sci Total Environ 368:456–464

Zeiger E (2006) The effect of air pollution on plants. In: Plant physiology. University of California, Los Angeles

Plants for Remediation: Uptake, Translocation and Transformation of Organic Pollutants

Giorgi Kvesitadze, G. Khatisashvili, T. Sadunishvili, and E. Kvesitadze

Contents

G. Kvesitadze (✉)
Durmishidze Institute of Biochemistry and Biotechnology, Agricultural University of Georgia, David Aghmashenebeli Alley, 13th km. 0159, Tbilisi, Georgia

Georgian National Academy of Sciences, 52 Rustaveli Ave., 0108, Tbilisi, Georgia
e-mail: kvesitadze@hotmail.com

G. Khatisashvili • T. Sadunishvili
Durmishidze Institute of Biochemistry and Biotechnology, Agricultural University of Georgia, David Aghmashenebeli Alley, 13th km. 0159, Tbilisi, Georgia

E. Kvesitadze
Department of Biotechnology, Georgian Technical University,
77 Kostava St., 0168, Tbilisi, Georgia

M. Öztürk et al. (eds.), *Plants, Pollutants and Remediation*,
DOI 10.1007/978-94-017-7194-8_12

Abstract Phytoremediation exploits plant physiological processes to decontaminate environment or to improve food chain safety by phytostabilisation of toxic elements. These phytotechnologies are based on plants remarkable absorption, transportation and metabolic capabilities that allow to uptake and transform environmental organic pollutants into nontoxic compounds or carry out their complete mineralization. Uptake of organic pollutants may occur through roots and leaves. Several factors such as cell wall permeability, temperature, pH, soil humidity affect pollutants uptake through roots. Thickness of leaf cuticle, its waxy layer serves as a rate-limiting barrier for penetration of organic pollutants in leaf cells. Translocation of organic pollutants uptaken by roots and leaves is carried out via transpiration stream and the flow of assimilates. Entering in cells pollutants undergo enzymatic transformations: functionalization, conjugation and compartmentation. Oxidases, reductases, dehalogenases, esterases and transferases, participating in these transformation processes are characterized according to their catalytic properties and regulation. Prolonged process of transformation of organic pollutants in plants that implies mostly deep oxidation is reflected on the cell normal metabolic processes. Enzymes, involved in catabolic processes leading to energy generation are indirectly participating in the detoxification process. Reversibilty of plants ultrastructural deviations and changes in enzyme activities, i.e. ability to overcome pollutants toxicity and proceed their detoxification highly depends on pollutants concentration and exposure time. Genetic engineering applied for improvement of plant remediation capacity involve: overexpression of existing or introduction of genes of enzymes catalyzing pollutants degradation; introduction of genes encoding biosynthetic pathway of biosurfactants to increase bioavailability of contaminants, etc.

Keywords Plants • Organic pollutants • Uptake • Translocation and transformation • Enzymes of pollutants transformation

1 Introduction

Plants are equipped with remarkable metabolic and absorption capabilities, as well as transport systems that allow them to uptake and transform environmental organic pollutants into nontoxic compounds or carry out their complete mineralization. Plant based remediation, i.e. phytoremediation is ecologically friendly, cost-effective technology having fast growing importance due to permanent increase of technogenic pollution, which is recognized as a serious global challenge. The chapter provides an overview of the fundamental aspects of pollutants uptake, translocation and transformation in plants. Enzyme systems involved in pollutants functionalization, conjugation and compartmentation as well as other plant processes and enzymes important for remediation are described. Latest achievements in engineering of plants with increased remediation ability are discussed.

2 Uptake of Organic Pollutants by Plants

Plants uptake the organic pollutants from water, soil and air by their roots and leaves. According to numerous experimental data, the penetration of different organic pollutants into plants physiologically is similar to processes that take place during entering of inorganic gases and nutrients into plants.

3 Uptake by Roots

Organic pollutants pass into the roots together with water, like ordinary nutrients, through cuticle-free unsuberized cell walls of young hairs of roots. After penetration, they move towards transport tissue of xylem along free intercellular space (apoplastic way) or cells (symplastic way). A comparatively small amount of pollutants move along the symplast, through cells and the plasmodesmata bridging cells, in the following way:

$$\text{Roots hairs} \rightarrow \text{intracellular spaces} \rightarrow \text{cell walls of cortical cells} \rightarrow$$
$$\rightarrow \text{endodermis} \rightarrow \text{get over the Casparian strip barrier} \rightarrow \text{xylem.}$$

The Casparian strip, which located around the central channel of the vessels and protects plants from water deficiency, creates additional resistance for the transportation of organic compounds. In the transport of hydrophobic compounds along the apoplast, the water-impermeable Casparian strip needs to be crossed, and for this an active transport must be used (Trapp and McFarlane 1995). Organic pollutants movement from root hairs to xylem is carried out mainly by apoplastic way. Penetration of compounds into the apoplast, a system of microcapillaries, takes place by diffusion. In contrast to symplastic transportation they easily move through these capillaries and do not meet membrane barriers on their way. Uptake of organic pollutants by roots takes place in two phases (Korte et al. 2000):

- In the first fast phase, pollutants diffuse from the surrounding medium into the root. Obviously, the rate of the process is directly proportional to pollutant concentration in the medium: soil or nutrient solution; besides, the intensity of the absorption process depends either on pollutant and soil physical-chemical chracteristics (solubility, lipophilicity, molecular mass, temperature, chemical content of soil, soil humidity) and some other factors (Ugrekhelidze et al. 1986; Ryan et al. 1988; Kritich and Schwarz 1989) as well as on soil morphology.
- In the second phase absorbed pollutants slowly accumulate in the tissue. The rate of this phase is determined by the course of the processes such translocation, transformation and deposition (compartmentalization) of absorbed pollutants, which will be discussed below.

A wide spectrum of hydrophilic and lipophilic organic molecules (aliphatic, aromatic and polycyclic aromatic hydrocarbons, alcohols, phenols, amines, etc.) can be uptaked by plant roots. Even substances with a very low solubility in water, such as polycyclic aromatic hydrocarbons and polychlorinated organics (among them PCBs and dioxins), were shown to be absorbed by roots (Dörr 1970; Ugrekhelidze and Durmishidze 1984; O'Connor et al. 1990; Webber et al. 1994; Kvesitadze et al. 2006; Inui et al. 2008). The most important property of pollutants for the process of absorption by roots is hydrophobicity, which usually is expressed as the 1-octanol/water partition coefficient (K_{ow}), or more often log K_{ow} (Dowdy and McKone 1997). Log K_{ow} spans over a wide range for different organic compounds. To describe the distribution of a chemical in soil the soil/water partition coefficient (K_d) is used. K_d is generally proportionally to the hydrophobicity of the compound and to the amount of soil organic matter (Briggs 1981, 1982, 1983; Korte et al. 1992; Sicbaldi et al. 1997). Hydrophobicity of organic pollutants significantly determines the degree of their sorption by soil. Hydrophobic compounds with K_{ow} >4, will be strongly sorbed and moderately hydrophobic compounds, K_{ow} 2–4, moderately sorbed (Trapp and McFarlane 1995). Organic pollutants with log K_{ow}>3.5, such as 1,2,4-trichlorobenzene, 1,2,3,4,5-pentachlorophenol, PAHs, PCBs, dioxins, etc., are well sorbed on soil granules or plant root surfaces and do not penetrate into the plant interior. Moderately hydrophobic pollutants with log K_{ow} between 1 and 3.5 (phenol, nitrobenzene, benzene, toluene, TCE, atrazine, etc.) are absorbed in large quantities and more easily penetrate into the plant. More hydrophilic pollutants with log K_{ow}<1 (aniline, hexahydro-1,3,5-trinitro-1,3,5-triazine (the explosive RDX), etc.) are slightly adsorbed and not intensively assimilated by plants (Schnoor and Dee 1997).

The cell wall is like a filter with pores that restricts the uptake or movement of organic molecules based on size. Low-molecular mass organics can readily move through the pores of the cell wall. High-molecular mass compounds are big to move through the cell wall pores. Besides, molecular mass of organic pollutants is the main limiting factor during their passage into roots. Organic substances with relative molecular mass not exceeding 1000 are easily absorbed by plants (Söchtig 1964). Larger molecules also penetrate into the roots; namely, it is shown that polyethylene glycol with such high molecular mass as between 4000 and 20,000 (Lawlor 1970; Janes 1974) are absorbed by plant roots. The amount of polyethylene glycol entering into the plant was inversely proportional to the polymer molecular mass. In case of root damage polyethylene glycol enters plants much faster and in significantly greater amounts. Absorbed by kidney bean and cotton seedlings polyethylene glycol was shown to translocate along the plant without changing its molecular characteristics (Andreopoulus et al. 1975). On contrary, it was reported that plant roots can absorb high molecular mass compounds only after partial degradation of the molecules (Führ and Sauerbeck 1967).

Using ^{14}C labeled humic acids have demonstrated that in sunflower (*Helianthus annuus*), wild radish (*Raphanus sativus*) and wild carrot (*Daucus carota*) major part of the high molecular mass humic acids is adsorbed on the surface of roots and partly penetrates into the cells of epidermis. Fulvic acids, smaller molecules penetrate more deeply and reach the central cylinder of the xylem, however, the labelled carbon of fulvic acids does not penetrate into the overground parts of plants. Experiments with polyurethane have shown that labelled carbon polymer is absorbed by tomato (*Lycopersicum esculentum*), cucumber and strawberry (*Fragaria vesca*) root systems after preliminary partial degradation of the polyurethane molecules in the soil (Führ and Mittelstaedt 1974). No recent data is available on the issue in literature.

One more essential factor of uptake of organic pollutants by roots is temperature. The temperature coefficient for diffusion processes, indicating how much the reaction rate increases for a 10 °C temperature rise, is comparatively low (1.2–1.4). Therefore, the diffusion process practically does not depend on temperature over plant-physiologically relevant temperature ranges. Though, passive diffusive absorption is followed by active transport controlled by transpiration, metabolic responses and the processes ofaccumulation of substrate. The temperature rise causes intensification of the transpiration stream and the rates of enzymatic reactions (temperature coefficient 1.3–5.0), resulting in corresponding intensification of the process of absorption of toxic compound (Korte et al. 2000).

Factors that determine mobility of pollutant molecules in soil (adsorption of pollutants by soil particles, degree of dissociation of ionogenic molecules) and permeability of absorptive root tissues in itself significantly depend on the pH of the soil or nutrient solution. Due to high content of humine and fulvic acids soil occurs in a role of anionite, and cations are more strongly bond in soil by the process of ion exchange, giving high K_d for simple bases and even several orders higher for doubly-charged organic cations (Trapp and McFarlane 1995). Therefore, uptake of ionzed compounds is also affected by pH besides properties such as hydrophobicity and organic matter. Uptake of weak acids normally increases when surrounding pH decrease (Briggs 1981). The non-ionic acid can pass the membrane dissociate to the anion in the plant compartment of higher pH and be trapped at that side of the membrane.

Good example of the influence of pH on the process of xenobiotics entering into plants roots is absorption of the insecticide picloram by the roots of oat (*Avena sativa*) and soybean (*Glycine max*) (Isensee et al. 1971). When soil acidity changes from pH 3.5 to pH 4.5, the amount of absorbed xenobiotic sharply decreases, while further change of pH in the range 4.5–9.5 only insignificantly influences the absorption process. Picloram is ionized at pH 3.5 by 20 %, while at pH 4.5 the degree of ionization exceeds 70 %. Therefore, plant roots absorb the insecticide predominantly in nonionic form. It was reported that many toxic compounds are predominantly assimilatedeither bt leaves or roots as undissociated molecules, i.e. without charge (Ugrekhelidze et al. 1986).

The processes of desorption of pollutants from particles of soil and their translocation in soil largely depends on absorbent acidity. Herbicide atrazine was shown is better extracted from weakly alkaline (pH 8.3) soil, while, chloramben and dicamba, are much better extracted under weakly acidic (pH 4.1) conditions (Lay and Casida 1976).

Since the organic pollutants enter the roots with the concentrate influx of water, the intensity of their absorption by the root system is also determined by soil humidity. With the reduction of the water potential in soil the amount of toxic compounds absorbed by plants is decreased. Soil moisture plays an important role in absorption and desorption of pollutants by soil. The addition of water to the organic solvent during the extraction of symmetric triazines, atrazine and chloramben from soil considerably enhances the process, and is a necessary condition for the complete extraction of pollutants (Lay and Casida 1976).

Plant processes, such as transpiration and metabolism, as well as mineral nutrition are significant factors influencing the process of organic compounds absorption by roots. Nutritional elements, their presence or absence differently influences the ingress of toxic compounds into roots (Kvesitadze et al. 2006).

4 Uptake by Leaves

Organic pollutants penetrate a leaf in two ways: through stomata or through the cuticle of epidermis which is covered by wax cuticle. Both pathways occur simultaneously in plants. Due to more simple mechanism of penetration, leaves absorb substances more selectively than roots.

The cuticle is a film-like wax layer covering almost all overground parts of higher plants, including the outer surfaces of the leaf cell epidermis. Generally, the cuticular layer is thicker on the upper (adaxial) side of leaves, and stomata are located on the lower (abaxial) side. Function of the cuticle is reduction of transpiration intensity and thus prevention of plant from dehydration. Besides, the waxy layer of cuticle serves as a rate-limiting barrier for penetration of organic pollutants into leaf cells. Principal component of the cuticle membrane is the lipid polyester cutin, which is a complex mixture of long-chain alkanes, alcohols, ketones, esters and carboxylic acids, synthesized by epidermal cells and deposited on the outer surface. Alkanes and esters predominate on the outside surface of a cuticle. Along with the components mentioned above, wax sometimes contains long-chain C_{29}–C_{33} diketones, triterpenoids (for example, ursolic acid), diterpenes, glycerides and phenolic compounds. The main mass of leaf wax is attributable to normal long-chain alkanes with an odd number of carbon atoms in the chain (C_{31}–C_{37}), in particular the *n*-alkanes $C_{29}H_{60}$ and $C_{31}H_{64}$, and esters of *n*-carboxylic acids with primary and secondary alcohols. In the cutin are incorporated structures of fibrillae and lamellae. The fibrillae are composed of polysaccharides, which may exhibit a distinctive reticulate pattern, while lamellae may contain wax compounds (Kirkwood 1999; Kvesitadze et al. 2006).

Thickness and chemical composition of cuticle varies and depends on plant species, age, location on the stem as well as on environmental factors such as temperature, humidity, etc. In young leaves the cuticle is usually thinner and less uniformly developed than in old ones. Cutin synthesis is terminated only after complete leaf greening (Kolattukudy 1980). Organic pollutants adsorbed on the lipophilic surface of leaf wax accumulate in the cuticle in a great amount and gradually penetrate into the leaf cells. The wax appears to be an active sorbent for lipophilic toxic compounds (Bucovac et al. 1990). Apparently, the molecules of the adsorbed organic pollutants together with individual wax components migrate from cuticle inside epidermal cells and are incorporated into intracellular membranes (Cassagne and Lessire 1975).

The penetration of organic pollutants through the cuticle greatly depends on the structure of pollutant. For example, pyrazon penetrates beet (*Beta vulgaris*) leaves comparatively easily, but phenmedipham and benzthiazuron penetrate rather slowly and in negligible amounts (Merbach and Schilling 1977). Leaf cuticle is also permeable for large molecules such as surface-active substances (Eynard 1974), long-chain fatty acids, (long-chain alkanes (Cassagne and Lessire 1975) and peptides (Shida et al. 1975).

Particles of some organic pollutants are sorbed on wax surface of leaf (dry deposition), or dissolve in water drops of wet leaf (wet deposition) and penetrate plant in these ways. It is shown that dry gaseous deposition was the principal pathway of Cl_4–Cl_6 dioxins and furans penetration in leaves of raygrass (*Lolium multiflorum*) (Welsch-Pausch et al. 1995). Similar results are obtained concerning deposition of polycyclic aromatic hydrocarbons in leaves (Simonich and Hites 1994; Nakajima et al. 1995).

The stomatal system serves as a major regulator of the penetration process of gaseous compounds in leaves due to constituent numerous apertures, which in case of need are resized. By changing the aperture diameter plants control the entry of different molecular mass compounds, including pollutants (Kvesitadze et al. 2006). Opening and closure of the stomata is controlled by movement of two modified kidney-shaped epidermal cells, so-called guard cells (Evert and Eichhorn 2006). The movement of these guard cells is regulated by the concentration of potassium ions – by increasing K^+ concentration stoma is opening. The degree of stomatal apertures opening depends on external environmental conditions such as light, temperature, humidity etc., and on internal factors such as the partial pressure of CO_2 in the intracellular space, plant hydration condition, ionic balance and presence of pheromones (Schroeder et al. 2001; Hetherington and Woodward 2003; Nejad and van Meeteren 2007; Shimazaki et al. 2007).

Penetration of liquids into leaves also occurs through stoma. If permeability for gases depends on the degree of opening of stomatal apertures, the permeability for liquids depends on moistening of the leaf surface, liquid surface tension and morphology of stomata. The majority of toxic compounds penetrate into a leaf as solutions (pesticides, air pollutants, liquid aerosols etc.). Surface tension of liquids is one of the important parameters that determine the ability of liquid penetration into pores (Starov 2004). Liquids with a surface tension less than 30 dyn/cm was shown to have a constant angle of contact with the surface of a leaf of zebrine (*Zebrina purpusii*) and instantly penetrate into the stomata. Liquids with surface tension more than 30 dyn/cm almost do not moisten plant leaves and hardly penetrate leaves via infiltration (Schönherr and Bukovac 1972).

It is experimentally proved that penetration of organic pollutants (2,4-D, α-naphthylacetic acid and its hydroxyderivative metabolites, such as 2-naphthoxyacetic acid and 2,2-dimethylhydrazide) into the stomata–rich lower surface leaves was considerably intensified under illumination (Greene and Bukovac 1977; Pemadasa 1979; Schönherr and Bukovac 1978).

Question arises: which way of organic pollutant penetration in leaves is preferable: through cuticle or through stomata? The question could be answered in part by the results of experiments obtained during the study of gaseous hydrocarbons methane and benzene absorption process by different plants with hypostomatous leaves (the leaves have stomata only on the lower surface) (Ugrekhelidze 1976; Ugrekhelidze et al. 1997; Kvesitadze et al. 2006). The leaves of the Field maple (*Acer campestre*), wild Caucasian pear (*Pyrus caucasica*), vine (*Vitis vinifera*) and narrow-leaved oleaster (*Elaeagnus angustifolia*) leaves were placed in an atmosphere containing ^{14}C-methane or [1–6-^{14}C] benzene. Leaf contact with labeled hydrocarbon occurred only from one side. The total radioactivity of the formed nonvolatile metabolites shows that the absorption of gaseous alkanes and vapours of aromatic hydrocarbons is carried out by leaves not only through stomata, but also through cuticle; At the same time, absorption through the stomata is preferable. Similar results were obtained for a number of herbicides (α-naphthylacetic acid, 2,4-D, picloram and derivatives of urea), applied in soluble form to leaves (Sharma and Vanden Born 1970; Sargent and Blackman 1972; Leece 1978). The above presented data clearly demonstrate that the abaxial side of a leaf, rich in stomata, absorbs the organic substances more intensively than the adaxial side. The cells of trichome (different outgrowths of the epidermis as filaments, warts, scales, setas etc.) can also participate in the absorption of toxic compounds. The number of radial trichomes on the adaxial surfaces of leaves has been correlated with the absorbed herbicide ^{14}C-triclopyr using young leaves of tanoak (*Lithocarpus densiflorus*) (King and Radosevich 1979).

5 Translocation of Organic Pollutants in Plants

Translocation of organic pollutants ubtaken by roots and leaves is carried out via two physiological processes such as the transpiration stream (transport of water and dissolved substances directed from roots to shoots, passing through vessels and tracheides located in the xylem), and the flow of assimilates (transport of substances from the leaves through sieve tubes located in the phloem to the parts of a plant located below (shoot axis, root) and above (shoot tops, fruits) the leaves) (Kvesitadze et al. 2006) (Fig. 1).

Organic pollutants in soil can bind to soil particles and this can be reversible or irreversible depending on the physical-chemical properties of the pollutants (particularly, on the value of K_{ow}, K_d, pK_a, etc.). Part of the pollutant can undergo microbiological transformations by rhizosphere microorganisms, that is paramount factor and often determines itensity of pollutant penetration and intensity of translocation in plants, as well as pollutant biodegradation extent. Pollutants and intermediates produced from pollutants transformation by microorganisms can penetrate into roots. Afterwards, these compounds together with water and dissolved nutrients are translocated by means of transpiration stream and distributed throughout the whole plant. Pollutants penetrated into the leaves through stomata or the cuticle or both, together with assimilates have reach the sieve tubes of phloem and translocate basipetally downwards to roots or acropetally upwards to top shoots and fruit or via both paths. There is a solid experimental data that confirms translocation of

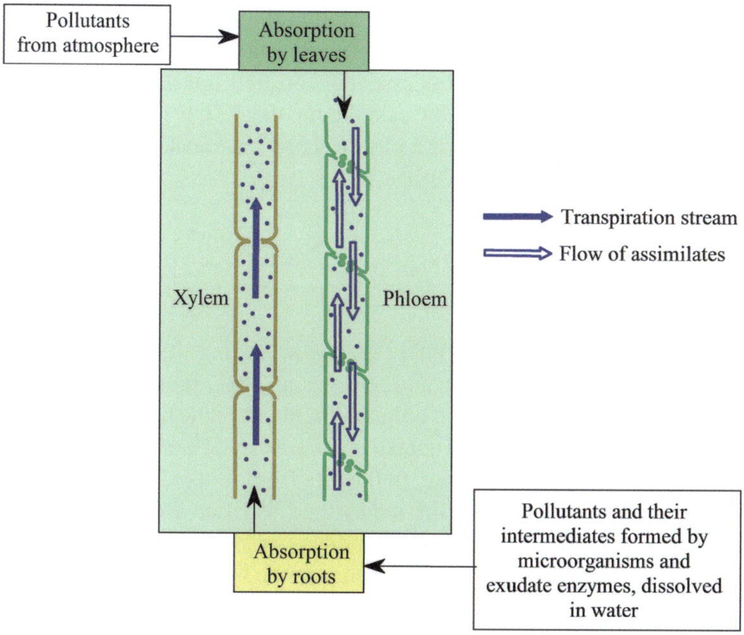

Fig. 1 Ways of environmental pollutants penetration and translocation in plants

toxic compounds in plants. Plants exposed to low concentrations of C_1–C_5 alkanes, cyclohexane, benzene, and toluene absorb these substances and carry out their deep oxidation. Experiments on 55 representatives of annual and perennial plants with using ^{14}C labeled hydrocarbons show that all tested plants uptake and transform alkanes and aromatic compounds with different intensity (Durmishidze et al. 1974a, b, c; Durmishidze and Ugrekhelidze 1975). Uptaken by the leaves hydrocarbons transformation products move along the stem to the roots, and uptaken and transformed hydrocarbons by the roots are transported to the leaves (Ugrekhelidze and Durmishidze 1984).

The transpiration stream is important in the absorption and translocation of environmental pollutants by plants (Beesley et al. 2010; Ma et al. 2012; Mitton et al. 2012). The formula, proposed for calculating the rate of pollutant assimilation from polluted soils, show that the rate of pollutants assimilation is dependent on the factors characterized transpiration stream (Schnoor and Dee 1997):

$$U = (TSCF)(T)(C) \tag{1}$$

U is the rate of pollutants assimilation (mg/day); *T* the rate of plant transpiration, (l/day); *C* the pollutants concentration in the water phase of soil (mg/l); *TSCF* the transpiration stream concentration factor, dimensionless, showing the ratio between the concentrations of organic pollutant in the liquid of the transpiration stream and in the environment (Paterson et al. 1994). The diffusion of organic pollutants from root symplast into xylem apoplast is essential for their translocation from roots to shoots, which can be quantified by calculating the TSCF (Verkleij et al. 2009; Nwoko 2010), which depends on the physical and chemical characteristics of the pollutant: (Burken and Schnoor 1998):

$$\text{TSCF} = 0.75\exp\left[-\frac{(\log K_{OW} - 2.50)^2}{2.4}\right] \quad (2)$$

The main parameter from the formula (2), characterizing the pollutant is K_{ow}, the partition coefficient between 1-octanol and water. Thus, most important property for transport of non-ionized organic pollutant in plants is the hydrophobicity, which predetermines the effectiveness of uptake and translocation of a pollutant in plants. Translocation of non-ionized organic pollutants from roots to shoots is an equilibrium process, rapidly attained and limited by the Casparian Strip. Movement across the membranes is optimal at log K_{ow} 1.8, and less for more polar or more lipophilic compounds (de Carvalho et al. 2007) To enhance the bioavailability of organic pollutants in soil, some amendments (e.g. Tween 80, citric and oxalic acids, biochar, and methylated-β-cyclodextrins) has been successfully applied in assisting phytoremediation of organic-polluted soils (Shen et al. 2009; Beesley et al. 2010; Mitton et al. 2012).

Organic pollutants uptaken by the leaves are translocated through the phloem together with the flow of assimilates formed in leaves. bi-directional flow is possible in the phloem, and through this vessel substances can penetrate both basipetally and acropetally. It has been shown that aryloxycarboxyl acid pesticides penetrate through the leaf cuticle in the form of undissociated molecules and are absorbed by the parenchymal cells. These xenobiotics reach phloem via symplast stream, penetrate into the sieve tubes, and through them enter leaves, growing tissues and reproductive organs. In kidney bean, herbicide 2,4-D and defoliant 2,4,5-T are translocated from the leaves to the rest of the plant both basipetally and acropetally (Long and Basler 1974). The herbicide mecoprop flows from leaves to roots with equal intensity in sensitive and resistant biotypes of common chickweed (*Stellaria media*) (Coupland et al. 1990).

For pesticides based on carbamates acropetal translocation is typical. Examples are carbofuran in seedlings of soybean and mung bean (*Vigna radiata*) (Talekar et al. 1977), methyl-2-benzimidazole carbamate in seedlings of peanut (Vias et al. 1976; Prasad and Ellis 1978) and safflower (*Carthamus tinctorius*) (Mathur and Jhamaria 1975). The pesticides phenmedipham and desmedipham, penetrating through the leaves of wild mustard (*Brassica kaber*), *Amaranthus* and sugar beet (*Beta vulgaris*), were translocated only acropetally (Hendrick et al. 1974).

It was found that plant resistance to pollutants depends on their translocation direction. For instance, the herbicide buthidazole, uptaken by leaves of sensitive to this herbicide plant *Amaranthus*, is translocated in both directions acropetally and basipetally, but in resistant maize leaves transport proceeds only basipetally (Hatzios and Penner 1980). This herbicide is insignificantly translocated along the apoplast in soybean leaves (Haderlie 1980). 4,4′-methylene-*bis*(2-chloroaniline) applied to different plant leaves is absorbed but not translocated.

Phloem flow of organic pollutants supports the hypothesis of intermediate permeability suggested by Tyree et al. (1979). This hypothesis takes into account the immediate proximity of the phloem and xylem vessels and proposes that: (i) any molecule with a high permeability through the membrane will be able to get into the phloem, but also can leave the phloem and the more rapidly transported to the xylem stream; (ii) any molecule with a low permeability through the membrane cannot reach a sufficiently high concentration in the phloem for effective transport; (iii) there must be an intermediate permeability between these extreme values, and substances having such permeability must be characterized by the highest phloem mobility.

Based on this hypothesis, and a broad range of experimental data of investigations of herbicide assimilation and transport by seedlings of castor bean, Kleier and coworkers have put forward a mathematical model enabling determination of the translocation of toxic compounds along plant transporting pathways (Hsu et al. 1988; Grayson and Kleier 1990; Hsu and Kleier 1990; Kleier 1994; Brudenell et al. 1995). The Kleier's model has been successfully used in the prediction of translocation of many important secondary metabolites, such as gibberelline A (O'Neill et al. 1986), salicylic acid (Yalpani et al. 1991), oligogalacturonides (Rigby et al. 1994) and glucosinolates (Brudenell et al. 1999).

Due to their ability to easily move along the transport pathways of plants, systemic herbicides are divided into phloem-mobile, xylem-mobile and ambimobile ones (the latter capable of penetrating into both phloem and xylem). Chemical compounds assignment to a particular class is predetermined by such physical-chemical parameters as the dissociation constant (pK_a) and hydrophobicity (K_{ow}). Herbicides with dissociation degree characteristic to strong and medium acids ($pK_a<4$) and having medium lipophilicity ($\log K_{ow}$ about 1 to 2.5–3) belong to the phloem-mobile type, while weaker acids with $pK_a>5$ and non-ionized compounds must be more polar to move well. Ability to translocate only along the xylem is characteristic for herbicides with the medium lipophilicity ($\log K_{ow}$ in the range 0–4) and with low degree of ionization ($pK_a>7$). Weak acids ($pK_a>7$), with high hydrophilicity ($\log K_{ow}<0$) are ambimobile. Highly lipophilic herbicides ($\log K_{ow}>4$), regardless of the value of pK_a, are often nonsystemic, because they cannot translocate into xylem or phloem (Bromilow et al. 1990).

Translocation of herbicide in phloem depends on the processes of biosynthesis of carbohydrate (predominantly sucrose) in tissues and streaming from mesophyllic cells to the complex of cells in phloem. There are at least two mechanisms of carbohydrate translocation in higher plants: translocation of sucrose through apoplast passing phloem and through symplastic connections between the cells of the mesophyll and the phloem. It was shown that in parallel to sucrose transportation, translocation of penetrated herbicides also occurs (Devine and Hall 1990). Sugars from phloem are translocated in the same symplastic and apoplastic ways.

Study of the uptake of some fungicides, herbicides and insecticides of different chemical classes by soybean roots and penetration through the xylem showed that the hydrophobicity of the organic pollutants greatly predetermines these processes and the maximum concentration of each pesticide in xylem juice is reached at log-K_{ow} ~ 3 (Sicbaldi et al. 1997).

Fungicide morpholine, despite the lipophilic character is systemic, i.e. it penetrates and is translocated throughout the plant. To explain this phenomenon, the process of assimilation and transport of labelled morpholine fungicides ^{14}C-dodemorph and ^{14}C-tridemorph at different pH have been studied. It was shown that at pH 5 the intensities of assimilation and translocation were insignificant, however at pH 8 the rates of the processes were increased by approximately two orders of magnitude. It has been stated that at pH 8 the more lipophilic tridemorphe is accumulated in large quantities by roots and moderately translocated to shoots. Dodemorphe is accumulated in roots in less quantity, but is translocated through the epidermis into the xylem very effectively and this picture practically does not change from 24 to 48 h. These data indicate that assimilation and translocation of toxic compounds in plants are the processes maintaining balance in plant cells, at least within definite periods of time (Chamberlain et al. 1998).

Plants can the partial release of some organic pollutants absorbed by roots or leaves, in unchanged form through the leaves or the root system, as though avoid chemical transformation of pollutants (Zaalishvili et al. 2000; Korte et al. 2000). This is the simplest pathway for moving of organic pollutants that have entered the plant. This pathway of pollutant elimination is rare and takes place only at high concentrations of highly mobile (phloem-mobile or ambi-mobile) compounds. Excretion of absorbed pollutants by plants differs from excretion in mammals, in which transformed as well as untransformed substances are eliminated from the body. Environmental pollutants absorbed by the roots are excreted via the leaves and vice versa, i.e. organic pollutants absorbed by the leaves are excreted via the roots. These two processes differ from each other in the translocation mechanisms of the absorbed compounds. Molecules of organic pollutants that initially penetrate through the roots are transported along the apoplast by the transpiration stream and have high xylem mobility; they are excreted by the leaf stomata. Phloem-mobile or ambi-mobile organic pollutants absorbed through leaves are translocated via the flow of assimilates and reach the roots and are excreted into soil or nutrient solution (Kvesitadze et al. 2006).

Existence of these two distinct mechanisms of excretion has been confirmed by many experiments. Pollutants absorbed by leaves, and capable of moving rapidly along the phloem, are often excreted by roots. Such excretion is not always conditioned by the concentration gradient but the process can also be directed against the gradient. [^{14}C] Alachlor applied to leaves of soybean and wheat (*Triticum aestivum*) is excreted via the roots into a nutrient solution containing a higher alachlor concentration than that in the roots (Chandler et al. 1974). These data indicate that the root excretion of phloem-mobile pollutants, translocated in the plant by the flow of assimilates is accomplished by an active transport mechanism. Excretion of organic pollutants via the roots is a functional process, characteristic of the higher plants. Besides the phloem-mobile compounds, ambimobile environmental pollutants absorbed by the leaf surface are sometimes excreted in untransformed form via the roots. Excretion via roots is especially characteristic for the phenoxyacetic acids (2,4-D, 2,4,5-T, etc.), dicamba, picloram and other systemic herbicides (Hallmen 1974; Schultz and Burnside 1980; Lingle and Suttle 1985).

Absorbed organic pollutants are more actively excreted by the root system as compared to leaves. For instance, the roots of *Ampelamus albidus* excrete approximately 37 % of the total amount of 2,4-D absorbed by the plant leaves over 8 days (Dexter et al. 1971).

The closer the pollutant-absorbing leaf is to the roots the higher is the rate of root excretion (Schultz and Burnside 1980). The intensity of excretion also rises in response to an increase in herbicide concentration applied to the leaf. There is no clear relation between excretion and plant resistance to herbicide, as the excretion often proceeds via the roots of both herbicide-sensitive and herbicide-resistant plants (Dexter et al. 1971).

Root system also excretes toxic compounds absorbed by the roots or by the stem of plant. For instance, cotton seedlings excrete backwards through the roots about 25–30 % of the herbicide bioxone, which was absorbed by the roots from herbicide-containing nutrient solution (Jones and Foy 1972). Such behaviour was observed after transfer of seedlings to nutrient solution free of herbicide for 2 days. In another example, 15 % of the total amount of the hydrazide of maleic acid injected over 30 days into the wood of saplings of rock maple (*Acer saccharinum*) and western plane (*Platanus occidentalis*) was excreted from the roots in untransformed form (Domir 1978).

The above presented data considers that most of the absorbed environmental pollutants are excreted without transformation by the root systems, although generally the amount of excreted xenobiotic varies between 0.1 and 2 %. The phenomenon of root excretion must be taken into account at treatment of plant surfaces with different pesticides, since excessive excretion of pollutants in unchanged form could become a source of serious chemical contamination of soil and groundwater.

Organic pollutants absorbed by the roots can be excreted via the leaves, although it occurs more rarely as compared to root excretion. An example is the excretion of phenol by the leaves of cane (*Scirpus lacustris* L.) plants kept on phenol solution (Seidel and Kickuth 1967). In this case the excretion proceeds so intensively that after 90 min the air near the leaves exhibits a positive qualitative reaction to the presence of phenol and after several hours its content in the air can be detected even by smell. Another example shows how leaves of tobacco and radish immersed with their petioles in a solution of 1,2-dibromoethane absorb and then rapidly excrete phenol into the atmosphere (Isaacson 1986). Thus, plants can excrete halogen derivatives of hydrocarbons absorbed from the soil or groundwater and gradually dilute them into the air. Further confirmation of the occurrence of this process is laboratory and field experiments on poplar hydrides removing TCE from artificially contaminated (260 mg/l) water and soil (Kassel et al. 2002). Only 10 % of the total TCE assimilated by the plant was evaporated via the poplar leaves without transformation and the rest was metabolized. This example indicates once more that excretion of absorbed organic pollutant proceeds at high concentrations and in this case, the organic pollutant penetrates rapidly along the xylem and is not transformed before it is evaporated through the stomata together with water.

To consider the process of excretion as a method of phytoremediation, it should be taken into account that the ability of higher plants to assimilate environmental pollutants via their root systems and excrete them in untransformed form via the leaves could be used for the decontamination of soils with high concentrations of organic pollutants. From the ecological point of view the shortcoming of excretion is that the pollutant does not undergo detoxification and returns to the environment with full retention of its toxic features (Kvesitadze et al. 2006).

6 Transformation of Organic Pollutants in Plants

To resist toxic action of environmental organic pollutants plants activate a definite set of biochemical and physiological processes. The process of detoxification of the organic pollutants is a sum of chemical reactions in separate cells. Mostly, plants totally or partially detoxify environmental organic pollutants entering their cells where they undergo enzymatic transformations leading to the decrease of pollutants toxicity. As the main pathways of uptaken organic pollutant transformation in plant cells resembles liver metabolism of drugs, the scheme illustrating the fate of organic pollutants entered in plant cells is known as "Green liver" model (Sandermann 1994, 1999). This scheme with little modification is represented on Fig. 2 (Kvesitadze et al. 2006).

Three successive phases of transformation of organic pollutants are considered to be significant (Sandermann 1999):

- The 1st phase is functionalization, when organic pollutant undergo enzymatic transformation as a result of which acquires new functional group;
- The 2nd phase is conjugation, when conjugation of formed intermediate with intracellular compounds takes place.
- The 3rd phase is compartmentalization, when formed in 2nd phase conjugates are deposited in vacuoles and/or cell walls.

Depending on chemical nature and concentration of the organic pollutant penetrated into plant cell, time of exposure and activities of plant detoxification enzymes, their complete detoxification via deep oxidation and formation of standard cell metabolites and carbon dioxide is possible (Kvesitadze et al. 2006).

Below is considered biological essence of each transformation phase, examples of occuring reactions and characterization of enzymes, participating in these reactions.

Processes of degradation of organic pollutants are closely related to many aspects of cellular metabolism. In prolonged detoxification processes quite a few enzymes are involved. Obviously these enzymes could be divided in two groups: enzymes directly and indirectly participating in detoxification processes (Kvesitadze et al. 2006).

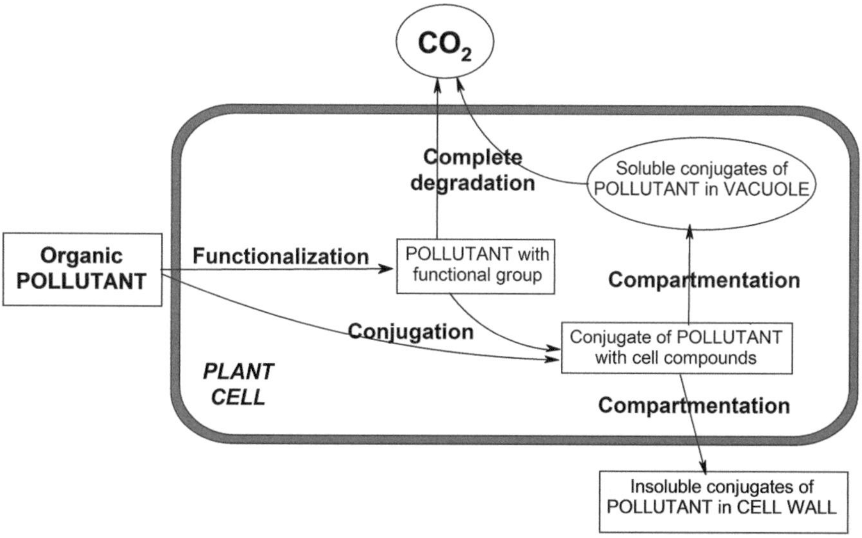

Fig. 2 The main pathways of organic pollutant transformation in plant cell according to Kvesitadze et al. (2006)

Reactions occurring during all three transformation processes (functionalization, conjugation and compartmentation) are of enzymatic nature. In the absence of xenobiotics these enzymes catalyze other reactions typical for plant cell regular metabolism. The main enzymes directly participating in the transformation of organic pollutants are listed in Table 1.

Prolonged process of transformation of organic pollutants in plants, that implies mostly deep oxidation is reflected on the cell normal metabolic processes, as cell requires in such cases additional energy to support these multi stage process of detoxification. Thus, enzymes, involved in catabolic processes leading to energy generation are thus indirectly participating in the detoxification process.

7 Functionalization

Functionalization is an initial rate limiting process of chemical modification of organic pollutants whereby a molecule of a hydrophobic organic xenobiotic acquires hydrophilic functional group (hydroxyl, amino, carboxyl, etc.) as a result of enzymatic transformations (oxidation, reduction, hydrolysis, etc.). Hydrophilic functional group promotes polarity and reactivity of the xenobiotic molecule, as a result of which its affinity to enzymes, catalyzing further transformation (conjugation or further oxidation) is increased. Finally, xenobiotic oxidative degradation proceeds to standard cell metabolites and mineralization to CO_2. This pathway of transformation enables not only full detoxification of the xenobiotic but also utilization of its

Table 1 Plant enzymes directly participating in the transformation of organic pollutants

1st phase (functionalization)	2nd phase (conjugation)	3rd phase (compartmentalization)
Oxidases:	**Transferases**:	**ATP-binding cassette (ABC) transporters** *(EC 3.6.3)*
Cytochrome P450-containing monooxigenases (EC 1.14.14)	*Glutathione S-transferase (EC 2.8.1.18)*	
Peroxidases (EC 1.11.1)	*O-glucosyl-transferase (EC 2.4.1.7)*	
Phenoloxidases (EC 1.14.18)	*N-glucosyltransferase (EC 2.4.1.71)*	
Laccases (EC 1.10.3)	*N-malonyltransferase (EC 2.3.1.114)*	
Reductases:	*Putrescine N-methyl-transferase (EC 2.1.1.53)*	
Nitroreductase (EC 1.6.6)		
Dehalogenases:		
Haloalkane dehalogenase (EC 3.8.1.1)		
Esterases:		
Arylesterase (EC 3.1.1.2)		
Lysophospholipase (EC 3.1.1.5)		
Acetylesterase (EC 3.1.1.6)		
Carboxylesterase (EC 3.1.1.1)		
Acid phosphatase (EC 3.1.3.2)		
Alkaline phosphatase (EC 3.1.3.1)		

carbon atoms for intracellular biosynthetic and energetic needs. The totality of such transformations is the essence of the detoxification process. Complete degradation of organic pollutants in a plant cell is however accomplished only at metabolizable concentrations of environmental pollutants, and time (in some cases the process lasts days or weeks) is needed for it (Kvesitadze et al. 2006). In case of high concentrations of pollutants full mineralization of xenobiotics is not usually achieved; typically only a small amount of organic pollutant present in the cell is mineralized, and the rest undergoes conjugation.

8 Hydroxylation

Hydroxylation is the most wide spread reaction of the functionalization phase, which proceeds almost similarly in all living organisms. As was mentioned above, in order to increase the reactivity of xenobiotics for their further transformations, the introduction of different hydrophilic groups is required. Introduction of a hydroxyl group into a xenobiotic molecule increases its polarity and hydrophilicity. Hydroxylation is considered in many cases as the primary detoxification reaction, followed by the processes of profound oxidation and conjugation (Sandermann

1994). Studies of the metabolites of xenobiotic alkanes and N-alkyl derivatives show that oxidative degradation of these molecules often begins with hydroxylation of the alkyl groups. Though it is not always possible to isolate and identify hydroxy derivatives of xenobiotics, the products of their further metabolism provide information on the chemical structures of the intermediates (Kvesitadze et al. 2006). The basic types of hydroxylation are represented in Table 2.

Table 2 The basic types of hydroxylation of organic pollutants

Pollutants	Types of reactions	Products
$H_3C{-}CH_3$ Ethane	Aliphatic hydroxylation	$H_3C{-}CH_2{-}OH$ Ethanol
 Benzene	Aromatic hydroxylation (via epoxide)	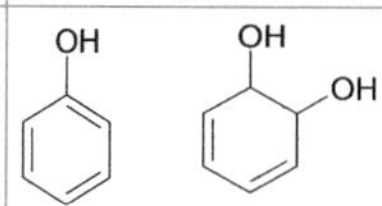 Phenol Catechol
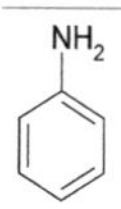 Aniline	Aromatic hydroxylation (via O-insertion)	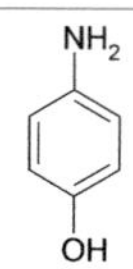 *p*-Hydroxyaniline
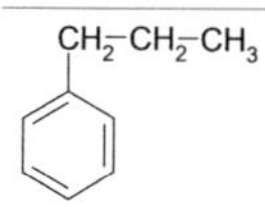 *n*-Propylbenzene	Aliphatic hydroxylation	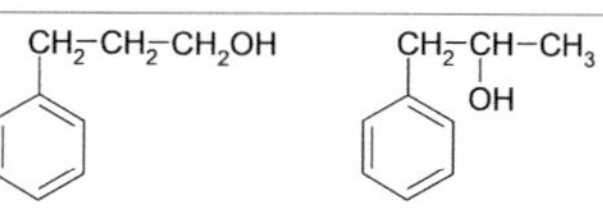 via ω-hydroxylation via (ω–1)-hydroxylation 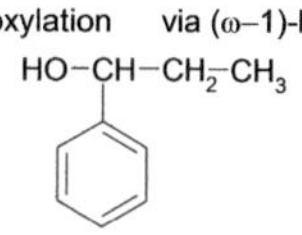 via α-hydroxylation
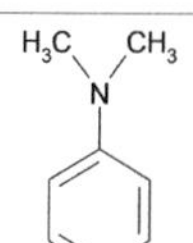 N,N-Dimethylaniline	N-Dealkylation	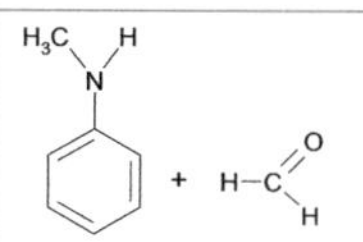 N-Methylaniline Formaldehyde
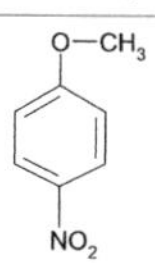 *p*-Nitroanisole	O-Dealkylation	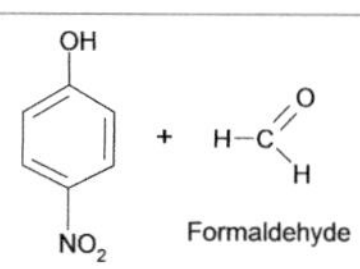 Formaldehyde *p*-Nitrophenol
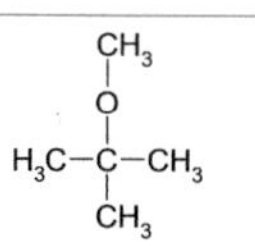 Methyl *tertyl*-butyl ether		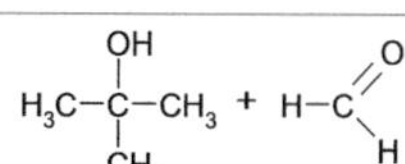 Formaldehyde *Tert*-butyl alcohol

(continued)

Table 2 (continued)

Pollutants	Types of reactions	Products
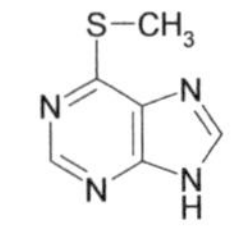 6-Methylmercapto-purine	S-Dealkylation	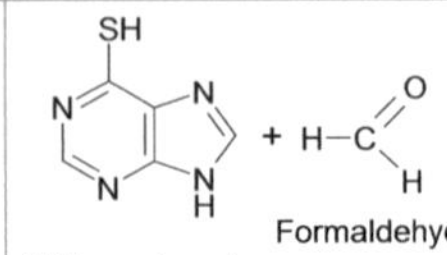 6-Mercaptopurine + Formaldehyde
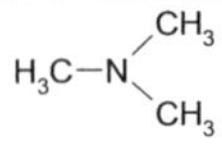 Trimethylamine	N-Oxidation	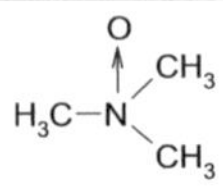 Trimethylamine oxide
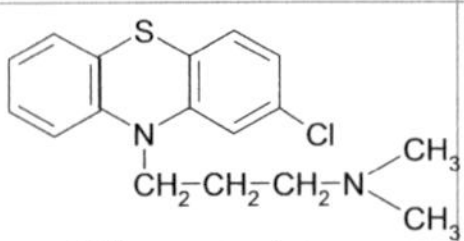 Chloropromazine	S-Oxidation	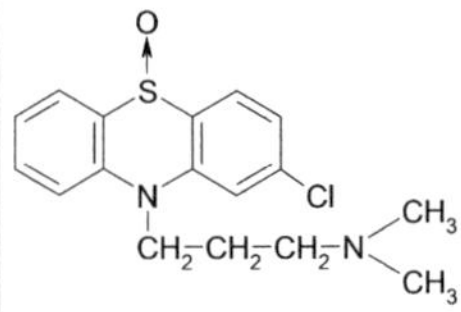 Chloropromazine sulphoxide
$R-CH_2-OH$ Alcohol	Alcohol oxidation	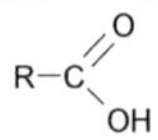 Carboxyl acid
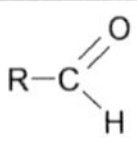 Aldehyde	Aldehyde oxidation	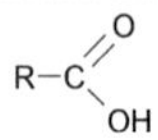 Carboxyl acid
R_1-NH-R_2 *Sec*-amine	N-Hydroxylation of sec-amines	$R_1-N(OH)-R_2$
$R-CH(NH_2)-CH_3$ Amine	Oxidative deamination	$R-C(=O)-CH_3$ + NH_3 Ketone Ammonia
$R_1-CH(Hal)-R_2$ Halogen-derivative of hydrocarbon	Oxidative dehalogenation	$R_1-C(=O)-R_2$ + HHal Ketone

8.1 *Cytochrome P450-Containing Monooxygenases*

(EC 1.14.14.1) belong to one of the major groups of enzymes that are responsible for detoxification of organic pollutants and are present in all forms of life (plants, bacteria, and mammals) (Morant et al. 2003; Meunier et al. 2004; Guengerich 2010). Cytochromes P450 are universally distributed and are detected in animals, plants and microorganisms. Cytochromes P450 are encoded by a highly divergent gene superfamily, and exhibit great diversity in reactive site and amino acid composition (Schuler 1996. This superfamily contains a spectrum of CYP gene families, which differ substantiallyaccording to their sequence, substrate specificity, genomic organization and inducibility. Over 120 cDNA and genomic DNA sequences for P450s of different plants have been identified (Schuler 1996): wheat, avocado (*Persea americana*), eggplant (*Solanum melongena* cv. Sinsadoharanasu), catmint (*Nepeta racemosa*), Madagaskar periwinkle, peppermint (*Mentha piperita*), pennycress (*Thlaspi arvense*), thale cress, maize, Jerusalem artichoke (*Helianthus tuberosus*), mung bean, alfalfa, sunflower, pea, flaxseed, guayule (*Parthenium argentatum*), petunia (*Petunia hybrida*), moth orchid (*Phalaenopsis* sp. hybrid SM9108), sorghum, barberry (*Berberis stolonifera*), field mustard (*Brassica campestris*), pigeon pea, tobacco, soybean, etc. The number of cytochrome P450 genes (*CYP*) in plant genomes is estimated to be up to 1 % of total gene annotations of each plant species (Mizutani and Ohta 2010). This implies that diversification within P450 gene superfamilies has led to the emergence of new metabolic pathways throughout land plant evolution. The conserved P450 families contribute to chemical defense mechanisms under terrestrial conditions and several are involved in hormone biosynthesis and catabolism. Plant P450s catalyze a wide variety of monooxygenation/hydroxylation reactions in secondary metabolism, and some of them are involved in unusual reactions such as methylenedioxy-bridge formation, phenol coupling reactions, oxidative rearrangement of carbon skeletons, and oxidative C-C bond cleavage (Mizutani and Ohta 2011). Low substrate specificity together with a preference for hydrophobic substances would suggest that these enzymes would not be particularly stereo-selective (Urlacher 2012).

Cytochromes P450 catalyze extremely diverse and often complex regio-specific and/or stereo-specific reactions in the biosynthesis or catabolism of plant regular and secondary metabolites (Morant et al. 2003). Cytochrome P450 is reported to plays a key role in over 20 physiologically important processes and reactions (Durst 1991; Schuler 1996). Among them biosynthesis of lignin monomers (Whetten and Sederoff 1995), anthocyanins (Holton and Cornish 1995), furanocoumarins (Berenbaum and Zangerl 1996), gibberellins (Jenings et al. 1993), isoflavonoid phytoalexins (Kochs and Grisebach 1986), alkaloids (Kutchan 1995), hydroxylation of fatty acids (Salaün and Helvig 1995), hydroxylation of limonene and geranoil (Hallahan et al. 1994), etc. are of vital importance for the plant cell.

Besides their physiological functions in the biosynthesis, some plant cytochrome P450-containing monooxygenases can play an important role in the hydroxylation of exogenous toxic compounds (environmental pollutants and other xenobiotics) after they penetrate into the plant cell (Sandermann 1994). Thus, cytochromes P450 enable plants to degrade toxic exogenous chemicals including pesticides and environmental pollutants, making them less phytotoxic. The recovery of an increasing number of plant P450 genes in recombinant form has enabled their use in experimentation, which has revealed their extraordinary potential for engineering herbicide tolerance, biosafening, phytoremediation and green chemistry (Werck-Reichhart et al. 2000). Cytochrome P450 monooxygenase superfamily are responsible for the Phase I metabolism of organic pollutants representing several classes of organic compounds. The majority of experimental evidence for P450 involvement in herbi-

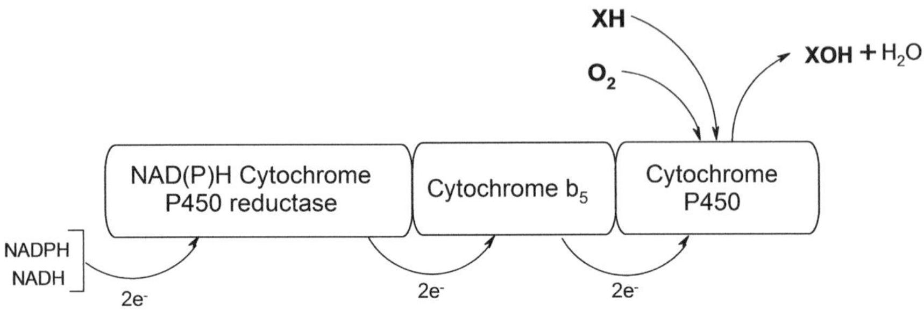

Fig. 3 Scheme of cytochrome P450 containing monooxygenase system. XH – nonpolar molecule of organic pollutant, XOH – polar product of hydroxylation

Fig. 4 Prosthetic group of cytochrome P450. An iron-(III) protoporphyrin-IX linked with a proximal cysteine ligand (Meunier et al. 2004)

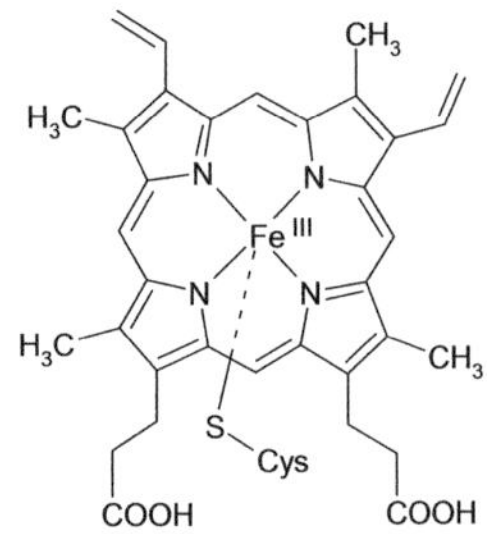

cide metabolism has been derived from *in vitro* studies in which the catalytic activity of plant microsomes towards herbicidal substrates was measured in the presence of various P450 inhibitors and activators (Siminszky 2006).

Cytochromes P450 monooxygenases are oxidases of mixed-function located in the membranes of the endoplasmic reticulum (microsomes) (Fig. 3). The cytochrome P450-containing monooxygenases use NADPH and/or NADH reductive equivalents for the activation of molecular oxygen and for the incorporation of one of its atoms into hydrophilic organic compounds (XH) that results in functionalized products (XOH) (Schuler 1996). In this case the second atom of oxygen is used for the formation of a water molecule.

The microsomal cytochrome P450 containing monooxygenase system is an electron transfer chain and contains the following components: the initial stage of electron transfer is a NADPH-cytochrome P450 reductase (EC 1.6.2.4); the intermediate carrier—cytochrome b_5, and the terminal acceptor of electrons—cytochrome P450. All these components are located in the membranes of the endoplasmic reticulum. When NADPH is used as the only source of reductive equivalents in this system, the existence of an additional carrier, a NADH-dependent flavoprotein is required. NADH may also be oxidized by the NADPH-dependent redox system. In the latter case cytochrome b_5 is not needed as the medium carrier (Hansikova et al. 1994).

Cytochrome P450 is cysteinato-heme enzymes and its prosthetic group is constituted of an iron(III) protoporphyrin-IX covalently linked to the protein by the sulfur atom of a proximal cysteine ligand (Meunier et al. 2004) (Fig. 4).

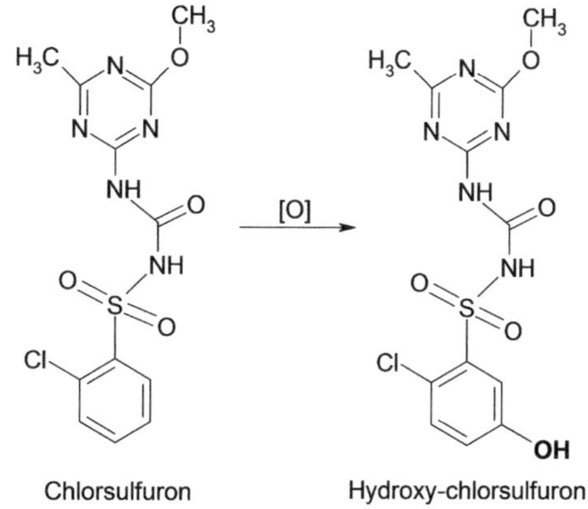

Fig. 5 Hydroxylation of aromatic ring of chlorsulfuron

Cytochromes P450 in plants participate in the reactions of C- and N-hydroxylation of aliphatic and aromatic compounds, N-, O-, and S-dealkylation, sulpho-oxidation, deamination, N-oxidation, oxidative and reductive dehalogenation, etc. (Schuler 1996). Resistance against many herbicides in plants is mediated by the rapid transformation of the herbicide into a hydroxylated, inactive product that is subsequently conjugated to carbohydrate moieties in the plant cell wall. For examples, N-demethylation and ring-methyl hydroxylation of the phenylurea herbicide chlorotoluron in wheat and maize are cytochrome P450-dependent processes (Fonné-Pfister and Kreuz 1990; Mougin et al. 1990). After hydroxylation both products undergo conjugation with glucose by the newly formed hydroxyl group (Mougin et al. 1990).

Sulphonylurea herbicides (primisulfuron, chlorsulfuron and triasulfuron) in wheat and maize under the action of cytochrome P450-dependent monooxygenase undergo hydroxylation in the aromatic ring (Fig. 5) (Schuler 1996).

Analogously, cytochrome P450-catalyzed hydroxylations are typical for other aromatic ring-containing herbicides. For example diclofop in wheat (McFadden et al. 1989), and bentazon in maize (McFadden et al. 1990) undergoes such transformations (Fig. 6). Similarly to chlorotoluron metabolites, after the hydroxylation the products of diclofop and bentazon are conjugated to O-glucosides.

For some phenylurea herbicides in the Jerusalem artichoke cytochrome P450-mediated N-demethylation is sufficient to cause partial or complete loss of phytotoxicity (Fig. 7) (Didierjean et al. 2002).

Hydroxylation of endogenous substrates and xenobiotics may be catalyzed by the same Cytochrome P450 may catalyze. Corroborations of this proposition are the oxidation of endogenous lauric acid and exogenous diclofop by the cytochrome P450-containing monooxygenase from wheat (Zimmerlin and Durst 1992), and hydroxylation of the endogenous *trans*-cinnamic acid and the exogenous *p-chloro-N*-methylaniline by a recombinant artichoke CYP73A1 (*trans*-cinnamic acid-4-hydroxylase) expressed in yeast (Pierrel et al. 1994). During the incubation of a

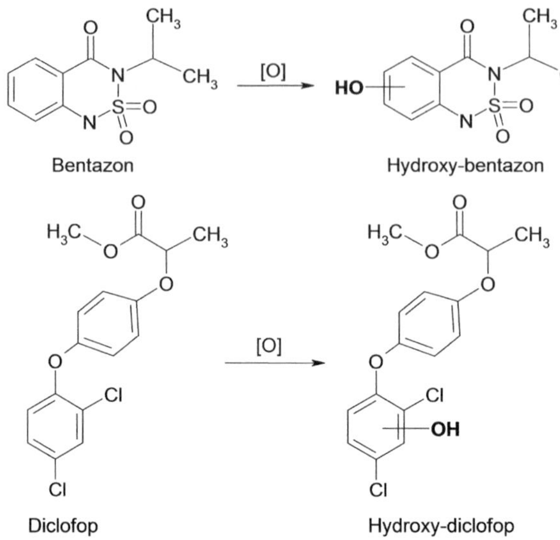

Fig. 6 Hydroxylation of the bentazon and diclofop aromatic rings

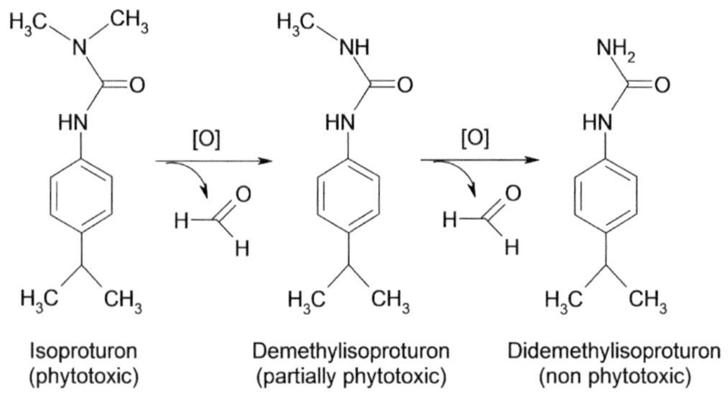

Fig. 7 N-demethylation of isoproturon by cytochrome P450 dependent monooxygenase

microsomal suspension (from etiolated soybean seedlings) with [1-^{14}C] *trans-cinnamic* acid (*qua* endogenous substrate) and N,N-dimethylaniline (*qua* model xenobiotic) simultaneously,the hydroxylation of the endogenous substrate was inhibited up to 70–80 % (Gordeziani et al. 1987). On the other hand, the demethylation of N,N-dimethylaniline in the presence of *trans*-cinnamic acid was inhibited by only 25–30 %. Besides N,N-dimethylaniline, the enzymatic transformation of cinnamic acid was also inhibited by other xenobiotics (ethylmorphine, *p-nitroanisole*, aniline and aminopyrine). The kinetics of the NADPH-dependent oxidation of cinnamic acid and xenobiotics revealed the competitive character of the inhibition of the cinnamic acid–hydroxylase activity by xenobiotics (Khatisashvili et al. 1997).

These results indicate that the switch over of cytochrome P450 to hydroxylation of xenobiotics enabled by and due to a decrease of its physiological function (hydroxylation of *trans*-cinnamic acid) takes place. The switch of an enzyme from biosynthesis to detoxification is determined by the polarity of the xenobiotic: the more hydrophobic the xenobiotic, the higher its affinity to cytochrome P450, the more universal the switch, and the faster the process of xenobiotic oxidation proceeds. Thus, it seems that after penetration of hydrophobic xenobiotics into the plant cell, switching of cytochrome P450 from an "endogenous" to an "exogenous" function regimen may take place. In essence, such switching is set into motion by the superior affinity of the xenobiotic for the enzyme compared to that of its natural substrates (Gordeziani et al. 1991).

In plants growing in a medium containing a xenobiotic, the concentration of cytochrome P450 increases. Nearly all xenobiotics examined have an inductive nature. The inductive abilities of xenobiotics such as phenobarbital, clofibrate, aminopyrine, and herbicides: 2,4-D, propanil, chloroacetamide, thiocarbamate, chlorotoluron, bentazon and others were clearly demonstrated (Salaün 1991). A cytochrome P450 (CYP76B1) isolated from Jerusalem artichoke is more actively induced by xenobiotics than other cytochrome P450-containing monooxygenases. This CYP76B1 metabolizes with high efficiency a wide range of xenobiotics, including alkoxycoumarins, alkoxyresorufins, and several herbicides of the phenylurea class (Robineau et al. 1998). CYP76B1 catalyzes also the removal of both N-alkyl groups of phenyl-ureas with turnover rates comparable for physiological substrates and produces non-phytotoxic compounds. Cytochrome P450-increased herbicide metabolism and tolerance can be achieved by ectopic constitutive expression of CYP76B1 in tobacco and *Arabidopsis* (Didierjean et al. 2002). Transformation with CYP76B1 resulted in a 20-fold increase in tolerance to the herbicide linuron and a tenfold increase in tolerance to the herbicides isoproturon or chlorotoluron in tobacco and *Arabidopsis*. Other than increased herbicide tolerance, the expression of CYP76B1 results in no other visible phenotype change in the transgenic plants. CYP76B1 can function as a selectable marker for plants that can be selected for the phytoremediation of contaminated sites.

Xenobiotics differ in their inductivity, namely the inductive effect of each particular xenobiotic depends on its chemical nature and/or the inductive potential of its intermediates. Some of these intermediates appear to be highly reactive and most of them cause inactivation of cytochrome P450 and its further conversion into cytochrome P420. Good examples of these intermediates are metabolites of 3,4-benzopyrene. Incubation of soybean and ryegrass with 3,4-benzopyrene causes the formation of epoxides, dioles and quinones (Trenck and Sandermann 1980; Sandermann 1988). The "aggressiveness" of these substances is expressed by the formation of active oxygen radicals that cause the irreversible conversion of cytochrome P450 to P420 (Khatisashvili et al. 1997). For instance, enhancement of the peroxidation of fatty acids also leads to the generation of oxygen radicals. According to hitherto unpublished data of the present authors, during cytochrome P450-mediated 3,4-benzopyrene oxidation in maize seedling microsomes, superoxide anion radicals are generated.

The expression of genetically engineered cytochrome P450 would be required for the low-cost production of several natural products, such as antineoplastic drugs (taxol or indole alkaloids), nutraceuticals (phytoestrogens) and antioxidants in plants (Morant et al. 2003). These compounds may have important functions in plant defence. Engineered cytochromes P450 could improve plant defence against insects and pathogens. These P450 may be tools to modify herbicide tolerance, and are selectable markers for phytoremediation.

8.2 Peroxidases

(EC 1.11.1.7) are ubiquitous enzymes found in all green plants, the majority of fungi and aerobic bacteria. The isoenzymes heterogenity of peroxidases appears to be result from *de novo* synthesis, as well as an array of physiological and ecological determinants including hormones, light, infection, etc. (Siegel 1993). Peroxidases have phylogenetically correlated similarities based on the chemical nature and redox potentials of the substrates, which they oxidize. Peroxidases often increase in response to stress, and one of principal roles of peroxidases appears to be the protection of cells from the hydrogen peroxide. The great catalytic versatility of the peroxidase is its predominant characteristic, and, therefore, no single role exists for this multifunctional enzyme.

The peroxidase catalyzes following reaction:

$$RH_2 + H_2O_2 \rightarrow 2H_2O + R$$

The enzyme is composed of a single peptide chain, and contains one haem (protoporphyrin IX). Horseradish peroxidase has two metal centers, one of iron heme group and two calcium atoms (Veitch 2004). The heme group has a planar structure with the iron atom held tightly in the middle of a porphyrin ring which is comprised of four pyrrole molecules (Fig. 8). Iron has two open bonding sites, one above and one below the plane of the heme group. The heme group has a histidine (enzyme) attached in the proximal histidine residue (His170) which is located below the heme group. The second histidine residue in the distal side of the heme group, above the heme group, is vacant in the resting state (Dunford and Jones 2010).

Plant peroxidases, different from those of animals, contain approximately 25 % of carbohydrates, which protect the enzyme from the action of proteolytic enzymes, and at the same time stabilize the protein conformation (Hu and Huystee 1989).

Peroxidases are known to catalyze a number of free radical reactions (Stahl and Aust 1995). The electron donors, such as veratryl alcohol (a free radical mediator) are oxidized by lignin peroxidase and generate cation radicals. As a result, the anion radical can catalyze the reduction of electron acceptors (cytochrome c, nitroblue tetrazolium, O_2). Similar reactions have been observed with Mn-dependent peroxidase in the presence of quinones. These reductive mechanisms may be involved in the metabolism of TNT in case of *Phanerochaete chrysosporium* action, but it is

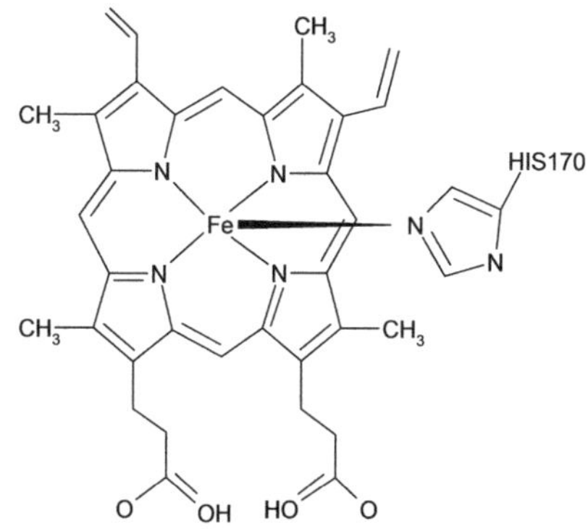

Fig. 8 Structure of heme in horseradish peroxidase

shown that these peroxidases do not participate in the initial reducing steps of this explosive (Stahl and Aust 1995).

Nearly all kinds of organic pollutants in plants are oxidized by peroxidases (Stiborova and Anzenbacher 1991). This notion is based on such facts as the ubiquitous occurrence of this enzyme in plants (the isozymes of peroxidase in green plants occur in the cell walls, plasmalemma, tonoplasts, and intracellular membrane systems of endoplasmic reticulum, plastids and cytoplasm), the high affinity of peroxidases from different plants to organic xenobiotics of different chemical structures, and their low substrate specificity. These facts indicate the universal character of the enzyme's action and the active participation of peroxidase isoenzymes in a wide variety of detoxification processes.

The formation of 2,4-dichloro-1,4-benzoquinone from 2,4,6-trichlorophenol (pesticide that has been used as a fungicide, herbicide, insecticide, antiseptic, defoliant and that has been found as environmental pollutant in fresh water lakes) is typical example of the peroxidase catalyzed degradation of the organic pollutants (Fig. 9) (Ortiz de Montellano 2010).

Several investigations indicate the participation of plant peroxidases in hydroxylation reactions of xenobiotics. For example, peroxidases from different plants are capable of oxidizing N,N-dimethylaniline (Shinohara et al. 1984), 3,4-benzpyrene, 4-nitro-*o*-phenylendiamine (Wilson et al. 1994), 4-chloroaniline (Laurent 1994), phenol, aminoflourene, acetaminophen, diethylstilbestrol, butylated hydroxytoluene, hydroxyanisoles, benzidine, etc. (Sandermann 1994). According to some data, horseradish peroxidase can oxidize methyl group of tritium-labeled [C^3H_3] TNT (Adamia et al. 2006).

8.3 *Phenoloxidase*

(EC 1.14.18.1), a copper-containing enzyme is characterized by universal distribution in plants, microorganisms, insects and animals (Mayer 1987; Sugumaran et al. 1999). Catalyzed by the enzyme is of great importance in such processes as

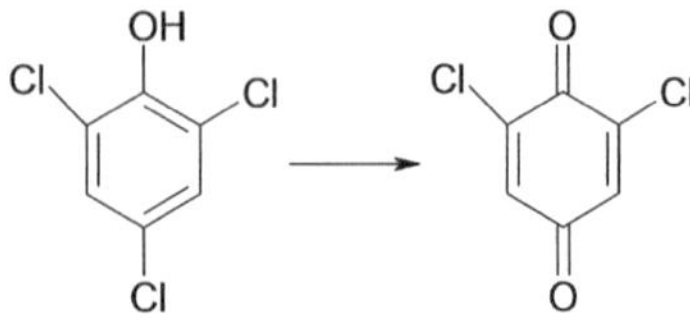

Fig. 9 Oxidation of 2,4,6-trichlorophenol by peroxidase

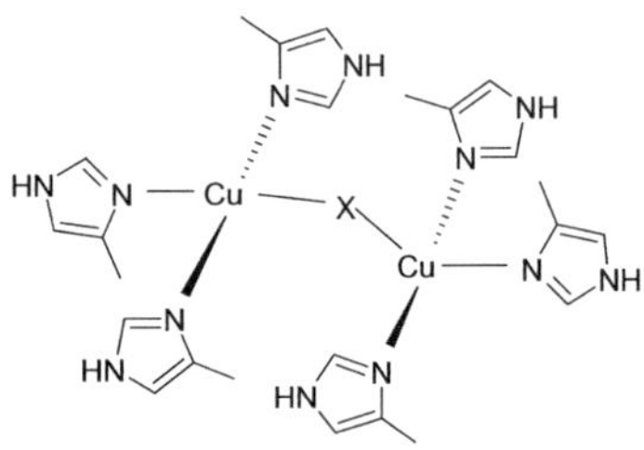

Fig. 10 The active site of phenoloxidase

vertebrate pigmentation and the browning of fruits and vegetables. Phenoloxidase exists in multiple forms in active and latent forms, and catalyzes both monooxygenase and oxygenase reactions: the *o*-hydroxylation of monophenols (monophenolase reaction) and the oxidation of *o*-diphenols to *o*-quinones (diphenolase reaction) (Sánches-Ferrer et al. 1994).

The enzyme has three forms, met-, deoxy- and oxy-, depending on the state of two copper ions of the binuclear site, where they are surrounded by six nitrogen atoms of histidine residues (Fig. 10) (Lu 2003). The met- and oxy- form copper ions are bivalent, and the deoxy- form copper ion is univalent. Besides, substrates (e.g. *o*-diphenol) bind to the met- and oxy- forms, but not to the deoxy- form. Oxygen can only bind to the free deoxy- form.

The catalytic cycles for the monophenolase and diphenolase activities are coupled not only to each other but also to non-enzymatic reactions involving *o*-quinone products (Rodrígez-López et al. 1992).

Type and rate of phenoloxidase activity in the same plants was shown to be depended on the molecular mass of multiple forms of the enzyme. In particular, low molecular forms of phenoloxidase (molecular mass 14, 21, 28, 35, 42, 55, and 70 kD) expose the abilities to act as both diphenolase and monophenolase; with increase of molecular mass (118 and 250 kD) only diphenolase activity is retained (Pruidze et al. 2003). This phenomenon is explained by the steric overlap of the enzyme active site upon association of the low molecular forms to create the high molecular forms, which prevents the binding of monophenols with the high molecular mass, i.e. oligomeric forms.

Besides catalyzing the oxidation of phenolic compounds, phenoloxidase also actively participates in the oxidation of xenobiotics of aromatic structure. In this process, actually both peroxidase and phenoloxidase are active, depending on the structure of the substrate. Phenoloxidase from spinach oxidizes aromatic xenobiot-

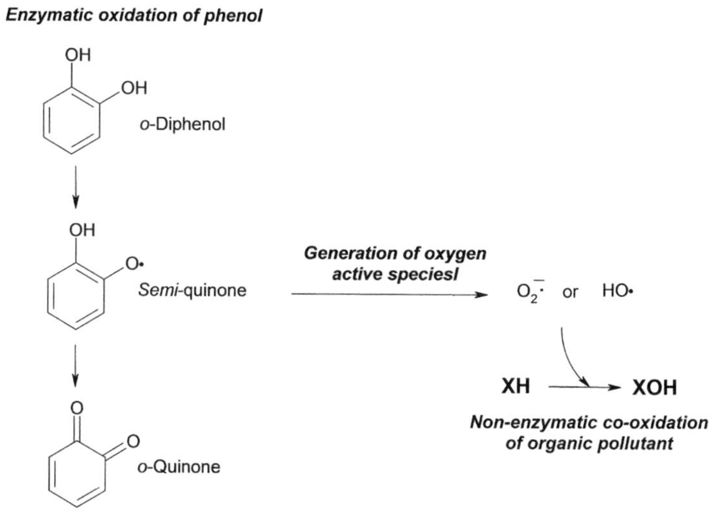

Fig. 11 Enzymatic oxidation of *o*-diphenol by phenoloxidase and non-enzymatic co-oxidation of organic pollutant

ics (benzene, toluene), and is active in their hydroxylation and further oxidation to quinone (Roy and Hanninen 1994). If the xenobiotic subjected to oxidation is not a substrate of the phenoloxidase, such a xenobiotic undergoes co-oxidation in the following manner: the enzyme oxidizes the corresponding endogenous phenol by forming quinones or semi-quinones or both, i.e. compounds with a high redox potential. These compounds activate molecular oxygen and form oxygen radicals, such as superoxide anion radical (O^{2-}) and hydroxyl radical (·OH) (Guillén et al. 1997, 2000; Solomon et al. 2001), which compounds have the capacity to oxidize the organic xenobiotics. In other words, the formation of these radicals enables the phenoloxidase to participate in detoxification processes by the co-oxidation mechanism presented below (Fig. 11).

Analogously, nitrobenzene is oxidized to *m*-nitrophenol, and the methyl group of [C^3H_3] TNT (Adamia et al. 2006) is oxidized by an enzyme preparation isolated from tea plant leaves.

Data indicating the participation of plant phenoloxidases in the oxidative degradation of xenobiotics are sparse (Ugrekhelidze et al. 1997), despite the fact that such activity should definitely be expected. Laccases of fungi have been better explored. Laccases biodegrade (oxidize) many aliphatic and aromatic hydrocarbons (Colombo et al. 1996), and also actively participate in the enzymatic oxidation of alkenes (Niku-Paavola and Viikari 2000). Crude preparations of laccase isolated from the white rot fungus *Trametes versicolor* oxidize 3,4-benzopyrene, anthracene, chrysene, phenanthrene, acenaphthene and other PAHs (Collins and Dobson 1997; Johannes and Majcherczyk 2000). The intensity of the oxidation of these environmental pollutants increases in the presence of such mediators as: phenol, aniline, 4-hydroxybenzoic acid and 4-hydroxybenzyl alcohol, which are substrates of laccase. The rate of PAH oxidation increases proportionally to the redox potential of the mediators in the range $E_h < 0.9$ V. The rate decreases with redox potential in the range $E_h > 0.9$ V (Johannes and Majcherczyk 2000). The natural substrates of laccase, methionine and cysteine, reduced glutathione, and others also stimulate the oxidation of xenobiotics. These data indicate that in the cases of laccase and *o*-diphenoloxidase, the oxidation of hydrocarbons is carried out by a co-oxidation mechanism (Ugrekhelidze 1976; Ugrekhelidze et al. 1997; Ugrekhelidze and Durmishidze 1984).

9 Reduction

Organic pollutants containing nitro groups (typically explosives such as TNT, RDX, HMX, etc.) in plant cell are transformed via reduction. For instance, TNT is initially reduced to ADNTs by most organisms. Nitroreductases are enzymes catalysing the reduction of the nitro groups in aromatic compounds such as explosives, namely TNT and are classified as EC 1.6.6 non-specific NAD(P)H-dependent nitroreductases (Esteve-Núñez et al. 2001). These enzymes are found in animals, plants and microorganisms.

The nitroreductase contains two monomers and binds two flavin mononucleotide prosthetic groups at the monomer interface (Haynes et al. 2002). The enzyme procures reducing equivalents from NADH and NADPH by means of two flavin mononucleotide cofactors (FMN). To catalyze its reaction, nitroreductase uses reduced pyridine nucleotides (both NADH and NADPH) as electron sources (Zenno et al. 1998). Free oxidized flavin has a planar configuration, the induced bend in the oxidized enzyme may favour reduction, and it may also account for the characteristic inability of the enzyme to stabilize one electron-reduced semi-quinone flavin, which is planar (Fig. 12).

The transformation of TNT is in many respects predetermined by its singular chemical structure. The polarization of the N–O bond, due to the greater electronegativity of oxygen compared with nitrogen, induces a partial positive charge on the nitrogen. This charge, combined with the high electronegativity of nitrogen, makes the nitro group easily reducible. On the other hand, the delocalized π electrons from the aromatic ring of TNT are removed by the electronegative nitro groups, which make the ring electrophilic (Preuss and Rieger 1995).

Two types of nitroreductases are known (Fig. 13) (Esteve-Núñez et al. 2001). Type I is found in animals, plants and a number of microorganisms (in bacterial

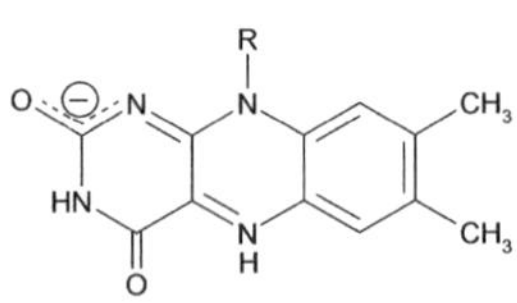

Fig. 12 Chemical structure of the flavin mononucleotide prosthetic group of nitroreductase fully reduced anionic form

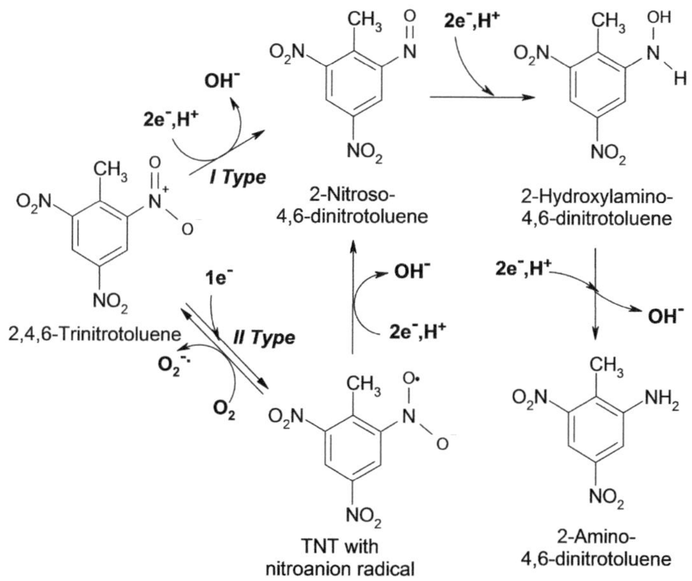

Fig. 13 Reduction of TNT by both types of nitroreductases

genera such as: *Bacillus*, *Staphylococcus*, *Pseudomonas*, etc. and actynomycetes). The enzyme reduces nitro group by two-electron transfers. This pathway is oxygen independent and no radicals are formed (Bryant et al. 1981). Therefore, the type I nitroreductases expose activity under both anaerobic and aerobic conditions. Type II are oxygen sensitive nitroreductases and they reduce nitro groups through single-electron transfers, forming a nitro-anion radical. Under aerobic conditions, an oxygen molecule reacts with a nitro-anion radical and forms a superoxide anion radical that makes the process of TNT transformation reversible. Such nitroreductases are found in rat liver microsomes, and in some strains of *Escherichia coli* (Peterson et al. 1979) and *Clostridium* (Angermaier and Simon 1983).

The first nitro group (in the 2- or 4- position) in TNT is generally much more rapidly reduced than that of the remaining groups. The transformation of the nitro to an amino group decreases the electron deficiency of the nitroaromatic ring, and consequently a lower redox potential is required to reduce the rest of the nitro groups from the molecule. Nitroreductases catalyze further transformations of the other nitro groups of TNT to amino groups. The removal of the nitro group from the o-position and subsequent reduction of the removed nitrite ion by nitrite reductase also takes place. As indicated above, the electron deficiency in the aromatic core of TNT induces nucleophilic attack. The hydride anion from the reduced pyridine nucleotides attacks the aromatic ring, and as a result a non-aromatic structure such as a Meisenheimer σ complex can be formed (Fant et al. 2001). Further, a nitrite anion is released from the Meisenheimer complex with the formation of

2,4-dinitrotoluene. Oxygen is not required for the formation of this compound, and thus this process is an alternative for the metabolism of nitroaromatic compounds when oxidative removal of the nitro groups is not possible.

The nitro, nitroso and hydroxylamino groups are responsible for the toxicity and mutagenic activity of TNT and its derivatives. It has been shown that complete reduction of the nitro groups to amino groups significantly decreases the mutagenic effect of this pollutant (Cash 1998).

It is important that plants that are used to phytoremediate explosive-contaminated soils and groundwater to contain highly active nitroreductase. The correlation between the plant nitroreductase activity and ability to absorb TNT from aqueous solutions has been demonstrated, with the corresponding increase of nitroreductase activity in parallel with the faster assimilation of TNT by the plant (Khatisashvili et al. 2003). These results support the hypothesis that plant nitroreductase activity may serve as a simple preliminary biochemical test to select plants with potential for the phytoremediation of areas contaminated by TNT and similar nitroaromatic compounds. Some plants actively absorb and transform TNT: yellow nutsedge (Palazzo and Leggett 1986), bush bean (Harvey et al. 1990), switch-grass (Peterson et al. 1998), parrot feather, stonewort, algae, ferns, monocotyledonous and dicotyledonous plants, aquatic and wetland species (Best et al. 1997), hybrid poplar (Thompson et al. 1998), soybean (Fialho and Bucker 1996). In the transformation of TNT by plants formation of the monoamino derivatives 2-amino-4,6-dinitrotoluene and 4-amino-2,6-dinitrotoluene takes place. A large proportion, sometimes about 60 %, of the metabolites seems to get involved in conjugation with insoluble biopolymers (Bhadra et al. 1999; Sens et al. 1999; Adamia et al. 2006) often with lignin and hemicellulose (Schoenmuth and Pestemer 2004). These conjugates are compartmented into the vacuoles and the cell wall.

10 Hydrolysis

Organic pollutants containing ester and ether bonds acquire functional groups via hydrolysis. Among them are such compounds as phthalate esters (chemical plasticizers), 2,4-D, diclofop-methyl, bromoxynil octanoate, binapacryl, aryloxyphenoxypropionate, pyrethrin, methyl paraoxon, malathion (pesticides), etc. (Kvesitadze et al. 2006). The functionalization of organic xenobiotics via hydrolysis are catalyzed by hydrolases such as carboxylesterases (EC 3.1.1.1), arylesterase (EC 3.1.1.2), lysophospholipase (EC 3.1.1.5), acetylesterase (EC 3.1.1.6), acylglycerol lipase (EC 3.1.1.23), acylcarnitine hydrolase (EC 3.1.1.28), palmitoyl-CoA hydrolase (EC 3.1.2.2), amidase (EC 3.5.1.4), aryl-acylamidase (EC 3.5.1.13), etc (Sandermann 1994; Krell and Sandermann 1985; Cummins et al. 2001; Cummins and Edwards 2004). These enzymes have a wide specificity and catalyze the hydrolytic cleavage of C–O bonds in organic pollutants. A broad specificity allows the esterases to actively participate in the functionalization phase of lipophilic xenobiotics.

One esterase out of 12 non-specific esterases in wheat shows a preference for a substrate with chain-length of 6–8 carbon atoms, and this form of esterase is active

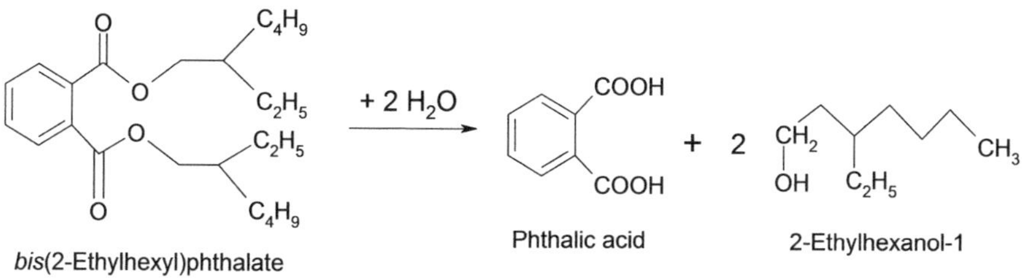

Fig. 14 Hydrolysis of *bis*(2-ethylhexyl)phthalate by esterase

with the toxic chemical plasticizer *bis*(2-ethylhexyl)phthalate (Cummins and Edwards 2004). The cleavage pathway of this molecule is presented in Fig. 14.

The enzymes also effectively hydrolyze model xenobiotics such as *p*-nitrophenyl acetate and α-naphthyl acetate. Comparison of various plant esterase activities showed that activity toward model xenobiotics was the highest in wheat, while esterases in weeds (wild oat, black-grass (*Alopecurus myosuroides*)) were more active in pesticide ester hydrolysis (diclofop-methyl, bromoxynil octanoate, binapacryl) (Cummins et al. 2001). This distinction is simply a consequence of the wide variety of esterases in plants. Weeds contain more basic esterases (pI > 5.0) with a high affinity towards pesticides, while the acidic esterase (pI = 4.6) from wheat has the greatest activity toward α-naphthyl acetate but no activity towards pesticides.

The esterase family in plants is important for the endogenous metabolism and herbicide bioactivation in crops and weeds. A member of the family of serine hydrolases (designated by GDSH), carboxyesterase, which activate aryloxyphenoxypropionate graminicides towards their bioactive herbicidal acids by hydrolyzing the respective ester precursors have been identified in black-grass, a problem weed of cereal crops in Northern Europe (Cummins and Edwards 2004). This enzyme (designated by *Am*GDSH1) was cloned and expressed in yeast (*Pichia pastoris*) as a secreted form of the enzyme. Expression was associated with activity towards aryloxyphenoxypropionate esters. *Am*GDSH1 was predicted to be glycosylated and exported to the apoplast of plants.

A special subclass of esterases is enzymes classified in a unified manner as phosphatases (EC 3.1.3), catalyzing the hydrolysis of ester (P–O–R) and anhydride (P–O–P) bonds by liberating phosphoric acid. Depending on different pH of action the existence of two types of phosphatases is recognized: alkaline phosphatases (EC 3.1.3.1) and acid phophatases (EC 3.1.3.2).

Acid phosphatases hydrolyze C–O–P bonds in substrates under acidic conditions. Acid phosphatases are detected in plants and microorganisms (Bylund et al. 1990; Gellatly et al. 1994; Wolfe and Hoehamer 2003). Recently increasing interest has been focused on plants and microorganisms phosphatases hydrolyzing the di- and triphosphate esters (Fig. 15).

Phosphatase from giant duckweed (*Spirodela oligorrhiza*), mung bean and slime mould (*Dictyostelium discoideum*) are capable of hydrolyzing such highly toxic

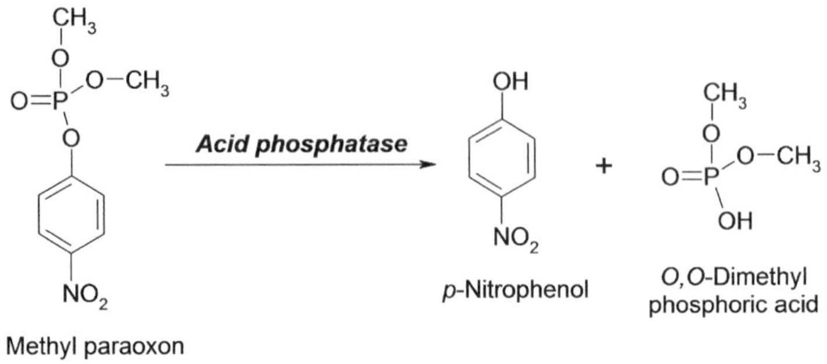

Fig. 15 Formation of p-nitrophenol from methyl paraoxon under the action of acid phosphatase (EC 3.1.3.2)

organophosphates as the nerve agents phosphofluoridic acid-(1-methylethyl) ester (DFP), soman and sarin (Hoskin et al. 1999). There are some other organophosphate compounds that are also hydrolyzed by enzymes, some of them being mixed-function oxidases (Bond and Bradley 1997; Gao et al. 2000), and flavin-containing monooxygenases (Levi and Hodgson 1992), catalyzing P=S bond oxidation. In this list carboxylesterase (EC 3.1.1.1) hydrolyzing the C–O bond should also be mentioned (Lan et al. 1983).

One should also note another group of enzymes, the organophosphate acid anhydrases, which specifically catalyze organophosphate hydrolysis (Yu and Sakurai 1995; Cheng et al. 1996).

Organic pollutants with ether bonds are transformed mainly at the ether site. When ether bonds are lacking, other easily oxidized side groups of the diphenyl ether system are transformed and only if the latter are absent, cleavage of ether bonds occurs in wheat seedlings (Jacobson and Shimabukuro 1984; Tanaka et al. 1990), oats (Jacobson and Shimabukuro 1984), oat cell suspension culture (Shimabukuro et al. 1987), and ryegrass (Shimabukuro and Hoffer 1991). Difenopenten-ethyl is de-etherified in soybean and wheat seedlings (Shimabukuro et al. 1989). The highly selective diphenyl ether herbicide AKH-7088 is metabolized in soybean by complete oxidation of the side chain (Kouji et al. 1990). However, the acifluorfen molecule is cleaved at the ether bond in soybean, to form the corresponding phenols (Fig. 16) (Frear et al. 1983). Similarly, fluorodifen, nitrofen (Shutte and Golfmann 1975), and other diphenyl ethers are cleaved into the corresponding phenols. All these phenols are directly transformed into corresponding conjugates.

11 Conjugation

Conjugation, 2nd phase of transformation of organic pollutants in plants, is a process when an organic pollutant is chemically coupled to cell endogenous compounds (proteins, peptides, amino acids, organic acids, mono-, oligo- and

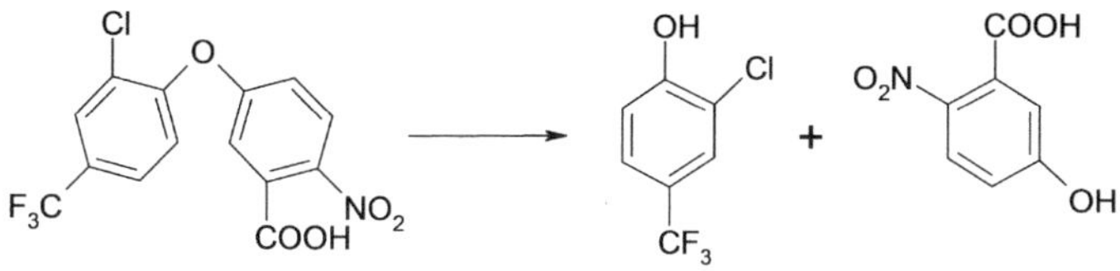

Fig. 16 Hydrolysis of acifluorfen by esterase

polysaccharides, lignin, etc.) by formation of peptide, ether, ester, thioether or other bonds of a covalent nature. Conjugation to biomolecules is regarded as a beneficial detoxification reaction (Schröder and Collins 2002). The toxicity of conjugates if compared with the toxicity of initial forms of toxic compounds is significantly decreased because of binding with non-toxic cellular compounds. Obviously, in such form, conjugates are kept in a cell for a definite interval without causing any pathological deviation in cell homeostasis. Intermediates of organic pollutant transformations, or pollutants already bearing functional groups capable of reacting with intracellular endogenous compounds are susceptible to conjugation. The formation of conjugates leads to the enhancement of the hydrophilicity of organic pollutants, and consequently to an increase in their mobility. Such characteristics simplify further compartmentation of the transformed organic pollutants. Being in conjugated form an organic pollutant in the plant cell is kept apart from vital processes and is therefore rendered harmless for the plant (Burken 2003).

Though conjugation is one of the most widely distributed pathways of plant self-defence against the toxicity of absorbed organic pollutants, it cannot be assumed that the process is energetically and physiologically advantageous for the plant (Burken 2003). Conjugation leads to the depletion of compounds of significant importance for the cell and their resulting deficiency decreases the capacity of plant resistance to prolonged contamination as well as to changes in environmental conditions. Unlike full mineralization, conjugation does not lead to complete detoxification of the organic pollutant, which preserves its basic molecular structure and hence only partially and provisionally loses its toxicity (Kvesitadze et al. 2006). Conjugation is not the most successful pathway of organic pollutant detoxification from the ecological point of view. Plant remains, containing the conjugated pollutants, actually become the toxicant carrier. Typically 70 % or more of the absorbed organic pollutants are accumulated in plants in the form of conjugates (Kvesitadze et al. 2001). This fact must be taken into account when considering the ultimate ecological fate of organic pollutants. Conjugates of organic pollutants are especially hazardous upon insertion into the food chain: enzymes of the digestive tract of warm-blooded animals can hydrolyze conjugates and release the organic pollutants or products of their partial transformation, which in some cases, due to increased reactivity, are more toxic than the initial organic pollutants. Therefore, it is highly desirable that plants applied to phytoremediation have a phenomenal capability to accomplish deep enzymatic degradation of organic pollutants. The selection of such plants, or the promotion of gene expression of enzymes participating in plant detoxification processes are the basic strategies of modern phytoremediation technologies (Kvesitadze et al. 2006; Abhilash et al. 2009).

Enzymes classified as transferases (EC 2) are responsible for catalyzing conjugation reactions of parent toxic compounds and the formation of intermediates with endogenous plant cell constituents. The participation of individual enzymes depends on the chemical nature of the intermediates and the existence of the cell constituents required. The process of conjugation is carried out by glutathione S-transferase (EC 2.8.1.18), O-glucosyl-transferase (EC 2.4.1.7), N-glucosyltransferase (EC 2.4.1.71), N-malonyltransferase (EC 2.3.1.114), putrescine N-methyl-transferase (EC 2.1.1.53) etc. (Sandermann 1994; Schröder and Collins 2002; Pilon-Smits 2005). Under normal conditions these enzymes participate in cell metabolism, and in case of organic pollutants (organic pollutants) penetrating into plant cells at appreciable (or indeed high) concentrations they are involved in organic pollutants conjugation. All cell constituents bound with organic pollutants in the conjugation processes have a hydrophilic nature and, thus, the hydrophobicity of the toxicants as a result of conjugation decreases. Therefore, the conjugates are more soluble in the cytoplasm than the parent toxicants and undergo compartmentation.

Among the known mechanisms of conjugation, glycosylation is one of the most widespread in higher plants (Kvesitadze et al. 2006). Organic pollutants having hydroxyl, caboxyl, amine, or other functional groups as a constituent part of their molecules are directly subjected to glycosylation. Often, after penetrating into the plant cell organic pollutants are exposed to minor initial transformations in which such groups are introduced into their molecules, significantly increasing the reactivity of the modified organic pollutants.

Glucosyltransferases catalyze the reaction between glucose and hydroxyl or amino groups (N-glucosyltransferases) of organic pollutants (Loutre et al. 2003). Both enzymes are inducible under the action of some herbicides and other organic pollutants (e.g. 3,4-DCA, 4-nitrophenol and 2,4,5-trichlorophenol) (Brazier et al. 2002).

In plants, organic pollutants undergo conjugation with the help of various transferases. For example, the herbicide 3,4-DCA is metabolized via N-malonyltransferase in soybean root cultures, but in *Arabidopsis thaliana* root cultures via N-glucosyltransferase (Fig. 17) (Gilbert and Crowley 1997).

2,2-*bis*(4-Chlorophenyl)-acetic acid, the first intermediate of the insecticide DDT metabolism in soybean, is conjugated by the formation of O-glucoside (Sandermann 1994). The conjugation capacity of soybean O-glucosyltransferase is 855 μg 2,2-*bis*(4-chlorophenyl)-acetic acid per hour per gram of cells fresh weight (Wetzel and Sandermann 1994).

Alcohols and phenols often undergo glycosylation in plants. This is illustrated in the literature by many examples indicating the formation of β-glucosides. Thus, formation of ethyl-β-glucoside was observed during cultivation of seedlings of mung bean in an ethanol-containing zone (Middleton et al. 1978).

The metabolic fate of [1–6-^{14}C] 3,4-dichloroaniline (DCA) was investigated in thale cress (*Arabidopsis thaliana*) root cultures and soybean plants after a 48 h treatment via the roots. DCA is rapidly taken up by both species and predominantly metabolized to N-malonyl-DCA in soybean and N-glucosyl-DCA in *Arabidopsis*.

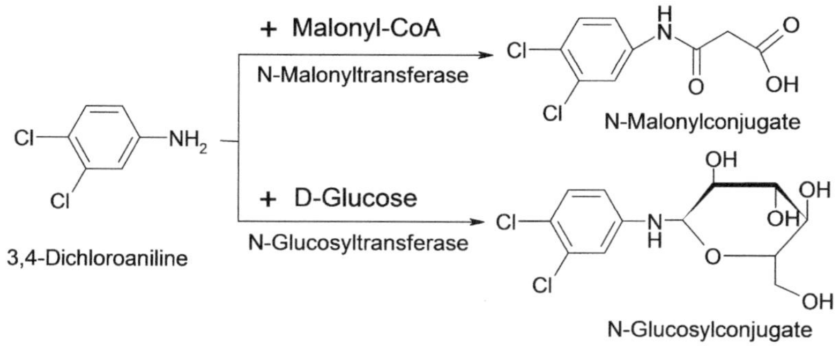

Fig. 17 Two pathways of 3,4-dichloroaniline conjugation

Synthesis is observed in roots and the respective conjugates are largely exported into the culture medium, with only a minority retained within the plant tissue. Once conjugated, the DCA metabolites from the medium are slowly taken up by the roots of both species. The difference in the routes of DCA detoxification in two plants could be explained partly by the relative activities of the conjugating enzymes, soybean having high DCA-N-malonyltransferase activity, while in *Arabidopsis* DCA-N-glucosyltransferase activity predominates (Lao et al. 2003).

DCA is rapidly conjugated by glucosylation in *Arabidopsis* root cultures by forming N-β-D-glucopyranosyl-DCA and excreted into the medium. The enzyme N-glucosyltransferase is responsible for this transformation (Loutre et al. 2003).

Pentachlorophenol is glucosylated in wheat and soybean plants via combination with malonic acid and is transformed to β-D-glucoside and O-malonyl-β-D-glucoside conjugates, which are found in plant tissues (Schmitt et al. 1985). Pridham (1958), in experiments with broad bean seedlings, showed that foreign mono-, di-, and triatomic phenols are easily (enzymatically) converted into the corresponding β-monoglucosides.

In some cases when phenols are glycosylated, the existence of di- and triglycosides has been shown. For instance, diglycoside (gentiobioside) and triglycosides are formed from exogenous hydroquinone in wheat embryos (Harborne 1977).

Often, hydroxyl derivatives are produced as one of the primary products of transformations of the organic pollutants in plant tissues, and are further subjected to rapid glycosylation. In this way a conjugate is formed from the oxidation product of the systemic fungicide etirimol. The aliphatic side chain (butyl group) of this herbicide is oxidized and the alcoholic hydroxyl formed is glycosylated in leaves of barley (Harborne 1977). The herbicide diphenamid is oxidized, i.e., the N-methyl group is hydroxylated, in pepper seedlings (Hodgson and Hoffer 1977) and in callus tissue of tobacco (Burrows and Leworthy 1976).

2,4-D glucose esters occur widely and in large amounts in herbicide-resistant wild wheat plants (*Triticum dicoccum*), timothy (*Phleum pratense*), and kidney bean (Chkanikov et al. 1976). Besides the carboxyl group, other acidic groups are also subjected to glycosylation in plants. Thus, the plant growth regulator ethephon

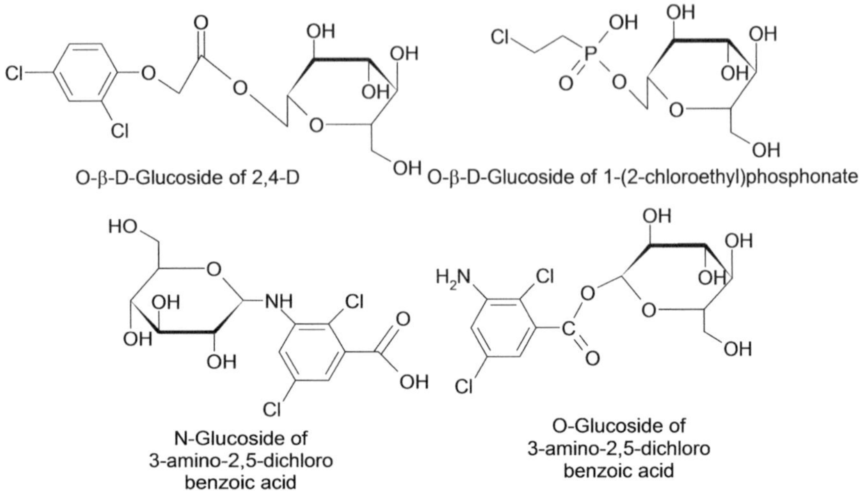

Fig. 18 Conjugates of pesticides with glucose

is glycosylated by formation of β-D-glucopyranoside-l-(2-chloroethyl)-phosphonate in bark cuts of hevea (*Hevea brasiliensis*) (Audley 1979) (Fig. 18). Glycosylation via blocking free amino groups of organic pollutants is another widespread mechanism of conjugation, since pollutants of different structures often contain such groups. Thus 3-amino-2,5-dichlorobenzoic acid is further transformed into the corresponding N- and O-glucosides after its initial transformation to the glucose ester (Fig. 18) in roots, shoots, and hypocotyls of *Setaria* sp. (Frear et al. 1978).

Amino acids, being multifunctonall secondary metabolites, participate in many vitally important for plant cells processes. Recently it was shown that in addition to all their previously known attributes of metabolic importance they effectively participate in the detoxification (conjugation) of a broad spectrum of organic pollutants.

Conjugation with amino acids is a widespread reaction of carboxyl group of organic pollutants in plants. A study of 2,4-D metabolism in soybean species has demonstrated that the glycoside conjugate of 4-oxy-2,5-dichlorophenoxyacetic acid is the primary metabolite in resistant species, but that in sensitive species forms conjugates with amino acids (White et al. 1990). 2,4-D formes conjugates with glutamic and aspartic acids (Fig. 19) in callus and differentiated root tissues of soybean (Davidonis et al. 1978), in tissue cultures of maize endosperm, and in the medullar parenchyma of tobacco, carrot, and sunflower (Feung et al. 1976).

The variety of peptides and proteins existing in plant cells, with a variety of functional groups, and their ability to couple with organic pollutants of different structures, creates several favourable opportunities for organic pollutants conjugation.

One of the most important pathways for maintaining the health of plants via detoxification is the conjugation of organic pollutants with the tripeptide –reduced glutathione (GSH) (-Glu–Cys–Gly).

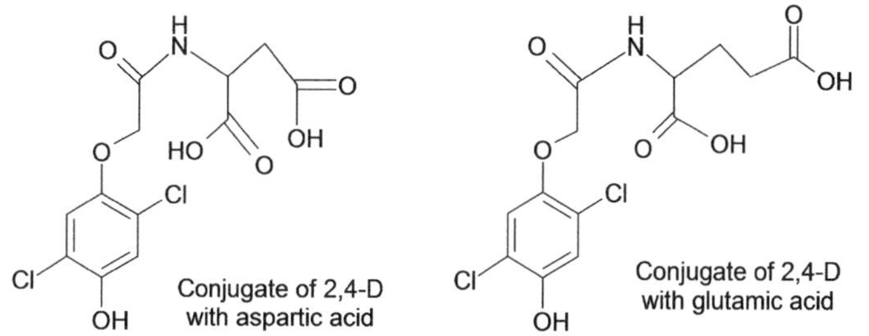

Fig. 19 Conjugates of 2,4-D hydroxy-derivatives with amino acids

Conjugation with GSH proceeds according to a widespread detoxification mechanism for organic pollutants in plants and mammals. Conjugation with glutathione has been demonstrated for chemical groups as diverse as chloroacetamides, triazines and sulphonyl-ureas (Leavitt and Penner 1979; Ezra and Stephenson 1985) and is regarded as an important contributor to the selectivity of herbicides in target crops. This detoxification pathway is most characteristic for symmetric triazines, chloroacetamides, and other halogen-containing compounds. A study on atrazine transformation in 53 herbaceous plant species (*Festucae*, *Avenae*, *Triticeae*, *Paniceae*, *Andropogonae*, *Eragrosteae*, *Chlorideae*) revealed that the herbicide formed conjugates with glutathione in all plants tested (Jensen et al. 1977).

Group of enzymes, called glutathione S-transferases (GSTs) have wide specificity and couple electrophilic organic pollutants and their metabolites with the GSH. In plants, a large and diverse gene family encodes the glutathione transferases. GSTs facilitate the reaction between the functional group of the pollutant intermediates and the SH-group of the glutathione cysteine residue. They participate in the conjugation of a wide spectrum of toxic compounds such as the herbicides flufenacet (FOE 5043), triflusulfuron, chlorimuron-ethyl, acetochlor, metolachlor, alachlor, atrazine (Bieseler et al. 1997), safeners (DeRidder et al. 2002), fluorodifen (Dixon et al. 2002), etc. In consequence, the toxicant is bound to intracellular compounds via a covalent bond to the sulphur atom (Fig. 20).

In the reaction presented in Fig. 20, R is an aliphatic, aromatic, heterocyclic, sulphate, nitrite or other group. The GSTs also catalyze the addition of aliphatic epoxides and arene oxides to glutathione; the reduction of polyol nitrate by glutathione to polyol and nitrite; certain isomerization reactions and disulphide exchange.

The best characterized among the plant proteins involved in detoxification are the GSTs from maize. Six isoenzymes differing in their DNA sequence, regulation and substrate specificity have so far been reported. Involvement in the detoxification of herbicides has been demonstrated for isoenzymes GST I, and GST III, that are constitutively expressed in maize, and for the safener-induced enzymes GST II and GST IV (Droog et al. 1993).

Another tripeptide, homoglutathione (differing from glutathione by containing alanine instead of glycine), may also participate in conjugation reactions with organic pollutants in plants. Formation of homoglutathione conjugates is character-

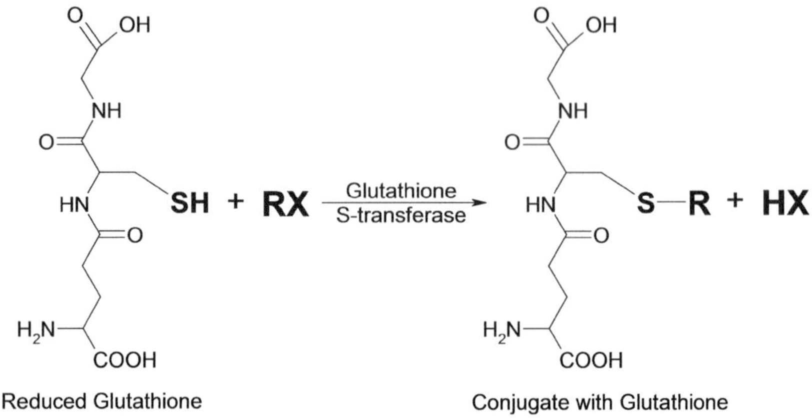

Fig. 20 The conjugation reaction of organic pollutants (RX) with reduced glutathione, catalyzed by GSTs

istic mainly for soybean. Thus, the herbicide propachlor forms a conjugate with homoglutathione in soybean seedlings (Lamoureux and Rusnes 1989). The same transformation occurs during chlorimuron-ethyl and thifensulfuron-methyl metabolism (Brown and Neighbors 1987; Brown et al. 1990). Homoglutathionic conjugates of acetochlor are formed also in other plants, particularly in soybean, mung bean, and alfalfa (Breaux 1987).

Another mechanism characteristic for the binding of organic pollutants to glutathione and homoglutathione is the reaction with alkylthio groups. The S-ethyldipropyl thiocarbamate conjugates with glutathione via an ethyl group in maize seedlings (Lay and Casida 1976; Carringer et al. 1978). It is supposed that in this particular case the herbicide initially is oxidized into the corresponding sulphoxide and then is conjugated with glutathione. The latter process is catalyzed by GST. Metribuzin binds with homoglutathione via a methylthio group in soybean (Frear et al. 1985). 3,4-Benzopyrene is oxidized by conjugation with glutathione in microsomes from parsley cell suspensions, soybean and primary leaves of pea seedlings (Trenck and Sandermann 1980).

Quite often organic pollutants are coupled with cell biopolymers, participating in formation of cell walls structure. Xylans (hemicelluloses), being widely presented in plant tissues and possessing many free carboxyl groups, actively participate in conjugation with amino or hydroxyl groups of foreign molecules. A good example of such combination is the conjugation of ADNTs, primary products of TNT reduction, with hemicellulose (Fig. 21) in the roots of hybrid willow (*Salix* sp.), and Norway spruce, trees used in dendroremediation of soils polluted by TNT (Schoenmuth and Pestemer 2004). Other example is metabolism of [^{14}C] TNT being absorbed by roots of kidney bean. TNT conjugates with lignin (20 %), hemicellulose (14 %), and pectin (5 %) (Sens et al. 1999).

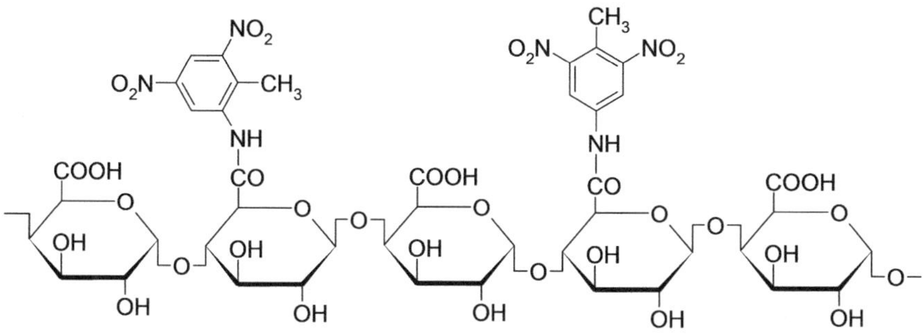

Fig. 21 Hemicellulose with bound residues of monoaminodinirtotoluenes (metabolites of TNT)

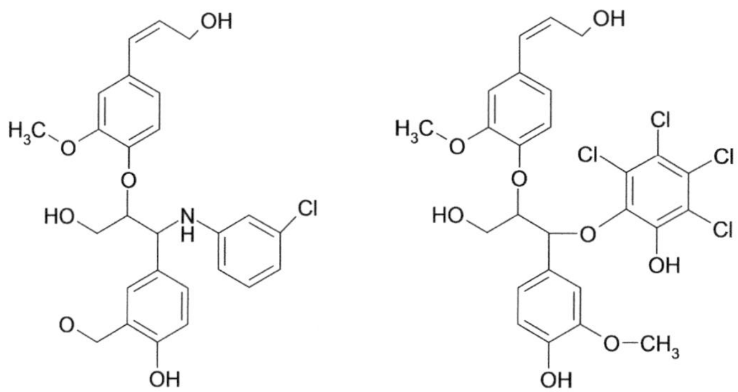

Fig. 22 Conjugates of 3-chloroaniline (*left*) and 1,2-dihydroxy-3,4,5,6-tetrachlorobenzene (*right*) with coniferyl alcohol

Lignin is a phenolic, structurally nonrepeating macromolecule, which is active in conjugation reactions, and often plays the role of a carrier of organic pollutants and their primary transformants (Sandermann 1994). Such compounds are incorporated into the lignin structure by being covalently coupled with the biopolymer. It has been shown that tautomeric forms of the lignin monomer coniferyl alcohol (quinone-metyd) couples organic pollutants with amino and hydroxyl groups, as for instance 3-chloroaniline (Fig. 22) (Sandermann et al. 1983). An analogous picture is observed in the case of pentachlorophenol. Coniferyl alcohol easily conjugates with 1,2-dihydroxy-3,4,5,6-tetrachlorobenzene, intermediate of pentachlorophenol hydroxylation (Fig. 22) (Sandermann 1987).

In summary, according to the existing experimental information, environmental pollutants may be conjugated directly with a biopolymers, or may be coupled with monomers and undergo copolymerization to form a modified biopolymer.

12 Compartmentation

Compartmentation is in most cases the Final step of conjugate processing. In this phase temporary (short or long term) storage of conjugates in defined compartments of the plant cell takes place. Soluble conjugates of toxic compounds (coupled with peptides, sugars, amino acids etc.) are accumulated in vacuoles, while insoluble conjugates (coupled with protein, lignin, starch, pectin, cellulose, xylan and other polysaccharides) are moved out of the cell via exocytosis and are accumulated in the apoplast or cell wall (Sandermann 1994). The step of compartmentation is analogous to mammalian excretion, essentially removing toxic compounds from metabolic tissues (Burken 2003). The major difference between detoxification processes in mammals (and other animals) and plants is that plants do not have a special excretion system for the removal of organic pollutant conjugates from the organism. Hence they use a mechanism of active transport for the removal of the toxic residues away from the vitally important sites of the cell (nuclei, mitochondria, plastids, etc.). This active transport is facilitated and controlled by the ATP-dependent glutathione pump (Martinova 1993). This process is also termed "storage excretion" (Merbach and Schilling 1977; Le Baron et al. 1988; Coleman et al. 1997).

The all three phases of metabolism of organic pollutants is well demonstrated for representatives of organochlorine pesticides (Sandermann 1987, 1988, 1994). For instance, insecticide DDT acquires hydrophilic carboxyl group instead of hydrophobic trichloromethyl group via hyrdoxylation, and then easily forms an ester with glucose enabling it to be stored in vacuoles according to the mechanism shown on the scheme (Fig. 23) (Sandermann 1994).

The biocide 2,3,4,5,6-pentachlorophenol also undergoes hydroxylation upon which the pentachlorophenol acquires a second hydroxyl group, this intermediate conjugates with lignin forming an insoluble compound, which is removed from the cell and stored in the cell wall (Fig. 24) (Sandermann 1994). But this organic pollutant has polar hydroxyl group, and is subjected to conjugation with β-D-glucose and/or O-malonyl-β-D-glucose, and forms soluble glucosides conjugates, which translocate and accumulate in vacuoles without the first phase of transformation (Fig. 25).

It could be stated that almost all organelles and compartments of plant cell directly or indirectly are involved in the processes that take place during transforma-

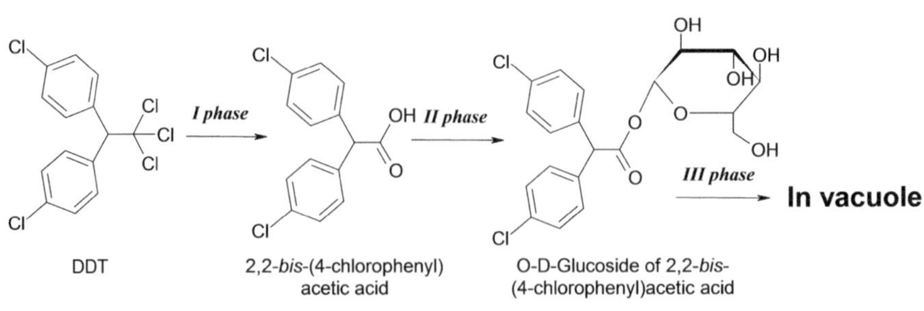

Fig. 23 Transformation of DDT with final deposition in vacuole

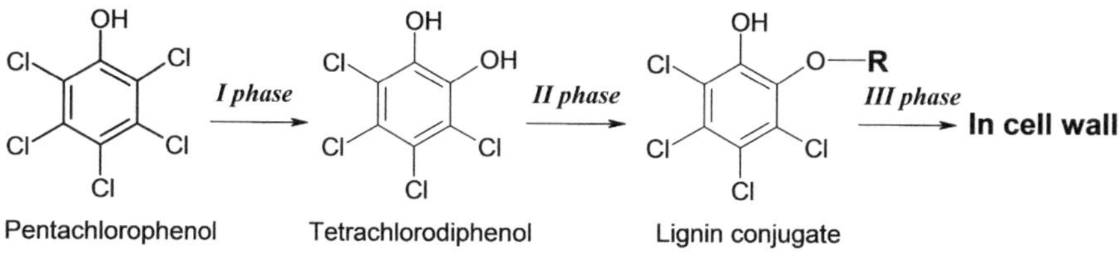

Fig. 24 Pentachlorophenol transformation for deposition in vacuoles and the cell wall

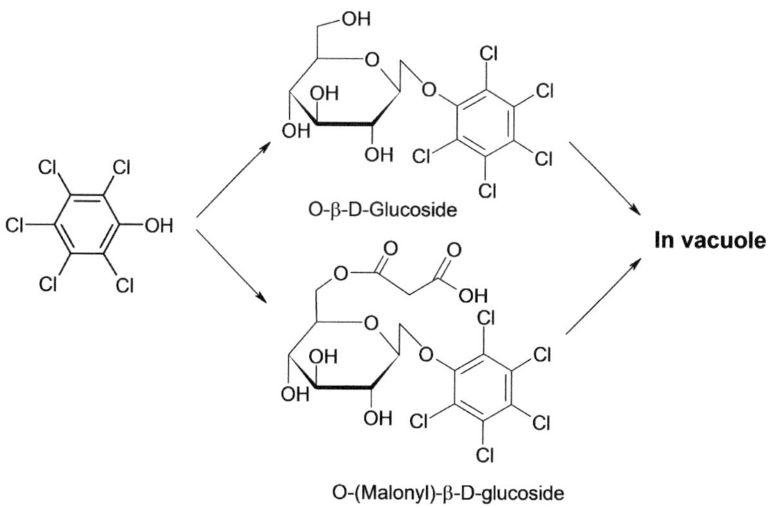

Fig. 25 Pentachlorophenol transformation for deposition in vacuoles

tion of organic pollutants. For example, functionalization phase predominately proceeds on membranes of endoplasmic reticulum, plasmalemma, peroxisomes and tonoplasts; the enzymes functioning in cytoplasm, participated in conjugation processes, mitochondria and plastids provide with energy separate phases of transformation; nuclei and ribosomes realize induction of the enzymes participating in detoxification processes; cell walls and vacuoles serve as compartments for storage of partially tranformed pollutants.

13 Complete Metabolism of Organic Pollutants in Plants

In recent years 4th phase of transformation of organic pollutants in plants, so-called further processing of organic pollutants by plants is considered (Edwards et al. 2011). When plants are transferred from a pollutant-containing environment to a pollutant-free medium, gradual degradation of deposited conjugates and mineralization of toxic residues of organic pollutants take place (Chrikishvili et al. 2006; Kvesitadze et al. 2006). Organic pollutants decomposition in plant cells is carried out mainly by oxidative enzymes. Basically, this process consists of several consecutive stages and terminates by the release of CO_2 (sometimes only in trace amounts however). Since carbon dioxide is considered to be an inorganic compound, the conversion of the organic pollutant into CO_2 is known as the process of mineralization. The verification of the concept of phytotransformation has been confirmed at different scales: in laboratory, greenhouse and field experiments.

Using organic pollutants with radioactively labeled carbon atoms in experiments on the uptake and transformation of organic pollutants the emission of $^{14}CO_2$ was demonstrated (Ugrekhelidze 1976), indicating that the initial transformations of organic pollutants is followed by the deep oxidation in plant cells.

Plants absorb alkanes and cycloalkanes from the environment and metabolize them. Experiments with [^{14}C] labeled hydrocarbons proved that sterile seedlings placed in an atmosphere containing low molecular mass alkanes (C_1–C_5) or cyclohexane uptake these compounds and further transform them by oxidation to the corresponding carboxyl acids. Alkanes undergo monoterminal oxidation, while cyclohexane is oxidized via ring cleavage. The emission of $^{14}CO_2$ in the dark during this process serves as evidence for the occurrence of mineralization and can be easily measured (depending on the time of exposure, the percent of mineralization was found to be as high as 30 %). Consequently, organic and amino acids are among the end products of this transformation and they can be used for further cell metabolism (Penner and Early 1973).

Methane, ethane, propane and pentane are metabolized by the formation of low molecular mass compounds largely composed by organic acids. Labeled fumaric, succinic, malonic, citric and lactic acids have been identified in plant leaves exposed to these low molecular mass alkanes, with most of the radioactivity incorporated into succinic and fumaric acids. Based on the fact that the carbon atoms originating from ethane are incorporated into these acids it is proposed that ethane in plants is oxidized monoterminally: if ethane was oxidized at both terminal carbon atoms, instead of one, the carbon atoms originating from ethane would be incorporated into glycolic, glyoxalic or oxalic acids. The oxidation of ethane at one terminal carbon atom leads to the formation of acetyl-CoA, which in turn is able to participate in the Krebs cycle (Durmishidze and Ugrekhelidze 1968a, b).

Propan oxidation at one terminal carbon atom leads to the formation of propionic acid, which successively undergoes -oxidation resulting in the formation of malonyl-CoA, and decarboxylation resulting in the formation of acetyl-CoA (Fig. 26) (Ugrekhelidze 1976). Acetyl-CoA is transferred to carboxyl groups of succinic acid. Based on the identified low molecular mass degradation products, it is suggested that propane is also oxidized monoterminally in plants into compounds that can be incorporated into the Krebs cycle. Pentane also may be oxidized monoterminally, forming valeric acid. Approximately the same organic acids are formed from pentane as from valeric acid (Ugrekhelidze 1976).

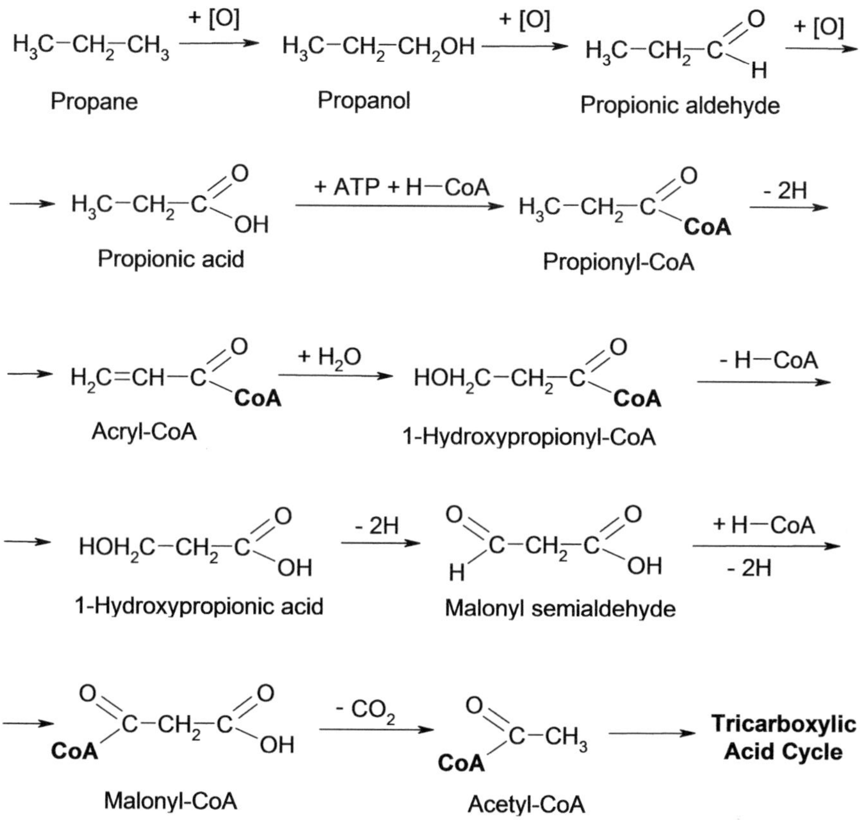

Fig. 26 Transformation of propane in higher plant cells

Long chain alkanes are subjected to transformations similar to those of short chain alkanes. For instance, after 40 min of incubation of leek leaves with an emulsion of exogenous [^{14}C] octadecane in water, 9.6 % of the total label is detected in esters, 6.4 % in alcohols and 4 % in organic acids (Cassagne and Lessire 1975). Following a similar experimental approach, it was firstly demonstrated that plants are also able to metabolize benzene and phenol via aromatic ring cleavage (Durmishidze et al. 1974a, b; Ugrekhelidze 1976). In this process the carbon atoms are incorporated into carboxyl acids and amino acids. Similar data were obtained for nitrobenzene and aniline (Mithaishvili et al. 2005), toluene (Jansen and Olson 1969; Tkhelidze 1969; Durmishidze et al. 1974b, c), *α-naphthol* (Ugrekhelidze and Kavtaradze 1970) and benzidine (Durmishidze et al. 1979).

Oxidation of benzene and phenol by crude enzyme extracts of plants forms muconic acid after ring cleavage, with catechol as an intermediate, according to the following scheme (Fig. 27) (Durmishidze et al. 1969).

Muconic acid may be further oxidized to fumaric acid. Muconic and fumaric acids are often found in plants exposed to benzene or phenol. Cleavage of the

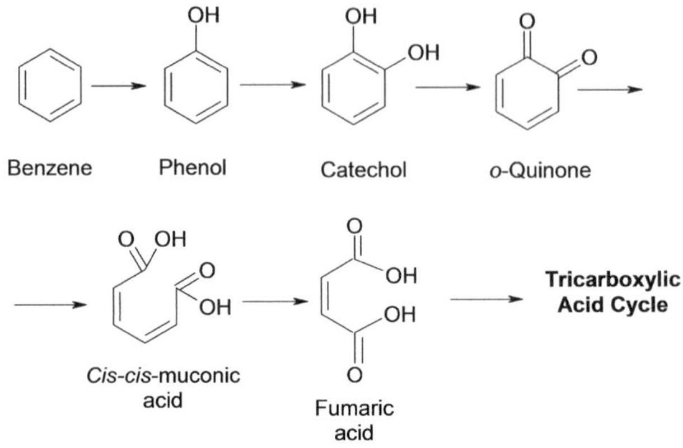

Fig. 27 Oxidative degradation of benzene in plant cells

aromatic ring of endogenous substrates proceeds by the same way, i.e. in the way that 3,4-dihydroxybenzoic acid is transformed into 3-carboxymuconic acid (Tateoka 1970).

The above information refers to the role of plants in maintaining a tolerable environment for humans and other fauna is already apparent from consideration of the circumstance that the burning of fuels for heating, electricity generation, and automotive transport (oil products, natural gas, coal, peat, etc.), formerly considered to be a fairly harmless activity, induces the release of carbon oxides into the environment in sufficient quantity to exterminate all life were plants not actively absorbing them from the atmosphere. The activity of green plants (together with certain microorganisms) directed in maintaining the atmosphere in tolerably good shape by absorbing and metabolizing carbon monoxide and dioxide and other pollutants is the main naturally beneficent ecological power. Yet the current preoccupation with global warming has made it starkly apparent that the biological self-cleaning potential of our planet is now inferior to the rate of accumulation of pollutants, as evidence not only by adverse changes of climatic parameters in the most damaged regions, but in the appearance of defective forms of plants, animals and microorganisms.

14 Enhancing of Phytoremediation Abilities of Plants

Phytoremediation implies application of plants solely or in combination with microorganisms for remediation of environment polluted with different inorganic or organic toxicants. The most important factor for successful phytoremediation is presence of suitable plants capable of actively assimilating contaminants. The process efficiency is determined mostly by the plant's ability to assimilate or accumulate (or both) organic and inorganic contaminants in cellular structures and to accomplish deep oxidative degradation of organic xenobiotics. Progress in the development of phytoremediation technologies applied to environments contaminated with organic contaminants is significantly ahead as compared to the developments in the assimilation of inorganic contaminants and radionuclides. This progress is achieved in selection of suitable plants with desirable features such as: adaptation to the particular soil-climatic zone, productivity (fast growing, large biomass formation), appropriate morphological (development of an appropriate leaf and root systems) and physiological (transpiration ability) characteristics, adaptation to field conditions, presence of enzymatic systems important for contaminants transformation, etc. (Kvesitadze et al. 2006).

These and possibly some other characteristics determine plants ability to deep degradation of organic contaminants. Definite progress has been achieved in the cloning of genes encoding enzymes important for contaminant transformation (Kvesitadze et al. 2006).

Although plants have the inherent ability to detoxify organic pollutants (Abhilash et al. 2009), as they acquire full set of enzymes, important for xenobiotic transformation. However complete detoxification of organic pollutants (mineralization or degradation to cell standard metabolites) may not be always achieved due to insufficient enzyme capacity, high persistence of pollutants to transformation (e.g. PAHs, PCBs and other POPs), or their high concentration. Incomplete detoxification of taken up by plants organic pollutants contains a risk that toxic compounds may be incorporated into food chain. This is a serious disadvantage of phytoremediation technologies. A direct method for enhancing the efficacy of phytoremediation is to overexpress in plants the genes involved in metabolism, uptake, or transport of specific pollutants. Detoxification of most classes of herbicides in major crops and weeds involves P450 enzymes. Only a few of them, mainly isoenzymes *CYP76B1* and *CYP71A10*, are characterised so far. They catalyse the dealkylation and/or the hydroxylation of phenylurea into non-phytotoxic products. The same reactions can be carried out by mammalian enzymes. Effective phytoremediation has so far only been achieved with a mammalian P450 transgene. *CYP2E1* expression allowed transgenic tobacco to take up and metabolise the pollutant trichloroethylene (Abhilash et al. 2009; Kumar et al. 2012).

In construction of transgenic plants for phytoremediation are successfully applied genes of enzymes catalyzing reactions of I and II phases transformations (Cytochrome P450s, peroxidases, laccases, nitroreductases, GSHs, etc.) from plants, bacteria, fungi, yeast, mammalians and humans (Abhilash et al. 2009). For example, the introduction of human and mammalian (rat, mouse, rabbit) cytochrome P450 isoenzymes genes (*CYP1*, *CYP1*, *CYP3*) in higher plants (*Arabidopsis thaliana*, *Oryza sativa*, *Nicotiana tabaccum*, *Solanum tuberosum*) by using *Agrobacterium tumefaciens*-mediated plant transformation or direct DNA gene transfer is successfully realized (Doty 2008; Kawahigashi 2009). Oryza sativa with mammalian (*Sus scrofa*) genes of Cytochrome P450 (*CYP2B22* and *CYP2C49*) revealed high tolerance to several herbicides (Kawahigashi et al. 2005); hybrid poplar (*Populus tremula x populous alba*) with Cytochrome P450 (*CYP4502E1*) from rabbit has increased ability to removal of organochlorine solvents such as trichloroethene, vinyl chloride, carbon tetrachloride, benzene and chloroform from hydroponic solution and air (Doty et al. 2007). Also, plants having yeast NADPH cytochrome P450 reductase have been obtained (Shiots et al. 2000; Schmidt et al. 2006).

In recent studies great attention is paid to genes of cytochrome P450 isoenzymes (Kumar et al. 2012). It was shown that the plant engineered with cytochrome P450 that participated in metabolism of a ω-fatty acid (*CYPBM3*), enhanced the selective hydroxylation of hydrocarbons of small to medium chain length (C_3-C_8 alkanes) (Glieder et al. 2002). Besides, *CYPBM3* was engineered for increased oxidation of aromatic hydrocarbons, such as benzene, propyl benzene, and naphthalene (Whitehouse et al. 2008). Cytochrome $P450_{cam}$ (*CYP101*) has been engineered using rational approach for efficient hydroxylation of alkanes and organochlorine toxicants (halogenated hexanes, PCBs), PAHs, and other organic pollutants (diphenylmethane, styrene, ethylbenzene, etc.) (Jones et al. 2001; Bell et al. 2001, 2002; Xu et al. 2005; Kumar et al. 2012). The engineered *CYPBM3* and *CYPCAM* enzymes can potentially be used to create transgenic plants for possible phytoremediation of organic pollutants. These engineered enzymes can be tested to examine whether they also show enhanced activity towards herbicides and other environmental contaminants. Furthermore, they can specifically be designed to metabolize herbicides, as well as other important petroleum aliphatic and aromatic hydrocarbons (Kumar et al. 2012).

By means of gene engineering manipulations gene of human cytochrome P450 (*CYP1A1*) was transferred to *Oryza sativa* for increase of plant tolerance to sulfonylurea (Hirose et al. 2005); for enhance of metabolism of chlorotoluron and norflurazon (Kawahigashi et al. 2008). Protoporphyrinogen IX oxidase (*Protox*) from *Bacillus subtilis* have been introduced to the same plant due to which plant acquired high resistance to herbicide oxyflufen high concentrations (Jung et al. 2008).

Besides the bacterial Cytochrome P450 (genes *XplA* and *XplB* from *Rhodococcus rhodochorus*) (Jackson et al. 2007), after transfer of Nitroreductase (*NfsA*) gene from *Escherichia coli* in this plant, initial transformation of TNT (reduction of nitro groups) have been increased 7–8 times (Kurumata et al. 2005); for removal of insecticide pentachlorophenol (PCP) gene of peroxidases from *Coriolus versicolor* (*Mn peroxidase*) are expressed (Iimura et al. 2002).

To increase phytoremediation properties expression of corresponding genes of the same plant is often used. Overexpression of glutamycysteine synthetase (*γ-ECS*) resulted in increased tolerance of *Populus trichocarpa* to chloroacetanilide herbicides (Gullner et al. 2001). Overexpression of UGTs genes Glycosyltransferases 743B4, 73C1 in *Arabidopsis thaliana* resulted in the enhanced detoxification of TNT and enhanced root growth (Gandia-Herrero et al. 2008). Similar results have been obtained by overexpression of γ-Glutamycysteine synthetase (*γ-ECS*) and Glutathione synthetase (*GS*) in *Brassica juncea* for enhanced tolerance to atrazine, 1-chloro-2,4-dinitrobenzene, phenanthrene and metolachlor (Flocco et al. 2004). The overexpression of Peroxidase gene (*tpxI*) in hairy roots of transgenic tomato *Lycopersicon esculentum* resulted in the enhanced removal of phenol (Oller et al. 2005).

Interesting experiments have been conducted on introduction of enzyme from one plant to another with the aim to increase plant tolerance to pesticides and organic contaminants. In particular, the introduction of cytochrome P450 (gene *CYP71A10*) from *Glycine max* to *Arabidopsis thaliana* and *Nicotiana tabaccum* was carried out, as a result of which tolerance of modified plant to phenyl urea herbicides have been significantly increased (Siminszky et al. 1999). Similar examples are the introduction of *Helianthus tuberosus* Cytochrome P450 genes (CYP76B1) in *Nicotiana tabacum* for increased tolerance to herbicides (Didierjean et al. 2002); the transfer of Glutathione S-transferases (GstI-6His) from maize to *Nicotiana tabacum* for forming of higher tolerance to alachlor (Karavangeli et al. 2005); transfer of Peroxidases genes (*tpx1* and *tpx2*) from *Lycopersicon esculentum* to *Nicotiana tabacum*, as a result of which hairy cultures of transgenic tobacco showed enhanced removal of phenol (Alderete et al. 2009).

Engineering of plants in connection with particular soil-climatic conditions and contaminants have also been conducted. A widely distributed explosive contaminant TNT has been chosen as contaminant for this purpose. In order to increase the degradability of TNT and similar compounds, several transgenic tobacco plants, containing the gene (*onr*) of the bacterial (*Enterobacter cloaceae*) enzyme pentaeritrole tetranitrate reductase (EC 1.6.99.7) have been created (French et al. 1999). This transgenic tobacco has been analyzed for its ability to assimilate the residues of TNT and trinitroglycerine. Later, novel transgenic tobacco with nitroreductase gene *NfsI* of the same bacteria have been obtained, which expressed significantly increased ability to uptake TNT from enviroenment (Hannink et al. 2001, 2007). Seedlings of the transgenic plants extracted explosives from liquid phases much faster, accomplishing denitration, than the seedlings of common forms of the same plants, in which growth was inhibited by the explosives (Hannink et al. 2001, 2007). Transgenic tobacco differed substantially from the common plant by its tolerance and fast uptake and assimilation of significant amounts of TNT. Analogous experimental results were obtained on other plants species (Kurumata et al. 2005).

Obtaining of transgenic plants with transformation ability of such persistent organic pollutants (POPs) as PCBs and dioxins are of extreme importance. With this aim a transgenic tobacco have been engineered in which gene *bphc* of PCB degrading bacteria encoding enzyme 2,3-dihydroxybiphenyl-1,2-dioxygenase, key enzyme of PCBs degradation is introduced (Chrastilova et al. 2007).

A comparatively novel genetic engineering approach is overexpression of secretory enzymes and other exudates that promote different pollutants degradation by specific microorganisms in rhyzosphere, i.e. phytoremediation *ex planta.* Different types of transgenic tobaco have been constructed with this purpose: (i) with fungal laccase (*LAC*) from *Coriolus versicolor.* Transgenic plant has ability to secretion of laccase into the rhizosphere and remove the pollutants bisphenol A and pentachlorophenol (Sonoki et al. 2005); (ii) with haloalkane dehydrogenase (*DhaA*) from *Terrabacter sp.*, enhanced detoxification of 1-chlorobutane in rhizosphere (Uchida et al. 2005); (iii) with biphenyl dioxygenase gene from *Burkholderia xenovorans*, catalyse the oxygenation of 4-chlorobiphenyls in rhizosphere (Mohammadi et al. 2007); (iv) with biphenyl catabolic enzymes (*bphC*) from *Pandoraea pnomenusa*, enhanced degradation of PCBs (Francova et al. 2003; Novakova et al. 2009). Modifications of *Arabidopsis thaliana* have also been obtained: (i) with root specific laccase (*LACI*) from *Cotton*, secretes laccase to the rhizosphere and has high resistance to chlorinated phenolic pollutants such as 2,4,6-trichlorophenol (Wang and Chen 2007); (ii) with aromatic-cleaving extradiol dioxygenase (*DbfB*) from *Terrabacter sp.*, enhanced detoxification of 2,3-dihydroxybiphenyl (Uchida et al. 2005).

Using plant exudation machinery to obtain transgenic plants with ability to excrete various detoxifying enzymes for degradation of different type pollutants seems prospective.

The above presented data concerns mainly engineering of plants with overexpression of genes, encoding enzymes involved in contaminants degradation. It should be noted that similar approaches are applied to engineering of microbes for rhizodegradation (Reineke 1998; Rugh et al. 1999).

Expression of genes of desired enzymes is not the only approach for obtaining transgenes for phyto and rhyzoremediation. There are examples of introducing genes encoding biosynthetic pathway of biosurfactants to increase bioavailability of contaminants; engineering plants to release specific exudates that can induce degradative pathways in rhizomicrobia; engineering plants with increased capacity for uptake, transport and sequestration of contaminants; adding genes to microbes and or plants to enhance resistance to contaminant or environmental stressors. The later is of extremely important as abiotic or biotic stressors are great challenge for *in situ* phytoremediation, despite many advantages that other remedial strategies do not provide (Gerhardt et al. 2009).

The stressors that affect phytoremediation in the field are variations in temperature, nutrients and precipitation, herbivory, plant pathogens, competition by weed species that are better adopted to the site (Nedunuri et al. 2000). For further development of transgenic approaches to engineer plants and microbes with enhanced phytoremediation efficiency, there is a need of better knowledge of the processes of contaminants uptake, translocation and transformation (chelation, degradation, and volatilization).

15 Other Processes and Enzymes Important for Remediation with Plants

Plant's abilities to absorb, deposit (conjugate), and deeply degrade pollutants, and to mineralize organic and to accumulate inorganic pollutants, within its cells determines the ecological potential of the plant. These abilities are the main technological parameters determining the application of plants in novel phytoremediation technologies. Plants integrate exposure to pollutants in their environment and respond at several levels of biological organization. Transmission and scanning electron microscopy, in combination with autoradiographic methods, with well-developed techniques of fixation of plant tissues and getting ultrathin sections, allows the deleterious effects of toxic contaminants to be revealed at the ultrastructural level, and the fate of toxicants in the plant cell to be followed. Complex morphological changes and alterations in the main metabolic processes of plant cell elicited by organic pollutants (pesticides, hydrocarbons, phenols, aromatic amines, etc.) were reported to be connected with partial or full destruction of cell ultrastructural architecture (Kumar and Subrash 1990; Allnuff et al. 1991; Buadze and Kvesitadze 1997; Buadze et al. 1998; Korte et al. 2000; Zaalishvili et al. 2000). The sequence and characteristics of the destruction of the plant cell organelles depend on the chemical nature, concentration and duration of action of the contaminant, the resistance of plant cell, and some other factors (Buadze et al. 1998).

For the first 30 min xenobiotics penetrate and accumulate in the subcellular organelles. Simultaneously, the induction of specific enzymes takes place, which participate in further oxidative transformations of the xenobiotics (Heggestad 1991; Kvesitadze et al. 2001). All the toxic compounds investigated so far changed the plant cell structure to a greater or lesser extent.

Attention should be paid to the processes promoting the detoxification of penetrated pollutants and their removal from the cell. Among such processes, deposition in the vacuole of xenobiotics that have penetrated into the cell must be emphasized. This phenomenon, observed practically in all cases where labelled organic compounds were used, allows the cell to at least temporarily resist the destructive action of the toxicant. Sending the xenobiotic to the vacuole excludes it from interfering with normal cell metabolism.

However, as soon as the cell is given a chance, the process of removing the toxic residues from the vacuoles to the extracellular and subsequently to the intercellular spaces begins. This phenomenon is observed after terminating the exposure of the plant to the toxicant. In such cases the periplasmic space of the cell is appreciably widened. The agranulation of the rough endoplasmic reticulum begins, then the cisternae of the smooth endoplasmic reticulum connect with vacuoles through which part of the conjugates are excreted from the cell (Kvesitadze et al. 2006).

Often a large number of ribosomes are visible in the plant cells under the influence of environmental pollutants. This phenomenon points to an increase in protein biosynthesis. Electron microscopy of ultrathin sections of soybean and maize roots apices, under the influence of nitrobenzene at different concentrations, clearly showed the appearance of cells darkened by numerous ribosomes (Zaalishvili et al. 2002).

The increase in protein biosynthesis could be explained by the induction of the enzymes participating in contaminants intracellular detoxification and supply of the diminished amount of protein during the conjugation process. Histochemical and biochemical analyses showed that concurrently induction of enzymes important for detoxification (peroxidases, cytochrome P450-containing monooxygenases and phenoloxidases) takes place. The content of these oxidative enzymes is significantly enhanced in the cell wall, on membranes of the plasmalemma, in endoplasmic reticulum, tonoplasts and vacuoles, i.e. where suitable conditions are created to detoxify the organic contaminants for their removal from the information-processing and energetic centres of the cell (Kvesitadze et al. 2001).

Presumably, cells attempt to minimize the destructive action of pollutants via their deep degradation, and this is expressed in the strong induction of enzymes participating in detoxification process, and in the creation of optimum conditions for consecutive and effective functioning of these enzymes.

Organic pollutants penetrated into plant cells cause significant changes across the whole range of intracellular metabolic processes. The cell is fighting against foreign and at the same time toxic compounds with all accessible means. This is firstly manifested in the activation of inductive processes directed to the synthesis of enzymes and enzymatic systems participating in xenobiotic detoxification. As a result of the progressive oxidation of the xenobiotics, standard cellular intermediates are formed, which then get entrained in the general metabolic cycle and are partially oxidized up to carbon dioxide. During the detoxification (especially oxidation) processes a significant amount of internal cell endogenous energy, as well as metabolites vitally important for the cell, are expended. Presumably the great majority of cell enzymes are mobilized and directly or indirectly involved in the xenobiotic degradation process.

Though collateral biochemical processes accompanying detoxification process in plants are not well investigated, there are in the literature quite a few examples indicating that the activities of the enzymes participating in different regular cellular processes are also influenced by xenobiotics that have penetrated into the cell. For instance, in alfalfa inhibition of glutamine synthetase activity (up to 50 %), and a simultaneous stimulation of glutamate dehydrogenase activity (40 %) took place as a result of phosphinotricine exposure (Bataynen et al. 1986). Toxicants that have penetrated into plant cell may furthermore affect the activity of the regulatory enzymes involved in the tricarboxylic acid (TCA or Krebs) cycle and in the process of oxidative phosphorylation. As a consequence, the process of biosynthesis of ATP and other energetically important adenosine nucleotides (ADP, AMP and GDP) are affected (Bataynen et al. 1986).

Plant responses depend on the stage of development, age and nutritional status of the plants (Kovács 1992).

Special investigations were carried out by the present authors to reveal the effects of different hydrocarbons and aromatic compounds on key enzymes of general metabolism, namely glutamine synthetase, glutamate and malate dehydrogenases.

During the study of aromatic compounds nitrobenzene and benzoic acid—the concentration-dependent effect, namely stimulation of glutamate dehydrogenase and malate dehydrogenase activities at lower (1 mM) and inhibition at higher (10 mM) concentrations was observed, it should be mentioned that after termination of the action of these aromatic compound, i.e. transfer of the plants from 10 mM xenobiotic-containing medium to a medium lacking nitrobenzene consecutive restoration of the enzyme activity took. Changes in cell ultrastructural organization under the influence of nitrobenzene and benzoic acid were also reversible (Sadunishvili et al. 2006).

The effect of other aromatic compounds—benzene and 3,4-benzopyrene—on the main metabolic enzymes of maize and kidney bean was studied in more detail, in particular the influence of different concentrations and duration of exposure as well as the presence and absence of light. In maize roots, grown under ordinary illumination under the influence of benzene and 3,4-benzopyrene (10 mM, exposure time 72 h) the increase of activities of all enzymes studied took place. The highest increase of activity was observed for glutamate dehydrogenase – fivefold in the case of benzene and threefold in the case of 3,4-benzopyrene. In etiolated seedlings inhibition of glutamine synthetase activity and a 50 % increase of glutamate dehydrogenase activity under the influence of 3,4-benzopyrene was noted. Upon an increase of the concentration of the aromatic compounds from 5 to 10 mM, a further (approximately 50 %) increase of glutamate dehydrogenase activities both in roots and leaves of maize seedlings was observed. More clearly manifested changes in activities of the studied enzymes were observed in ryegrass seedlings.

The picture was different in case of kidney bean – there a slight (20 %) increase of glutamate dehydrogenase activity in roots at high concentrations (10 mM) of benzene and stimulation of malate and glutamate dehydrogenase activities in leaves were observed.

Not the least important factor is the duration of plant exposure to aromatic compounds. Experiments carried out in this direction have demonstrated that at shorter exposures (2 h) in leaves and roots of maize seedlings none of the studied enzyme activities changed. Upon increasing the exposure duration (24 h) an insignificant stimulation of glutamate dehydrogenase activity both in leaves and roots of maize seedlings was observed. The picture is drastically changed at relatively longer (72 h) exposures. In leaves of maize seedlings benzene causes inhibition of glutamate dehydrogenase activity by 40 %, malate dehydrogenase by 72 %, and glutamine synthetase by 32 %. On contrary, in roots a significant fourfold stimulation of glutamate dehydrogenase activity and comparatively low stimulation (22 %) of malate dehydrogenase were observed.

These studies indicate that aromatic compounds at the given concentrations are toxic for plants upon continued exposure. Thus, high concentrations of benzene and 3,4-benzopyrene are somehow critical for the studied plants. However, upon prolonged exposure plants elaborate a defense mechanism against the toxic effect of aromatic compounds that is expressed via ultrastructural reorganization of the cell and mobilization of energetic resources. The increase of glutamate dehydrogenase activity and number of mitochondria, as well as intensification of their contacts with chloroplasts and endoplasmic reticulum was evidence for it (Kvesitadze et al. 2006).

The response of enzymes participating in energy generation was observed in plants exposed to TNT and RDX (Unpublished data of authors).

Benzoic acid and its esters occur naturally in many plant and animal species. However benzoic acid a toluene oxidation product is found in vehicle exhausts (Kawamura et al. 1985). Transformation of [1-^{14}C]- and [7-^{14}C] benzoic acids in sterile seedlings of maize (*Zea mays*) and pea (*Pisum sativum*) was studied. The tested labeled compounds were supplied to plants through roots as water solutions. The bigger part of the assimilated benzoic acid form conjugates with plant low molecular weight peptides. After removal of plants from labeled benzoic acids containing medium the amount of conjugation products gradually decreases and the process is accompanied by emission of labeled carbon dioxide, indicating on the degradation of conjugation products and the oxidation of their radioactive component carbon atoms to carbon dioxide. Parallel to conjugation reaction smaller part of entered in plant benzoic acid is transformed via oxidation, as a result of which an aromatic ring is cleaved and the obtained aliphatic fragment is incorporated into cell regular metabolism.

Benzoic acid, radioactive label of which is detected in plant subcellular organelles and finally deposits in vacuoles, affects cell ultrastructural organization (Figs. 28, 29 and 30).

Benzoic acid significantly influenced enzymes: glutamate dehydrogenase, malate dehydrogenase and glutamine synthetase in maize roots. It was demonstrated that benzoic acid at concentration 1 mM cause the increase of glutamate dehydrogenase deamination activity, also of malate dehydrogenases and glutamine synthetase activities to different extent. The highest 60 % increase of glutamate dehydrogenase activity is observed at the beginning of incubation. It seems that the energy necessary for xenobiotic detoxification is generated in result of intensification of catabolic processes. Intensification of glutamic acid catabolic degradation leads to ammonia release, which is assimilated by glutamine synthetase and promotes synthesis of amino acids to restore their deficiency in cell after their catabolic degradation (Chrikishvili et al. 2006). However in case of ammonia excess, as shown during the study of kinetics of externally supplied ammonium ^{15}N incorporation in amino acids in kidney bean, prevention of its toxicity is carried out via activation of reductive amination activity of glutanmate dehydrogenase (Sadunishvili et al. 1993).

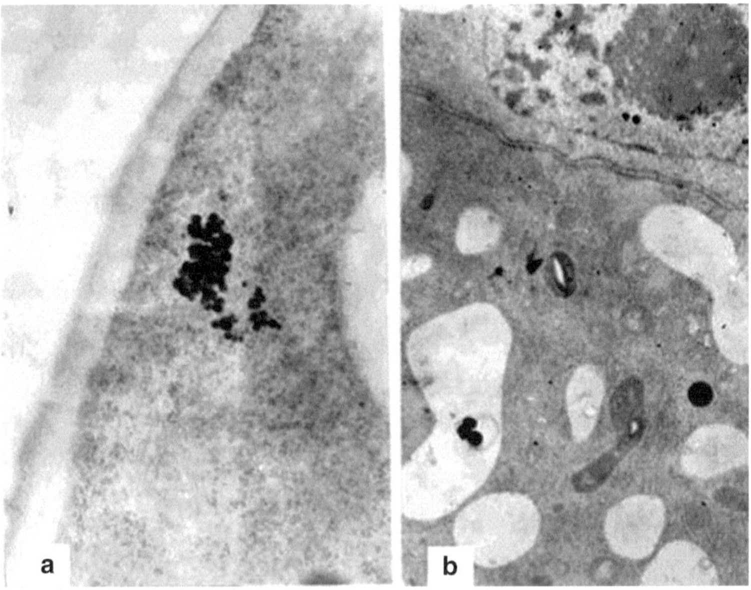

Fig. 28 Autoradiograph of root tip cap cells of maize seedlings incubated on 3 mM [1-^{14}C]benzoic acid containing medium, exposure time 1 h. (**a**) Radioactive label in nucleus and vacuole (×12000). (**b**) Radioactive label in peripheral cytoplasm (×36000)

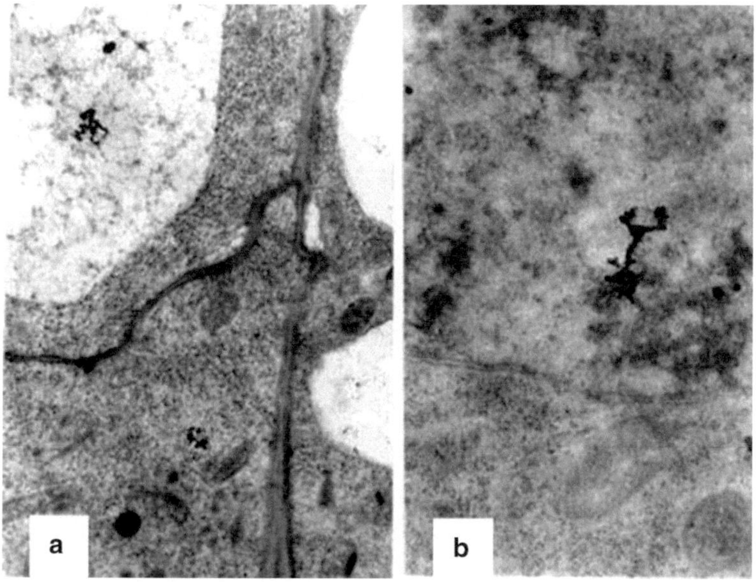

Fig. 29 Autoradiograph of root tip cap cells of maize seedlings incubated on 3 mM [1-^{14}C]benzoic acid containing medium, exposure time 24 h. (**a**) Radioactive label in nucleus (×27000). (**b**) Radioactive label in vacuole and cytoplasm. Big vacuoles with osmiophilic insertions (×20000)

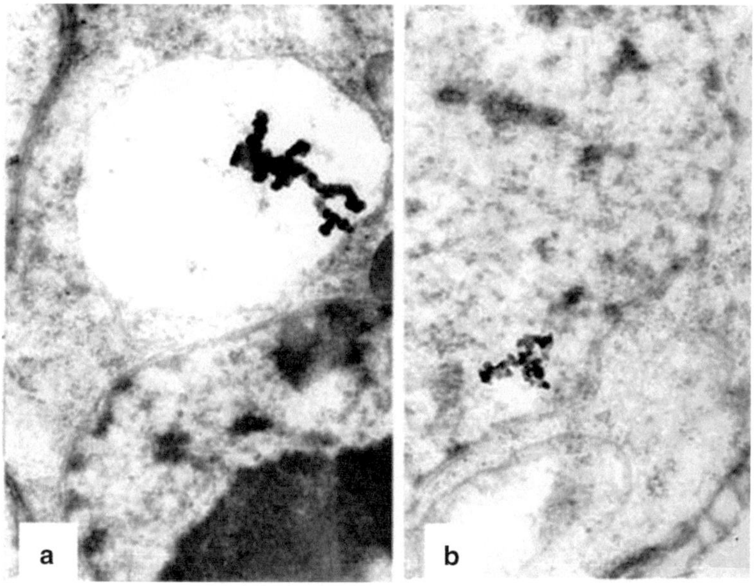

Fig. 30 Autoradiograph of root tip cap cells of maize seedling incubated 3 mM [1-^{14}C]benzoic acid containing medium, exposure time 72 h. (**a**) Radioactive label in nucleus. Multiple contacts of endoplasmic reticulum with mitochondria (×48000). (**b**) Radioactive label in vacuole. Multiple contacts of endoplasmic reticulum with vacuoles (×25000)

Apparently, aliphatic products, resulting from benzoic acid transformation are oxidized in Tricarboxylic Acid Cycle, evidence for this is activation of malate dehydrogenase.

Multiple contacts of endoplasmic reticulum with mitochondria indicates on correlation of biosynthetic process in cell necessary for induction of enzymes for benzoic acid transformation via conjugation and oxidation and activation of tricarboxylic acid cycle, leading to energy generation (Fig. 30).

Benzoic acid at high concentration – 10 mM, causes the inhibition of all the enzymes studied. A correlation between the rate of inhibition and duration of incubation was observed. When xenobiotics action was terminated after the transfer of seedlings to Knopp's solution, which didn't contain benzoic acid activation caused by low concentration of benzoic acid (1.0 mM) gradually disappeared and 48 h. later the activities of the studied enzymes return to the initial level. Inhibition caused by high concentration of benzoic acid (10 mM) practically is of irreversible character. The irreversible inhibition of the enzymes caused by high concentration of benzoic acid, probably indicate the highly toxic dose of xenobiotic concentration.

Obviously, energy necessary for xenobiotic detoxification is generated as a result of intensification of catabolic processes degradation (Chrikishvili et al. 2006).

Presented data clearly indicates the correlation between the penetration of organic contaminants in plant cells and the corresponding changes in the activities of the enzymes participating in energy and nitrogen metabolism. This correlation is highly affected by the xenobiotics concentration in the cell.

The influence of organic contaminants on plant cells, due to wide spectrum of their action, is a rather complicated process that cannot be conveniently expressed in compact form. However, in spite of the shortage of experimental data, it can be suggested that all subcellular organelles and key enzymes of general metabolism are involved in the process of xenobiotics transformation.

More investigations are important for elucidation of the mechanism of involvement of cell regular metabolism enzymes in organic pollutants detoxification. Plant response, expressed in changes of cell metabolic processes, key enzyme activities would serve as an indication for the objective evaluation of plant vitality in polluted environments. Such investigations are important also for understanding plant rehabilitation processes and elaboration of a proper strategy for agricultural crop cultivation in contaminated soils with a guarantee of safe harvest.

Existing knowledge and information do not allow to answer all questions concerning the action of environmental contaminants on plant cell structure-function organization. However based on our results on deviations in cell structure and activities of regular metabolism enzymes we suggest that all kinds of detoxification processes in higher plants are closely related to cell metabolism. Apparently, not only the absence of particular enzymes or enzymatic systems directly participating in conjugation/oxidation of xenobiotics are rate limiting in the remediation process, but other metabolic enzymes, determining cell energetic potential indirectly participate in detoxification on pollutants and plant survival.

References

Abhilash PC, Jamil S, Singh N (2009) Transgenic plants for enhanced biodegradation and phytoremediation of organic xenobiotics. Biotechnol Adv 27:474–488

Adamia G, Ghoghoberidze M, Graves D, Khatisashvili G, Kvesitadze G, Lomidze E, Ugrekhelidze D, Zaalishvili G (2006) Absorption, distribution and transformation of TNT in higher plants. Ecotoxicol Environ Saf 64:136–145

Alderete LGS, Talano MA, Ibannez SG, Purro S, Agostini E (2009) Establishment of transgenic tobacco hairy roots expressing basic peroxidases and its application for phenol removal. J Biotechnol 139:273–279

Allnuff FCT, Ewy R, Renganathan M, Pan RS, Dilley KA (1991) Nigericin and hexylamine effects on localized proton gradients in thylacoids. Biochim Biophys Acta Bioenerg 1059:28–36

Andreopoulos-Renaud U, Glas J, Falgoux D, Schiedecker D (1975) Absorption D'un polyethyleneglycol par de jeunes plantes de haricot et de cotonnier. Dosage par chromatographie on phase gaze use sur exsudat de tige. Cr Acad Sci D 280:2333–2340

Angermaier L, Simon H (1983) On nitroaryl reductase activities in several clostridia. Hoppe-Seylers Z Physiol Chem 364:1653–1663

Audley BG (1979) Structure and properties of 2-chloroethylphosphonic acid (ethephon) metabolite from *Hevea brasiliensis* bark. Phytochemistry 18: 53–60

Bataynen N, Kopacz SI, Lee CP (1986) The modes of action of long chain alkali compounds on the respiratory chain-linked energy transducing system in submitochondrial particles. Arch Biochem Biophys 250:476–487

Beesley L, Moreno-Jiménez E, Gomez-Eyles JL (2010) Effects of biochar and greenwaste compost amendments on mobility, bioavailability and toxicity of inorganic and organic contaminants in a multi-element polluted soil. Environ Pollut 158:2282–2287

Bell SG, Harford-Cross CF, Wong LL (2001) Engineering the CYP101 system for in vivo oxidation of unnatural substrates. Protein Eng 14: 797–802

Bell SG, Stevenson JA, Boyd HD, Campbell S, Riddle AD, Orton EL, Wong LL (2002) Butane and propane oxidation by engineered cytochrome P450cam. Chem Commun (Camb) 7: 490–491

Berenbaum MR, Zangerl AR (1996) Physical-chemical diversity: adaptation or random variation? Recent Adv Phytochem 30:1–12

Best EPH, Zappi ME, Fredrickson HL, Sprecher SL, Larson SL, Ochman M (1997) Screening of aquatic and wetland plant species for the phytoremediation of explosives-contaminated groundwater from the Iowa Army Ammunition Plant. Ann NY Acad Sci 829:179–194

Bhadra R, Wayment DG, Hughes JB, Shanks JV (1999) Confirmation of conjugation processes during TNT metabolism by axenic plant roots. Environ Sci Technol 33:446–452

Bieseler B, Fedtke C, Neuefeind T, Etzel W, Prade L, Reinemer P (1997) Maize selectivity of FOE 5043: Degradation of active ingredient by glutathione-S-transferases. Pflanzenschutz-Nachrichten Bayer 50: 117–140

Bond JA, Bradley BP (1997) Resistance to malathion in heat-shocked Daphnia Magna. Environ Toxicol Chem 16:705–712

Brazier M, Cole DJ, Edwards R (2002) O-Glucosyltransferase activities toward phenolic natural products and xenobiotics in wheat and herbicide-resistant and herbicide-susceptible black-grass (Alopecurus myosuroides). Phytochemistry 59:149–156

Breaux EJ (1987) Initial metabolism of acetochlor in tolerant and susceptible seedlings. Weed Sci 35:463–468

Briggs GG (1981) Theoretical and experimental relationships between soil adsorption, octanolwater partition coefficients, water solubilities, bioconcentration factors, and the parachor. J Agric Food Chem 29:1050–1059

Briggs GG, Bromilow RH, Evans AA (1982) Relationships between lipophilicity and root uptake and translocation of non-ionized chemicals by barley. Pestic Sci 13:495–504

Briggs GG, Bromilow RH, Evans AA, Williams M (1983) Relationships between lipophilicity and the distribution of non- ionised chemicals in barley shoots following uptake by the roots. Pestic Sci 14:492–500

Bromilow RH, Chamberlain K, Evans AA (1990) Physicochemical aspects of phloem translocation of herbicides. Weed Sci 38:305–314

Brown HM, Neighbors SM (1987) Soybean metabolism of chlorimuron ethyl: physiological basis for soybean selectivity. Pestic Biochem Physiol 29:112–120

Brown HM, Wittenbach VA, Forney DR, Strachan SD (1990) Basis for soybean tolerance to thifensulfuron methyl. Pestic Biochem Physiol 37:303–313

Brudenell AJP, Baker DA, Grayson BT (1995) Phloem mobility of xenobiotics, tabular review of physicochemical properties governing output of the Kleier model. J Plant Growth Regul 16:215–231

Brudenell AJP, Griffiths H, Rossiter JT, Baker DA (1999) The phloem mobility of glucosinolates. J Exp Bot 50:745–756

Bryant DW, McCalla DR, Leelsma M, Laneuville P (1981) Type I nitroreductases of Escherichia coli. Can J Microbiol 27:81–86

Buadze O, Kvesitadze G (1997) Effect of low-molecular weight alkanes on the cell photosynthetic apparatus. Ecotoxicol Environ Saf 38:36–44

Buadze O, Sadunishvili T, Kvesitadze G (1998) The effect of 1,2-benzanthracene and 3,4-benzpyrene on the ultrastructure on maize cells. Int Biodeterior Biodegrad 41:119–125

Bukovac MT, Petracek PD, Fader RG, Morse RD (1990) Sorption of organic compounds by plant cuticles. Weed Sci 38:289–298

Burken JG (2003) Uptake and metabolism of organic compounds: green liver model. In: McCutcheon SC, Schnoor JL (eds) Phytoremediation. Transformation and control of contaminants. Wiley-Interscience, Hoboken, pp 59–84

Burken JG, Schnoor JL (1998) Predictive relationships for uptake of organic contaminants by hybrid poplar trees. Environ Sci Technol 32:3379–3385

Burrows WJ, Leworthy DP (1976) Metabolism of N,N-diphenylurea by cytokinin-dependent tobacco callus identification of the glucoside. Biochem Biophys Res Comm 70: 1109–1117

Bylund JE, Dyer JK, Feely DE, Martin EL (1990) Alkaline and acid phosphatases from the extensively halotolerant bacterium Halomonas elongata. Curr Microbiol 20:125–131

Carringer RD, Rieck CE, Bush LP (1978) Metabolism of EPTC in corn (Zea mays). Weed Sci 26:157–163

Cash GG (1998) Prediction of chemical toxicity to aquatic microorganism: ECOSAR vs. Microtox assay. Environ Toxicol Water Qual 132:211–216

Cassagne C, Lessire R (1975) Studies on alkane biosynthesis in epidermis of Allium porrum L. leaves. 4. Wax movement into and out of the epidermal cells. Plant Sci Lett 5:261–266

Chamberlain K, Patel S, Bromilow RH (1998) Uptake by roots and translocation to shoots of two morpholine fungicides in barley. Pestic Sci 54:1–7

Chandler JM, Basler E, Santelman PW (1974) Uptake and translocation of alachlor in soybean and wheat. Weed Sci 22:253–259

Cheng TC, Harvey SP, Chen GL (1996) Cloning and expression of a gene encoding a bacterial enzyme for decontamination of organophosphorous nerve agents and nucleotide sequence of the enzyme. Appl Environ Microbiol 62:1636–1641

Chkanikov DI, Makeev AM, Pavlova NN, Artemenko EN, Dubovoi VP (1976) The role of 2,4-D metabolism in plant resistance to this herbicide (in Russian). Agrokhimiya 2:127–132

Chrastilova Z, Mackova M, Novakova M, Macek T, Szekeres M (2007) Transgenic plants for effective phytoremediation of persistent toxic organic pollutants present in the environment. Abstracts/J Biotechnol 131:S38

Chrikishvili D, Sadunishvili T, Zaalishvili G (2006) Benzoic acid transformation via conjugation with peptides and final fate of conjugates in higher plants. Ecotoxicol Environ Saf 64:390–399

Coleman JOD, Mechteld MA, Kalff B, Davies TGE (1997) Detoxification of xenobiotics in plants: chemical modification and vacuolar compartmentation. Trends Plant Sci 2:144–151

Collins PJ, Dobson ADW (1997) Regulation of laccase gene transcription in Trametes versicolor. Appl Environ Microbiol 63:3444–3450
Colombo JC, Cabello MN, Arambarri AM (1996) Biodegradation of aliphatic and aromatic hydrocarbons by natural soil microflora and pure culture of imperfect and ligninolytic fungi. Environ Pollut 94:355–362
Coupland D, Lutman PJW, Heath C (1990) Uptake, translocation, and metabolism of mecoprop in a sensitive and a resistant biotype of Stellaria media. Pestic Biochem Physiol 36:61–67
Cummins I, Edwards R (2004) Purification and cloning of an esterase from the weed black-grass (Alopecurus myosuroides), which bioactivates aryloxyphenoxypropionate herbicides. Plant J 39:894–904
Cummins I, Burnet N, Edwards R (2001) Biochemical characterisation of esterases active in hydrolysing xenobiotics in wheat and competing weeds. Physiol Plant 113:477–485
Davidonis GH, Hamilton RH, Mumma RO (1978) Metabolism of 2,4-dichlorophenoxyacetic acid in soybean root callus and differentiated soybean root cultures as a function of concentration and tissue age. Plant Physiol 62: 80–86
De Carvalho RF, Bromilow RH, Greenwood R (2007) Uptake and translocation of non-ionised pesticides in the emergent aquatic plant parrot feather Myriophyllum aquaticum. Pest Manag Sci 63:798–802
DeRidder BP, Dixon DP, Beussman DJ, Edwards R, Goldsbrough PB (2002) Induction of glutathione S-transferases in Arabidopsis by herbicide safeners. Plant Physiol 130:1497–1505
Devine MD, Hall LM (1990) Implication of sucrose transport mechanism for the translocation of herbicides. Weed Sci 38:299–304
Dexter AG, Slife FW, Butler HS (1971) Detoxification of 2,4-D by several plant species. Weed Sci 19:721–729
Didierjean L, Gondet L, Perkins R, Lau S-MC, Schaller H, O'Keefe DP, Werck-Reichhart D (2002) Engineering herbicide metabolism in tobacco and *Arabidopsis* with CYP76B1, a cytochrome P450 enzyme from Jerusalem artichoke. Plant Physiol 130:179–189
Dixon DP, Lapthorn A, Edwards R (2002) Plant glutathione transferases. Genome Biol 3:3004.1–3004.10
Domir SC (1978) Translocation and metabolism of injected maleic hydrazide in silver maple and American sycamore seedlings. Physiol Plant 42:387–396
Dörr R (1970) Die Aufnahme von 3,4-Benzpyren durch Pflanzenwurzeln. Landwirtsch Forsch 23:371–376
Doty SL, Shang QT, Wilson AM, Moore AL, Newman LA, Strand SE (2007) Enhanced metabolism of halogenated hydrocarbons in transgenic plants contain mammalian P450 2E1. Proc Natl Acad Sci U S A 97:6287–6291
Doty SL (2008) Enhancing phytoremediation through the use of transgenic plants and entophytes. New Phytol 179:318–333
Dowdy DL, McKone TE (1997) Predicting plant uptake of organic chemicals from soil or air using octanol/water and octanol/air partition ratios and a molecular connectivity index. Environl Toxicol Chem 16:2448–2456
Droog FJN, Hooykaas PJJ, Libbenga KR, van der Zaal EJ (1993) Proteins encoded by an auxin-regulated gene family of tobacco share limited but significant homology with glutathione S-transferases and one member indeed shows in vitro GST activity. Plant Mol Biol 21:965–972
Dunford HB, Jones PA (2010) Peroxidases and Catalases: Biochemistry, Biophysics, Biotechnology and Physiology. 2nd edn. Wiley, New York
Durmishidze S, Ugrekhelidze D (1968a) Absorption and conversion of butane by higher plants (in Russian). Dokl Akad Nauk SSSR 182:214–216
Durmishidze S, Ugrekhelidze D (1968b) Oxidation of ethane, propane and pentane by higher plants (in Russian). Bull Georgian Acad Sci 50:661–666
Durmishidze S, Ugrekhelidze D (1975) Absorption and transformation of methane by plants (in Russian). Fiziol Rast 22:70–73

Durmishidze S, Ugrekhelidze D, Djikiya A, Tsevelidze D (1969) The intermediate products of enzymatic oxidation of benzene and phenol (in Russian). Dokl Akad Nauk SSSR 184:466–469

Durmishidze S, Ugrekhelidze D, Djikiya A (1974a) Absorption and transformation of benzene by higher plants (in Russian). Fiziologiya i Biochimiya Kulturnikh Rastenii 6:217–221

Durmishidze S, Ugrekhelidze D, Djikiya A (1974b) Uptake of benzene by fruits from atmosphere (in Russian). Appl Biochem Microbiol 10:472–476

Durmishidze S, Ugrekhelidze D, Djikiya A (1974c) Absorption and transformation of toluene by higher plants (in Russian). Appl Biochem Microbiol 10:673–676

Durmishidze S, Djikiya A, Lomidze E (1979) Uptake and transformation of benzidine by plants in sterile conditions (in Russian). Dokl Akad Nauk SSSR 247:244–247

Durst F (1991) Biochemistry and physiology of plant cytochrome P-450. In: Ruckpaul K (ed) Frontiers in biotransformation, vol 4. Academic-Verlag, Berlin, pp 191–232

Edwards R, Dixon DP, Cummins I, Brazier-Hicks M, Skipsey M (2011) New perspectives on the metabolism and detoxification of synthetic compounds in plants. In: Schröder P, Collins CD (eds) Organic xenobiotics and plants: from mode of action to ecophysiology, plant ecophysiology, vol 8. Springer, Dordrecht New York, pp 125–148

Esteve-Núñez A, Caballero A, Ramos JL (2001) Biological degradation of 2,4,6-trinitrotoluene. Microbiol Mol Biol Rev 65:335–352

Evert RF, Eichhorn SE (2006) Esau's plant anatomy: meristems, cells, and tissues of the plant body: their structure, function, and development. Wiley, Hoboken

Eynard I (1974) Determination on foliage of surfactant solution by a radioisotope technique. Allionia 17:131–135

Ezra G, Stephenson GR (1985) Comparative metabolism of atrazine and EPTC in proso millet (Panicum miliaceum L.) and corn. Pestic Biochem Physiol 24:207–212

Fant FA, de Sloovere A, Matthijse K, Marle C, el Fantroussi S, Werstraete W (2001) The use of amino compounds for binding 2,4,6-trinitrotoluene in water. Environ Pollut 111:503–507

Feung C, Hamilton RH, Mumma RO (1976) Metabolism of 2,4-dichlorophenoxyacetic acid: 10. Identification of metabolites in rice root callus tissue cultures. J Agric Food Chem 24: 1013–1019

Fialho RC, Bucker J (1996) Changes in levels of foliar carbohydrates and myoinositol before premature leaf senescence of Populus-Nigra induced by a mixture of O_3 and SO_2. Can J Bot 74:965–970

Flocco CG, Lindblom SD, Smits EAHP (2004) Overexpression of enzymes involved in glutathione synthesis enhances tolerance to organic pollutants in *Brassica juncea*. Int J Phytoremediation 6:289–304

Fonné-Pfister R, Kreuz K (1990) Ring-methyl hydroxylation of chlortoluron by an inducible cytochrome P450-dependent enzyme from maize. Phytochemistry 29:2793–2796

Francova K, Sura M, Macek T, Szekeres M, Bancos S, Demnerova K (2003) Preparation of plants containing bacterial enzyme for degradation of poly chlorinated biphenyls. Fresenius Environ Bull 12:309–313

Frear DS, Swanson HR, Mansager ER, Wien RG (1978) Chloramben metabolism in plants: isolation and identification of glucose ester. J Agric Food Chem 26:1347–1354

Frear DS, Swanson HR, Mansager ER (1983) Acifluorfen metabolism in soybean: diphenylether bond cleavage and the formation of homoglutathione, cysteine, and glucose conjugates. Pestic Biochem Physiol 20:299–310

Frear DS, Swanson HR, Mansager ER (1985) Alternate pathways of metribuzin metabolism in soybean: formation of N-glucoside and homoglutathione conjugates. Pestic Biochem Physiol 23:56–65

French CE, Hosser SJ, DaviesGJ NS, Bruce NC (1999) Biodegradation of explosives by transgenic plants expressing pentaerythritol tetranitrate reductase. Nat Biotechnol 17:491–494

Führ F, Mittelstaedt W (1974) Versuche mit polyurethan-Hartschaum auf Basis von ^{14}C-markiertem Diphenylmethan-diisocyanat (MDI) zu Fragen der Mineralisierung und Aufnahme durch Planzen. Ber Kernforschungsanlage Jülich 1062:30–42

Führ F, Sauerbeck D (1967) The uptake of colloidal organic substances by plant roots as shown by experiments with ^{14}C-labelled humus compounds. In: Report FAO/IAEA meeting, Viena. Pergamon Press, Oxford, pp 73–82

Gandia-Herrero F, Lorenz A, Larson T, Graham IA, Bowles J, Rylott EL (2008) Detoxification of the explosive 2,4,6- trinitrotoluene in Arabidopsis: discovery of bifunctional O and C-glucosyltransferases. Plant J 56:963–974

Gao JA, Garrison WA, Mazur C, Hoehamer CF, Wolfe NL (2000) Uptake and phytotransformation of organophosphorous pesticides by axenically cultivated aquatic plants. J Agric Food Chem 48:6121–6127

Gellatly KS, Moorhead GBG, Duff SMG, Lefebvre DD, Plaxton WC (1994) Purification and characterization of a potato tuber acid phosphatase having significant phosphotyrosine phosphatase activity. Plant Physiol 106:223–232

Gerhardt KE, Huang X-D, Glick BR, Greenberg BM (2009) Phytoremediation and rhizoremediation of organic soil contaminants: potential and challenges. Plant Sci 176:20–30

Gilbert ES, Crowley DE (1997) Plant compounds that induce polychlorinated biphenyl biodegradation by *Arthrobacter* sp. Strain B1b. Appl Environ Microbiol 63:1933–1938

Glieder A, Farinas ET, Arnold FH (2002) Laboratory evolution of a soluble, self-sufficient, highly active alkane hydroxylase. Nat Biotechnol 20:1135–1139

Gordeziani M, Durmishidze S, Khatisashvili G, Adamia G, Lomidze E (1987) The investigation of biosynthetic and detoxification ability of plant cytochrome P-450 (in Russian). Dokl Akad Nauk SSSR 295:1491–1493

Gordeziani M, Khatisashvili G, Kvesitadze G (1991) Free and coupled with xenobiotic hydroxylation oxidation of NADPH (in Russian). Dokl Akad Nauk SSSR 320:467–470

Grayson BT, Kleier DA (1990) Phloem mobility of xenobiotics. IV. Modeling of pesticide movement in plants. Pestic Sci 30:67–79

Greene DW, Bukovac MJ (1977) Foliar penetration of naphthalene-acetic acid enhancement by light and role of stomata. Am J Bot 64:96–104

Guengerich FP (2010) Cytochrome P450 enzymes. In: McQueen CA (ed) Comprehensive toxicology, vol 4: biotransformation. Elsevier, New York, pp 41–76

Guillén F, Gómez-Toribio V, Martýnez MJ, Martýnez AT (2000) Production of hydroxyl radical by the synergistic action of fungal laccase and aryl alcohol oxidase. Arch Biochem Biophys 382:142–147

Guillén F, Martýnez MJ, Muñoz C, Martýnez AT (1997) Quinone redox cycling in the ligninolytic fungus Pleurotus eryngii leading to extracellular production of superoxide anion radical. Arch Biochem Biophys 339:190–199

Gullner G, Komives T, Rennenberg H (2001) Enhanced tolerance of transgenic poplar plants overexpressing gamma-glutamylcysteine synthetase towards chloroacetanilide herbicides. J Exp Bot 52:971–979

Haderlie RC (1980) Absorption and translocation of buthidazole. Weed Sci 28:352–360

Hallahan DL, Lau SM, Harder PA, Smiley DW, Dawson GW, Pickett JA, Christoffersen RF, O'Keefe DP (1994) Cytochrome P-450-catalysed monoterpenoid oxidation in catmint (Nepeta racemosa) and avocado (Persea americana) evidence for related enzymes with different activities. Biochim Biophys Acta 1201:94–100

Hallmen U (1974) Translocation and complex formation of picloram and 2,4-D in rape and sunflower. Physiol Plant 32:78–83

Hannink N, Rosser SJ, French CE, Basran A, Murray JA, Nicklin S, Bruce NC (2001) Phytodetoxification of TNT by transgenic plants expressing a bacterial nitroreductase. Nat Biotechnol 19:1168–1172

Hannink NK, Subramanian M, Rosser SJ, Basran A, Murray JAH, Shanks JV (2007) Enhanced transformation of TNT by tobacco plants expressing a bacterial nitroreductase. Int J Phytoremediation 9:385–401

Hansikova H, Frei E, Anzenbacher P, Stiborova M (1994) Isolation of plant cytochrome P450 and NADPH:cytochrome P450-reductase from tulip bulbs (Tulipa fosteriana). Gen Physiol Biophys 13:149–169

Harborne JB (1977) Introduction to ecological biochemistry. National Academic Press, London/New York/San Francisco
Harvey SD, Fellows RJ, Cataldo DA, Bean RM (1990) Analysis of 2,4,6-trinitrotoluene and its transformation products in soils and plant tissues by high-performance liquid chromatography. J Chromatogr 518:361–374
Hatzios KK, Penner D (1980) Site of uptake and translocation of 14C-buthidazole in corn (Zea mays) and redroot pigweed (Amaranthus retroflexus). Weed Sci 28:285–291
Haynes CA, Koder RL, Miller A-F, Rodgers DW (2002) Structures of nitroreductase in three states. J Biol Chem 277:11513–11520
Heggestad HE (1991) Origin of Bel-W3, Bel-C and Bel-B tobacco varieties and their use as indicators of ozone. Environ Pollut 74:264–291
Hendrick LW, Meggitt WF, Penner D (1974) Basis for selectivity of phenmedipham and desmodipham on wild mustard, redroot pigweed, and sugar beet. Weed Sci 22:179–186
Hetherington AM, Woodward FI (2003) The role of stomata in sensing and driving environmental change. Nature 424:901–908
Hirose S, Kawahigashi H, Ozawa K, Shiota N, Inui H, Ohkawa H (2005) Transgenic rice containing human CYP2B6 detoxifies various classes of herbicides. J Agric Food Chem 53:3461–3467
Hodgson RH, Hoffer BL (1977) Diphenamid metabolism in pepper and an ozone effect. 1. Absorption, translocation, and the extent of metabolism. Weed Sci 25:324–330
Holton TA, Cornish EC (1995) Genetics and biochemistry of anthocyanin biosynthesis. Plant Cell 7:1071–1083
Hoskin FCG, Walker JE, Mello CM (1999) Organophosphorous acid anhydrolase in slime mold, duckweed and mung bean: a continuing search for a physiological role and a natural substrate. Chem Biol Interact 119/120:399–404
Hsu FC, Kleier DA (1990) Phloem mobility of xenobiotics. III. Sensitivity of unified model to plant parameters and application to patented chemical hybridizing agents. Weed Sci 38:315–323
Hsu FC, Kleier DA, Melander WH (1988) Phloem mobility of xenobiotics. II. Bioassay testing of the unified model. Plant Physiol 86:811–816
Hu C, Huystee RB (1989) Role of carbohydrate moieties in peanut peroxidases. Biochem J 263:129–135
Iimura Y, Ikeda S, Sonoki T, Hayakawa T, Kajita S, Kimbara K (2002) Expression of a gene for Mn-peroxidase from Coriolus versicolor in transgenic tobacco generates potential tool for phytoremediation. Appl Microbiol Biotechnol 59:246–251
Inui H, Wakai T, Gion K, Kim YS, Eun H (2008) Differential uptake for dioxinlike compounds by zucchini subspecies. Chemosphere 73:1602–1607
Isaacson P (1986) Uptake and transpiration of 1,2-dibromoethane by leaves. Plant Soil 95:431–434
Isensee AR, Jones GE, Turner BC (1971) Root absorption and translocation of picloram by oats and soybeans. Weed Sci 19:727–736
Jackson EG, Rylott EL, Fournier D, Hawari J, Bruce NC (2007) Exploring the biochemical properties and remediation applications of the unusual explosive-degrading P450 system XplA/B. Proc Natl Acad Sci U S A 104:16822–16827
Jacobson A, Shimabukuro RH (1984) Metabolism of diclofop-methyl in root-treated wheat and oat seedlings. J Agric Food Chem 32:742–748
Janes BE (1974) The effect of molecular size, concentration in nutrient solution, and exposure time on the amount and distribution of polyethylene glycol in pepper plants. Plant Physiol 54:226–236
Jansen EF, Olson AC (1969) Metabolism of carbon-14-labelled benzene and toluene in avocado fruit. Plant Physiol 44:786–791
Jenings JC, Coolbaugh RC, Nakata DA, West CA (1993) Characterization and solubilization of kaurenoic acid hydroxylase from Gibberella fujikura. Plant Physiol 101:925–930

Jensen KIN, Stephenson GR, Hunt LA (1977) Detoxification of atrazine in three Gramineae subfamilies. Weed Sci 25:212–218

Johannes C, Majcherczyk A (2000) Natural mediators in the oxidation of polycyclic aromatic hydrocarbons by laccase mediator systems. Appl Environ Microbiol 66:524–528

Jones DW, Foy CL (1972) Absorption and translocation of bioxone in cotton. Weed Sci 20:116–124

Jones JP, O'Hare EJ, Wong LL (2001) Oxidation of polychlorinated benzenes by genetically engineered CYP101 (cytochromeP450(cam)). Eur J Biochem 268:1460–1467

Jung S, Lee HJ, Lee Y, Kang K, Kim YS, Grimm B (2008) Toxic tetrapyrrole accumulation in protoporphyrinogrn IX oxidase overexpressing transgenic rice plants. Plant Mol Biol 67:535–546

Karavangeli M, Labrou NE, Clonis YD, Tsaftaris A (2005) Development of transgenic tobacco plants overexpressing glutathione S-transferase I fro chloroacetanilide herbicides phytoremediation. Biomol Eng 22:121–128

Kassel AG, Ghoshal D, Goyal A (2002) Phytoremediation of trichloroethylene using hybrid poplar. Physiol Mol Biol Plants 8:1–8

Kawahigashi H, Hirose S, Ohkawa H, Ohkawa Y (2005) Transgenic rice plants expressing human CYP1A1 remediate the triazine herbicides atrazine and simazine. J Agric Food Chem 53:8557–8564

Kawahigashi H, Hirose S, Ohkawa H, Ohkawa Y (2008) Transgenic rice plants expressing human P450 genes involved in xenobiotic metabolism for phytoremediation. J Mol Microbiol Biotechnol 15:212–219

Kawahigashi H (2009) Transgenic plants for phytoremediation of herbicides. Curr Opin Biotechnol 20:225–230

Kawamura K, Ng LL, Kaplan IR (1985) Determination of organic acids (C1–C10) in the atmosphere, motor exhausts, and engine oils. IPCS guidelines for the monitoring of genotoxic effects of carcinogens in humans. Environ Sci Technol 19:1082–1086

Khatisashvili G, Gordeziani M, Kvesitadze G, Korte F (1997) Plant monooxygenases: participation in xenobiotic oxidation. Ecotoxicol Environ Saf 36:118–122

Khatisashvili G, Adamia G, Pruidze M, Kurashvili M, Varazashvili T, Ananiashvili T (2003) TNT uptake and nitroreductase activity of plants. In: 8th international FZK/TNO conference on contaminated soil. ConSoil 2003, Gent, Belgium, pp 2482–2485

King MG, Radosevich R (1979) Tanoak (Lithocarpus densiflorus) leaf surface characteristics and absorption of triclopyr. Weed Sci 27:599–605

Kirkwood RC (1999) Recent developments in our understanding of the plant cuticle as a barrier to the foliar uptake of pesticides. Pestic Sci 55:69–77

Kleier DA (1994) Phloem mobility of xenobiotics. V. Structural requirements for phloem-systemic pesticides. Pestic Sci 42:1–11

Kochs G, Grisebach H (1986) Enzymatic synthesis of isoflavonoids. Eur J Biochem 155:311–318

Kolattukudy PE (1980) Cutin suberin and waves. In: Stumpf PK (ed) The biochemistry of plants: a comprehensive treatise, vol 4, Lipids: structure and function. National Academic Press, New York, pp 571–645

Korte F, Behadir M, Klein W, Lay JP, Parlar H, Sceunert I (1992) Lehrbuch der ökologischen chemie. Grundlagen und Konzepte fur die ökologische Beurteilung von Chemikalien. Georg Thieme Verlag, Stuttgart/New York

Korte F, Kvesitadze G, Ugrekhelidze D, Gordeziani M, Khatisashvili G, Buadze O, Zaalishvili G, Coulston F (2000) Review: organic toxicants and plants. Ecotoxicol Environ Saf 47:1–26

Kouji H, Masuda T, Matsunaka S (1990) Mechanism of herbicidal action and soybean selectivity of AKH-7088, a novel diphenyl ether herbicide. Pestic Biochem Physiol 37:219–226

Kovács M (1992) Herbaceous (flowering) plants. In: Kovács M (ed) Biological indicators in environmental protection. Ellis Horwood, New York

Krell HW, Sandermann H (1985) Plant biochemistry of xenobiotics. Purification and properties of a wheat esterase hydrolyzing the plasticizers chemical, bis(2-ethylhexyl)phthalate. Eur J Biochem 143:57–62

Kristich MA, Schwarz OJ (1989) Characterization of 14C-naphthol uptake in excised root segments of clover (*Trifolium pratense* L.) and fescue (*Festuca arundinaceae* Screb.). Environ Monit Assess 13:35–44

Kumar VN, Subrash C (1990) Influences of SO2 on seed germination and seedling growth of *Eulalipsis binate* (Rets) Hubbord. Geobios (India) 17:190–191

Kumar S, Jin M, Weemhoff JL (2012) Cytochrome P450-mediated phytoremediation using transgenic plants: a need for engineered cytochrome P450 enzymes. J Pet Environ Biotechnol 3:127. doi:10.4172/2157-7463.1000127

Kurumata M, Takahashi M, Sakamotoa A, Ramos JL, Nepovim A, Vanek T (2005) Tolerance to and uptake and degradation of 2, 4, 6-trinitrotoluene (TNT) are enhanced by the expression of a bacterial nitroreductase gene in *Arabidopsis thaliana*. Z Naturforsch C 60:272–278

Kutchan TM (1995) Alkaloid biosynthesis – the basis for metabolic engineering of medicinal plants. Plant Cell 7:1059–1061

Kvesitadze G, Gordeziani M, Khatisashvili G, Sadunishvili T, Ramsden JJ (2001) Some aspects of the enzymatic basis of phytoremediation. J Biol Phys Chem 1:49–57

Kvesitadze G, Khatisashvili G, Sadunishvili T, Ramsden JJ (2006) Biochemical mechanisms of detoxification: basis of phytoremediation. Springer, Berlin/Heidelberg

Lamoureux GL, Rusnes DG (1989) Propachlor metabolism in soybean plants, excised soybean tissues, and soil. Pestic Biochem Physiol 34:187–204

Lan PT, Main R, Motoyama N, Dauterman WC (1983) Hydrolysis of malathion by rabbit liver oligomeric and monomeric carboxylesterases. Pestic Biochem Physiol 20:232–237

Lao S-H, Loutre C, Brazier M, Coleman JOD, Cole DJ, Edwards R, Theodoulou FL (2003) 3,4-Dichloroaniline is detoxified and exported via different pathways in Arabidopsis and soybean. Phytochemistry 63:653–661

Laurent FMG (1994) Chloroaniline peroxidation by soybean peroxidases. Pestic Sci 40:25–30

Lawlor DW (1970) Absorption of polyethylene glycols by plants and their effects on plant growth. New Phytol 69:501–506

Lay MM, Casida JE (1976) Dichloroacetamide antidotes enhance thiocarbamate sulfoxide detoxification by elevating corn root glutathione and glutathione S-transferase activity. Pestic Biochem Physiol 6:442–446

Le Baron HM, McFariand JE, Simoneaux BJ, Ebert E (1988) Metolachlor. In: Kearney PC, Kaufman DD (eds) Herbicides: chemistry, degradation, and mode of action, vol 3. Dekker, New York, pp 335–381

Leavitt JR, Penner D (1979) In vitro conjugation of glutathione and other thiols with acetanilide herbicides and EPTC sulfoxide and the action of the herbicide antidote R-25788. J Agric Food Chem 27:533–536

Leece DR (1978) Foliar absorption in Prunus domestica L. I. Nature and development of the surface wax barrier. Aust J Plant Physiol 5:749–752

Levi PE, Hodgson E (1992) Metabolism of organophosphorous compounds by the flavin-containing monooxygenase. In: Chambers JE, Levi PE (eds) Organophosphorous: chemistry, fate, and effects. National Academic Press, San Diego, pp 141–145

Lingle SE, Suttle JC (1985) A model system for the study of 2,4-D translocation in leaf of spurge. Can J Plant Sci 65:369–377

Long JW, Basler E (1974) Patterns of phenoxy herbicide translocation in bean seedlings. Weed Sci 22:18–24

Loutre C, Dixon DP, Brazier M, Slater M, Cole DJ, Edwards R (2003) Isolation of a glucosyltransferase from Arabidopsis thaliana active in the metabolism of the persistent pollutant 3,4-dichloroaniline. Plant J 34:485–493

Lu Y (2003) Cupredoxins. In: McCleverty JA, Meyer TJ (eds) Comprehensive Coordination Chemistry II: From Biology to Nanotechnology. Vol 8: Lawrence QJ, Tolman WB (eds) Biocoordination Chemistry. Elsevier, New York, pp 91–122

Ma T, Luo Y, Christie P, Teng Y, Liu W (2012) Removal of phthalic esters from contaminated soil using different cropping systems: a field study. Eur J Soil Biol 50:76–82

Martinova E (1993) An ATP-dependent glutathione-S-conjugate "export" pump in the vacuolar membrane of plants. Nature 364:247–249

Mathur AK, Jhamaria SL (1975) Translocation studies of two systemic fungicides. Pesticides 9:40–46

Mayer A (1987) Polyphenoloxidases in plants-recent progress. Phytochemistry 26:11–20

McFadden JJ, Frear DS, Mansager ER (1989) Aryl hydroxylation of diclofop by a cytochrome P-450 dependent monooxygenase from wheat. Pestic Biochem Physiol 34:92–100

McFadden JJ, Gronwald JW, Eberlein CV (1990) In vitro hydroxylation of benzaton by microsomes from naphtallic anhydride-treated corn shoots. Biochem Biophys Res Comm 168:206–211

Merbach W, Schilling G (1977) Ursachen der Unempfindlichkeit von Beta vulgaris L. gegemiber Pyrazon, Phenmedipham und Benzthiazuron. Biochem Physiol Pflanz 171:187–190

Meunier B, de Visser SP, Shaik S (2004) Mechanism of oxidation reactions catalyzed by cytochrome P450. Enzymes Chem Rev 104:3947–3980

Middleton W, Jarvis BC, Booth A (1978) The effect of ethanol on rooting and carbohydrate metabolism in stem cuttings of *Phaseolus aureus* Roxb. New Phytol 81:2790–2797

Mithaishvili T, Scalla R, Ugrekhelidze D, Tsereteli B, Sadunishvili T, Kvesitadze G (2005) Degradation of aromatic compounds in plants grownunder aseptic conditions. Z Naturforsch C Biosci 60:97–102

Mitton FM, Gonzalez M, Peña A, Miglioranza KS (2012) Effects of amendments on soil availability and phytoremediation potential of aged p, p'-DDT, p, p'-DDE and p, p'-DDD residues by willow plants (Salix sp.). J Hazard Mater 203–204:62–68

Mizutani M, Ohta D (2010) Diversification of P450 genes during land plant evolution. Annu Rev Plant Biol 61:291–315

Mizutani M, Ohta D (2011) Unusual P450 reactions in plant secondary metabolism. Arch Biochem Biophys 507:194–203

Mohammadi M, Chalavi V, Novakova-Sura M, Laiberte JF, Sylvestre M (2007) Expression of bacterial biphenyl-chlorobiphenyl dioxygenase genes in tobacco plants. Biotechnol Bioeng 97:496–505

Morant M, Bak S, Moller BL, Werck-Reichhart D (2003) Plant cytochromes P450: tools for pharmacology, plant protection and phytoremediation. Curr Opin Biotechnol 2:151–162

Mougin C, Cabanne F, Canivenc M-C, Scalla R (1990) Hydroxylation and N-demethylation of chlortoluron by wheat microsomal enzymes. Plant Sci 66:195–203

Nakajima DJY, Suzuki J, Suzuki S (1995) Seasonal changes in the concentration of polycylic aromatic hydrocarbons in azalea leaves and inrelationship to the atmospheric concentration. Chemosphere 30:409–418

Nedunuri KV, Govindaraju RS, Banks MK, Schwab AP, Chen Z (2000) Evaluation of phytoremediation for field-scale degradation of total petroleum hydrocarbons. J Environ Eng 126:483–490

Nejad AR, van Meeteren U (2007) The role of abscisic acid in disturbed stomatal response characteristics of Tradescantia virginiana during growth at high relative air humidity. J Exp Bot 58:627–636

Niku-Paavola ML, Viikari L (2000) Enzymatic oxidation of alkenes. J Mol Cat 10:435–444

Novakova M, Mackova M, Chrastilova Z, Viktorova J, Szekeres M, Demnerova K (2009) Cloning of the bacterial bphC gene into Nicotiana tabacum to improve the efficiency of PCB phytoremediation. Biotechnol Bioeng 102:29–37

Nwoko CO (2010) Trends in phytoremediation of toxic elemental and organic pollutants. Afr J Biotechnol 9:6010–6016

O'Connor GA, Kiehl D, Eiceman GA, Ryan JA (1990) Plant uptake of sludge-borne PCBs. J Environ Qual 19:113–118

Oller ALW, Agostini E, Talano MA, Capozucca C, Milrad SR, Tigier HA (2005) Overexpression of a basic peroxidase in transgenic tomato (*Lycopersicon esculentum* Mill. cv. Pera) hairyroots increases phytoremediation of phenol. Plant Sci 169:1102–1111

O'Neill SD, Keith B, Rappaport L (1986) Transprot of gibberellin A1 in cowpea membrane vesicles. Plant Physiol 80:812–817

Ortiz de Montellano PR (2010) Catalytic mechanisms of heme peroxidases. In: Torres E, Ayala M (eds) Biocatalysis based on heme peroxidases. Peroxidases as potential industrial biocatalysts. Springer, Berlin Heidelberg, pp 79–107

Palazzo AJ, Leggett DC (1986) Effect and disposition of TNT in a terrestrial plant. J Environ Qual 15:49–52

Paterson S, Mackay D, McFarlane C (1994) A model of organic chemical uptake by plants from soil and the atmosphere. Environ Sci Technol 28:2259–2266

Pemadasa MA (1979) Stomatal responses to two herbicidal auxins. J Exp Bot 30:267–276

Penner D, Early RW (1973) Effect of alachlor, butylate and chlorbromuron on carbofuran distribution and metabolism in barley and corn. Weed Sci 21:360–366

Peterson FJ, Mason RP, Horspian J, Holtzman JL (1979) Oxygen-sensitive and insensitive nitroreduction by Escherichia coli and rat hepatic microcosomes. J Biol Chem 254:4009–4014

Peterson MM, Horst GL, Shea PJ, Comfort SD (1998) Germination and seedling development of switchgrass and smooth bromegrass exposed to 2,4,6-trinitrotoluene. Environ Pollut 99:53–59

Pierrel MA, Batard Y, Kazmaier M, Mignotte-Vieux C, Durst F, Werck-Reichhart D (1994) Catalytic properties of the plant cytochrome P450 CYP73 expressed in yeast. Substrate specificity of a cinnamate hydroxylase. Eur J Biochem 224:835–842

Pilon-Smits E (2005) Phytoremediation. Annu Rev Plant Biol 56:15–39

Prasad S, Ellis BE (1978) In vivo characterization of catechol ring- cleavage in cell cultures of Glycine max. Phytochemistry 17:187–193

Preuss A, Rieger PG (1995) Anaerobic transformation of 2,4,6-TNT and other nitroaromatic compounds. In: Spain JC (ed) Biodegradation of nitroaromatic compounds. Plenum Press, New York, pp 69–85

Pridham JB (1958) Metabolism of phenolic compounds by the broad bean Vicia faba. Nature 182:795–800

Pruidze GK, Mchedlishvili NI, Omiadze NT, Gulua LK, Pruidze NG (2003) Multiple forms of phenol oxidase from Kolkhida tea leaves (Camellia sinensis L.) and Mycelia sterilia IBR 35219/2 and their role in tea production. Food Res Int 36:587–595

Reineke W (1998) Development of hybrid strains for the mineralization of chloroaromatics by patchwork assembly. Annu Rev Microbiol 52:287–331

Rigby NM, McDougall AJ, Needs PW, Selvendran RR (1994) Phloem translocation of a reduced oligogalacturonide in Ricinus communis L. Planta 193:536–541

Robineau T, Batard Y, Nedelkina S, Cabello-Hurtado F, LeRet M, Sorokine O, Didierjean L, Werck-Reichhart D (1998) The chemically inducible plant cytochrome P450 CYP76B1 actively metabolizes phenylureas and other xenobiotics. Plant Physiol 118:1049–1056

Rodrígez-López JN, Tudela J, Varón R, Fenoll LG, García-Carmona F, García-Cánovas F (1992) Analysis of a kinetic model for melanin biosynthesis pathway. J Biol Chem 267:3801–3810

Roy S, Hanninen O (1994) Pentachloroohenol: uptake/elimination kinetics and metabolism in aquatic plant Eichornia crassipes. Environ Toxicol Chem 13:763–773

Rugh CL, Senecoff JF, Meagher RB, Hrywna Y (1999) Construction and characterization of two recombinant bacteria that grow on *ortho-* and *para*-substituted chlorobiphenyls. Appl Environ Microbiol 65:2163

Ryan JA, Bell RM, Davidson JM, O'Connor GA (1988) Plant uptake of non-ionic organic chemicals from soils. Chemosphere 12:2299–2323

Sadunishvili T, Gvarliani M, Nutsubidze N, Kvesitadze G (1993) Enzymatic mechanism of ammonia excess detoxication in kidney bean. Fresenius Environ Bull 2:534–539

Sadunishvili T, Kuprava N, Aplakov V, Betsiashvili M, Zaalishvili G, Mithaishvili T, Kvesitadze G (2006) Effect of nitrobenzene on ammonia assimilation enzymes and cell ultrastructure in maize and soybean. Proc Georgian Acad Sci Biol Ser B 4:16–22

Salaün JP (1991) Metabolization de xenobiotiques par des monooxygenases a cytochrome P-450 chez les plantes. Oceanis 17:459–474

Salaün JP, Helvig C (1995) Cytochrome P450-dependent oxidation of fatty acids. In: Durts F, O'Keef DP (eds) Drug metabolism and drug interactions. Freund Publishing House, London, pp 12–49

Sánches-Ferrer A, Rodrígez-López JN, García-Cánovas F, García-Carmona F (1994) Tyrosinase: a comprehensive review of its mechanism. Biochim Biophys Acta 1247:1–11

Sandermann H (1987) Pestizid-Rückstände in Nahrungspflanzen. Die Rolle des pflanzlichen Metabolismus. Naturwissenschaften 74:573–578

Sandermann H (1988) Mutagenic activation of xenobiotics by plant enzymes. Mutat Res 197:183–194

Sandermann H (1994) Higher plant metabolism of xenobiotics: the "green liver" concept. Pharmacogenetics 4:225–241

Sandermann H, Scheel D, Trenck T (1983) Metabolism of environmental chemicals by plants. Copolymerization into lignin. J Appl Polymer Sci 37:407–420

Sandermann H (1999) Plant metabolism of organic xenobiotics. Status and prospects of the 'Green Liver' concept. In: Altman A, Ziv M, Izhar S (eds) Plant biotechnology and in vitro biology in the 21st century. Current plant science and biotechnology agriculture, vol 36, Kluwer Academic, Dordrecht, pp 321–328

Sargent JA, Blackman GE (1972) Studies on foliar penetration. 9. Patterns of penetration of 2,4-dichlorophenoxyacetic acid into the leaves of different species. J Exp Bot 23:830–839

Schmidt B, Joußen N, Bode M, Schuphan I (2006) Oxidative metabolic profiling of xenobiotics by human P450s expressed in tobacco cell suspension cultures. Biochem Soc Trans 34:1241–1245

Schmitt R, Kaul J, Trenck T, Schaller E, Sandermann H (1985) β-D-glucosyl and O-malonyl-β-D-glucosyl conjugates of pentachlorophenol in soybean and wheat: identification and enzymatic synthesis. Pestic Biochem Physiol 24:77–85

Schnoor JL, Dee PE (1997) Phytoremediation. Technology evaluation report TE-98-01. Ground-Water Remediation Technologies Analysis Center. Ser E. Iowa City

Schoenmuth BW, Pestemer W (2004) Dendroremediation of trinitrotoluene (TNT) part 2: fate of radio-labelled TNT in trees. Environ Sci Pollut Res 11:331–339

Schönherr J, Bukovac MJ (1972) Penetration of stomata by liquids. Dependence on surface tension, wettability, and stomatal morphology. Plant Physiol 49:813–823

Schönherr J, Bukovac MJ (1978) Foliar penetration of succinic acid-2,2-dimethylhydrazide: mechanism and rate limiting step. Physiol Plant 42:243–249

Schröder P, Collins C (2002) Conjugating enzymes involved in organic pollutant metabolism of organic organic pollutants in plants. Int J Phytoremediation 4:247–265

Schroeder JI, Kwak JM, Allen GJ (2001) Guard cell abscisic acid signalling and engineering drought hardiness in plants. Nature 410:327–330

Schuler MA (1996) Plant cytochrome P450 monooxygenases. Crit Rev Plant Sci 15:235–284

Schultz ME, Burnside OC (1980) Absorption, translocation and metabolism of 2,4-D and glyphosate in hemp dogbane (Apocynum cannabinum). Weed Sci 28:13–19

Seidel K, Kickuth R (1967) Exkretion von Phenol in der Phylosphäre von Scirpus lacustris L. Naturwissenschaften 52:517–525

Sens C, Sheidemann P, Werner D (1999) The distribution of 14C-TNT in different biochemical compartments of the monocotyledoneous Triticum aestivum. Environ Pollut 104:113–119

Sharma MP, Vanden Born WH (1970) Foliar penetration of picloram and 2,4-D in aspen and balsam poplar. Weed Sci 18:57–65

Shen C, Tang X, Cheema SA, Zhang C, Khan MI (2009) Enhanced phytoremediation potential of polychlorinated biphenyl contaminated soil from e-waste recycling area in the presence of randomly methylated-beta-cyclodextrins. J Hazard Mater 172:1671–1676

Shida T, Homma Y, Misaio T (1975) Absorption, translocation and degradation of N-lauoryl-L-valine in plants. 6. Studies on the control of plant diseases by amino acid derivatives. J Agric Chem Soc Jpn 49:409–418

Shimabukuro RH, Hoffer BL (1991) Metabolism of diclofop-methyl in sensitive and resistant biotypes of Lolium rigidum. Pestic Biochem Physiol 39:251–260

Shimabukuro RH, Walsh WC, Jacobson A (1987) Aryl-O-glucoside of diclofop: a detoxification product in wheat shoots and wild oat suspension diclofop. J Agric Food Chem 35:393–399
Shimabukuro RH, Walsh WC, Hoffer BL (1989) The absorption, translocation and metabolism of difenopenten-ethyl in soybean and wheat. Pestic Biochem Physiol 33:57–68
Shimazaki K, Doi M, Assmann SM, Kinoshita T (2007) Light regulation of stomatal movement. Annu Rev Plant Biol 58:219–247
Shinohara A, Kamataki T, Ichimura Y, Opochi H, Okuda K, Kato R (1984) Drug oxidation activities of horse-redish peroxidase, myoglobin and cytochrome P-450cam reconstituted with synthetic hemes. Jpn J Pharmacol 45:107–114
Shiots N, Kodama S, Inui H, Ohkama H (2000) Expression of cytochrome P450 1A1 and P450 1A2 as fused enzymes with yeast NADPH-cytochrome P450 oxidoreductase in transgenic tobacco plants. Biosci Biotechnol Biochem 64:2025–2033
Shutte R, Golfmann GP (1975) Metabolism of herbicides: derivatives of diphenyl ether in oat. In: Mechanism of plant herbicide and synthetic growth regulators actions and their further fate in biosphere (in Russian). Materials of XI Symp, Pushchino, pp 129–133
Sicbaldi F, Sacchi GA, Trevisan M, Del Re AAM (1997) Root uptake and xylem translocation of pesticides from different chemical classes. Pestic Sci 50:111–119
Siegel BZ (1993) Plant peroxidases – an organismic perspective. J Plant Growth Regul 12:303–312
Siminszky B, Corbin FT, Ward ER, Fleischmann TJ, Dewey RE (1999) Expression of a soybean cytochrome P450 monooxygenase cDNA in yeast and tobacco enhances the metabolism of phenylurea herbicides. Proc Natl Acad Sci U S A 96:1750–1755
Siminszky B (2006) Plant cytochrome P450-mediated herbicide metabolism. Phytochem Rev 5:445–458
Simonich SL, Hites RA (1994) Vegetation-atmosphere partioning of polycyclic aromatic hydrocarbons. Environ Sci Technol 28:939–943
Söchtig H (1964) Beeinlussung des Stoffwechsels der Planzen durch Humus und seine Bestandteile und die Auswirkung auf Wachstum und Ertrag. Landbauforsch Völkenrode 14:9–15
Solomon EI, Chen P, Metz M, Lee SK, Palmer AE (2001) Oxygen binding, activation, and reduction to water by copper proteins. Angew Chem Int Ed 40:4570–4590
Sonoki T, Kajita S, Ikeda S, Uesugi M, Tatsumi K, Katayama Y (2005) Transgenic tobacco expressing fungal laccase promotes the detoxification of environmental pollutants. Appl Microbiol Biotechnol 67:138–142
Stahl JD, Aust SD (1995) Biodegradation of 2,4,6-trinitrotoluene by the white rot fungus Phanerochaete chrysosporium. In: Spain JC (ed) Biodegradation of nitroaromatic compounds. Plenum Press, New York, pp 117–134
Starov VM (2004) Surfactant solutions and porous substrates: spreading and imbibition. Adv Colloid Interface Sci 111:3–27
Stiborova M, Anzenbacher P (1991) What are the principal enzymes oxidizing the xenobiotics in plants: cytochrome P-450 or peroxidase? Gen Physiol Biophys 10:209–216
Sugumaran M, Duggaraju R, Generozova F, Ito S (1999) Insect melanogenesis. II. Inability of Manduca phenoloxidase to act on 5,6-dihydroxyindole-2-carboxylic acid. Pigm Cell Res 12:118–125
Talekar NS, Lee EM, Sun LT (1977) Absorption and translocation of soil and foliar applied 14C-carbofuran and 14C-phorate in soybean and mung bean seeds. J Econ Entomol 70:685–688
Tanaka FS, Hoffer BL, Simabukuro RH (1990) Identification of the isomeric hydroxylated metabolites of methyl-2-[4-(2,4-dichlorophenoxy)phenoxy] propanoate (diclofop-methyl) in wheat. J Agric Food Chem 38:599–603
Tateoka TN (1970) Studies on the catabolic pathway of protocatechuic acid in mung bean seedlings. Bot Mag (Tokyo) 83:49–54
Thompson PL, Ramer LA, Schnoor JL (1998) Uptake and transformation of TNT by hybrid poplar trees. Environ Sci Technol 32:975–980

Tkhelidze P (1969) Oxidative transformation of benzene and toluene in vine grapes (in Russian). Bull Georg Acad Sci 56:697–700
Trapp M, McFarlane C (1995) Plant contamination. Lewis Publisher, Boca Raton
Trenck T, Sandermann H (1980) Oxygenation of benzo[a]pyrene by plant microsomal fractions. FEBS Lett 119:227–231
Tyree MT, Peterson CA, Edgington LV (1979) A simple theory regarding ambimobility of xenobiotics with special reference to the nematicide oxamyl. Plant Physiol 63:367–374
Uchida E, Ouchi T, Suzuki Y, Yoshida T, Habe H, Vamaguchi I (2005) Secretion of bacterial xenobiotic-degrading enzymes from transgenic plants by an apoplastic expressional system: an applicability for phytoremediation. Environ Sci Technol 39:7671–7677
Ugrekhelidze D (1976) Metabolism of exogenous alkanes and aromatic hydrocarbons in plants (in Russian). Metsnieraba, Tbilisi
Ugrekhelidze D, Durmishidze S (1984) Penetration and detoxification of organic xenobiotics in plants (in Russian). Metsniereba, Tbilisi
Ugrekhelidze D, Kavtaradze L (1970) The question of metabolism of α-naphthol in higher plants (in Russian). Bull Georg Acad Sci 57:465–469
Ugrekhelidze D, Phiriashvili V, Mithaishvili T (1986) Uptake of salicylic acid and aniline by pea roots (in Russian). Fiziol Rast 33:165–170
Urlacher VB (2012) 7.13 Oxidation: stereoselective oxidations with cytochrome P450 monooxygenases. In: Turner NJ (ed) Comprehensive chirality. Vol 7: Synthetic methods VI: enzymatic and semi-enzymatic. Elsevier, New York, pp 275–294
Veitch NC (2004) Horseradish peroxidase: a modern view of a classic enzyme. Phytochemistry 65:249–259
Ugrekhelidze D, Korte F, Kvesitadze G (1997) Uptake and transformation of benzene and toluene by plant leaves. Ecotoxicol Environ Saf 37:24–28
Verkleij JAC, Golan-Goldhirsh A, Antosiewisz DM, Schwitzguébel J-P, Schröder P (2009) Dualities in plant tolerance to pollutants and their uptake and translocation to the upper plant parts. Environ Exp Bot 67:10–22
Vias SC, Sinha OK, Josh LK (1976) Systemic uptake and translocation of bavistin and calixin in peanut (*Arachis hypogaea* L.). Pesticides 10:32–38
Wang G-D, Chen X-Y (2007) Detoxification of soil phenolic pollutants by plant secretory enzyme. In: Willey N (ed) Phytoremediation: methods and reviews. Methods in biotechnology, vol 23. Springer, Humana Press, Totowa pp 49–57
Webber MD, Pietz RI, GranatoTC SML (1994) Plant uptake of PCBs and other organic contaminants from sludge-treated coal refuse. J Environ Qual 23:1019–1026
Welsch-Pausch K, McLachlan MS, Umlauf G (1995) Determination of the principal pathway of plychlorinated dibenzo-*p*-dioxins an dibenzofuranes to *Lolium multiflorum*. Environ Sci Technol 29:1090–1098
Werck-Reichhart D, Hehn A, Didierjean L (2000) Cytochromes P450 for engineering herbicide tolerance. Trends Plant Sci 5:116–123
Wetzel A, Sandermann H (1994) Plant biochemistry of xenobiotics: isolation and characterization of a soybean O-glucosyltransferase of DDT metabolism. Arch Biochem Biophys 314:323–328
Whetten R, Sederoff R (1995) Lignin biosynthesis. Plant Cell 7:1001–1013
White RH, Liebel RA, Hymowitz T (1990) Examination of 2,4-D tolerance in perennial Glycine species. Pestic Biochem Physiol 38:153–161
Whitehouse CJ, Bell SG, Tufton HG, Kenny RJ, Ogilvie LC, Wong LL (2008) Evolved CYP102A1 (P450BM3) variants oxidise a range of non-natural substrates and offer new selectivity options. Chem Commun (Camb) 28:966–968
Wilson L, Williamson T, Gronowski J, Gentile GI, Gentile JM (1994) Characterization of 4-nitro-o-phenylendiamine activities by plant systems. Mutat Res 307:185–193
Wolfe NL, Hoehamer CF (2003) Enzymes used by plants and microorganisms to detoxify organic compounds. In: McCutcheon SC, Schnoor JL (eds) Phytoremediation. Transformation and control of contaminats. Wiley-Interscience, Hoboken, pp 159–188

Xu F, Bell SG, Lednik J, Insley A, Rao Z, Wong LL (2005) The heme monooxygenase cytochrome P450cam can be engineered to oxidize ethane to ethanol. Angew Chem Int Ed Engl 44:4029–4032

Yalpani N, Silverman P, Wilson MA, Kleier DA, Raskin I (1991) Salicylic acid is a system signal and an inducer of pathogenesis-related proteins in virus-infected tobacco. Plant Cell 3:809–818

Yu MH, Sakurai S (1995) Diisopropylfluorylphosphate (DFP)-hydrolyzing enzymes in mung bean (Vigna radiata) seedlings. Environ Sci (Tokyo) 3:103–111

Zaalishvili G, Lomidze E, Buadze O, Sadunishvili T, Tkhelidze P, Kvesitadze G (2000) Electron microscopic investigation of benzidine effect on maize root tip cell ultrastructure, DNA synthesis and calcium homeostasis. Int Biodeterior Biodegradion 46:133–140

Zaalishvili G, Sadunishvili T, Scala R, Laurent F, Kvesitadze G (2002) Electron microscopic investigation of nitrobenzene distribution and effect on plant root tip cells ultrastructure. Ecotoxicol Environ Saf 52:190–197

Zenno S, Kobori T, Tanokura M, Saigo K (1998) Conversion of NfsA, the major Escherichia coli nitroreductase, to a flavin reductase with an activity similar to that of Frp, a flavin reductase in Vibrio harveyi, by a single amino acid substitution. J Bacteriol 180:422–425

Zimmerlin A, Durst F (1992) Aryl hydroxylation of the herbicide diclofop by a wheat cytochrome P-450 monooxygenase. Plant Physiol 100:868–873

Potential Use of Licorice in Phytoremediation of Salt Affected Soils

Dilfuza Egamberdieva and Nazim A. Mamedov

Contents

Abstract The salinisation of lands has become a major environmental issue and has been recognized as the most important economic, social and environmental problem in many regions of the world. Salt and drought tolerant halophytes may help to restore abandoned saline lands for sustainable use for crop production. Licorice has been considered as salt tolerant plant which could be used for remediation of abandoned salt affected soils. Phytoremediation of saline soils with nitrogen-fixing leguminous licorice can improve the soil nitrogen content, increase the soil organic matter, stimulate soil biological activity and improve soil water-holding capacity. This paper focuses on the potential use of licorice for the phytoremediation of salt affected soils.

Keywords Phytoremediation • Salinity • Licorice • Salt tolerance • Nutrients

D. Egamberdieva
Institut für Landschaftsbiogeochemie, Leibniz-Zentrum für Agrarlandschaftsforschung e. V., Eberswalder, Straße 84 15374, Müncheberg
e-mail: Dilfuza.egamberdieva@zalf.de

N.A. Mamedov (✉)
Medicinal Plants Program, School of Agriculture, University of Massachusetts at Amherst, Amherst, MA, USA
e-mail: mamedov@cas.umass.edu

M. Öztürk et al. (eds.), *Plants, Pollutants and Remediation*,
DOI 10.1007/978-94-017-7194-8_13

1 Introduction

There is considerable evidence for several treats of climate change such as increased temperature, salinity and drought with intervening arid episode. As a result of climate changes, there is a negative impact on agricultural production and livelihood of rural population in across regions of Asia, and Africa (Nellemann et al. 2009). A crop loss due to negative impact of climate change is a major area of concern to cope up with the increasing food requirements which affects nearly one billion people around the world (Munns and Tester 2008; Beddington et al. 2011; Hakeem et al. 2012). Salinity alone affects 33 % of the potentially arable land area of the world, whereas 950 million ha of salt-affected lands occur in arid and semi-arid regions (UNEP 2008). According Qadir et al. (2007) one of the major factor that increase soil salinity include irrigation of cultivated lands with saline water, and on average, 24 % of the world's irrigated lands have been affected by salinity. There are several approaches management practices developed to cope with salinity and improve plant growth including identification of saline tolerant crops, crop diversification, reduced tillage, manure application, crop rotation, adaptation of crops and phytoremediation (Kushiev et al. 2005; Soltani et al. 2012; Adesemoye and Egamberdieva 2013). Planting perennial and annual halophytes in salt affected soils may also help to restore abandoned lands for crop production (Toderich et al. 2008). Plant biomass could be used efficiently for the removal of pollutants from contaminated sites through extraction, immobilization, degradation of contaminants and this approach is called phytoremediation. This emerging technology has lower cost, environmentally friendly and holds a great promise for the decontamination of pollutants (Greenberg et al. 2006). However, salt cannot be degraded, therefore the use of salt tolerant plants is prerequisite for removing salt ions from the soil. The phytoremediation approaches aimed at the rehabilitation of salt affected soils have been carried out in various parts of the world (Ghaly 2002; Kushiev et al. 2005; Ravindran et al. 2007; Hakeem et al. 2015).

Several plants has been used for phytoremediation of salt affected soils and wetlands e.g. *Bassia indica* (kochia) (Shelef et al. 2012), *Tetragonia tetragonoides* (New Zealand spinach) *Portulaca oleracea* (Purslane) (Ben Asher et al. 2012), *Sesbania* (agati sesban) (Ilyas et al. 1993), *Medicago sativa* (alfalfa) (Cuartero et al. 2002), *Atriplex* (saltbush) (Malcolm et al. 1988), *Chenopodium album* (lamb's-quarters) (Hamidov et al. 2007), *Suaeda maritime* (seablite) and *Sesuvium portulacastrum* (sea purslane) (Ravindran et al. 2007). Licorice (*Glycyrrhiza glabra* L.) has been also considered as salt tolerant plant which could be used for remediation of abandoned salt affected soils. In addition licorice was considered to be an alternative income source for farmers in salt affected arid regions (Abe et al. 2005).

This chapter focuses on the potential use of licorice for the phytoremediation of salt affected soils. After discussing information on the negative effect of salinity on plant growth, we address the overview of licorice, its biology, and the role of plant in remediation of salt affected soils.

2 Soil Salinity and Plant Growth

Salinization of soil and water resources is recognized as the main threat to environmental resources and almost affects 33 % of the potentially arable land area of the world (UNEP 2008). The irrigation of cultivated lands with saline water, poor cultural practices considered as a major factors increasing salinity and is widely responsible for increasing the salt concentration in soil (Rengasamy 2006; Egamberdiyeva et al. 2007).

Salt-affected soils differ in their chemical composition and the concentration of Ca^{+2}, K^+, Na^+, CO_3^{2-}, and Cl^- were influenced significantly by soil salinity (Egamberdieva et al. 2010). The Ca^{2+} was the dominant (>60 %) salt associated cation in all soils followed by Mg^{+2}, K^+, and Na^+, respectively. Significantly higher concentration of Ca^{2+}, K^+, and Na^+ were associated with CO_3^{-2} and Cl^- and reflected a dominance of CO_3^{-2} and Cl^- in irrigation-induced saline soils and its probably responsible for higher E_C (Garcia and Hernandez 1996). The salinity negatively affect on soil microbial activities and organic matter dynamics (Davranova et al. 2013; Egamberdieva 2011).

Many studies have demonstrated that salinity inhibits seed germination and growth of various crop plants e.g. wheat (Egamberdieva 2009), rice (Xu et al. 2011), maize (Khodarahmpour et al. 2012), and sugar beet (Jamil et al. 2006). The negative effects are assumed to be caused by the osmotic stress (Shirokova et al. 2000), disturbance of the hormonal balance (Prakash and Parthapasenan 1990), and the toxicity of Na^+ and Cl^- ions on uptake of N, P, K, and Mg by the plant (Heidari and Jamshid 2010). Only salt tolerant halophytes might be a good candidate to cultivate in salt-affected soils because they are perennial, and deep rooted plants. According Nobel (1999) in saline soils halophytes are subjected to accumulation of excessive inorganic ions in cell walls. Halophytes can grow in the presence of high concentration of salt (500 mM) (Zhu 2003). Manousaki and Kalogerakis (2011) explained this tolerance as an adaptation of halophytes to extreme environments. In that respect halophytes can be divided to salt-excluders, salt-includers and salt-accumulators (Breckle 2002).

3 Phytoremediation

The phytoremediation technology has shown promise as an effective low-cost and economically sustainable approach to detoxify metals and organic chemicals through plant species capable of hyperaccumulating target ionic species in their shoots (Wiltse et al. 1998; Schwitzguébel 2001; Qadir and Oster 2004). In salt affected soils, the phytoremediation is achieved through the increase of the dissolution rate of calcite by plant roots which are resulted in enhanced levels of Ca+2 in soil solution to effectively replace Na from the cation exchange complex (Qadir et al. 2007). That is the most economically feasible technology in large salt affected area where plants remove salts from soil through their deep and wide root system. It is known that plant roots play an important role in facilitating the process of leaching Na+ (Qadir et al. 2007). The halophytic plants will also improve soil fertility through improvement of soil biological activity and increase the availability of nutrients. In addition they improve hydraulic properties of sodic soils (Akhter et al. 2004). It has been reported that highly salt-resistant species such as halophytes may accumulate quite high levels of salts and Na+ in their shoots and considered as potential tool for phytoremediation of saline-sodic soils (Qadir et al. 2007). However, their efficiency highly variable, whereas highly salt tolerant and deep rooted plants have been found to be efficient in phytoremediation of saline soils (Kaur et al. 2002).

Ben Asher et al. (2012) reported that lettuce (*Lactuca sativa*) absorbed more salt at higher salt concentration compared to lower salt concentration and the percentage of salt absorbed by plant was about 3–5 % of the salt applied. Similar results they observed for New Zealand spinach (*Tetragonia tetragonioides*), which also showed a high ability to extract salts from the soil. The Na concentration of plants increased from 1.8 to 2.4 % in accordance with increased salinity levels from 0.65 up to 3.5 dS/m (Ben Asher et al. 2012). Lyas et al. (1993) studied alfalfa, sesbania – wheat rotation – alone and in conjunction with the application of gypsum for remediation of saline sodic soil and found increased saturated hydraulic conductivity (Ks) by alfalfa. In following investigations alfalfa would contribute to only 1–2 % of the total Na removed during phytoremediation of sodic soils (Qadir et al. 2003). The potential capacity of halophytes *Chenopodium album* and *Apocynum lancifolium*, has been evaluated for its ability to remove of chloride, sodium, magnesium and calcium ions from the salt-affected soils of Uzbekistan (Hamidov et al. 2007). The authors observed that *Chenopodium album* was most effective in removing chloride ions (104.5 mg g^{-1} dry biomass), and sodium (33.6 mg g^{-1} dry biomass). It has been also reported that licorice adapted to dry saline lands can be used for remediation of salt affected soils. In other study Ravindran et al. (2007) reported that *Suaeda maritime* and *Sesuvium portulacastrum* exhibited greater accumulation of salts in their tissues as well as higher reduction of salts in the soil medium.

4 The Potential of Licorice for Phytoremediation of Salt Impacted Soils

The genus *Glycyrrhiza* is a perennial herb belongs to the leguminous family (*Fabaceae* L.) and about 30 species are accepted up to today including *G. aspera*, *G. bucharica*, *G. echinata*, *G. eurycarpa*, *G. glabra*, *G. iconica*, *G. inflata*, *G. korshinskyi*, *G. lepidota*, *G. macedonica*, *G. pallidiflora*, *G. squamulosa*, *G. triphylla*, *G. uralensis* and *G. yunnanensis* (Nomura et al. 2002; Fiore et al. 2005). Most widely distributed species *Glycyrrhiza glabra* is found in Spain, Italy Turkey, the Caucasus, Central Asia, and the western part of China whereas *Glycyrrhiza uralensis* is distributed form Central Asia to Mongolia and China (Hayashi et al. 2003).

In traditional medicine licorice roots have been used against treating various disease including lung diseases, bronchial asthma, catarrhs of the upper respiratory

Fig. 1 The effect of NaCl concentrations on growth of *Glycyrrhiza uralensis*

tract, and certain viral infections (Anon 2005; Sharma et al. 2013). Licorice has deep root, often reach 17 m in arid areas in order to access deep ground water and may yield 10–15 t/ha of fresh roots after 3 years growth. The plant has a fusiform root system with numerous suckers that are often more than 1 m in length (Kushiev et al. 2005). Licorice is generally salt tolerant, well-adapted to grow in saline soils and used in rehabilitating abandoned saline fields (Lu et al. 2013; Kushiev et al. 2005). *G. uralensis* fix N_2 from the atmosphere in the symbiosis with highly specific rhizobia belonging to the species of *Mezorhizobium* (Wei et al. 2008; Li et al. 2012). Wei et al. (2008) studied the diversity of *Mezorhizobium* associated with *G. uralensis* and selected salt tolerant strain *Mezorhizobium* sp. CCNWGX035. The strain was able to grow in the presence of 60 mg mL^{-1} NaCl (6 % NaCl).

Lu et al. (2013) studied the response of *Glycyrrhiza inflata* to NaCl stress and observed that the higher concentration of NaCl (>200 mmo L^{-1}) inhibited growth and development of licorice. In our study we have observed that licorice *Glycyrrhiza uralensis* germinated at potting soil irrigated with 50 and 100 mM NaCl nutrient solution. However, 100 mM NaCl concentration totally inhibited plant growth after 1–2 weeks (unpublished data, Fig. 1).

Phytoremediation of saline soils with nitrogen-fixing legumes would be an optimal technique for restoration of soil fertility since legumes can also improve the soil nitrogen content, increase the soil organic matter, stimulate soil biological activity and improve soil water-holding capacity (Ashraf and McNeilly 2004). There are several reports on the use of *Glycyrrhiza glabra* in the reclamation of saline soils and the subsequent restoration of irrigated cropping systems (Kerbabaev 1971; Badalov 1996). For example Lu et al. (2013) reported on the increased uptake of K^+, Ca^{2+} and Mg^{2+} by licorice with increasing NaCl concentration and proline synthesis which increase osmoregulation ability through which plant adapt to high saline environment. Kushiev et al. (2005) studied the potential use of *Glycyrrhiza glabra* for the remediation of abandoned saline areas over a 4 year period before being returned to a cotton/wheat crop rotation. They have observed that above ground biomass ranged from 3.55, 5.6 to 8.55 t ha in following 3 years respectively. Another interesting observation was that the total dissolved salt (TDS) of the ground water in the control plot increased from 5.19 to 6.11 g l^{-1} over 3 years period, but under the licorice treatment the TDS declined over the same period from 6.35 to 3.99 g l^{-1}, indicating that salts had been decreased from the system. In addition Mg^{2+}, Ca^{2+}, $SO4^{2-}$, N^{+}, and Cl^{-} declined markedly in licorice grown field when compared to the field where no licorice was introduced (Kushiev et al. 2005). In general this study clearly indicates declined salt content from 215 t ha (first year) to 185 t ha (third years) in the top 2 m depth of soil under licorice plot. The yield of cotton and wheat cultivated after licorice was also increased compared to that of the local regional average (Kushiev et al. 2005).

The high plant biomass and deeper root system is important for successful phytoremediation. It is already reported that salinity inhibit plant growth, especially root development (Egamberdieva et al. 2013), thus decrease plants ability to extract salts from the soil. Since amelioration occurs throughout the root zone, it is important to improve root growth and development in highly saline soils. One of the biotechnological approaches for stimulation of plant root system and biomass under saline condition is to use plant growth promoting rhizobacteria (PGPR) (Egamberdieva and Lugtenberg 2014) and arbuscular mycorrhizal fungi (AMF) (Hameed et al. 2014). This strategy play key role in enhancing the efficacy of phytoremediation (Glick 2003; Chaudhry et al. 2005). Selvaraj and Sumithra (2011) studied the effect of *Glomus aggregatum* (AMF) and PGPR on the growth, nutrient uptake and biomass of *Glycyrrhiza glabra* and they found that microbial consortium enhanced nutrient uptake and plant biomass compared to untreated control plants. Recently we have also observed that shoot dry weight and length, N uptake and nodulation of *G. uralensis* was clearly improved when the plant was inoculated with its salt tolerant PGPR strains compared with the uninoculated plants irrigated with 75 mM NaCl (unpublished data). The beneficial root associated bacteria may alleviate plant stress by lowering stress ethylene, stimulate root growth and development by supplying additional phytohormones to plants and facilitates mineral uptake of plants (Egamberdieva 2011, 2012; Jabborova et al. 2013; Lugtenberg and Kamilova 2009). Considering above beneficial traits of root associated microbes, the development of salt-tolerant symbioses is an absolute necessity to enable cultivation of plants in salt-affected soils.

5 Conclusion

The soil salinity, drought and desertification have become a major environmental issue and have been recognized as the most important economic, social and environmental problem in many regions of the world. Halophytes are naturally adapted to cope with environmental stresses and considered as potential plants for the remediation of salt-impacted soils. Licorice is salt and drought tolerant halophytic plant and widespread on salt affected lands. Plant has several advantages in saline environment: (a) tolerant to salinity and drought, (b) deep root system that enables plants to access water and a high accumulation of ions, (c) improve soil fertility through stimulation of microbial biomass and organic matter, (d) allow the land to be returned to a high productive farmland, (e) generate income for farmers through the production of high quality forage for feeding to livestock, (f) root material can be used in drinks and medicinal preparations.

References

Abe J, Araki H, An P, Shimizu H, Li J, Guo Y, Inanaga S (2005) Combating desertification and rehabilitating degraded arid lands in Alashan, Inner Mongolia, China. Root Res 14(2):51–58 (in Japanese)

Adesemoye AO, Egamberdieva D (2013) Beneficial effects of plant growth promoting rhizobacteria on improved crop production: the prospects for developing economies. In: Maheshwari DK et al (eds) Bacteria in agrobiology: crop productivity. Springer, Berlin/Heidelberg

Akhter J, Murray R, Mahmood K, Malik KA, Ahmed S (2004) Improvement of degraded physical properties of a saline-sodic soil by reclamation with kallar grass (*Leptochloa fusca*). Plant Soil 258:207–216

Anon (2005) *Glycyrrhiza glabra*. Altern Med Rev 10:230–237

Ashraf M, McNeilly T (2004) Salinity tolerance in *Brassica* oilseeds. Crit Rev Plant Sci 23(2):157–174

Badalov MA (1996) Multiplication of liquorice naked and its cultivation. Fan Publisher, Tashkent

Beddington J, Asaduzzaman M, Fernandez A, Clark M, Guillou M, Jahn M, Erda L, Mamo T, Van Bo N, Nobre CA, Scholes R, Sharma R, Wakhungu J (2011) Achieving food security in the face of climate change: summary for policy makers from the Commission on Sustainable Agriculture and Climate Change. CGIAR Research Program on Climate Change, Agriculture and Food Security (CCAFS). Copenhagen, Denmark. Available online at: www.ccafs.cgiar.org/commission

Ben Asher J, Beltrao J, Aksoy U, Anac D, Anac S (2012) Controlling and simulating the use of salt removing species. Int J Energy Environ 6(3):360–370

Breckle SW (2002) Salinity, halophytes and salt affected natural ecosystems. In: Läuchli A, Lüttge U (eds) Salinity: environment-plants-molecules. Kluwer, Dordrecht, pp 53–77

Chaudhry MS, Batool Z, Khan AG (2005) Preliminary assessment of plant community structure and arbuscular mycorrhizas in rangeland habitats of Cholistan Desert, Pakistan. Mycorrhiza 15(8):606–611

Cuartero J, Pulido JM, Gómez-Guillaumón ML, Alvarez M (2002) Salt removal potential of barley, alfalfa, *Atriplex patula* and *A. prostate*. Acta Horicult 573:387–396

Davranova N, Egamberdieva D, Ismatov Z, Wirth S (2013) Impact of crop management practice on soil microbial populations in a semi arid soil of Uzbekistan. Soil Water J 2(2):921–927

Egamberdieva D (2009) Alleviation of salt stress by plant growth regulators and IAA producing bacteria in wheat. Acta Physiol Plant 31:861–864

Egamberdieva D (2011) Survival of Pseudomonas extremorientalis TSAU20 and P. chlororaphis TSAU13 in the rhizosphere of common bean (Phaseolus vulgaris) under saline conditions. Plant Soil Environ 57(3):122–127

Egamberdieva D (2012) *Pseudomonas chlororaphis*: a salt-tolerant bacterial inoculant for plant growth stimulation under saline soil conditions. Acta Physiol Plant 34:751–756

Egamberdieva D, Lugtenberg B (2014) PGPR to alleviate salinity stress on plant growth. In: Miransari M (ed) Use of microbes for the alleviation of soil stresses, vol 1. Springer, New York, pp 73–96
Egamberdieva D, Renella G, Wirth S, Islam R (2010) Secondary salinity effects on soil microbial biomass. Biol Fertil Soils 46(5):445–449
Egamberdieva D, Berg G, Lindström K, Räsänen LA (2013) Alleviation of salt stress of symbiotic *Galegaofficinalis* L. (*Goat's Rue*) by co-inoculation of rhizobium with root colonizing *Pseudomonas*. Plant Soil 369(1):453–465
Egamberdiyeva D, Gafurova L, Islam KR (2007) Salinity effects on irrigated soil chemical and biological properties in the Syr Darya basin of Uzbekistan. In: Lal R, Sulaimanov M, Stewart B, Hansen D, Doraiswamy P (eds) Climate change and terrestrial C sequestration in Central Asia. Taylor-Francis, New York, pp 147–162
Fiore C, Eisenhut M, Ragazzi E, Zanchin G, Armanini D (2005) A history of the therapeutic use of liquorice in Europe. J Ethnopharmacol 99:317–324
Garcia C, Hernandez T (1996) Influence of salinity on the biological and biochemical activity of a calciorthird soil. Plant Soil 178:255–263
Ghaly FM (2002) Role of natural vegetation in improving salt affected soil in northern Egypt. Soil Tillage Res 64:173–178
Glick BR (2003) Phytoremediation: synergistic use of plant and bacteria to clean up the environment. Biotechnol Adv 21(5):383–393
Greenberg BM, Huang XD, Gurska J, Gerhardt KE, Lampi MA, Khalid MA, Isherwood D, Chang P, Wang W, Wang H, Dixon DG, Glick BR (2006) Development and successful field tests of a multi-process phytoremediation system for decontamination of persistent petroleum and organic contaminants in soils. In: Tisch B, Zimmerman K, White P, Beckett P, Guenther L, Macleod A, Rowsome S, Black C (eds) Reclamation and remediation: policy and practice. Canadian Land Reclamation Association, Calgary, pp 124–133
Hakeem KR, Khan F, Chandna R, Siddiqui TO, Iqbal M (2012) Genotypic variability among soybean genotypes under NaCl stress and proteome analysis of salt tolerant genotype. Appl Biochem Biotechnol 168:2309–2329
Hakeem KR, Sabir M, Ozturk M, Mermut A (2015) Soil remediation and plants: prospects and challenges. Academic Press, Elsevier, New York, p 724
Hameed A, Egamberdieva D, Abd-Allah EF, Hashem A, Kumar A, Ahmad P (2014) Salinity stress and arbuscular mycorrhizal symbiosis in plants. In: Miransari M (ed) Use of microbes for the alleviation of soil stresses, vol 1. Springer, New York, pp 139–159
Hamidov A, Khaydarova N, Khamidov M, Neves MA, Beltrao J (2007) *Apocynum Lancifolium* and *Chenopodium Album*- potential species to remediate saline soils. WSEAS Trans Environ Dev 7(3):123–128
Hayashi H, Hattori S, Inoue K, Khodzhimatov O, Ashurmetov O, Ito M, Honda G (2003) Field survey of *Glycyrrhiza* plants in Central Asia: chemical characterization of *G. glabra* collected in Uzbekistan. Chem Pharm Bull 51(11):1338–1340
Heidari M, Jamshid P (2010) Interaction between salinity and potassium on grain yield, carbohydrate content and nutrient uptake in pearl millet. J Agric Biol Sci 5:39–46
Ilyas M, Miller RW, Qureshi RH (1993) Hydraulic conductivity of saline-sodic soil after gypsum application and cropping. Soil Sci Soc Am J 57:1580–1585
Jabborova D, Egamberdieva D, Räsänen L, Liao H (2013) Salt tolerant *Pseudomanas* strain improved growth, nodulation and nutrient uptake of soybean grown under hydroponic salt stress condition. In: XVII International plant nutrition colloquium and boron satellite meeting proceedings book. Sabanci University, Istanbul, pp 260–261
Jamil M, Lee DB, Jung KY, Ashraf M, Lee SC, Rhal ES (2006) Effect of salt (NaCl) stress on germination and early seedling growth of four vegetables species. J Cent Eur Agric 7:273–282
Kaur B, Gupta SR, Singh G (2002) Bioamelioration of a sodic soil by silvopastoral systems in northwestern. India Agrofor Syst 54:13–20

Kerbabaev BB (1971) Protection of liquorice naked production resources. In: Proceeding book of conference on protection of flora of Central Asian Republics and Kazakhstan. Fan publisher, Tashkent, pp 252–256

Khodarahmpour Z, Ifar M, Motamedi M (2012) Effects of NaCl salinity on maize (*Zea mays* L.) at germination and early seedling stage. Afr J Biotechnol 11:298–304

Kushiev H, Noble AD, Abdullaev I, Toshbekov V (2005) Remediation of abandoned saline soils using *Glycyrrhiza glabra*: a study for the Hungry Steppes of Central Asia. Int J Agric Sustain 3:102–113

Li L, Sinkko H, Montonen L, Wei G, Lindstrom K, Rasanen L (2012) Biogeography of symbiotic and other endophytic bacteria isolated from medicinal *Glycyrrhiza* species in China. FEMS Microbiol Ecol 79:46–68

Lu JH, Lu X, Liang YC, Lin HL (2013) Salt tolerance of *Glycyrrhiza inflata* seedlings in Xinjiang and its ion response to salt stress. Chin J Plant Ecol 37(9):839–850

Lugtenberg B, Kamilova F (2009) Plant-growth-promoting-rhizobacteria. Annu Rev Microbiol 63:541–556

Malcolm CV, Clarke AJ, D'Antuono MF, Swaan TC (1988) Effects of plant spacing and soil conditions on the growth of five Atriplex species. Agric Ecosyst Environ 21:265–279

Manousaki E, Kalogerakis N (2011) Halophytes – a new trend in phytoremediation. Int J Phytorem 13:959–969

Munns R, Tester M (2008) Mechanisms of salinity tolerance. Annu Rev Plant Biol 59:651–681

Nellemann C, MacDevette M, Manders T, Eickhout B, Svihus B, Prins AG, Kaltenborn BP (eds) (2009) The environmental food crisis – the environment's role in averting future food crises. A UNEP rapid response assessment. United Nations Environment Programme, GRID-Arendal, www.grida.no

Nobel PS (1999) Physicochemical and environmental plant physiology. Academic, San Diego

Nomura T, Fukai T, Akiyama T (2002) Chemistry of phenolic compounds of licorice (*Glycyrrhiza* species) and their estrogenic and cytotoxic activities. Pure Appl Chem 74(7):1199–1206

Prakash L, Parthapasenan G (1990) Interactive effect of NaCl salinity and gibberelic acid on shoot growth, content of absisic acid and gibberelin like substances and yield of rice (*Oruza sativa*). Plant Sci 100:173–181

Qadir M, Steffens D, Yan F, Schubert S (2003) Sodium removal from a calcareous saline-sodic soil through leaching and plant uptake during phytoremediation. Land Degrad Dev 14:301–307

Qadir M, Oster JD (2004) Crop and irrigation management strategies for saline-sodic soils and waters aimed at environmentally sustainable agriculture. Sci Tot Environ 323:1–19

Qadir M, Oster JD, Schubert S, Noble AD, Sahrawat KL (2007) Phytoremediation of sodic and saline-sodic soils. Adv Agron 96:197–247

Ravindran KC, Venkatesan K, Balakrishnan V, Chellapan KP, Balasubramanian T (2007) Restoration of saline land by halophytes for Indian soils. Soil Biol Biochem 39:2661–2664

Rengasamy P (2006) World salinization with emphasis on Australia. J Exp Bot 57:1017–1023

Schwitzguébel JP (2001) Hype of hope: the potential of phytoremediation as an emerging green technology. Remediation 11(4):63–78

Selvaraj T, Sumithra P (2011) Effect of *Glomus aggregatum* and plant growth promoting rhizomicroorganisms on growth, nutrition and content of secondary metabolites in *Glycyrrhiza glabra* L. Indian J Appl Pure Biol 26(2):283–290

Sharma V, Agrawal RC, Pandey S (2013) Phytochemical screening and determination of antibacterial and anti-oxidant potential of *Glycyrrhiza glabra* root extracts. J Environ Res Dev 7(4):1552–1558

Shelef O, Gross A, Rachmilevitch S (2012) The use of *Bassia indica* for salt phytoremediation in constructed wetlands. Water Res 46(13):3967

Shirokova Y, Forkutsa I, Sharafutdinova N (2000) Use of electrical conductivity instead of soluble salts for soil salinity monitoring in Central Asia. Irrig Drain Syst 14:199–205

Soltani A, Khodarahmpour Z, Jafari AA, Nakhjavan S (2012) Selection of alfalfa (*Medicagosativa* L.) cultivars for salt stress tolerance using germination indices. Afr J Biotechnol 31:7899–7905

Toderich K, Tsukatani T, Shoaib I, Massino I, Wilhelm M, Yusupov S, Kuliev T, Ruziev S (2008) Extent of salt-affected land in Central Asia: biosaline agriculture and utilisation of salt-affected resources. Discussion paper No. 648. Kyoto Institute of Economic Research, 34 pp

UNEP (2008) In dead water. Merging of climate change with pollution, over-harvest, and infestations in the world's fishing grounds. UNEP/GRID-Arendal, Arendal, Norway. Available online at: http://www.grida.no/_res/site/file/publications/InDeadWater_LR.pdf

Wei GH, Yang XY, Zhang ZX, Yang YZ, Lindsröm K (2008) Strain *Mesorhizobium* sp. CCNWGX035: a stress-tolerant isolate from Glycyrrhiza glabra displaying a wide host range of nodulation. Pedosphere 18(1):102–112

Wiltse CC, Rooney WL, Chen Z, Schwab AP, Banks MK (1998) Greenhouse evaluation of agronomic and crude oil-phytoremediation potential among alfalfa genotypes. J Environ Qual 27:169–173

Xu GY, Rocha PS, Wang ML, Xu ML, Cui YC, Li LY, Zhu YX, Xia X (2011) A novel rice calmodulin-like gene, OsMSR2, enhances drought and salt tolerance and increases ABA sensitivity in Arabidopsis. Planta 234:47–59

Zhu JK (2003) Regulation of ion homeostasis under salt stress. Curr Opin Plant Biol 6:441–445

Effect of Nutrient Enrichment on Metal Accumulation and Biological Responses of Duckweed (Lemnaceae) Spread in Turkey

Ahmet Aksoy and Zeliha Leblebici

Contents

Abstract Wetlands have been proposed as sites for the phytoremediation of metals. The fate of metals within plant tissues is a critical issue for the effectiveness of this process. This study was intended to test the hypothesis that nutrient enrichment (P, NO_3^--N and SO_4^{2-}) enhances the metal tolerance and biological responses of floating macrophytes. To test this hypothesis, duckweed species which spread in Turkey (*Lemna minor* L., *Lemna gibba* L., *Lemna trisulca* L., *Lemna turionifera* Landolt and *Spirodela polyrhiza* (L.) Schleid.) were exposed to heavy metals (Pb, Ni, Cd) in the absence and presence of nutrients for 7 days under laboratory conditions. Metal accumulation, relative growth rates (RGR) and photosynthetic pigments (chlorophyll a) were measured. It was determined that metal and nutrient concentration in water decreased throughout the experiments. The highest Pb accumulation was seen at a dose of 50 mg l^{-1} in *L. gibba* (22,596 μg g^{-1}), after 7 days. Relative growth rates were negatively correlated with metal exposure, but

A. Aksoy
Department of Biology, Akdeniz University, Antalya 07058, Turkey

Z. Leblebici (✉)
Faculty of Science and Arts Department of Molecular Biology and Genetics, Nevsehir University, 50300 Nevsehir, Turkey
e-mail: zleblebici@nevsehir.edu.tr

M. Öztürk et al. (eds.), *Plants, Pollutants and Remediation*,
DOI 10.1007/978-94-017-7194-8_14

nutrient addition was found to suppress this effect. Photosynthetic pigment level was found to be negatively correlated with metal exposure, and nutrient addition attenuated chlorophyll decrease in response to metal exposure. Levels of chl *a* decreased in a Pb concentration-dependent and time-dependent manner, with a minimum value of 0.386 mg g^{-1} in the 50 mg l^{-1} on *L. gibba*. The study concluded that nutrient enrichment increases the tolerance of duckweeds to metals. However, it was determined that nutrient enrichment decreased metal accumulation at 5 mg l^{-1} nutrient addition. Our finding may be useful for the phytoremediation of water polluted with heavy metals.

Keywords Nutrient enrichment • Duckweed • Heavy metal • Photosynthetic pigment

1 Introduction

Duckweed is a floating aquatic macrophyte belonging to the botanical family Lemnaceae, which can be found world-wide on the surface of fresh waters. The Lemnaceae family consists of two genera (*Lemna* L., *Spirodela* Schleiden) and five species in Turkey. Compared to most other plants, duckweed has low fiber content (about 5 %), since it does not require structural tissue to support leaves and stems. Of these, applications of (duckweed) in wastewater treatment was found to be very effective in the removal of nutrients, soluble salts, organic matter, heavy metals and in eliminating suspended solids, algal abundance and total and fecal coliform densities. In particular, species of *Lemna* are reported to accumulate toxic metals and therefore are being used as experimental model systems to investigate heavy metal induced responses (Halaimi et al. 2014; Leblebici et al. 2010, 2011). Bioavailability and bioaccumulation of various heavy metals in aquatic and wetland ecosystems is gaining tremendous significance globally (Bianconi et al. 2013; Leblebici and Aksoy 2011).

Water contamination with heavy metals is a very important problem in the current world. Occurrence of toxic metals in pond, ditch and river water affect the lives of local people that depend upon these water sources for their daily requirements (Rai et al. 2002). A number of methods are available to remove toxic metals from water: ion exchange, chemical precipitation, membrane filtration, reverse osmosis, electrolysis, and adsorption. The latter is by far the most versatile and widely used. However, the methods present different efficiencies for different metals and they can be very expensive, especially if large volumes, low metal concentration and high standards of cleaning are required. The main processes by which heavy metals are removed from aquatic environment are physical, biological and biochemical and they take place in water, biota and suspended solids. The predominance of one of them will depend on the composition of the system, the pH, the redox condition and the nature of the pollutant (Miretzky et al. 2004).

Phytoremediation is considered an effective, low cost, preferred cleanup option for moderately contaminated areas (Hakeem et al. 2015). Wetlands are often considered sinks for contaminants, and there are many cases in which wetland plants are utilized for removal of pollutants, including metals. Chang and Corapcioglu (1998) describe how plants promote, through various processes, the remediation of toxic chemicals. These processes include: (1) modifying the physical and chemical properties of contaminated soils; (2) releasing root exudates, thereby increasing organic carbon; (3) improving aeration by releasing oxygen directly to the root zone, as well as increasing the porosity of the upper soil zones; (4) intercepting and retarding the movement of chemicals; (5) effecting co-metabolic microbial and plant enzymatic transformations of recalcitrant chemicals; (6) decreasing vertical and lateral migration of pollutants to ground water.

Five main subgroups of phytoremediation have been identified: *Phytoextraction*, in which metal accumulating plants are used to transport and concentrate metals into the harvestable parts of roots and aboveground shoot (Kumar et al. 1995). *Rhizofiltration*, in which plant roots absorb, precipitate, and concentrate toxic metals from polluted effluents (Dushenkov et al. 1995); *Phytostabilization*, in which mobility of heavy metals is reduced through the use of tolerant plants (Dushenkov et al. 1995). *Phytotransformation/phytodegradation*, in which a contaminant can be eliminated via phytodegradation or phytotransformation by plant enzymes or enzyme, co-factors (Susarla et al. 2002). *Phytovolatilisation*: Volatilisation of pollutants into the atmosphere via plants (Burken and Schnoor 1999).

Plant species with potential for phytoremediation should possess the following properties: (1) they should extract and accumulate, transform, degrade, or volatilise contaminants at levels that are toxic to ordinary plants; and (2) The plant species must have fast growth and high yield. Additionally, a good phytoremediation species should be able to remediate multible pollutant simultaneously because pollution rarely occurs with a single chemical (Mkandawire and Dudel 2007).

Currently, a few plants species are known to possess the properties that qualify them to be good phytoremediation species for terrestrial and aquatic environments (Hume et al. 2002; Cossu et al. 2002; Sooknah and Wilke 2004; Duman et al. 2009). Among species identified for aquatic phytoremediation are species from the genus *Lemna,* a free-floating tiny macrophyte (Chaudhuri et al. 2014; Salt et al. 1995; Wang et al. 2002; Mkandawire et al. 2004a; Goulet et al. 2005; Stout and Nusslein 2005). *Lemna* species commonly grow naturally in wetlands including some highly contaminated water bodies (Landolt 1982). *Lemna* species are highly advocated for application in wastewater treatment facilities, in constructed wetlands and even in the restoration of contaminated water bodies (Wang et al. 2002). Their relatively simple but advanced anatomical and physiological structure has scientific and engineering significance. These properties allow easy handling, and manipulating under laboratory conditions. Consequently, they are considered to be a model plant-representative of higher plants-for a number of chemical and biogeochemical studies involving the regulation of element assimilation in higher plants. Apart from phytoremediation studies and use, the *Lemna* spp. are among the most standardized test organisms in aquatic ecotoxicology (EPA 1996; DIN 2000; ISO 2001; OECD 2002).

2 Pytoremediation Studies of Lemnaceae

Lemnaceae has received a lot of attention from scientists because of its potential accumulation capacity of contaminants. There are several studies that have shown that most *Lemna* spp. show an exceptional capability and potential for the uptake and accumulation of heavy metals, radionuclides as well as metalloids, surpassing that of algae and other aquatic macrophytes (Chaudhary and Sharma 2014; Körner et al. 1998; Szabo et al. 1999; Axtell et al. 2003; Zimmo et al. 2004). Table 1 presents some selected metals reported in the literature which shows high accumulation capacity in some *Lemna* species. For example, the zinc concentration in frond tissue was 2700 times higher than that of its medium (Sharma and Gaur 1994). Under experimental conditions, *L. minor* is a good accumulator of Cd, Se, and Cu, but a moderate accumulator of Cr and a relatively poor accumulator of Ni and Pb (Zayed et al. 1998). *Lemna* spp. have also shown potential in the attenuation of uranium as well as arsenic in surface waters of decommissioned uranium mining (Mkandawire et al. 2004a, b; Mkandawire and Dudel 2005). The uptake rates of all by *Lemna* spp. is estimated between 0.8 and 17 mg g^{-1} d^{-1} (Goulet et al. 2005). The effects of

Table 1 Selected bioaccumulation and transfer factors of some heavy metals in some *Lemna* spp. As reported in the literature (Charpentier et al. 1987; Steveninck et al. 1992; Miranda and Ilangovan 1996; Mkandawire et al. 2004a; Leblebici et al. 2010; Leblebici and Aksoy 2011)

Species	Metal	Bioaccumulation (mg kg^{-1}dry biomass)	Bioaccumulation coefficient[a]
L. gibba	As	1000–1500	>500
	Cu	745–1050	~10,000
	U	850–1100	~3000
	Cr	900–1710	~7000
L. minor	Zn	212–1010	>1000
	Cd	14,200	~20,000
	Cu	200–800	>500
	Cd	14,200	~12,000
	Co	200–2000	>1500
	Pb	>750	>500
	Cr	13	~250
	Ba	226	~800
	Al	1700–4560	>4000
L. trisulca	Cd	130–1200	>3500
	Al	19,238	~18,000
	Cr	1555	~2000
	Cu	217	~300
	Ba	107	>500
	Zn	1308	~700
	Pb	233	>500

[a]Bioaccumulation coefficients were estimated from values from the same literature source

Pb was conducted on *L. minor* L. and *Spirodela polyrhiza* (L.) Schleid. *L. minor* accumulated 561 mg g^{-1} Pb and *S. polyrhiza* accumulated 330 mg g^{-1} Pb after 7 days (Leblebici and Aksoy 2011). The high Cd and Cu accumulation were seen on *L. gibba* 6415 and 2351 mg g^{-1} respectively (Leblebici et al. 2010).

Sekomo et al. (2012) explored the use of algae and duckweed ponds as a post-treatment for textile wastewater. The study was conducted using the hypothesis that differing conditions such as pH, redox potential and dissolved oxygen in these ponds would lead to different heavy metal removal efficiencies. The authors indicated that both treatment systems are not very suitable as a polishing step for removing these heavy metals. Despite the significant differences in terms of physico-chemical conditions, differences in metal removal efficiency between algal and duckweed ponds were rather small.

Lahive et al. (2012) investigated the effects of zinc on photosynthetic performance in the three species over the 7 days. The maximum quantum efficiency of photosystem II, *Fv/Fm*, the effective quantum efficiency, *Y(II)*, and photochemical quenching, *qP*, were measured in mature and young fronds, as well as along a developmental gradient within fronds. *Fv/Fm* and *Y(II)* in young, emerging *Landoltia punctata* fronds were more severely impacted by zinc than developed, mature fronds. The authors explained that younger proximal sections of *L. punctata* fronds were more impacted than older distal frond sections. Overall, *Fv/Fm* and *Y(II)* also tended to be more affected by zinc in young, compared to mature, *Lemna gibba* and *Lemna minor* fronds. Single colony, time-point or leaf-zone analyses may not, therefore, show the full biological picture of the impact of a toxicant, while species-specific differences need also to be considered.

Zhang et al. (2011) investigated arsenic (As) accumulation and tolerance of duckweed *Spirodela polyrhiza* L. and its potential for Asphytofiltration. *S. polyrhiza* was able to survive in high concentrations of As(V) solution. This study suggested that this floating aquatic plant has some potential for as phytofiltration in contaminated water bodies or paddy soils.

Sasmaz and Obek (2012) investigated *L. gibba's* capacity to remove silver (Ag) and gold (Au) from secondary effluents. *L. gibba* accumulated significant amounts of Ag and Au for 6 days from the initiation of the experimental study. The highest accumulations were 2303 % for Ag and 247 % for Au. However, after 6 days, the rate of Ag and Au accumulation in *L. gibba* declined, as saturation levels had been reached in the plant tissues. The authors showed that metal accumulating property of *L. gibba* can also be commercially exploited to recover Au and Ag from wastewater and mining wastes.

Appenroth et al. (2008) investigated the modification of chromate toxicity by sulphate in duckweeds. The authors explained that the sulphate influences the toxicity of chromate, mainly by chromate uptake, with negligible impact on other physiological processes.

Greater duckweed (*S. polyrhiza* L.) was tested for arsenic accumulation under laboratory conditions by Rahman et al. (2007) and Rahman et al. (2008) to investigate arsenic uptake efficiency and mechanisms. The results showed that *S. polyrhiza* L. accumulated higher amount of arsenic from As(V) solution compared to that from DMAA solution. They also observed that As(V) uptake into *S. polyrhiza* L. was negatively correlated with phosphate uptake and positively correlated with iron uptake. The facts were explained by the competitive uptake inhibition of As(V) by phosphate and adsorptive affinity of As(V) on iron oxides of root surfaces. In contrast, DMAA uptake in *S. polyrhiza* L. was neither affected by phosphate nor correlated with iron (Rahman et al. 2007). Thus, it has been proposed that *S. polyrhiza* L. might use different mechanisms for As(V) and DMAA uptake. However, the arsenic uptake ability of *S. polyrhiza* L. suggests that this macrophyte would be a good option for the phytoremediation of contaminated water.

Since the early 1970s, considerable work has been done on the use of *Lemna* spp. as a means of treating wastewater of both agricultural and domestic origin (Cheng et al. 2002; Zimmo et al. 2004; Goulet et al. 2005). Almost a decade ago, Koles et al. (1987) described the guidelines for the use of *Lemna* spp. to remove ammonia and phosphorus from water. A *Lemna*-covered wastewater treatment system, in practice, works optimally within depths between 30 and 150 cm. Smith and Moelyowati (2001) have also developed guidelines for designing a *Lemna* spp. based wastewater treatment system. The guidelines have a design program that suggests that a combination of anaerobic ponds, *Lemna* spp.-based treatment system and maturation ponds can minimize land requirements associated with wastewater treatment using only phytoremediation procedure. Vatta et al. (1995) developed models for *L. gibba*-based wastewater treatment plants. They developed a comprehensive process model which simulates the behaviour of a waste-water treatment system based on *L. gibba*. The model accounts for the main chemical and biochemical phenomena involved in a natural waste-water treatment system. Their predictions are quite reliable, especially in mini-ponds and in real-size treatment. *Lemna* spp. grows very densely in nutrient-rich environments in which layers of fronds grow one on top of another to form a mat that can be as much as 10 cm thick. This thick mat creates an anaerobic environment in the water on which this mat floats, which promotes anaerobic digestion and denitrification of wastewater (Landesman 2000; Cheng et al 2002). Therefore, *Lemna* spp. can also be part of constructed wetland systems, either in the wastewater-receiving or in polishing ponds in wetland-treated effluents. Polishing is one of the last steps in wastewater treatment used where residual nutrient, organic and suspended solids are removed either aerobically or facultatively. The elimination capacity for organic material in terms of biological oxygen demand (BOD) and chemical oxygen demand (COD) is lower in comparison to other vascular plants and rich in cellulose but emerge growing macrophytes in constructed wetland. Nitrogen removal is at the same level or even higher (Vymazal 2005). P-elimination is higher in halophyte than *Lemna*-dominated treatment systems because phosphates are usually fixed on the gravel beds in the benthic zone.

Generally, *Lemna* species are considered to be very fast growing, thereby providing a high turnover and yield (Landolt 1986). Most *Lemna* species have a mean specific growth rate range of 0.2–0.3 d^{-1} with a doubling time in the ranging between 0.7 and 2 days (Landolt 1986; Cheng et al. 2002). However, *L. minor* and *L. gibba* can reach a specific growth rate of about 0.6 d^{-1} under ideal conditions which are rich in nutrients (Mkandawire et al. 2004a). Maximum relative growth rates of 0.73–0.79 d^{-1} have been observed in *Lemna aequinoctialis* Welw and *Wolffia microscopica* (Griffith) Kurz, which correspond to doubling times between 20 and 24 h (Körner et al. 2003). The lowest maximal growth rates are observed in submerged species (Landolt and Kandeler 1987). In general, Körner et al. (2003) find the RGR values of *Lemna* sp. comparable to angiosperm herbaceous plants which range between 0.03 and 0.37 d^{-1}, whereas algae grow at rates between 0.26 and 2.84 d^{-1}.

Landolt and Kandeler (1987) estimate the annual mean yield for *Lemna* species to be 73 tons ha^{-1} $year^{-1}$ dry biomass. Some yield of above 180 tons ha^{-1} $year^{-1}$ dry biomass have been recorded. The yield of *Lemna* spp., when compared to algae in aquatic systems is relatively high. The average yield of *Lemna* reported in literature lies between 25 and 50 g m^{-2} d^{-1} dry biomass in natural uncontaminated water bodies, even though a daily yield of close to 200 g m^{-2} d^{-1} have been estimated under laboratory cultures and in some tropical regions. Thus, *Lemna* species are estimated to have 41 and 75 % of biomass-related extraction potential for metals (Mkandawire and Dudel 2007). For instance, Mkandawire et al. (2004a) estimated that *L. gibba* biomass can extract arsenic and uranium in the magnitude of 751.9 ± 250 and 662.7 ± 203 kg ha^{-1} $year^{-1}$ representing an extraction potential of 48.3 ± 15.1 and 41.4 ± 11.9 % under ideal laboratory condition-optimal steady state condition with unlimited growth. *Lemna* species are capable of growing throughout the year and thereby can provide the required biomass to take up contaminants from the aquatic system.

3 Materials and Methods

3.1 *Plant Material and Treatment Conditions*

Lemna is a genus of monocotyledonous free-floating aquatic macrophytes in the Lemnaceae family, which is commonly known as duckweed. They commonly grow in stagnant or slow-flowing, nutrient-enriched waters through-out tropical and temperate zones. Their growth conditions include temperatures range of 6–33 °C, a wide pH range with optimal growth between pH 5.5 and 7.5 (Mkandawire and Dudel 2005). Unlike most terrestrial and aquatic angiosperms, *Lemna* spp. reproduces almost exclusively asexually despite being flowering plants, thereby allocating almost all their resources to vegetative growth (Landolt and Kandeler 1987). Anatomically, they a diffuse unit known as a frond which is composed of leaflets and a root-like structure. From the phylogenetic point of view, *Lemna* spp. are in the evolutionary path of secondary simplification of a former complex and highly differentiated vascular plants (Les et al. 1997).

Lemna species were obtained from Soysalli-Kayseri, (38° 23′ 500″ N, 035° 21′ 919″ E, 1075 m), and Beyşehir-Yeşildağ Town Konya, (37° 33, 348 N 031° 29, 331

E, 1123 m) Turkey. The chemical composition of Soysallı and Beyşehir waters were (mean ± standard deviation): pH = 6.4 ± 0.1, conductivity = 92 ± 8 μS cm^{-1}, NH_4^+-N = 0.021 ± 0.01 mg l^{-1}, NO_3^--N = 0.02 ± 0.001 mg l^{-1}, NO_2^--N = 0.002 ± 0.001 mg l^{-1}, SO_4^{2-} = 0.2 ± 0.03 mg l^{-1}; pH = 6.5 ± 0.3, iletkenlik = 72 ± 8 μS cm^{-1}, NH_4^+-N = 0.021 ± 0.01 mg l^{-1}, NO^{-3}-N = 0.02 ± 0.001 mg l^{-1}, NO_2^--N = 0.002 ± 0.001 mg l^{-1}, SO_4^{-2} = 0.09 ± 0.03 mg l^{-1}. Before metal treatment, plants were acclimatized for 5 days under laboratory conditions (23 °C and 14 h photoperiod, 350 μmol m^2 s^{-1}). In this study, lead chloride ($PbCl_2$) nickel chloride ($NiCl_2 6H_2O$) and cadmium chloride ($CdCl_2$) were used without further purification for experimental treatments. Plants were treated with different concentrations of Pb (0, 5, 10, 25, 50 mg l^{-1}), Ni (1–5–10–20 mg l^{-1}) and Cd (0.5–1–2–4 mg l^{-1}) maintained in double deionized water in 500 ml conical flasks under the aforementioned conditions for periods of 1, 3, 5 and 7 days. Plant growth rates in response to metal exposures were compared with exposures enriched with 5 mg l^{-1} P (KH_2PO_4), 5 mg l^{-1} NO_3^--N (KNO_3) and SO_4^{2-} (K_2SO_4) (Hadad et al. 2007). Flasks without metals grown alongside each set of experimental groups served as controls. After harvesting, plants were washed with double deionized water. Plants were placed on blotting paper and allowed to drain for 5 min before weighing. Water samples were filtered through for nutrient determinations. Nitrate, phosphate and sulphate were determined by Hach Lange DR 2800 spectroscopy. All treatments were carried out in triplicate.

Lemna species relative growth rates were calculated in each group according to Hunt's equation

$$R = lnW2 - lnW1/T2 - T1,$$

Where R is the relative growth rate (gg^{-1} d^{-1}), W_1 and W_2 are the initial and final fresh weights, respectively, and (T_2–T_1) is the experimental period (Hunt 1978).

3.2 Pb, Ni and Cd Quantification

Harvested plants were washed thoroughly with double deionized water, blotted and oven dried at 80 °C. Each sample was then digested with 10 ml pure HNO_3, using a CEM-MARS 5 (CEM Corporation Matthews, NC, USA) microwave digestion system (maximum power: 1200 W, power: 100 %, ramp: 20:00 min, pressure: 180 psi, temperature: 210 °C and hold time: 10:00 min). After digestion, the volume of each sample was adjusted to 25 ml using double deionized water. Determinations of Pb Ni and Cd concentrations in plant samples were carried out by inductively coupled plasma optical emission spectroscopy (Varian-Liberty II, ICP-OES) (Duman et al. 2009). Peach leaves (NIST, SRM-1547) were used as reference material; also all analytical procedures were performed for reference material.

3.3 *Plant Growth Parameters*

Plant biomass was measured on the basis of fresh weight. Photosynthetic pigment of treated and untreated plants (100 mg) were extracted in 80 % chilled acetone in the dark. After centrifugation at 10,000 × g for 10 min, absorbance of the supernatant was taken at 450, 645 and 663 nm. The content of chlorophyll a was estimated as previously described (Witham et al. 1971).

3.4 *Statistical Analysis*

Two-way analysis (ANOVA) was done with all the data to confirm the variability of data and validity of results, and Duncan's multiple range test (DMRT) was performed to determine the significant difference between treatments. Statistical Package for the Social Sciences (SPSS) statistical program was used for statistical analysis (Kinnear and Gray 1994).

4 Results and Discussion

4.1 *Accumulation of Heavy Metal and Its Effect on Growth of Plant*

Pb bioaccumulation was measured in *Lemna* fronds. The highest Pb accumulation was seen at a dose of 50 mg l^{-1} in *L. gibba* (22,596 μg g^{-1}), after 7 days (Fig. 1). In the groups enriched with nutrients, the highest Pb accumulation was seen at a dose of 50 mg l^{-1} in *S. polyrhiza* (14,589 μg g^{-1}), after 7 days (Fig. 1). Bioaccumulation of Ni was measured in *Lemna* fronds. The plants were found to accumulate high amounts of Ni in a concentration-time dependent manner. The highest Ni accumulation was observed at a dose of 20 mg l^{-1} in *L. gibba* (2899 μg g^{-1}), after 7 days (Fig. 2). In the groups enriched with nutrients, the highest Ni accumulation was seen at a dose of 20 mg l^{-1} in *L. gibba* (2383 μg g^{-1}), after 7 days (Fig. 2). Cd bioaccumulation was measured in *Lemna* fronds. The highest Cd accumulation was seen at a dose of 4 mg l^{-1} in *L. gibba* (6428 μg g^{-1}), after 7 days (Fig. 3). In the groups enriched with nutrients, the highest Cd accumulation was seen at a dose of 4 mg l^{-1} in *L. gibba* (4249 μg g^{-1}), after 7 days (Fig. 3). Metal concentrations in plants increased with metal concentration as well as over time. In the present study, a high accumulation of heavy metal was observed in *Lemna* species over a 7-day period and *L. gibba* was more effective in the accumulation of heavy metal than other species

The relative growth rates of *Lemna* species decreased in the presence of Pb in a concentration dependent manner (Fig. 4). However, the treatments enriched with nutrients did not show a similar correlation. The highest decline of RGR was seen

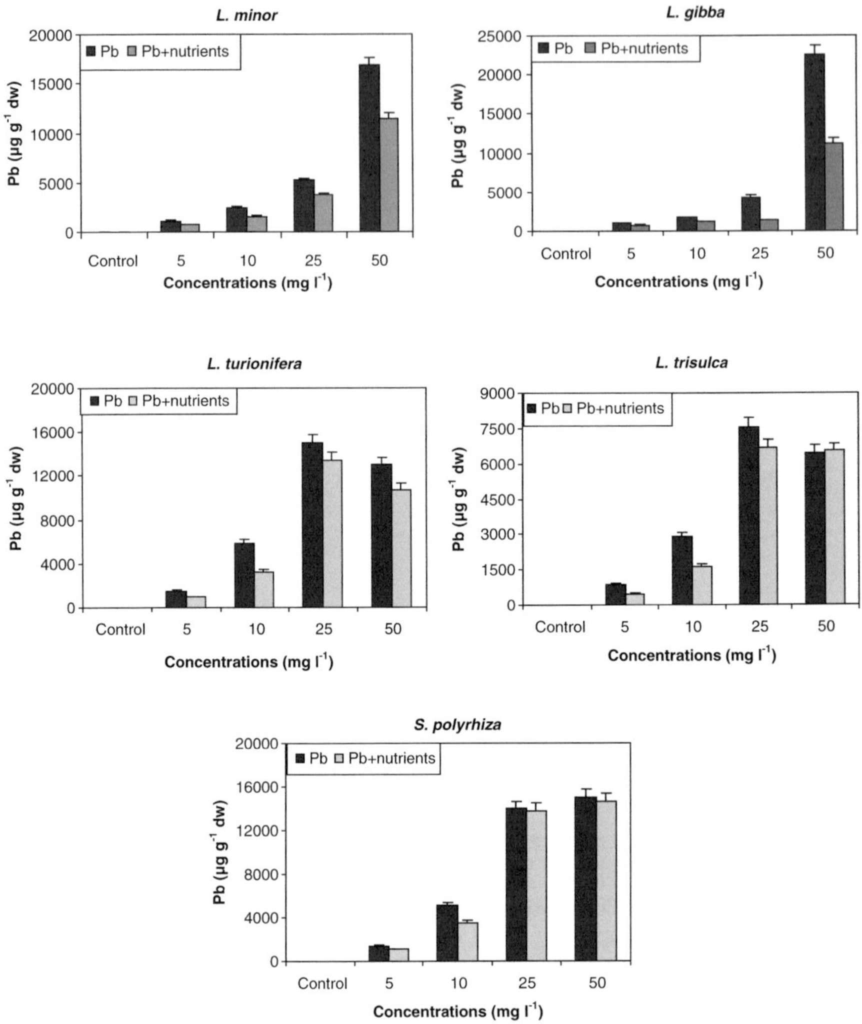

Fig. 1 Accumulation of Pb and Pb+nutrients by *Lemna minor*, *Lemna gibba*, *Lemna trisulca*, *Lemna turionifera* and *Spirodela polyrhiza* exposed to different concentrations over various periods of time. All values are means of triplicates ± S.D. ANOVA significance was set at $p \leq 0.05$

at 20 mg l^{-1} Ni exposure in *L. turionifera*, after 7 days (Fig. 5). The Growth rates of the *Lemna* species declined with increasing heavy metal concentrations. However, nutrient enrichment led to increased growth rates at heavy metal concentrations that impaired growth in the non-enriched groups (Fig. 6).

Appenroth et al. (2008) investigated the modification of chromate toxicity by sulphate in duckweeds. The authors explained that the sulphate influences the toxicity of chromate mainly by chromate uptake, with negligible impact on other

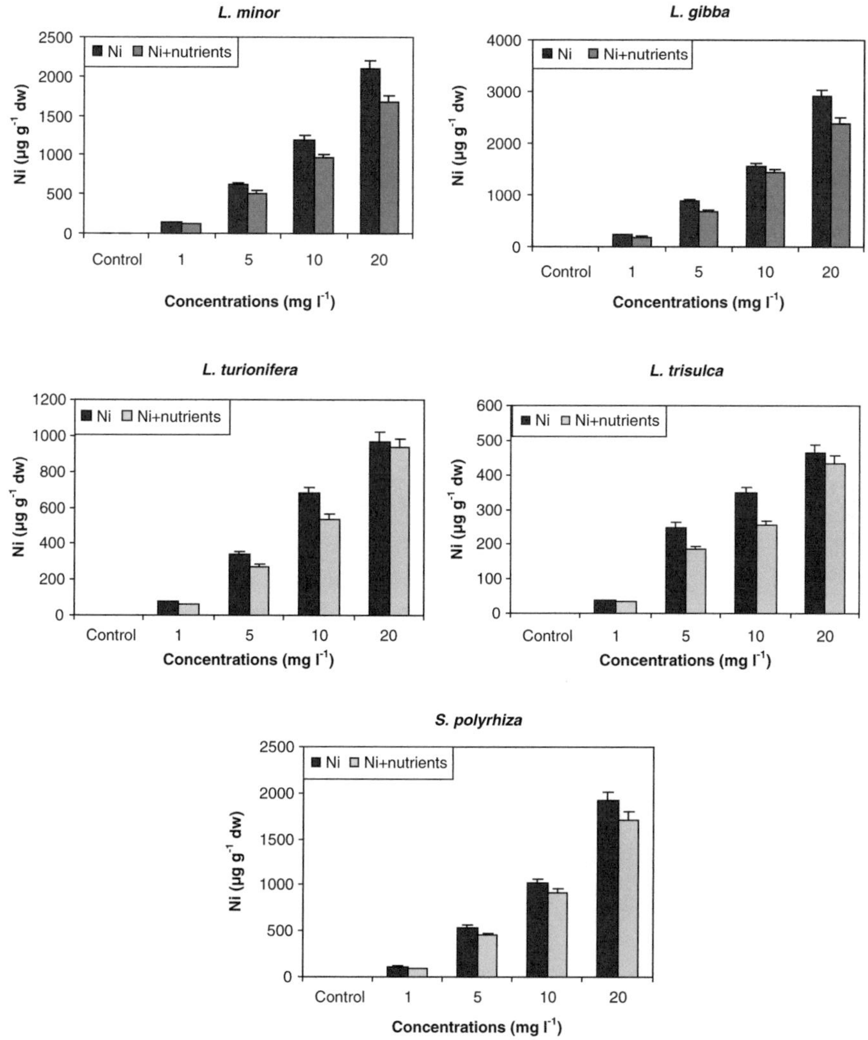

Fig. 2 Accumulation of Ni and Ni + nutrients by *Lemna minor*, *Lemna gibba*, *Lemna trisulca*, *Lemna turionifera* and *Spirodela polyrhiza* exposed to different concentrations over various periods of time. All values are means of triplicates ± S.D. ANOVA significance was set at $p \leq 0.05$

physiological processes. Rahman et al. (2008) reported on the uptake of arsenate in *S. polyrhiza* and its interactions with PO_4^{3-} and Fe ions. Their study found that arsenate uptake in *S. polyrhiza* occurred through the phosphate uptake pathway and by physico-chemical adsorption on Fe-plaques of plant surfaces as well. In line with our results, Hadad et al. (2007) found that nutrient enrichment enabled growth of *Salvinia hergozii* at Zn and Ni exposures that impaired growth in plants without

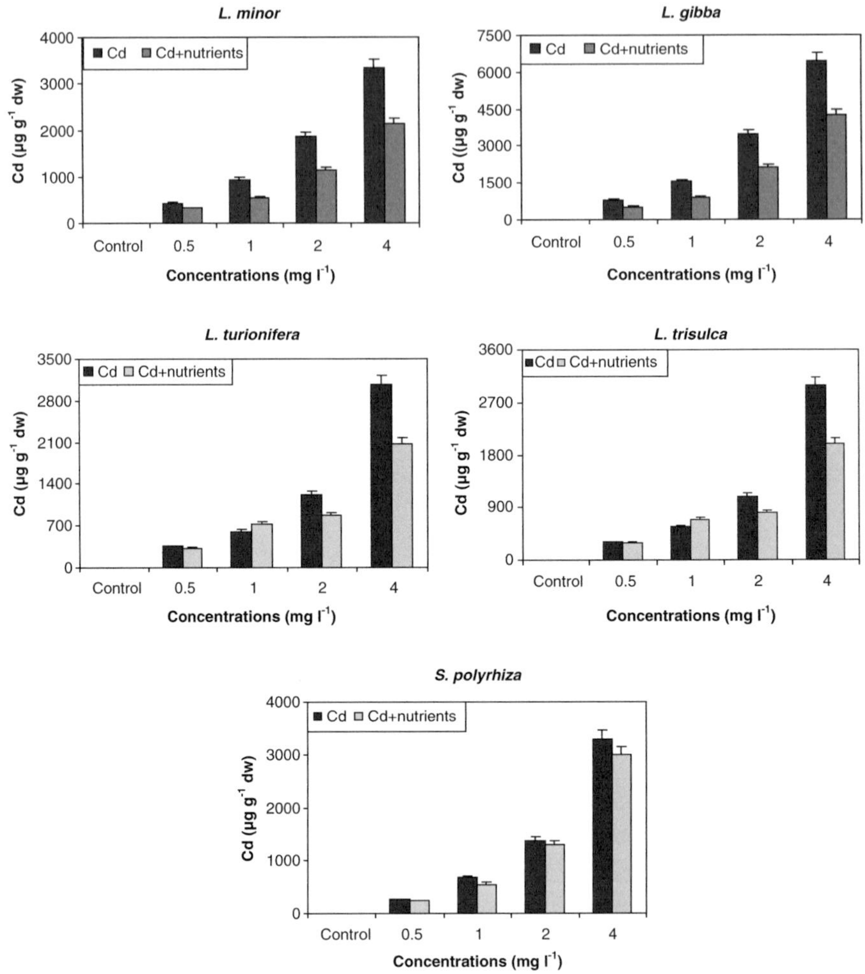

Fig. 3 Accumulation of Cd and Cd+nutrients by *Lemna minor*, *Lemna gibba*, *Lemna trisulca*, *Lemna turionifera* and *Spirodela polyrhiza* exposed to different concentrations over various periods of time. All values are means of triplicates ± S.D. ANOVA significance was set at $p \leq 0.05$

nutrient addition. Göthberg et al. (2004) found high metal concentrations in *Ipomea aquatica* cultivated for human consumption in freshwater courses near Bangkok that were receiving variable amounts of cultural nutrient loads. The authors proposed fertilization as a means to attenuate metal accumulation. Their experimental work concurs with our findings, and showed that nutrient enrichment led to increased *I. aquatica* tolerance to cadmium, lead, and mercury. A decrease in lead and mercury accumulation was observed with increasing concentrations of nutrients, as was observed for lead, nickel and cadmium in the present study.

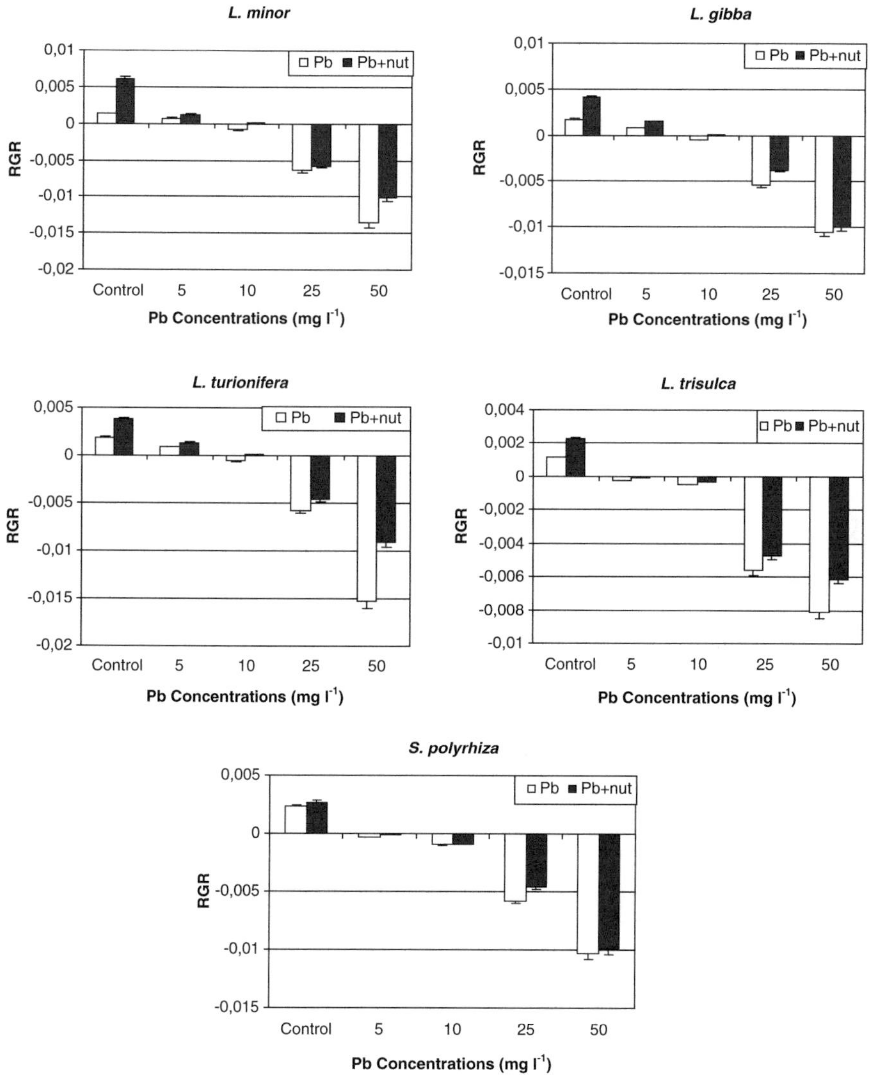

Fig. 4 Relative growth rates of Pb treat by *Lemna minor*, *Lemna gibba*, *Lemna trisulca*, *Lemna turionifera* and *Spirodela polyrhiza*

4.2 Effect of Metals on Photosynthetic Pigments

Chlorophyll concentration in the *Lemna* species were negatively correlated with heavy metal exposures (Figs. 7, 8, and 9). Nutrient enrichment attenuated the observed decrease in chlorophyll concentration by heavy metal exposures. Levels of chl *a* decreased in a Pb concentration-dependent and time-dependent manner, with

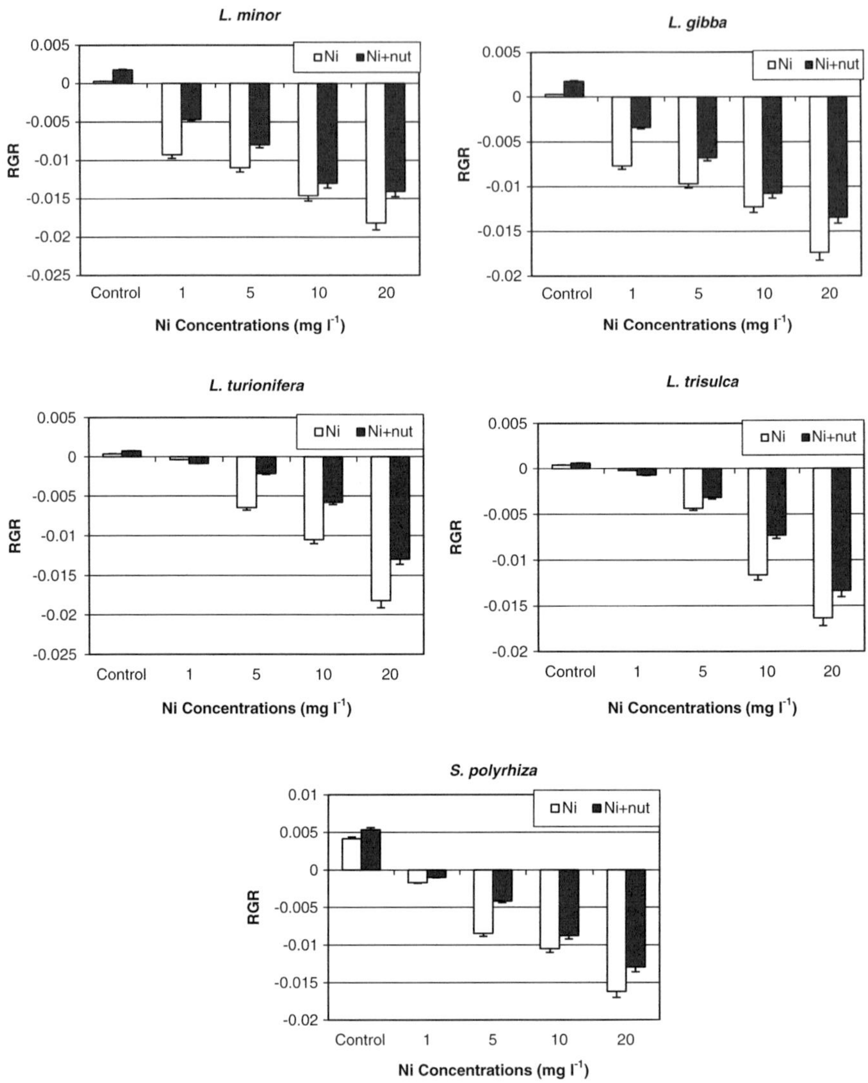

Fig. 5 Relative growth rates of Ni treat by *Lemna minor*, *Lemna gibba*, *Lemna trisulca*, *Lemna turionifera* and *Spirodela polyrhiza*

a minimum value of 0.386 mg g^{-1} in the 50 mg l^{-1} on *L. gibba* (Fig. 7b). In the nutrient-enriched *L. gibba*, 50 mg l^{-1} Pb group, chl *a* levels were reduced to a minimum value of 0.565 mg g^{-1} fresh weight on day 7 (Fig. 7a). Chlorophyll concentration in *Lemna* species were negatively correlated with Ni exposures (Fig. 8). Nutrient enrichment attenuated the observed reduction in chlorophyll concentration caused by Ni exposures. When *L. gibba* fronds were exposed to Ni concentrations

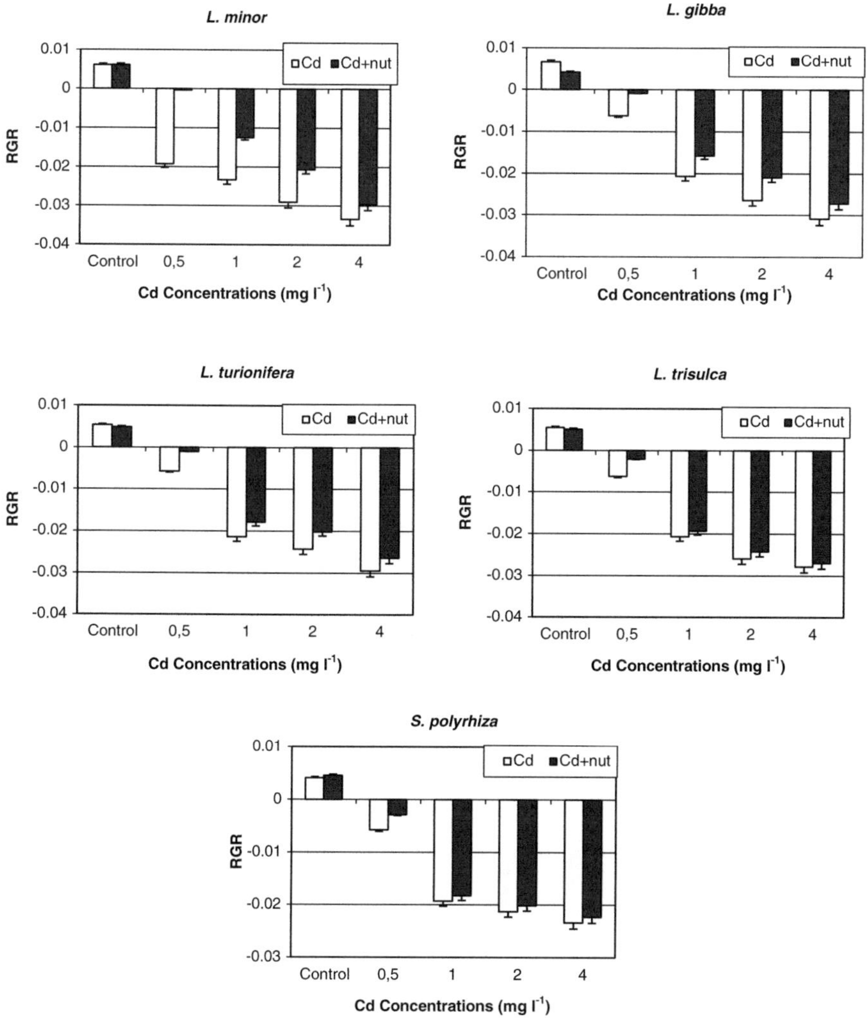

Fig. 6 Relative growth rates of Cd treat by *Lemna minor*, *Lemna gibba*, *Lemna trisulca*, *Lemna turionifera* and *Spirodela polyrhiza*

of 1 mg l^{-1} or higher, a dose-dependent decrease of chlorophyll pigments was also observed, with a minimum chl *a* value of 0.514 mg g^{-1} fresh weight on day 7 at 20 mg l^{-1} compared to 1.601 mg g^{-1} in controls (Fig. 8b). On day 7, the 20-mg l^{-1} Ni-treated groups with nutrient enrichment of chl *a* reached a minimum value of 0.845 mg g^{-1} fresh weight (Fig. 8a). Levels of chl *a* decreased in a Cd concentration-dependent and time-dependent manner, with a minimum value of 0.321 mg g^{-1} in the 4 mg l^{-1} on *L. minor* (Fig. 9b). In the nutrient-enriched *L. minor*, 4 mg l^{-1} Cd

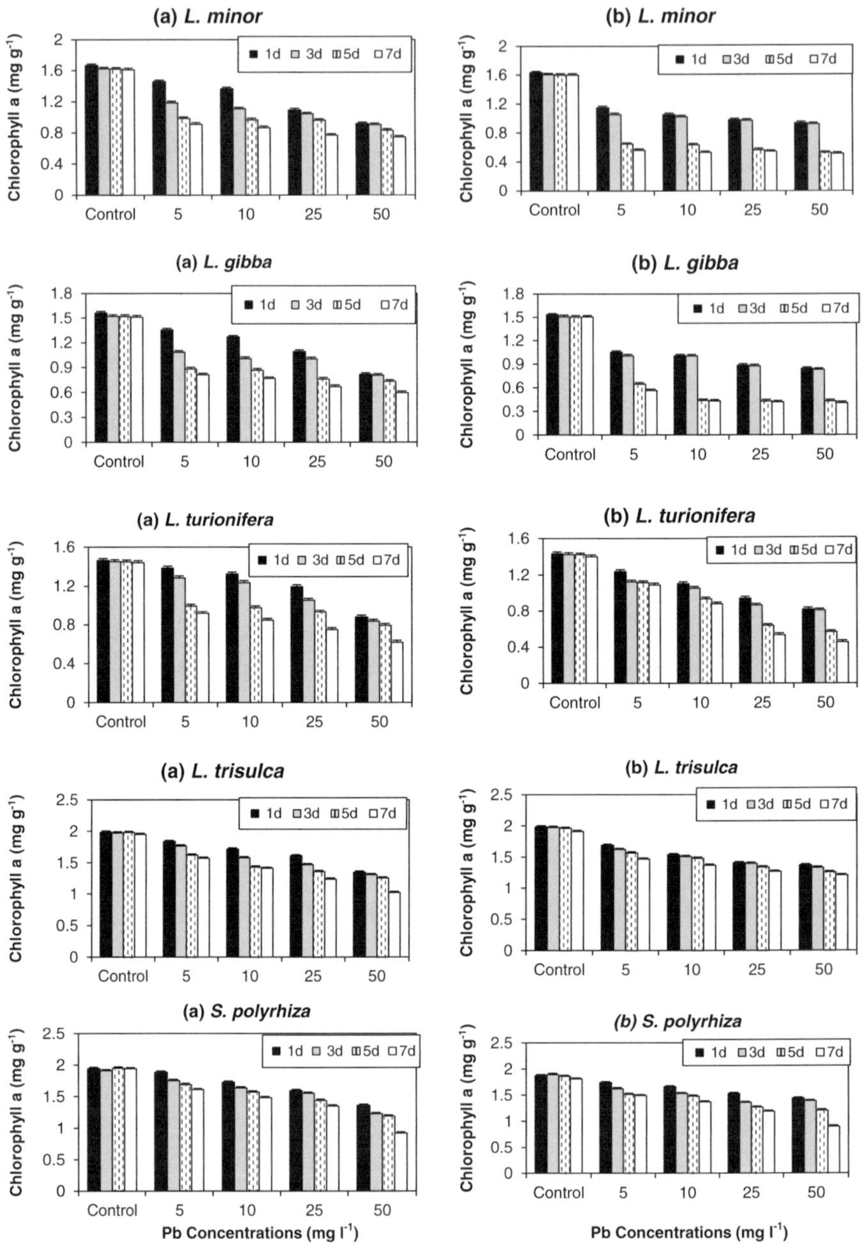

Fig. 7 Effect of different concentrations of Pb+nutrients and Pb (**a** and **b**) on chlorophyll *a* contents of *Lemna minor, Lemna gibba, Lemna trisulca, Lemna turionifera* and *Spirodela polyrhiza*. All values are means of triplicates ± S.D. ANOVA significance was set at $p \leq 0.05$

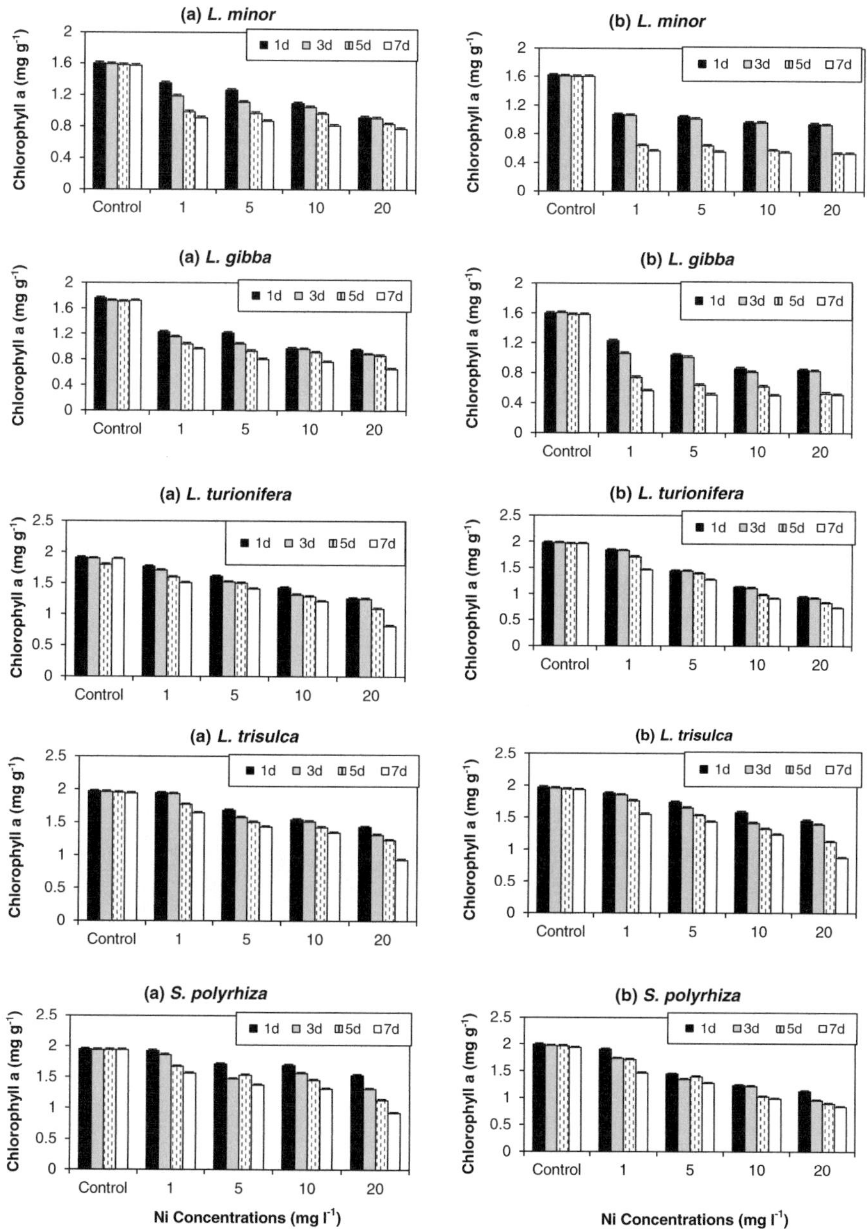

Fig. 8 Effect of different concentrations of Ni + nutrients and Ni (**a** and **b**) on chlorophyll *a* (**a** and **b**), contents of *Lemna minor, Lemna gibba, Lemna trisulca, Lemna turionifera* and *Spirodela polyrhiza*. All values are means of triplicates ± S.D. ANOVA significance was set at $p \leq 0.05$

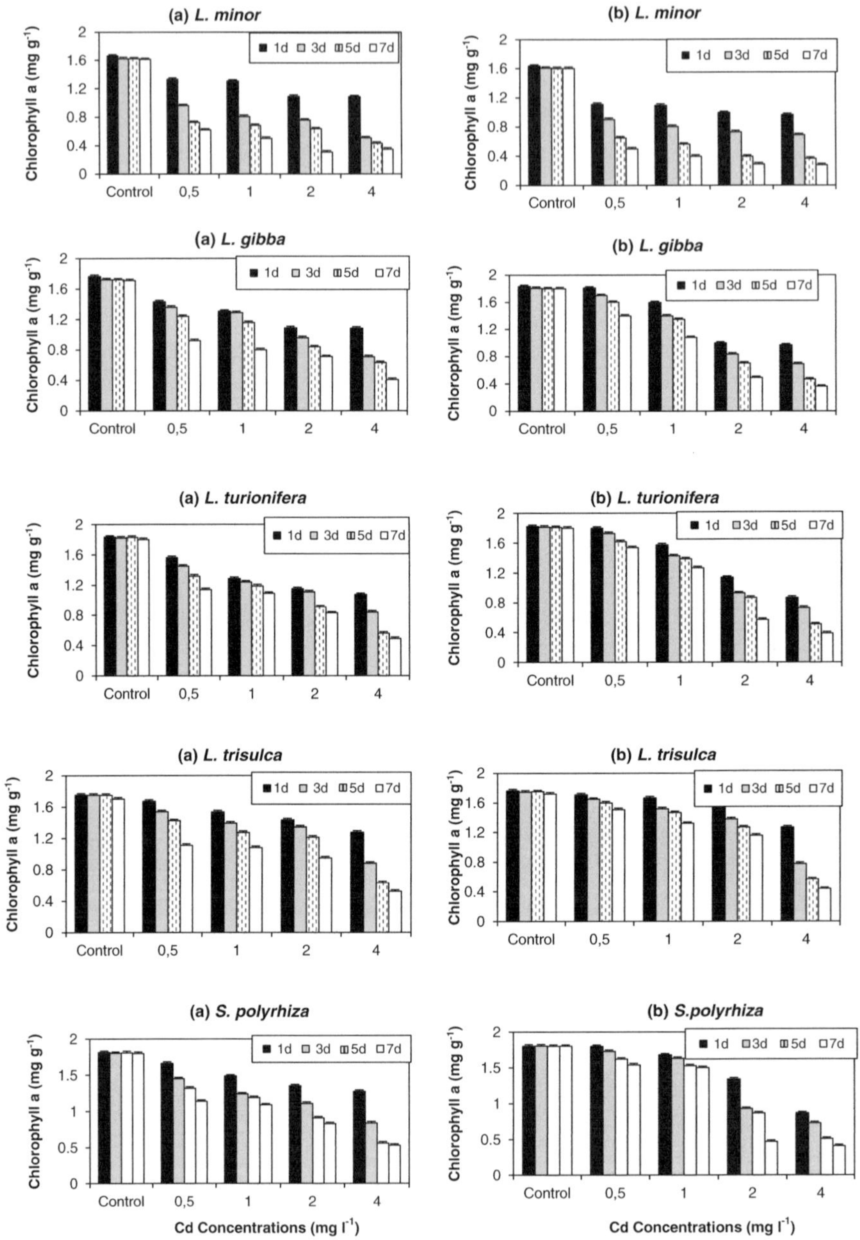

Fig. 9 Effect of different concentrations of Cd + nutrients and Cd (**a** and **b**) on chlorophyll *a* (**a** and **b**), contents of *Lemna minor, Lemna gibba, Lemna trisulca, Lemna turionifera* and *Spirodela polyrhiza*. All values are means of triplicates ± S.D. ANOVA significance was set at $p \leq 0.05$

group, chl *a* levels were reduced a minimum value of 0.425 mg g^{-1} fresh weight on day 7 (Fig. 9a).

Chlorophyll concentrations in *Lemna* species were negatively correlated with heavy metal accumulation, but nutrient enrichment mitigated the decrease in chlorophylls. In fronds treated with Pb, a concentration of 5 mg l^{-1} was sufficient to cause a decrease in pigment molecules, indicating that although lead is a non-essential metal ion, it was toxic for the growth and development of plants and at high levels could be a strong inhibitor of photosynthesis (Frankart et al. 2002; Vavilin et al. 1995). The loss in chlorophyll content could be due to peroxidation of chloroplast membranes or replacement of magnesium in chlorophyll molecules by Pb ions (Mal et al. 2002; Sandmann and Boger 1980).

5 Conclusions

The *Lemna* species are important in the treatment of domestic and industrial wastewater and effluents as well as in the restoration of decommissioned mining sites. They have most of the properties of an ideal phytoremediaton species. In this study, we investigated the toxic effects of lead, nickel and cadmium on the *Lemna* species. Our results showed that nutrient enrichment raised the tolerance of *Lemna* species to metal contamination. This effect has important implications in the use of constructed wetlands for industrial wastewater treatment. Many processes in the metallurgical industry produce wastewater containing high concentrations of metal ions. A higher tolerance would be useful for wastewater treatment as it allows macrophyte growth at metal concentrations that would otherwise impair their development. Nutrient addition will thus aid metal removal by increasing macrophyte production, leading to higher metal uptake by the macrophyte biomass, and thereby enhancing overall biological activity, to reach higher metal retention levels in the detritus fractions.

References

Appenroth KJ, Luther A, Jetschke G et al (2008) Modification of chromate toxicity by sulphate in duckweeds (*Lemnaceae*). Aquat Toxicol 89:167–171

Axtell NR, Stenberg PK, Claussen K (2003) Lead and nickel removal using *Microspora* and *L. minor*. Bioresour Technol 89:41–48

Bianconi D, Pietrini F, Massacci A et al. (2013) Uptake of cadmium by Lemna minor, a (hyper?-) accumulator plant involved in phytoremediation applications E3S web of conferences. In: Proceedings of the 16th international conference on heavymetals in the environment, vol 1, p 4

Burken JG, Schnoor JL (1999) Distribution and volatilisation of organic compounds following uptake by hybrid poplar trees. Int J Phytoremediation 1:139–151

Chang YY, Corapcioglu MY (1998) Plant-enhanced subsurface bioremediation of nonvolatile hydrocarbons. J Environ Eng 124:162–169
Charpentier S, Garbier J, Flaugnatti R (1987) Toxicity and bioaccumulation of cadmium in experimental cultures of duckweed *Lemna minor* L. Bull Environ Contam Toxicol 38:1055–1061
Chaudhary E, Sharma P (2014) Assessment of heavy metal removal efficiency of *Lemna minor*. Int J Innov Res Dev 3:176–178
Chaudhuri D, Majumder A, Misra AK et al (2014) Cadmium removal by *Lemna minor* and *Spirodelapolyrhiza*. Int J Phytoremediation 16:1119–1132
Cheng J, Landesman L, Bergmann BA et al (2002) Nutrient removal from swine lagoon liquid by *Lemna minor* 8627. Trans ASAE 45:1003–1010
Cossu R, Haarstad K, Lavagnolo MC et al (2002) Removal of municipal solid waste COD and NH4-N by phytoreduction: a laboratory-scale comparison of terrestrial and aquatic species at different organic loads. Ecol Eng 16:459–470
DIN (2000) Duckweed growth inhibition test: determination of non-poisonous effect of water constituentandwastewatertoduckweed(*Lemnaminor,Lemnagibba*).DeutschesInstitutfürNormunge.v. DurchBeuthVerlag, Berlin
Duman F, Leblebici Z, Aksoy A (2009) Bioaccumulation of nickel, copper, and cadmium by *Spirodelapolyrhiza* and *Lemnagibba*. J Fresh Ecol 24:177–179
Dushenkov V, Kumar PBAN, Motto H et al (1995) Rhizofiltration- the use of plants to remove heavy metals from aqueous streams. Environ Sci Technol 29:1239–1245
EPA (1996) Aquatic plant toxicity test using Lemna spp.: tiers I and II-OPPTS 850.4400. United States Environmental Protection Agency Prevention, Pesticides and Toxic Substances Unit, New York
Frankart C, Eullaffroy P, Vernet G (2002) Photosynthetic responses of *Lemna minor* exposed to xenobiotics, copper and their combinations. Ecotoxicol Environ Saf 53:439–445
Göthberg A, Greger M, Holm K et al (2004) Influence of nutrient levels on uptake and effects of mercury, cadmium and lead in water spinach. J Environ Qual 33:1247–1255
Goulet RR, Lalonde JD, Munger C et al (2005) Phytoremediation of effluents from aluminum smelters: a study of Al retention in mesocosms containing aquatic plants. Water Res 39:2291–2300
Hadad HR, Maine MA, Natale GS et al (2007) The effect of nutrient addition on metal tolerance in *Salvinia herzogii*. Ecol Eng 31:122–131
Hakeem KR, Sabir M, Ozturk M, Mermut A (2015) Soil remediation and plants: prospects and challenges. Academic/Elsevier, New York, p 724
Halaimi FZ, Kellali Y, Couderchet M et al (2014) Comparison of biosorption and phytoremediation of cadmium and methylparathion, a case-study with live *Lemnagibba* and *Lemnagibba* powder. Ecotoxicol Environ Saf 105:112–120
Hume NP, Fleming MS, Horne AJ (2002) Denitrification potential and carbon quality of four aquatic plants in wetland microcosms. Soil Sci Soc Am J 66:1706–1712
Hunt R (1978) Plant growth analysis. Studies in biology. Edward Arnold Ltd, London
ISO (2001) Water quality – determination of the toxic effect of water constituents and waste water to duckweed (*Lemna minor*) – Duckweed growth inhibition test. ISO TC 147/Sc 5 N, International Organisation for Standardisation, Geneva
Kinnear PR, Gray CD (1994) SPSS for windows made simple. Lawrence Erlbaum Associates, Mahwah
Koles SM, Petrell RJ, Bagnall LO (1987) Duckweed culture for reduction of ammonia, phosphorus and suspended solids from algal-rich water. In: Reddy KR, Smith WH (eds) Aquatic plants for water treatment and resource recovery. Magnolia Publishing, Orlando
Körner S, Lyatuu GB, Vermaat JE (1998) The influence of *Lemnagibba* L. on the degradation of organic material in duckweed-covered domestic wastewaters. Water Res 23:3092–3098
Körner S, Das SK, Veenstra S et al (2003) The capacity of duckweed to treat wastewater: ecological considerations for a sound design. J Environ Qual 32:1583–1590

Kumar PBAN, Dushenkov V, Motto H et al (1995) Phytoextraction- the use of plants to remove heavy metals from soils. Environ Sci Technol 29:1232–1238
Lahive E, Halloran JO, Jansen MAK (2012) Frond development gradients are a determinant of the impact of zinc on photosynthesis in three species of *Lemnaceae*. Aquat Bot 101:55–63
Landesman L (2000) Effects of herbivory and competition on growth of Lemnaceae in systems for wastewater treatment and livestock feed production. University of Louisiana at Lafayette, Loisiana
Landolt E (1982) Distribution patterns within the family Lemnaceae. In: Symoens JJ, Hooper SS, Compere P (eds) Studies on aquatic vascular plants. Royal Botanic Society of Belgium, Brussels
Landolt E (ed) (1986) The family of Lemnaceae: a monographic study (Vol I). Biosystematic investigation in the family of Duckweeds. Geobotanischen Institutes, StiftungRübel, Zürich
Landolt E, Kandeler R (eds) (1987) The family Lemnaceae: a monographic study (Vol 2). Phytochemistry, physiology, application, bibliography. StiftungRübel, Zürich
Leblebici Z, Aksoy A (2011) Growth and heavy metal accumulation capacity of *Lemna minor* and *Spirodelapolyrhiza* (Lemnaceae): interactions with nutrient enrichment. Water Air Soil Pollut 214:175–184
Leblebici Z, Aksoy A, Duman F (2010) Influence of nutrient addition on growth and accumulation of cadmium and copper in *Lemnagibba*. Chem Speciat Bioavailab 22:157–164
Leblebici Z, Aksoy A, Duman F (2011) Influence of salinity on the growth and heavy metal accumulation capacity of *Spirodelapolyrhiza*(Lemnaceae). Turk J Biol 35:215–220
Les DH, Landolt E, Crawford DJ (1997) Systematics of the Lemnaceae (duckweeds): inferences from micromolecular and morphological data. Syst Evol 204:161–177
Mal TK, Adorjan AL, Corbertt AL (2002) Effect of copper on growth of an aquatic macrophytes *Elodea canadensis*. Environ Pollut 120:307–311
Miranda MG, Ilangovan K (1996) Uptake of lead by *Lemnagibba* L.: influence on specific growth rate and basic biochemical changes. Bull Environ Contam Toxicol 56:1000–1007
Miretzky P, Saralegui A, Cirelli AF (2004) Aquatic macrophytes potential for the simultaneous removal of heavy metals (Buenos Aires, Argentina). Chemosphere 57:997–1005
Mkandawire M, Dudel EG (2005) Accumulation of arsenic in *Lemnagibba* L. (duckweed) in tailing water of two abandoned uranium mining sites in Saxony, Germany. Sci Total Environ 336:81–89
Mkandawire M, Dudel EG (2007) Are *Lemna* spp. effective phytoremediation agents? Bioremediation, biodiversity and bioavailability. Glob Sci Books 1:56–71
Mkandawire M, Taubert B, Dudel EG (2004a) Capacity of *Lemnagibba* L. (duckweed) for uranium and arsenic phytoremediation in mine tailing waters. Int J Phytorem 6:347–362
Mkandawire M, Lyubun YV, Kosterin PV et al (2004b) Toxicity of arsenic species to *Lemnagibba* L. and the influence of phosphate on arsenic bioavailability. Environ Toxicol 19:26–35
OECD (2002) OECD guidelines for the testing of chemicals: *Lemna* sp. Growth inhibition test. Organisation for Economic Corporation and Development, Berlin
Rahman MA, Hasegawa H, Ueda K et al (2007) Arsenic accumulation in duckweed (*Spirodelapolyrhiza* L.): a good option for phytoremediation. Chemosphere 69:493–499
Rahman MA, Hasegawa H, Ueda K et al (2008) Arsenic uptake by aquatic macrophyte *Spirodelapolyrhiza* L.: interactions with phosphate and iron. J Hazard Mater 160:356–361
Rai UN, Tripathi RD, Vajpayee P et al (2002) Bioaccumulation of toxic metals (Cr, Cd, Pb and Cu) by seeds of *Euryale ferox* Salisb (Makhana). Chemosphere 46:267–272
Salt DE, Blaylock M, Kumar NPBA et al (1995) Phytoremediation: a novel strategy for the removal of toxic metals from the environment using plants. Biotechnology 13:468–474
Sandmann G, Boger P (1980) Copper-mediated lipid peroxidation processes in photosynthetic membranes. Plant Physiol 66:797–800
Sasmaz A, Obek E (2012) The accumulation of silver and gold in *Lemnagibba* L. exposed to secondary effluents. ChemErde Geochem 72:149–152

Sekomo CB, Rousseau DPL, Saleh SA et al (2012) Heavy metal removal in duckweed and algae ponds as a polishing step for textile wastewater treatment. Ecol Eng 44:102–110
Sharma SS, Gaur JP (1994) Potential of *Lemnapolyrrhiza* of removal of heavy metals. Ecol Eng 4:37–43
Smith MD, Moelyowati I (2001) Duckweed based wastewater treatment (DWWT): design guidelines for hot climates. Water Sci Technol 43:291–299
Sooknah RD, Wilkie AC (2004) Nutrient removal by floating aquatic macrophytes cultured in anaerobically digested flushed dairy manure wastewater. Ecol Eng 22:27–42
Steveninck RFM, Steveninck ME, Fernando DR (1992) Heavy-metal (Zn, Cd) tolerance in selected clones of duckweed (*Lemna minor*). Plant and Soil 146:271–280
Stout LM, Nusslein K (2005) Shifts in rhizoplane communities of aquatic plants after cadmium exposure. Appl Environ Microbiol 71:2484–2492
Susarla S, Medina VF, McCutcheon SC (2002) Phytoremediation: an ecological solution to organic chemical contamination. Ecol Eng 18:647–658
Szabo SL, Braun M, Borics G (1999) Elemental flux between algae and duckweed (*Lemnagibba*) during competition. ArchivfürHydrobiologie 146:355–367
Vatta G, Rota R, Boniardi N et al (1995) Dynamic modelling of treatment plants based on *Lemnagibba*. Chem Eng J Biochem Eng J 57:37–48
VavilinDV PVA, Matorin DN et al (1995) Sublethal concentration of copper stimulates photosystem II photoinhibition in *Chlorella pyrenoidosa*. J Plant Physiol 146:609–614
Vymazal J (ed) (2005) Natural and constructed Wetlands: nutrient, metal and management. Backhuys Publishers, Prague
Wang Q, Cui Y, Dong Y (2002) Phytoremediation of polluted waters potentials and prospects of wetland plants. Acta Biotechnol 22:199–208
Witham FH, Blaydes DF, Deulin RM (1971) Experiments in plant physiology. Van Nostrand Reinhold Company, New York
Zayed A, Gowthaman S, Terry N (1998) Phyoaccumulation of traces elements by Wetland plants: I. Duckweed. J Environ Qual 27:715–721
Zhang X, Hu Y, Liu Y et al (2011) Arsenic uptake, accumulation and phytofiltration by duckweed(*Spirodela polyrhiza* L.). J Environ Sci 23:601–606
Zimmo OR, Van der Sten NP, Gijzen HJ (2004) Nitrogen mass balance across pilot-scale algae and duckweed-based wastewater stabilisation ponds. Water Res 38:913–920

Temporal Variation of Biological Oxygen Demand (BOD), Chemical Oxygen Demand (COD), and pH Values in Surface Waters of Gölcük-Kocaeli, Turkey

Funda Dökmen

Contents

Abstract In this study, eight different surface water resources located in five villages of Gölcük, Kocaeli-Turkey, were tested between 2007 and 2008 to determine the biological oxygen demand (BOD_5), chemical oxygen demand (COD) and pH values. Temporal variation in these parameters reflects changes in precipitation. The test results showed that the biological oxygen demand (BOD_5), chemical oxygen demand (COD) and pH levels varied between 2 and 11 mg L^{-1}, 11–22 mg L^{-1} and 8.33–5.15, respectively. Data showed a strong relationship between biological oxygen demand (BOD_5), chemical oxygen demand (COD) and pH concentrations and amount of precipitation. Based on this fact, water quality of the surface water resources was also examined.

Keywords Drinking water • Irrigation water • Rainfall • Seasonal precipitation • Spring water • Water quality • Water classification

F. Dökmen (✉)
Vocational School of Food and Agricultural, University of Kocaeli,
Campus of Arslanbey, 41285 Arslanbey-Kartepe/Kocaeli, Türkiye
e-mail: f_dokmen@hotmail.com; fun@kocaeli.edu.tr

M. Öztürk et al. (eds.), *Plants, Pollutants and Remediation*,
DOI 10.1007/978-94-017-7194-8_15

1 Introduction

As the world population grows and the demand for food increases, agriculture in arid areas is required to improve its productivity, while its development is severely restricted by water availability (Watanable et al. 2006).

Pollution of environment is one of the most important problems of the world. Determination of pollution sources is very important to understand amount and results of water pollution.

Water quality is related to its surrounding natural elements and phenomena, such as soils, cover of plants, and flora as well as meteorological, hydrological, geological conditions of the region. Any change in these conditions, which may result from global climate change, inevitably affects the quality of water resources.

Changes in precipitation quality affect surface water resources depending on seasons. The precipitation amount varies temporally and spatially. The pH is one of the very important factors which fix water quality. Generally, pH of natural waters has fixed concentration of soluble carbonate, bicarbonate and carbon dioxide (Dökmen and Kurtuluş 2009).

Biochemical Oxygen Demand (BOD_5) is a chemical procedure for determining how fast biological organisms use up oxygen in a body of water. It is used in water quality management and assessment, ecology and environmental science (Sawyer et al. 2003). Similarly, the Chemical Oxygen Demand (COD) test is commonly used to indirectly measure the amount of organic compounds in water. Most applications of chemical oxygen demand (COD) determine the amount of organic pollutants found in surface water, making chemical oxygen demand (COD) a useful measure of water quality (Sawyer et al. 2003).

The aim of this work was to study the influence of precipitation events and quantity on the surface water resources in terms of biological oxygen demand (BOD_5), chemical oxygen demand (COD), and pH.

2 Materials and Methods

The research area is located to the east of Istanbul between latitudes 41° 00′ and 41° 45′ N and between longitudes 29° 00′ and 29° 45′ E. Gölcük is located near the İzmit-Bursa Government Highway 17 km south of İzmit town of Kocaeli City in Turkey. The north of Gölcük is surrounded by the İzmit Gulf. The investigation area is a mountainous area surrounded by valleys (Fig. 1).

The investigation area has the Marmara Climate. The summers are hot and dry, and the winters are cold and rainy. In the research area, annual mean temperature is 23.6 °C and annual mean of precipitation is 808.4 mm $year^{-1}$. In the long period years (1975–2006), total annual potential evaporation is 540.5 mm water $year^{-1}$ and total annual real evaporation of research area is 476.9 mm water. The maximum and the minimum temperatures of the last 20 years are 15.5 °C and 13.7 °C, respectively (Anonymous 2007).

Fig. 1 The map of research area

In this study, eight different surface water resources located in five different villages in the vicinity of Gölcük-Kocaeli in Turkey were investigated. The content of biological oxygen demand (BOD5), chemical oxygen demand (COD), and pH were examined for the relation to amount of rainfall. This work was accomplished analyzing eight water samples every month from the sources in Ümmiye, Mamuriye, Ferhadiye, Nüzhetiye and in Yeniköy villages in the period of 2007–2008 years. Standard sampling methods were used in the work, and the samples were analyzed according to standard methods (APHA 1985).

3 Results and Discussion

The 13 month average values of biological oxygen demand (BOD_5), chemical oxygen demand (COD), and pH in the samples taken from eight different surface water sources in Kocaeli-Gölcük and surrounding villages are given in Table 1. The names, places and characteristics of the water sources are shown in Table 1. The average values of the samples for 13 months were as follows: biological oxygen demand (BOD_5) 5.37–3.62 mg L^{-1}, chemical oxygen demand (COD) 14.25–16.30 mg L^{-1} and pH 7.15–7.34.

Table 1 The water sources and some of their characteristics

Spring no.	Location	Name of the spring	Flow (m s^{-1})	Soil characteristics	pH	BOD_5 (mg L^{-1})	COD (mg L^{-1})
1	Ümmiye	Şelale	0.5	Sandy	7.34	4.75	16.3
2	Mamuriye	Altınoluk	1.0	Sandy-Lime	7.31	3.87	14.87
3	Ferhadiye	Çürükbayır	0.5	Clayish-Lime	7.15	3.87	14.50
4	Nüzhetiye	Karanlıkdere	0.3	Lime	7.21	3.87	14.50
5	Nüzhetiye	Değirmendere	0.9	Lime-Sandy	7.29	3.62	14.25
6	Nüzhetiye	Sakarbıçkı 1	0.5	Sandy-Lime	7.23	5.37	16
7	Nüzhetiye	Sakarbıçkı 2	0.5	Clayish-Lime	7.28	4	14.50
8	Yeniköy	Havuzlubahçe	0.9	Clayish-Lime	7.27	4	15.37
Mean			0.63		7.26	4.16	15.03

The biological oxygen demand (BOD_5) values were 2–7 mg L^{-1} between January and December, 2008, and increased in the winter period proportionally with the precipitation rate for all of the water resources (Fig. 2). The value of biological oxygen demand (BOD_5) fluctuates between 2 and 4 mg L^{-1} (number 3, 4, 5, 7) and 4–10 mg L^{-1} (number 1, 2, 6, 8). According to standards, sources 3, 4, 5 and 7 are first quality of water (maximum limit is 4 mg L^{-1}) and sources 1, 2, 6 and 8 are second quality. The quantity of water resources increases and mixes solid and organic matter as intensive due to rainfall in the winter and autumn seasons. For this reason, the value of biological oxygen demand (BOD_5) can be very high than summer months. The loadings of biological oxygen demand (BOD_5) were much higher in the wet season (June-August) than in the dry season (September-May). Generally, of the annual biological oxygen demand (BOD_5) loading more than 50 % occurred in the wet season, whereas only 30–40 % in the dry season.

Related to the amount of precipitation and biological oxygen demand (BOD_5) concentrations were studied on nine streams in New Jersey by Dunlap (1976). Their relationships were determined so closely related to biological oxygen demand (BOD_5) load. Biological oxygen demand (BOD_5) loadings in the lower part of the Nakdong River, Mulgum, Korea Republic were investigated at weekly and biweekly intervals from June 1994 to June 1998 by SangKyun et al. (1999) Inter-annual variability of biological oxygen demand (BOD_5) loadings was remarkable during this period. When annual precipitation was high (1348 mm) in 1997, the loadings of biological oxygen demand (BOD_5) were also high (BOD_5 133,599 t $year^{-1}$). When the annual precipitation was exceptionally low (676 mm) in 1994, biological oxygen demand (BOD_5) loadings were also low (BOD_5 61,065 t $year^{-1}$). Therefore, yearly variability of nutrient loadings could be largely affected by the magnitude of rainfall during the summer of each year.

An increase of 15.0–20.0 mg L^{-1} was detected in chemical oxygen demand (COD) concentration in the water resources in spring and autumn depending on rainfall as depicted in Fig. 3. The chemical oxygen demand (COD) concentration decreased when the amount of rainfall was 110.0 kg m^{-2} in December 2007 and

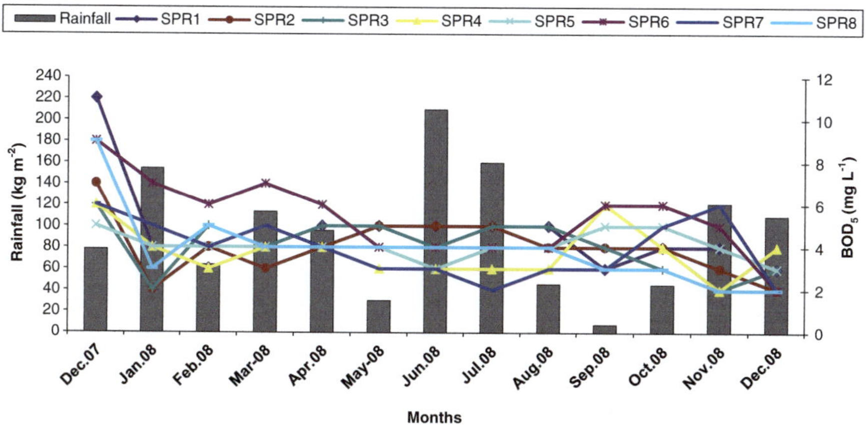

Fig. 2 The change of BOD_5 values from the water source depending on rainfall

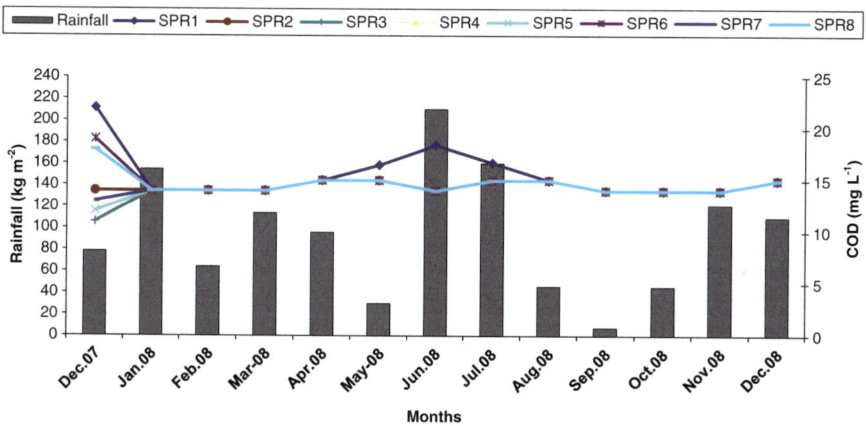

Fig. 3 The change of COD values from the water source depending on rainfall

January 2008. All of the water resources were shown so closely related to chemical oxygen demand (COD) concentration. According to classification, all the water sources are of first class (maximum limit is 25 mg L^{-1}). These values indicate that the amount of the chemical and biochemical oxidized material is very little.

All the water sources satisfy the first class water quality for pH. The pH values of all the water sources are suitable for drinking water and irrigation water in accordance with TSE 266 (pH 7.0–8.5) (Anonymous 1986). The result of 13 months precipitation rate shows that the pH values between December 2007 and June 2008 are 8–8.5 for all the water resources and decreased to 5–6 from July 2008 to October 2008 due to precipitation rate of 130.0 kg m^{-2} (Fig. 4). The acidic concentration of the water has increased due to the amount of organic and inorganic materials. In

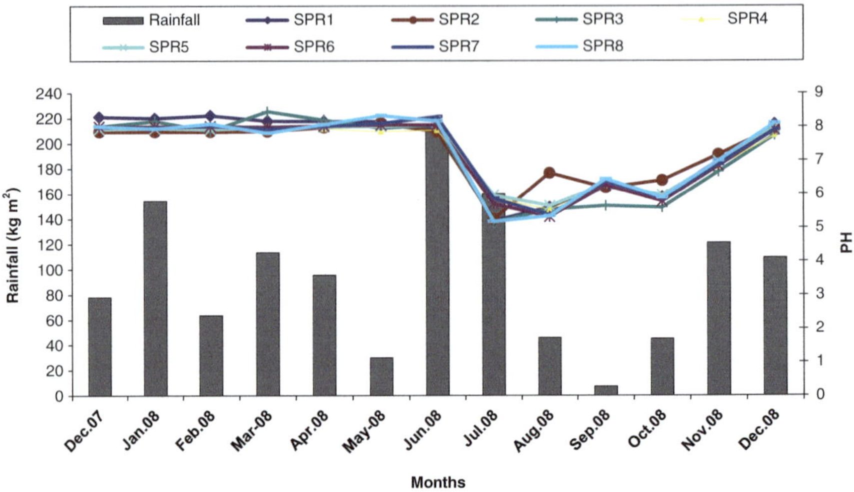

Fig. 4 The change of pH values from the water source depending on rainfall

summer, since the amount of organic and inorganic materials mixed with the water decreases, the pH value of the water changes. This causes the pH value of the water sources to be more than 7.

4 Conclusion

In this study, to analyzed the relationship between biological oxygen demand (BOD_5), chemical oxygen demand (COD), and pH values in surface waters and also variations of parameters' concentration were related to precipitation. Not only the rainfall change impacts, but also implication of water quality in local environment is to be discussed. Based on the local knowledge on living with the resources and environment of the region, conditions for sustainable water quality and management are to be suggested. In practice, quality of surface water resources related to environmental problems and amount of precipitation are described including their aspect, reason, and measures in the past and present, especially with special reference to climate changes. Surface water is naturally replenished by precipitation and naturally lost through discharge to evaporation, and sub-surface seepage into the groundwater (Fetter 2009). The assessment of water quality has vital importance for agricultural production soil and water management, and sustainable use of natural resources one of the most important natural resources is surface water in a plain and watershed (Özbek 2004). Furthermore, understanding hydrological and hydrochemical processes has become increasingly more challenging with high rates of population growth and the subsequent alteration of the environment (Serengil and Özyuvaci 2000). It observed that the rainfall during winter and spring periods affects the pH values to increase in alkalinity, and during summer and autumn periods to decrease in acidity. The monthly total amount (kg m^{-2}) of rainfall in winter and summer showed some differences. There was no significant change on pH values, and the values were similar for all water resources. The reason of the pH value decrease in acidity is existence of different chemical substances and their concentrations of sulphide and carbon dioxide in summer rainfalls. Biological oxygen demand (BOD_5) is similar in function to chemical oxygen demand (COD), in that both measure the amount of organic compounds in water. However, chemical oxygen demand (COD) is less specific, since it measures everything that can be chemically oxidized, rather than just levels of biologically active organic matter (Anonymous 2006). The soil erosion which occurs in the rainy months plays an important role in the value of biological oxygen demand (BOD_5) and chemical oxygen demand (COD). The smaller biological oxygen demand (BOD_5) values indicate that the water is clear and the microorganisms do not exhaust the organic materials.

References

Anonymous (1986) İçme suları standartı (standard of drinking water), TSE 266/UDK 663–6:543, Nisan-1986, Ankara

Anonymous (2006) Chemical methods in the determination of BOD7 in a pulp and paper Mill's wastewaters. J Autom Methods Manag Chem (Hindawi Publishing Corporation) 2006, (Article ID 903384):1–5. doi:10.1155/JAMMC/2006/90384

Anonymous (2007) Kocaeli meteoroloji istasyonu kayıtları (recorders of Kocaeli wheather station), Kocaeli

APHA (1985) Standard methods for the examination of water and wastewater. American Public Health Association, Washington, DC

Dökmen F, Kurtuluş C (2009) Temporal variation of nitrate, chlorine and pH values in surface waters. J Food Agric Environ 7(2):689–691

Dunlap VD (1976) The influence of precipitation on biochemical oxygen demand on nine streams in New Jersey, PhD. thesis, 1976, http://geog.rutgers.edu/dissertation_abstract/1976

Fetter CW (2009) Applied hydrogeology, 4th ed. http://en.wikipedia.org/wiki/Surface_water, Last modified: 27 March 2009

Özbek AK (2004) Assessment of some soil quality parameters for Daphan plain soils. International Soil Congress (ISC) on "Natural Resource Management for Sustainable Development", Abstract Book, ISBN: 975-96629-2-2, June 7–10, Erzurum, Turkey, p 47

SangKyun L, SangHo C, HyunWoo K, Kyong H, GeaJae J (1999) Inter-annual variability of nutrient loading in the lower Nakdong River, Mulgum, Korea. Acta Hydrobiol Sinica 23:17–23

Sawyer CN, McCarty PL, Parkin GF (2003) Chemistry for environmental engineering and science, 5th edn. McGraw-Hill, New York. ISBN 0-07-248066-1

Serengil Y, Özyuvacı N (2000) Water quality management in Balabandere creek catchment, international conference on the future of the mediterranean rural environment: prospects for sustainable land use and management, Abstract Book, Cranfield University, School of Agriculture Food and Environment, 8–11 May 2000, Menemen, Turkey, p 82

Watanable T, Nagona T, Kanber R (2006) Innovated cross-disciplinary approach to impact assessment of climate change on agricultural production systems in arid areas, international symposium on water and land management for sustainable irrigation agriculture, Keynote papers and abstracts, Çukurova University, April 4–8, Adana-Turkey, p 79–88

Ambient Ozone Levels in the Eastern Mediterranean Region and Assessment of Its Effect on the Forested Mountain Areas of Southern Turkey

Evrim Akkoyunlu, Rukiye Tıpırdamaz, Saime Başaran, Halil Sarıbaşak, Dudu Özkum, and Gülen Güllü

Contents

Abstract Ambient ozone measurements were conducted from the beginning of May 2003 to the end of October 2004 in the forest ecosystems of west Mediterranean mountains of Turkey. The ozone concentrations were estimated using a passive sampling method from the bottom of the valley (altitude 10 m) to the top of the mountain (1950 m), over 20 sites distributed all over the study area. Active continuous measurements of ozone were done at one of the monitoring site which helped to calibrate the concentrations of ozone obtained by passive method. The results

E. Akkoyunlu • G. Güllü
Department of Environmental Engineering, Hacettepe University, 06532 Ankara, Turkey

R. Tıpırdamaz (✉)
Department of Biology, Hacettepe University, 06532 Ankara, Turkey
e-mail: tuz@hacettepe.edu.tr

S. Başaran • H. Sarıbaşak
Ministry of Environment and Forestry, West Mediterranean Research Institute, Ankara, Turkey

D. Özkum
Near East University, Mersin 10, Nicosia, Northern Cyprus, Turkey

M. Öztürk et al. (eds.), *Plants, Pollutants and Remediation*,
DOI 10.1007/978-94-017-7194-8_16

indicated that ozone concentrations were in the range of 19–410 μg m^{-3}. The yearly average of ozone was 89.37 ± 71.25 μg m^{-3}. Generally, maximum ozone concentrations were measured at high altitude stations (1950 amsl) during spring and summer seasons, and minimum concentrations were measured at the locations near to the road traffic. The symptoms of probable ozone injury in the vicinity of passive ozone samplers were examined by collecting needle and leaf samples from the main native plants. According to the visual inspection of leaves and measurement of photosynthetic pigments of control plants and symptomatic leaves, out of 41 species of native plants, 11 species were identified as potential bioindicators of ozone. Ozone concentrations in the west Mediterranean part of Turkey appear to be high enough and of sufficient duration to cause foliar injury in a wide variety of native plants.

Keywords Ozone • Bioindicators • Forest • Mediterranean • Photosynthetic pigment

1 Introduction

Ozone has been proved to cause foliar injury in a variety of native forest species in different Southern European countries as a consequence of the particular air pollution chemistry and dynamics of this region (Guderian et al. 1989; Steubing et al. 1989). Due to high solar radiation, widespread emission of the precursors of photochemical oxidants (e.g. NOx, VOCs, NMHC) and prolonged drought periods, the Mediterranean forests become potentially sensitive to climatic fluctuations and changes (Ozturk and Secmen 1981; Ozturk 1989; Çepel 1989; Butkovic et al. 1990). Owing to the complex relationships existing between species sensitivity, ozone exposure and doses, length of the vegetative periods, the impact areas on the Mediterranean coastal areas have not been identified yet (Barnes et al. 2000). Among the Mediterranean countries, only in Spain, Israel and Greece, poor tree condition attributed to the action of ozone have been reported (Gimeno et al. 1995; Velissariou et al. 1996). Recent surveys carried out in southern Switzerland reported ozone-like symptoms on a variety of broadleaved species (Bussotti and Ferretti 1998).

The results of a 7 year monitoring across a 16 × 16 km network in Southern Turkey, Antalya, show that defoliation is increasing (Güllü et al. 2001). During 1991 survey, 40 % of the trees were in the moderately damaged class, whereas, in 1997 survey, it was determined that 80 % of the trees were in the moderately damaged class. According to the results of this study, as there is an extensive neutralization of acidity by local soil and long range trasported Saharan dust which are rich in carbonates, the reasons of forest decline in the study region cannot be attributed to the acidic precipitation (Güllü et al. 2001; Al-Momani et al. 1998). The acidic deposition could not be accepted as the significant factor contributing to forest health problems. From October 1994 to January 1996, a continuous hourly ozone sampling was performed on the Mediterranean coast of Turkey (30.34 °E, 36.47 °N). In this study, the ozone levels in the region show values over the critical level of 10 ppm.h which is the level accepted by European Community as a threshold for the forests (Güllü et al. 2003). As the measured ambient ozone levels are higher than critical level, it has been proposed that ozone may be one of the antropogenic stress factors causing forest decline in Southern Turkey observed during this study.

The main purpose of this study was to determine if ozone may contribute to forest decline in Southern Turkey. To accomplish this goal, ambient ozone concentrations were measured with passive and active monitors and vegetation near the passive samplers and active monitor was evaluated for probable ozone injury symptoms during May and early September. Native plants, growing in situ, are preferred for vegetation surveys. These established detector bioindicators visibly respond to ozone only when sufficient soil moisture and climatic conditions allow uptake of enough ozone for a sufficient time to allow inactivation of antioxidant defense mechanisms (Manning et al. 2002 and Davison and Barnes 1998).

2 Materials and Methods

2.1 *Ozone Monitoring*

During three weekly sequences from May 2003 to October 2004, passive ozone samplers (Passam) and a continuous active monitor (Environement 41 M-LCD) were used for monitoring sites in forested areas, along the slopes of Taurus mountains in Antalya (Fig. 1). Samplers were placed in open field stations within the forest area, at a height of 3 m above ground level. The samplers were provided with rain shelters, which also protected them from the wind.

Co-located measurements of O_3 with the photometric analyzer and passive samplers indicate good agreement between the two methods ($p < 0.05$, $R^2 = 0.891$). Some data periods were excluded from the data set due to malfunctioning of the equipments.

2.2 *Collection and Storage of Foliar Samples for Physiological Measurements*

In the vicinity of the location where the ozone passive sampler is installed, foliar samples for physiological measurements were collected from main tree species within 0.5 km of the ozone monitoring stations during late summer (September, 2003) in the case of broadleaved species, and spring (May, 2003) for conifers. Samples were taken from the upper sun exposed part of the crown, for conifers, second and third year of the foliar parts were sampled. Symptomatic leaf or needle

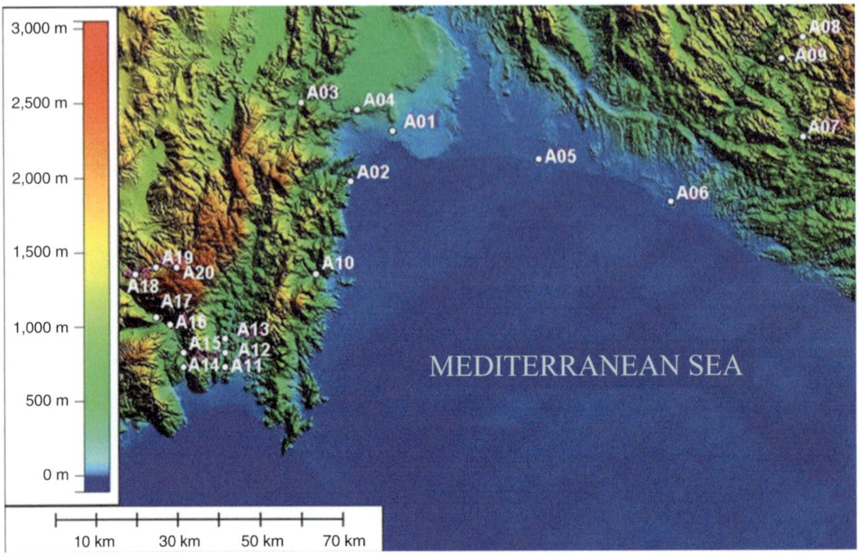

Fig. 1 The West Mediterranean study area in southern Turkey, showing the relative position of the ozone passive and continuous monitoring sites and their altitudes

material was collected for later analyses in the laboratory. Asymptomatic material was also sampled on the same individuals in a nearby branch for control.

Samples were collected 1–2.5 m above ground from the sun side, stored in paper bags at room temperature, and moistened three or four times daily. After identifying all tree, bush and perennial species growing inside the plots and along the edges and examining the foliage up to 2 m for ozone-like injury, symptomatic and asymptomatic branch and leaf samples were photographed with Canon Powershot SD600 digital camera and then dried in a plant press.

2.3 Symptom Identification and Injury Scoring

Visible ozone like symptoms was identified according to the instructions given in Submanual for the Assessment of Ozone Injury on European Forest Ecosystems (http://www.icp-forests.org/Manual.htm). The identification of visible ozone injury was based on the examples given in the "Submanual", on the current list of sensitive species (Skelly et al. 1999). The ozone injury criterion for coniferous species was the presence of photo bleaching and/or mottling on the light-exposed portion of the needles. Among angiosperm species, various symptoms were observed on the adaxial and light-exposed side of leaves. These include light-green or reddish to brownish stippling between leaf second-order veins, variable bronzing, whole leaf reddening or intercostal band necroses. Additional characteristics for discriminating between ozone injury and other abiotic stress factors were the increase in degree of injury with leaf age, with lower position on the branch, and with light exposure. Biotic origins of the symptoms were excluded by close examination of leaves and needles using a 12_ magnifying lens and looking for fungal fruit-bodies and for insect and mite individuals or evidence of them such as sucking injury, scats and eggs. If symptoms were found, the number of affected plants/species was determined and an estimate of severity of incidence was made, using a simple 0–4 scale, where 0 = no symptoms, 1 = 1–5 %, 2 = 6–25 %, 3 = 26–50 %, and 4 = 50–100 %.

2.4 *Quantification of Photosynthetic Pigments*

Chl *a* and *b* and carotenoid concentrations were measured in leaf extracts with 100 % acetone. The absorbance of the extracts was measured with a spectrophotometer at 662, 645 and 470 nm. The individual levels of Chl *a*, Chl *b* and Car($x+c$) were calculated by means of Lichtenthaler equations (Lichtenthaler 1987).

2.5 *Statistical Analyses*

An ANOVA analysis for each parameter and species was performed to evaluate O_3 effects on plant health related parameters. Also, a combined analysis involving all the assessed species was performed to assess whether plant sensitivity to O_3 exposure could be related to plant family and their location; therefore a three-way ANOVA analyses was carried out considering ozone exposure and family as factors. When significant differences ($p<0.05$) were detected, the differences between means were assessed using the least significant difference (LSD) test. All statistical analyses were carried out using Statgraphics Plus v.3.1 software.

3 Results and Discussion

Across the study region, maximum ozone concentrations were observed during August, 2004, 410.90 μg m^{-3}, and minimum in December, 2003, 19.03 μg m^{-3}. Monthly average ozone concentrations show a marked seasonal variation with maximum concentrations occurring from July to November (Fig. 2). During this period the average ozone concentration generally exceeds 60 μg m^{-3}. In other months the average concentration is somewhat lower. Observed summer-time peak cycle is due to the effect of continental emissions causing an ozone reduction in winter by titration with NO and ozone excess in summer due to photochemical formation. Summer-time peak ozone cycle has an adverse effect on vegetation as the time of the ozone increase often coincides with the growing season.

The diurnal variation is also pronounced at Antalya (Fig. 2). Here the mean of the hourly average concentrations for May 2003–October 2004 show the concentration increases from about 15 μg m^{-3} before sunrise to an average maximum of about 55 μg m^{-3} at 15:00. There after the concentration decreases rapidly to sunset at about 19:00, it is followed by a more gradual decrease to the minimum at sunrise. The average concentration is above 40 μg m^{-3} for about 11 h.

It is well-known that ozone concentrations are highly dependent on meteorological variables, increasing with higher air temperature and solar radiation and decreasing in cloudy and rainy periods (e.g. Sanz et al. 2007). There is a significant correlation between ozone and temperature ($r = 0.745$) and solar radiation ($r = 0.877$). This reveals that warm and sunny weather enhances the ozone concentration because of the emission of volatile hydrocarbons which increase with temperature, higher solar radiation increases photochemical processes, and high temperature results in more rapid chemical ozone formation (Fiala et al. 2003).

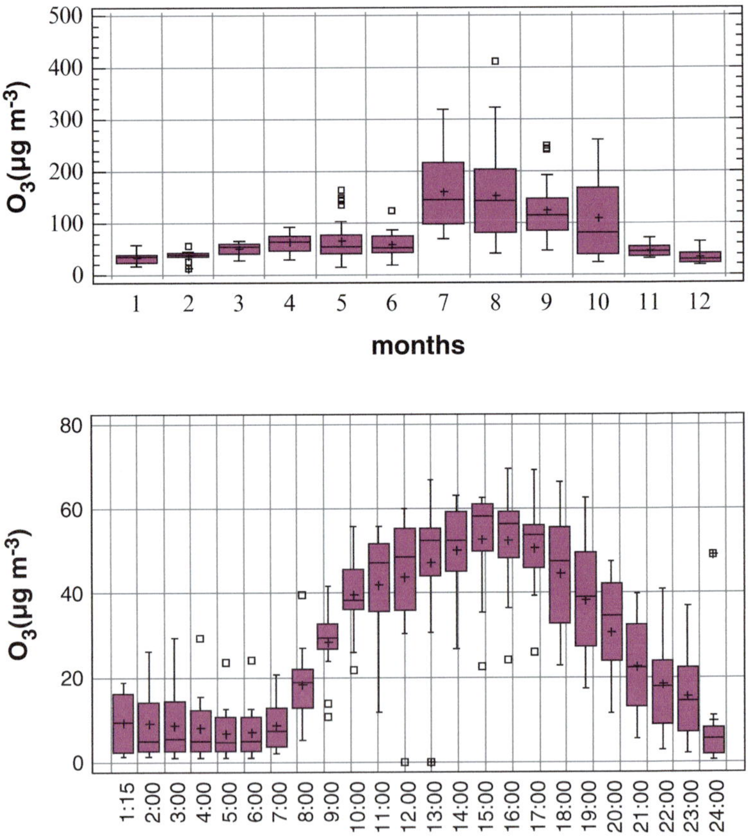

Fig. 2 Mean monthly ozone concentration at Antalya for 2003–2004 (*top*) and mean hourly concentrations in 2003–2004 (*bottom*)

Table 1 Mean ozone concentrations over the region and at the stations far away and near to the roads during different seasons (μg m^{-3})

	Mean annual ozone	Summer (April–September)	Winter (October–March)
	Mean ± std	Mean ± std	Mean ± std
Stations near to the roads	85.59 ± 62.78*	101.69 ± 63.49*	61.32 ± 53.65
Stations far away from the roads	100.59 ± 75.09*	122.68 ± 76.74*	62.65 ± 54.36
All	89.37 ± 71.25	113.25 ± 71.68	62.00 ± 53.79

*Statistically significant difference P < 0.05 observed pairs

It has been tested whether the ozone concentrations varied between the geographical locations of the stations, their altitudes and closeness to roads. There is a statistically significant difference of ozone concentrations between the locations of stations, closeness to road and their altitudes ($p < 0.05$). The maximum ozone concentrations were observed on the high altitude stations. It has been found that ozone concentration increased gradually with an increase in altitude. Similar altitudinal gradients have been identified at a regional scale in rural areas of the Mediterranean, mountain–top locations by several workers.

When the closeness of the stations to roads is taken as a factor for ozone variation, it has been found that, as the distance to roads increases, the ozone concentrations also increases. The difference is more pronounced during summer season (April to September). From the previous studies, it has been reported that, elevated NOx emission sources act as a local sink for ozone (ICP 1996). For this reason, in areas with heavy traffic as in town centres, ozone levels are usually lower than outside city areas. As can be seen in Table 1, due to an increase in the traffic activities during summer time in the study area, the ozone concentrations are lower at the stations near the roads.

Our results indicate that concentrations of ozone in southern Turkey are similar and even higher to those measured in forested parts of Romanian and Ukranian Carpacian mountains and Sumava mountains (Bytnerowicz et al. 2004), which all suffer from ozone phytotoxicity. With an annual mean ozone level of 89.37 μg m^{-3} (summer time mean ozone, 113.25 μg m^{-3}), forest ecosystems in the southern Turkey are under an influence of anthropogenic sources and chronic exposure to ambient ozone concentrations is probably one of the primary reasons for forest decline.

3.1 *Relation Between Ozone Exposure and Plant Injury*

The most common method of assessing plant injury induced by air pollutants, involves visual estimation of the percent leaf area that is injured and variation of chlorophyll content.

Band shape of yellowing on the light-exposed upper side of the needles with increasing needle age is a characteristic symptom of ozone (Schmieden and Wild 1995). Based on visual injury damage percentages, it has been observed that, visual injury increases with needle age on the coniferous species. For the whole region, the percent of visual injury for 2nd class of severity index (6–25 % injury) on symptomatic needles for the current year is around 37 % and the previous year's is around 50 %.

The most common coniferous species in the study area is *P. brutia*. In addition to this we also find *Pinus nigra, Pinus pinea, Abies cilicica, Cedrus libani, Cupressus sempervirens* and *Juniperus oxycedrus* in the region. When the maximum ozone concentrations observed in the stations were compared with the visual injury damage of the previous year needles of *P. brutia* of the same station, it has been seen that, there is a statistically significant increase in the visual damage if the maximum ozone concentrations are higher than 60 µg m^{-3} ($p = 0.015 < 0.05$).

There were 36 different natural decidiuous tree species throughout the study area; *Amelonchier paruiflora, Berberis, Capparis ovata, Ceratonia siliqua, Cistus crateagus, Cotinus coggyria, Crataegus, Daphne mezereum, Daphne sericea, Ficus carica, Fontanesia phylliraeoides, Myrtus communis, Nerium oleander, Olea europea, Osyris alba, Paliurus spina-christi, Phillyrea latifolia, Pistacia lentiscus, Pistacia terebinthus, Platanus orientalis, Pyrus, Quercus, Quercus coccifera, Rhamnus, Rhus coriaria, Rosa, Ruscus oculeatus, Salvia, Smilax, Smilax aspera, Styrax officinalis, Thymelea tartonraira, Vitex agnus-castus.* For the deciduous species investigated in the study area, 26 % of the trees do not show any visual damage, whereas 60 % had 1–5 % of the leaf area damage and the rest had more than 6 % of the leaf area damage similar to current year needle visual damage.

The change in chlorophyll content has been used in many studies investigating the effects of ozone on plants (Knudson et al. 1977; Köllner and Krause 2000; Robinson and Wellburn 1991; Della Torre et al. 1998). The results, however, of such investigations do not always lead to the same conclusions concerning the pattern of effects. Knudson et al. (1977) found higher reduction in chl *a* in ozone-exposed plants of *Phaseolus vulgaris*, and *Spinacia oleracea*. Robinson and Wellburn (1991) also observed reduction in the chl *a*/*b* ratio in *Picea abies* plants, due to summer ozone exposures. Price et al. (1990) found no differences in the relative reduction of chlorophylls *a* and *b*, although both chlorophyll forms decreased by 20–40 % in barley exposed to 200 ppb of O_3.

On the basis of results of chlorophyll content of the coniferous species, control groups have more than 18 % Chl *a*, 22 % Chl *b*, 17 % Chl $a + b$ and 10 % carotenoid content compared to symptomatic needles, the values however change from species to species. For deciduous species the Chl *a*, Chl *b* and Chl $a + b$ content of symptomatic foliars is less than 31, 34 and 32 % respectively, compared to their control groups. For carotenoids, no significant difference was observed for deciduous species.

For common coniferous species the chlorophyll content variation for symptomatic and control groups is presented in Fig. 3. The difference between symptomatic and control group is significant for *Pinus pinea* and *Cupressus sempervirens* ($p < 0.05$). No difference has been observed in the case of *Pinus brutia* and *Pinus*

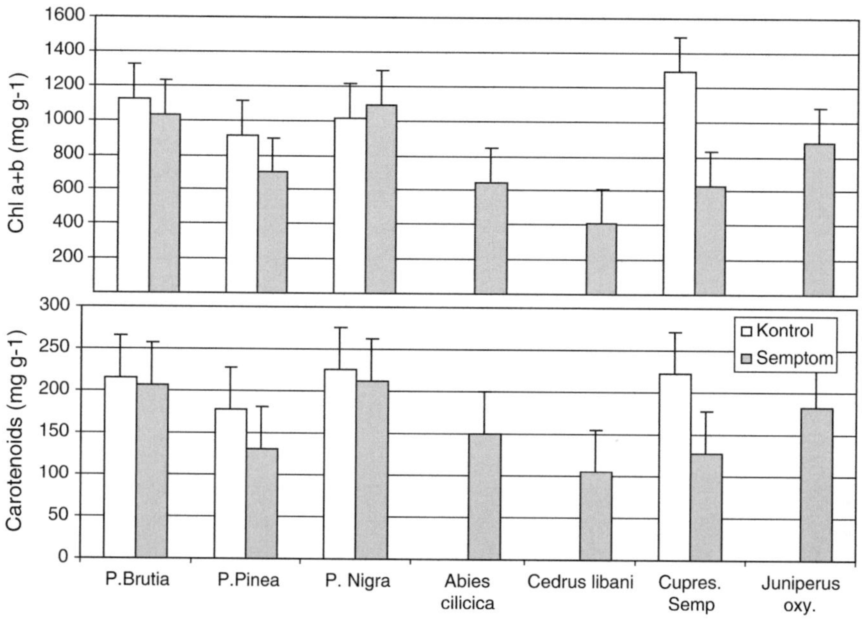

Fig. 3 Variation of Chl $a+b$ and Carotenoid content of symptomatic and control group of coniferous species

nigra. Since there was no control group for *Abies cilicica, Cedrus libani* and *Juniperus oxycedrus*, the variation of pigment content could not be followed.

For different deciduous species, the variation of pigment content from symptomatic to control group foliars is high. The most significant reduction in Chl *a* content on the symptomatic leaves was observed in *Vitex agnus castus, Paliurus spina cristii, Phillyrea latifolia, Thymelea tartonraira* with 51, 44, 42 and 36 % reduction, respectively. For the same species, Chl *b*, Chl $a+b$ reduction varies from 30 % to 70 % (Fig. 4). Although, for whole deciduous species there is no significant variation in the carotenoid content, a statistically significant carotenoid reduction has been recorded in *Crateagus, Quercus coccifera, Phillyrea latifolia* and *Vitex agnus-castus*. These reductions are 81, 65, 55 and 36 % respectively. It has been observed that the reduction of chlorophyll content on foliar parts is increasing with increase in ozone concentrations in the region. When ozone concentrations are around 40–60 µg m^{-3}, the reduction of pigment content is 2–10 % for coniferous species and 15 % for deciduous species, whereas the reduction of pigment content increases 30 % for coniferous, 45 % for deciduous species at the ozone concentrations higher than 80 µg m^{-3}.

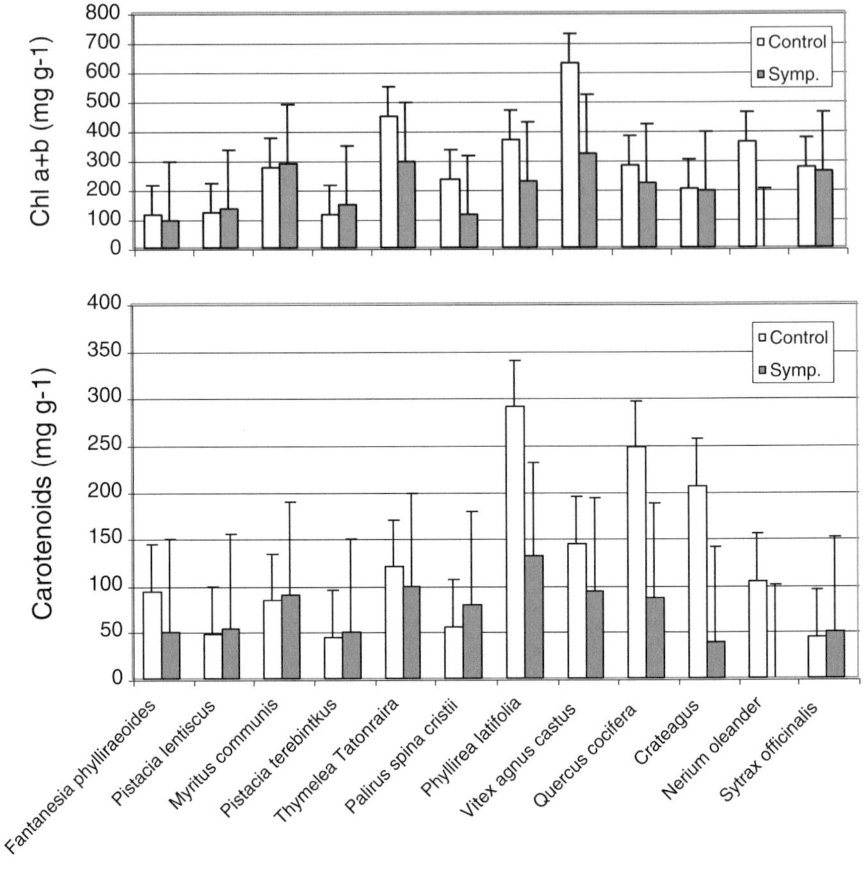

Fig. 4 Variation of Chl $a + b$ and Carotenoid content of symptomatic and control group of deciduous species

4 Conclusion

The results of our study indicate that ambient ozone is prevalent in forested regions of southern Turkey at concentrations high enough to injure sensitive plants. With respect to ozone increase, visual injury and reduction of chlorophyll pigment content of the foliar samples was observed. Based on visual injury indexes and pigment losses, out of 41 investigated natural species, 11 were identified as bioindicators for ozone injury: *Vitex agnus castus, Palirus spina cristii, Phyllirea latifolia, Thymelea Tatonraira, Crateagus, Quercus cocifera, Salvia, Berberis, Ceratonia siliqua, Pinus pinea and Cupressus sempervires.* Out of these identified sensitive species *Crateagus, Berberis* and *Pinus pinea* have already been identified as sensitive species to ozone in Europe.

There is a clear need for development of long-term O_3 monitoring networks in mountain forest ecosystems of the southern Mediterranean Region. Such networks should consist of active and passive samplers.

Acknowledgement We acknowledge financial support from the Scientific and Technological Research Council of Turkey (No. İÇTAG-Ç062 102I049). We thank many people we worked with in the field for their collaboration. Organizations and individuals include: Dr. Gürdal Tuncel, Middle East Technical University, Turkey, Yusuf Cengiz, Dr. Mehmet ali Başaran, West Mediterranean Forest Research Center, Antalya, Turkey

References

Al-Momani IF, Aygün S, Tuncel G (1998) Wet deposition of major ions and trace elements in the eastern Mediterranean basin. J Geophys Res Atmos 103(D7):8287–8299

Barnes J, Gimeno B, Davison A, Dizengremel P, Gerant D, Bussotti F, Velissariou D (2000) Air pollution impacts on pine forests in the Mediterranean basin. In: Ne'eman G, Trabaud L (eds) Ecology, biogeography and management of Pinus halephensis and P. brutia forest ecosystems in the Mediterranean basin. Backhuys Publishers, Kerkwerve, pp 391–404

Bussotti F, Ferretti M (1998) Air pollution, forests condition and forest decline in Southern Europe. Environ Pollut 10:49–65

Butkovic V, Cvitas T, Klasinc L (1990) Photochemical ozone in the Mediterranean. Sci Total Environ 99(1–2):145–151

Bytnerowicz A, Godzik B, Grodzinska K, Fraczek W, Musselman R, Manning W, Badea O, Popescu F, Fleisceher P (2004) Ambient ozone in forests of the Central and Eastern European mountains. Environ Pollut 130:5–16

Çepel N (1989) Pollution problems in Turkish forests. In: Ozturk MA (ed) Plants and pollutants in developed and developing countries. Ege University Press, Izmir, pp 207–212

Davison AW, Barnes JD (1998) Effects of ozone on wild plants. New Phytol 139:135–151

Della Torre G, Ferranti F, Lupattelli M, Pocceschi N, Figoli A, Nali C, Lorenzini G (1998) Effects of ozone on morpho–anatomy and physiology of *Hedera helix*. Chemosphere 36:651–656

Fiala J, Cernikovsky L, de Leeuw F, Kurfuerst P (2003) Air pollution by ozone in Europe in summer 2003. Overview of exceedances of EC ozone threshold values during the summer season April–August 2003 and comparison with previous years, European Environmental Agency, Copenhagen (EEA Topic Report 3/2003)

ICP Forests, (International Co-operative Programmes and the Mapping Programme under the Working Group on Effects) (1996) Convention on long range transboundary air pollution "Effects of O_3 and NO_2"

Gimeno BS, Penuelas J, Porcuna JL, Reinert RA (1995) Biomonitoring ozone phytotoxicity in eastern Spain. Water Air Soil Pollut 85:1521–1526

Guderian R, Klumpp G, Klumpp A (1989) Effects of SO_2, O_3 and NO_x singly and in combination on forest species. In: Ozturk MA (ed) Plants and pollutants in developed and developing countries. Ege University Press, Izmir, pp 231–268

Güllü G, Şirin G, Örtel E, Tuncel G (2001) The effects of air pollution on crown and foliar conditions of forests in the Southern Turkey. In: Proceedings of the 11th international symposium on environmental pollution and its impact on life in the Mediterranean Region, October 6–10, Limmasol, Cyprus

Güllü G, Şirin G, Tuncel G (2003) Forest decline evidence in Southern Turkey and its possible dependence on ozone trends. Water Air Soil Pollut Focus 3(5–6):263–275

Knudson LL, Tibbitts TW, Edwards GE (1977) Measurement of ozone injury by determination of leaf chlorophyll concentration. Plant Physiol 60:606–608

Köllner B, Krause GHM (2000) Changes in carbohydrates, leaf pigments and yield in potatoes induced by different ozone exposure regimes. Agr Ecosyst Environ 78:149–158

Lichtenthaler HK (1987) Chlorophylls and carotenoids: pigments of photosynthetic biomembranes. Methods Enzymol 148(1987):351–382

Manning WJ, Godzik B, Musselman R (2002) Potential bioindicator plant species for ambient ozone in forested mountain areas of central Europe. Environ Pollut 119:283–290

Ozturk M (ed) (1989) Plants and pollutants in developed and developing countries. Ege University Press, Izmir, 759 pp

Ozturk M, Secmen O (1981) Air pollutants and plants. Nature Man 3:23–27

Price A, Young A, Beckett P, Britton G, Lea P (1990) The effects of ozone on plant pigments. In: Balscheffsky M (ed) Current research in photosynthesis, vol 4. Kluwer Academic Publishers, Dordrecht, pp 595–598

Robinson DC, Wellburn AR (1991) Seasonal changes in the pigments of Norway spruce, *Picea abies*. Karst, and the influence of summer ozone exposures. New Phytol 119:251–259

Sanz MJ, Calatayud V, Sanzhez-Pena G (2007) Measures of ozone concentrations using passive sampling in forest of South Western Europe. Environ Pollut 145:620–628

Schmieden U, Wild A (1995) The contribution of ozone to forest decline. Physiol Plant 94:371–378

Skelly JM, Innes JL, Snyder KR, Savage JE, Vanderheyden D, Zhang J, Sanz MJ (1999) Observation and confirmation of foliar ozone symptoms of native plant species of Switzerland and southern Spain. Water Air Soil Pollut 116:227–234

Steubing L, Fangmeier A, Biermann J, Both R (1989) Effects of SO_2, NO_2 and O_3 on native plants and plant communişties in a beech forest. In: Ozturk MA (ed) Plants and pollutants in developed and developing countries. Ege University Press, Izmir, pp 269–276

Velissariou D, Gimeno BS, Badiani M, Fumigalli I, Davison AW (1996) Records of ozone visible injury in the ECE Mediterranean region. In: Kärenlampi L, Skärby L (eds) Critical levels for ozone in Europe: testing and finalizing the concepts UNECE workshop report Department of Ecology and Environmental. University of Kuopio, Kuopio, pp 343–354

Determination of Environmental Activities and Perspectives of Plants: A Field Research in Kayseri

Arzum Büyükkeklik, Murat Toksarı, and Tuba Bekiş

Contents

Abstract The plants have great effects on environmental pollution and other related environmental problems that threat life. Therefore, civil society organizations, governments and consumers impose responsibility to plants in order to decrease or solve environmental problems. So, in all stages, plants have to consider the effects of production activities on environment and they should study in order to decrease these negative effects. The aim of this field research is to determine the environmental activities and management perspectives of medium and big-sized plants operating in Industrial Zone of Kayseri. Also the activities of plants about decreasing environmental effects are examined along the process from product design till completion of its life cycle. Moreover, the aims of these activities such as reducing raw material, energy and water costs, gaining competitive advantages, contribution to the decrease of environmental pollution, obeying the laws and regulations about the environment are investigated. Consequently, it is found out -contrary to expectations- that most of plants do not evaluate environmental

A. Büyükkeklik • M. Toksarı (✉) • T. Bekiş
Faculty of Economics and Administrative Sciences, University of Niğde, 51100 Niğde, Turkey
e-mail: mtoksari@nigde.edu.tr

M. Öztürk et al. (eds.), *Plants, Pollutants and Remediation*,
DOI 10.1007/978-94-017-7194-8_17

management as a cost driver. They integrate environmental subjects into production activities.

Keywords Environmental problems • Environmental activities of plants • Environmental perspectives of plants • Medium and big-sized plants

1 Introduction

Since the late 1980s, there has been an increase in awareness about environmental issues around the world and the impact of business on the environment has become an important issue. Especially in developed countries the negative effects of industrial products threaten the future of generations.

Accordingly, civil society organizations, governments, consumers, suppliers, financial institutions and employees have focused on this issue of increasing concern. The reflection of this concern has been many legislations and regulations at both international and national levels. So there has been an increasing pressure for plants to improve their environmental activities. Table 1 shows general forms of these pressures and their impacts. However, intensity of these pressures may vary by country or industry.

The plants should develop environmental activities under these pressures. The most important factor that effects plants to develop environmental activities is environmental management perspectives of them. These perspectives force the plants to take an environmental management certification, form a written environmental policy and establish a separate environmental department and all of them are important orijin points for a plant to develop and apply environmental activities. So in this study, environmental management perspectives and environmental activities of

Table 1 Environmental pressures and their impacts

Forms of pressure	Environmental impact
Government	Encompassing environmental regulations
Consumer and supplier	Better informed and more aware of the environmental content and impact of consumer products
	Acceptance of green products by industry and end customers
Investor	Examining the environmental record of potential companies
Employees	Employment implications, high skill requirements to operate complex pollution abatement equipment
Local community	Complaints associated with pollution
Financial	Use of environmental risk surveys by banks and insurance companies

Source Gupta (1995: 38)

medium and big-sized plants operating in industrial zone of Kayseri are determined.

In the rest of the paper first, there are definitions about perspectives of the environmental management, environmental management system and environmental policy. Second, the methodology of research is presented. Third, results of research in Kayseri are presented by frequency tables. In the conclusion, the findings of research are evaluated and suggestions for plants are developed.

2 The Perspectives of the Environmental Management

The perspectives of plants to the environmental management is considered under three categories in literature; such as passive, active and proactive perspectives (Sarkis 1998: 159; Yüksel 2008: 51). In passive perspective, plants regard environment issues as a cost driver and they do not apply environmental management. In active perspective, environmental management is regarded as a tool for the environmental laws and regulations that sector needs. However, active perspective is at the minimum level and insufficient for developing environmental management (Akdoğan 2003: 253). Moreover, in this perspective environment issues are regarded as cost drivers and environmental management is applied as an obligation by an impulsive force of laws and regulations. In proactive perspective, unlike the others, the main aim is to gain competitive advantage by developing well planned environmental management appropriate to the plant's strategy (Sarkis 1998: 159). In this perspective, the meaning of environment may imply harmonizing corporate environmental performance with stockholders' expectations as well as constituting a significant new source of competitive advantage, such as lower costs and/or expanded market share (Gupta 1995: 36). Thus, environment effects the operations of a plant.

3 Environmental Management System

An environmental management system is a formal system and database that integrates processes for training of personnel, monitoring and reporting of specialized environmental performance information to stakeholders of the plant (Melynk et al. 2003: 332). And it reduces or clears off the plants' effects to the environment and is used to improve and monitor the environmental performance of them. Environmental management effects various aspects of operations of a plant, from the purchase of various inputs through process control and changes to the output itself (Gupta 1995: 43).

One of the extensively used environmental management systems is ISO 14001. ISO 14001 is a system certification issued by International Standarts Organization in 1996. ISO 14001 defines the organization's structure, planning activities, procedures, processes for an effective environmental management system. Also it describes developing, applying and revision levels of environmental policy and guides for the application (Sarkis 1998: 163). The plants, that have ISO 14001 standard, evaluate environmental issues in all stages from the design of products and processes, to production, distribution and disposal of the products (Yüksel 2008: 52). Another management system certification is EMAS (Environmental Management and Auditing Scheme) developed by European Union. European Union describes the objective of EMAS as promoting continual improvements in the environmental performance of organizations (Hertin et al. 2007: 261).

4 Environmental Policy

Environmental policy is a declaration that explains intentions and principles about general environmental perspectives of top managers' perspectives (Sarkis 1998: 163). Written environmental policy indicates that the plant accepts the "environment" as an important factor in its activities and decision processes (Gadenne et al. 2009: 61). In environmental policy, the plant clearly expresses its main goal. Environmental policy can be accepted as an orijin to emplace the environmental sensitivity into the plant's culture. Throughout this orijin, plant intents to reduce or clear off its affect to the environment during its activities and form its processes on the basis of this apprehension. Additionally, environmental policy is an important indicator for stakeholders (employees, customers, rivals etc.) to evaluate the plant.

5 Environmental Management Department

To form a environmental management department is so vital in terms of attaching importance to environmental management and the results that they want to achieve (Gadenne et al. 2009: 48).

6 Methodology

This study is a part of comprehensive research project that explores environmental management and innovation applications of medium and big-sized plants operating in industrial zone of Kayseri. This is a descriptive study that explores the perspectives and activities of these plants.

The research data has been collected by a survey questionnaire designed based on literature (Akdoğan 2003; Melynk et al. 2003; Frondel et al. 2008; Yüksel 2008) review. The questionnaires were answered by top managers of plants.

Because of the thought that medium and big-sized plants' environmental activities are more than small plants' activities, the population of the study is determined as medium and big-sized plants in industrial zone of Kayseri. To determine the number of plants in population, it is benefited from Chamber of Industry Work Guidebook of Kayseri 2008 and Sectoral Guidebook 2007 prepared by Industrial Zone of Kayseri District Office. According to these resources and the information received from managers of Chamber of Industry of Kayseri, there are 250 medium and big-sized plants in Kayseri. Because of limitations about time and costs, it has been required to study on a sample rather than the whole population.

The sample of the research is determined as 96 plants, at the level of $e=0.1$ and $\alpha=0.05$ (95 % confidence interval $z=1.96$). But 96 plants are more than the 0.05 of population ($96/250 \geq 0.05$), it is multiplied by adjustment factor $\left(\frac{N-n}{N-1}\right)$ and is determined that 60 plants will represent the population (Kurtuluş 1998: 235). Although 60 plants are accepted sufficient to represent the population, the data collected from 75 plants selected by judgmental sampling. In this sampling method the samples are determined by researchers by considering the contribution to the research (Nakip 2003: 184).

The findings of research consist of descriptive statistics obtained from SPSS (Statistical Package For Social Sciences) 15.0 programme.

7 Findings

7.1 Environmental Management System Certification

The participants of the research are asked whether they have ISO 14001 environmental management system certification, only 5 of them (6.7 %) stated that they have ISO 14001 environmental management system certification. 8 of them (10.7 %) stated that they are at the planning stage of applying this management system. Remaining 62 plants (82.7 %) stated that they have no environmental system certification (see Table 2). Any of the participants has no EMAS. Only 1 plant stated that it is at the planning stage of getting this environmental system certification.

In this content, the 67 % of the participants have environmental management system certification and they are at the beginning stage of environmental management.

Table 2 Environmental management system certification

	ISO 14001			
Size of the plant	Yes	No	At the planning stage	Total number of plants
Medium sized (50–250)	0	57	6	63
Big sized (251 and more)	5	5	2	12
Total	5 (6.7 %)	62 (82.7 %)	8 (10.7 %)	75 (100 %)

Table 3 Written environmental policy

Written environmental policy	Number of plants	Percentage of plants
Yes	18	24.0
No	57	76.0
Total	**75**	**100**

Table 4 Environmental department

	Department about environmental issues		
Size of the plant	Yes	No	Number of plants
Medium sized (50–250)	48	15	63
Big sized (251 and more)	12	0	12
Total	60 (80 %)	16 (20 %)	75 (100 %)

7.2 *Written Environmental Policy*

Participants answered the question about written environmental policy. 18 of them (24 %) stated that they have a written environmental policy, 57 of them (76 %) stated that they have not got any written environmental policy (see Table 3). In literature environmental policy is accepted as an orijin to form environmental management and it is regarded as an indicator of environmental sensibility, It can be said that this sensibility hasn't been adopted by the most participants (76 %).

7.3 *Environmental Department*

The participants of the research are asked if they have department about environmental issues. And if they have, it is asked the name of it. 60 of 75 participants (80 %) stated that they have department about environmental issues (see Table 4). 48 of the medium sized plants and all of the big sized plants have department about environmental issues.

The departments of plants responsible for environmental issues is seen on the Table 5. With reference to this table, production, R&D (research and development) and quality control departments are responsible for environmental activities. And any of the participants don't have a separate environmental department that has the word of "environment" in it. Moreover in some plants (3 of medium sized plants) general managers are directly responsible for environmental activities. However, the perspective about "environment" should be similar to the perspective about "quality".

As there is a department named "quality department", "quality control department" or "quality assurance department" to determine, apply, control and monitor the activities about quality, there should be a separate department named "environ-

Table 5 Departments of plants about environmental issues

The name of the department about environmental issues	The number of plants	Percent
Production	21	28
R&D	18	24
Quality control	15	20
General manager	3	4
Exploitation management	2	2.6
Purchasing department and logistics	1	1.3
Total	**60**	**80**

Table 6 The perspectives of the plants to environmental management

The perspectives of plants to environmental management	Number of the firms	Percent of the firms
Environmental issues are cost drivers for our plant.	6	8
Environmental practices comply with regulations are sufficient for our firm.	31	41.3
Gains after the environmental management programmes meet the costs for these programs	17	22.7
Environmentally conscious business practices help our plant to find opportunities for gaining competitive advantage.	19	25.3
Total	**73**	**97.3**

mental management department", "job security&environmental management department" or "quality&environmental management department".

7.4 *Environmental Management Perspectives*

It is asked the participants to select one of the judgement about environmental perspectives of plants (see Table 6). This part of the questionnaire is answered by 73 plants. With reference to Table 6, 6 of the plants (8 %) evaluate the environmental activities as a cost driver. 31 of them (41.3 %) states that they perform their environmental activities to adapt environmental regulations. 17 of them (22.7 %) state that the gains after the environmental management programs meet the costs for these programs. 19 of them (25.3 %) mentions that environmentally conscious business practices help their plant to find opportunities for gaining competitive advantage. In this content, after evaluating the environmental perspectives of plants in the research sample it is understood that, only 8 % of them have passive perspective; 41.3 % of them have active perspective and rest of them (48 %) have proactive perspective. Hence most of the plants (48 %) evaluate the environmental issues as a competitive advantage, it is hopeful in terms of developing environmental applications.

Table 7 Environmental activities of plants

Environmental activities	Low level	–	–	–	High level	Mean
1. We study in order to change our inputs harmless or less harmful to the environment	4	18	18	16	2	2.90
2. We study in order to reduce the amount of our inputs	2	0	16	31	9	3.78
3. We study in order to use the transformed (secondary) material as the input of our production process	1	4	12	26	15	**3.86**
4. We study in order to reduce or clear off the waste occurred during our production process.	0	2	12	24	20	**4.07**
5. We study in order to re-use the waste occurred during production	1	7	12	25	13	3.72
6. We study for recovery of our products turned back by any reason	0	2	19	27	10	3.78
7. We study in order to design our production technology by using less water and energy	2	0	12	30	14	**3.93**
8. We study related to the clear energy resources such as steam energy, solar energy, wind energy	7	12	11	21	7	3.16
9. We study in order to reduce carbon dioxide occurred during production	3	11	23	16	5	3.16
10. We study in order to eliminate the product's potential environmental effect in product design	2	9	18	25	4	3.34
11. We study in order to develop some scales (measurement of waste amount, rate of re-cycled product) to monitor environmental sensibility of our product.	10	13	13	16	6	2.91
12. We study with our suppliers and customers to develop environmental sensibility	19	16	11	10	2	2.31

7.5 Environmental Activities

The part of the questionnaire about environmental activities-comprised 12 judgements- is answered by 58 plants (77.3 %). Participants were asked to indicate their judgement about environmental activities using a five point Likert scale; 1:low level – 5:high level (see Table 7).

According to the answers, it is seen that the mean of all of the judgements are higher than 2. This indicates that the plants develop environmental activities. When the environmental activities at low and high levels are investigated, it is seen that the environmental activities at the high levels are (i) the activities that reduce or clear off the waste occurred during production process (mean=4.07), (ii) the activities that design production technology by using less water and energy (mean=3.93) and (iii) the activities that use the transformed (secondary) material as the input of production process (mean=3.86). So it is understood that the plants focused on environmental activities in production processes.

The environmental activities that the plants perform at low levels are (i) studying with suppliers and customers to develop environmental sensibility (mean=2.31), (ii) changing inputs becoming harmless or less harmful to the environment (mean=2.90) and (iii) developing some scales (measurement of waste amount, rate of recycled product) to monitor environmental sensibility of the product (mean=2.91). These are the activities required common studies with the members in supply chain such as suppliers and customers. So these judgements indicate that the plants yet perform the environmental activities on their own and they can not develop integrated applications.

8 Conclusions

The research determining the environmental perspectives and activities of the medium and big-sized plants that have operations in industrial zone of Kayseri and its results are summarized below;

- The main determining factor that the plants develop their environmental activities are the environmental management perspectives of them. When a plant accept the environmental management as a cost driver and obligation, it may not perform it despite its fines. Or it perform the environmental management at minimum level just because of its fine. In such a case, environmental management transforms to cost driver. However, because of environmental pollution and its effects, environment management become a competition tool that will be developed complied with competitive strategies, and it requires proactive perspective. In this research 48 % of the medium and big-sized plants operating in Kayseri stated that they exhibit proactive perspective to environmental management and this perspectives triggers them to develop environmental activities. In addition to this, the transformation of plants' perspectives from passive/active to proactive is necessary to reduce the costs and gain competitive advantage.
- In literature, environmental system certification, written environmental policy and separate environmental department are regarded as an origin point of developing and applying environmental activities. According to the findings of the research a few plants have environmental system certification, written environmental policy and separate environmental department. All these show that, most of the plants couldn't perform a formal environmental management practice.
- In this study, the levels of the plants' environmental activities are explored. These environmental activities mentioned by 12 judgements on the questionnaire. It is understood that plants perform environmental activities in production processes at higher level. However they perform the other activities that needs common studies with members of supply chain-such as suppliers and customers- at lower level. So it can be said that although integrated applications provide competitive advantage for plants, they perform environmental activities on their own and they can not develop integrated applications.

References

Akdoğan A (2003) Çevreye Duyarlı Yönetim ve İşletmecilik, Kayseri Ticaret Odası Yayınları, No 48, Kayseri

Frondel M, Horbach J, Rennings K (2008) What triggers environmental management and innovation? Emprical evidence for Germany. Ecol Econ 66(1):153–160

Gadenne DL, Kennedy J, McKeiver C (2009) An empirical study of environmental awareness and practices in SMEs. J Bus Ethics 64:45–63

Gupta MC (1995) Environmental management and its impact on the operations function. Int J Oper Prod Manage 15(8):34–51

Hertin J, Berkhout F, Wagner M, Tyteca D (2007) Are EMS environmentally effective? The link between environmental management systems and environmental performance in European companies. J Environ Plan Manage 51(2):259–283

Kurtuluş K (1998) Pazarlama Araştırmaları, İ.Ü. İşletme Fakültesi Yayın No: 274, Genişletilmiş ve Gözden Geçirilmiş 6. Baskı, İstanbul

Melynk SA, Sroufe RP, Calantone R (2003) Assessing the impact of environmental management systems on corporate and environmental performance. J Oper Manage 21:329–351

Nakip M (2003) Pazarlama Araştırmaları Teknikler ve (SPSS Destekli) Uygulamalar. Seçkin Yayıncılık, Ankara

Sarkis J (1998) Evaluating environmentally conscious business practices. Eur J Oper Res 107:159–174

Yüksel H (2008) An empirical evaluation of cleaner production practices in Turkey. J Cleaner Prod 16(S1):50–57

An Approach for Sustainable Management of The Balikligol Lakes, Turkey

Bulent Armagan, Nurettin Besli, and Deniz Ucar

Contents

Abstract The Balikligol Lakes in Sanliurfa, Turkey (Lake Ayn-i Zeliha and Lake Halil-ur Rahman) are freshwater lakes, which possess not only environmental value but also touristic value due to their natural aquarium look and their historical and sacred status in the past and present. The fish deaths have been encountered in these lakes from time to time. Deteriorating water quality can harm the health of the fish in the water. Therefore, the water quality in both the lakes needs to be monitored and proper management strategies should be developed. The pollution in the lakes exceeding the acceptable levels endangers the sustainable management of the biodiversity. With the advent of the measurement technology, it is now possible to set-up permanent monitoring and management systems in a cost-effective way. The objective of this study is to establish an online water quality monitoring and management system to protect the water quality in the Balikligol Lakes. In all eight measuring stations were selected at random for keeping track of the pollution levels. The applicable system units were determined. Each station will make predeter-

B. Armagan (✉)
Faculty of Engineering, Department of Environmental Engineering, Istanbul University, Avcilar, Istanbul, Turkey
e-mail: bulentarmagan@gmail.com

N. Besli • D. Ucar
Department of Computer Engineering, Harran University, Sanliurfa, Turkey

M. Öztürk et al. (eds.), *Plants, Pollutants and Remediation*,
DOI 10.1007/978-94-017-7194-8_18

mined measurements and send data to the Environmental Administration Center of Sanliurfa Governorship via wireless internet connection. The closeness of these Lakes to the city center makes the wireless internet access available. Moreover, the measuring station locations were geocoded on a digital map, and information tables for each station linked to the digital map using Geographic Information System (GIS) software. The basic approach of the system is that the collected data at the Environmental Administration Center will be analyzed with the support of GIS enabled software, and the action plan will be determined according to the guidelines established for the surface water quality standards. This whole process will create a management system for "The Balikligol Lakes" which receives real-time data continuously and respond promptly according to the guideline.

Keywords Water quality • Monitoring • The Balikligol Lakes • GIS

1 Introduction

Seasonal changes in surface water quality are an important aspect for evaluating temporal variations of surface water resources. The quality of the surface water resources is evaluated with the help of some physical and chemical parameters (Yucel et al. 1995; Ozturk et al. 1996, 2005, 2011; Ouyanga et al. 2006; Sakcali et al. 2009). Some important primary physical and chemical parameters can be listed as turbidity, color, temperature, conductivity, dissolved oxygen, chemical oxygen demand and heavy metals (Ozturk et al. 1991). Although all pollutants in water above certain concentrations pose a health threat (Dogan and Ozturk 1991; Yucel et al. 1995; Yesilnacar and Uyanik 2005), heavy metals are among the most toxic pollutants in groundwater and industrial wastewater (Ozturk and Secmen 1986; Ozturk 1989). The source of heavy metals in water systems can be mainly attributed to manmade sources (Ozturk et al. 2004, 2010, 2012; Karamanis et al. 2008; Altay and Ozturk 2012; Ashraf et al. 2015). The heavy metal concentrations may show variations, depending on the nature of pollution. Even low metal concentrations (e.g. 1.0 g/m^3) can cause illness and even death in living beings (Dogan and Ozturk 1991). The heavy metals may accumulate within algae cells and fish body. Hence, monitoring of water quality and development of appropriate management strategies are necessary to protect water resources (Eckenfelder 2000).

Water monitoring is used to help water resource managers to understand and avert potential negative impacts of anthropogenic or natural factors on water resources (Baltacı and Onur 2006; Nas et al. 2006).

Consistent and comparable long term water quality and quantity monitoring data are needed in order to:

1. Describe the status and trends of a water resource,
2. Identify existing and emerging water quality issues, and
3. Determine compliance with regulations.

The basic principles of a monitoring program include understanding the system to be monitored, designing the monitoring program to meet set objectives, paying attention to details, monitoring source activities, and building an ongoing program evaluation processes (Annan 2001; Jung et al. 2004). Some researchers have investigated integrated water quality models (Richards et al. 1996; Ning et al. 2001; Beck 2005; Lindenschmidt et al. 2005; Yang et al. 2008) and environmental management systems based on hydrological modelling (Chau et al. 2002; Mujumdar and Saxena 2004; Zacharias et al. 2005), but these systems are not connected with any online monitoring system. Even in emergency cases of water pollution, no feasible management scheme can be worked out in timely manner (Thoms and Swirepik 1998; Rauch and Harremoes 1999; Huang and Xia 2001; Quinn 2003). These problems justify the development of an online water quality monitoring and management system that can provide an early warning of water-pollution events (Yang et al. 2008).

Even though considerable progress has been made in recent years to develop online water quality monitoring capability, these installations still only complement laboratory testing, which is not yet a fully viable alternative (Yang et al. 2008; Drage et al. 1998). This paper describes the Balıklıgol online water quality monitoring and management system, which uses modern data transmission and support from GIS which enables software techniques to monitor the lake's water environment and hydrological–environmental models to forecast the potential environmental water demand. This combination of techniques allows optimal allocation of water using information acquired from the monitoring system and estimates from the water environment models.

In this way we can;

1. provide an overview of present water quality in the Balikligol Lakes, Turkey and
2. compile all the data sent from the field into GIS database.

The statistical analyses of the data is to be employed. Information sharing and integration with other systems in the Governorship of Sanliurfa needs to be established to ensure more effective working of the system.

2 Study Site

The Balikligol Lakes consist of two lakes and a canal carrying the water that overflows from those lakes (Cetin et al. 2000). The small lake is known as Ayn-i Zeliha (1537 m^2) and the big one as Halil-ur Rahman (3060 m^2). The depth of both lakes is around 1.5 m. Periodic samples were collected from both lakes (Fig. 1).

The lakes are fed with spring water. The water from the surface drains into an underground lake and springs up at the bottom of the lakes. The spring water levels change with regard to seasons and rainfall. The Karakoyun stream merges with the Culap stream which in turn merges with the Euphrates River within the borders of Syria (Ozturk 1995; Ozturk et al. 2004). The monitoring stations consist of 17 sampling locations (Fig. 1).

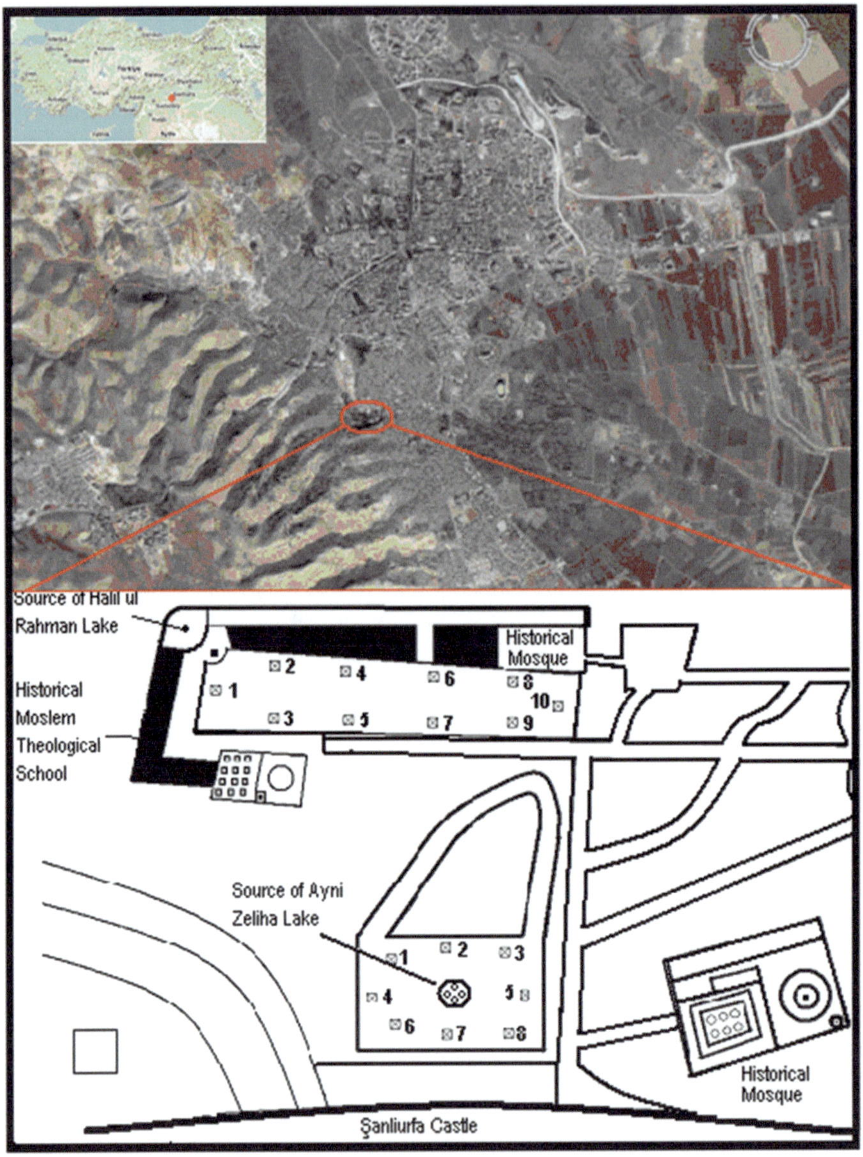

Fig. 1 The Map of Turkey and Sanliurfa City with The Balikligol Lakes

3 Laboratory Studies

The samples were collected 15 cm below the surface and the parameters such as temperature, pH, conductivity, chemical oxygen demand (COD), dissolved oxygen (DO), Iron, Sodium, Chromium, Copper, Nickle and Lead were determined in situ.

Table 1 Seasonal distribution of measured water quality parameters with WHO Standarts

Parameters	Unit	Spring	Summer	Autumn	Winter	Average	WHO (1993)
pH	–	7.32	7.22	7.69	7.66	7.47	6.5–9.5
Conductivity	μS/cm	323.81	306.90	322.95	301.52	313.80	250
Temperature	°C	20.30	26.76	20.75	16.37	21.05	–
DO	mg/l	4.28	3.93	4.88	5.25	4.59	–
COD	mg/l	63.36	67.63	58.69	53.60	60.82	–
Fe	mg/l	0.25	0.55	1.57	0.54	0.73	300
Cr	mg/l	0.88	0.81	0.37	0.00	0.52	2000(max)
Na	mg/l	2.44	2.57	2.31	2.00	2.33	200

The analyses were performed at the laboratory of the Environmental Engineering Department of Harran University. Standard methods were used to determine the chemical and physical features of the water samples (STM 2005). As to the chemical parameters, COD and DO were determined according to the closed reflux and iodimetric titration methods respectively. Heavy metal cited above were analysed with a Varian AA 140. As for the physical parameters, temperature, conductivity, turbidity were measured with turbidimeter and the pH was determined with a portable pH meter (Table 1). The samples were filtered with a 0.45 mm filter paper before the analyses on the atomic adsorption to prevent the clogging of the machine. The analyses was compared with standards (Armagan et al. 2008).

4 Establishment of The Surface Water Monitoring System

4.1 Mobile Software and Hardware

The software used in the system is capable of being integrated with other software used for analyzing and collecting data. It is also compatible with the industry standards. As for the mobile hardware, there is measurement equipment, which is installed at each station, a server and wireless modem (Infotech 2002).

4.2 Wireless Communication Network

The Mobitex wireless communication network is used between the mobile units and the Environment Management Center. The communication network is supported by GSM-SMS, Orbcomm, INMARSAT C, D+ in case of the future requirements. The Infotech ITS-system is used here, which has features like 512 KB package message infrastructure, less communication more information, alternatives custom for the needs of the communication network and a developed location sensitiveness (2–10 m).

4.3 Storage and Evaluation of Data

The communication with the Environment Management Center is being carried out with Mobitex wireless communication network. The software, which is capable of sending the data to different applications, stores the data and oversees the user access. The subcomponents of the central software are communication gateway, digital maps, application server, database and the integration module. The digital maps used in the projects are produced by aerial photography based on photogrametric method; kinematic navigation system is used for 3D evaluation. Furthermore the locations of the analysis stations are geocoded as a layer for the project used.

4.4 Reporting of Data

Information sharing and integration with other software in the Metropolitan Municipality environment will be established to ensure a more effective working of the system. A management information system (MIS) was set up for The Balikligol Lakes in Sanliurfa to coordinate network and manage regional pollution data coming from monitoring system of The Balıklıgol Lakes Region. Therefore Sanlıurfa Governorship Environment Administration Center will be connected to the Sanlıurfa Municipality Control Center, Harran University Environmental Engineering Department and interested offices after evaluating the data. In this way the management information system (MIS) will take the necessary required steps. This system will send information warning to the interested units by e-mail, SMS message or radio wireless. MIS will send the report to the interested units from time to time.

4.5 Sustainable Management and Action Plan

Based on the explanations given above a model (Fig. 2) for the management of lakes has been developed. The model has the following steps:

1. Data from measurement stations will be collected,
2. Important control parameters (dissolved oxygen, organic concentration, metal concentration and fresh water flow rate) will be compared with the legislations or the values predetermined by the authorities and
3. If the values are out of the desired ranges, some special actions and warnings will be proposed.

With the algorithm developed, the management of the lakes will be easier.

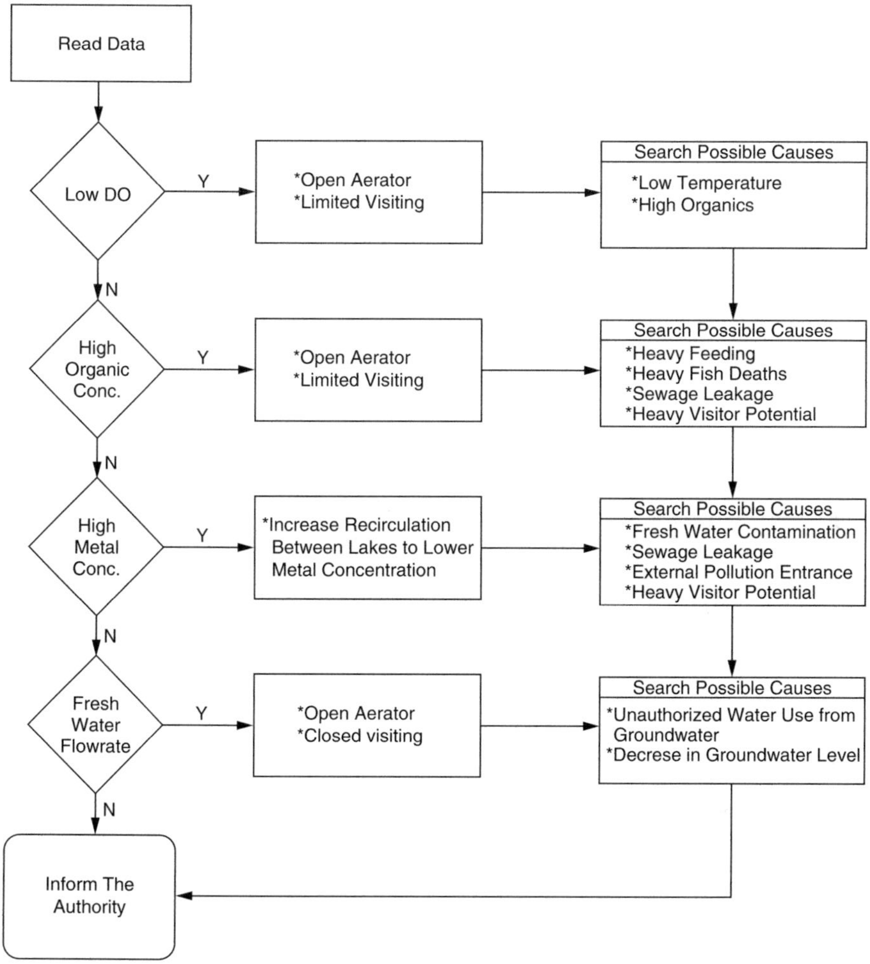

Fig. 2 Sustainable management and action scenarios for the Balikligol Lakes, Turkey

5 Conclusion

The Balikligol Lakes in Sanliurfa are located in the Southeastern Anatolian Region, Turkey. The Balikligol Lakes (Lake Ayn-i Zeliha and Lake Halil-ur Rahman) are freshwater lakes that have not only environmental value but are also of touristic value, due to their historical and sacred status in the past and present and their natural aquarium look (Cetin et al. 2000). From time to time, fish deaths have been encountered in these lakes. Deteriorating water quality harms the health of the fishes in the water.

GIS being a powerful tool for supporting environmental activities allows the managers to save, manipulate and display information electronically and make efficient decisions. Considerable improvements have been made during the last couple of years in improving the quality of water in the Balıklıgol Lakes. The goal of this study is to establish a preemptive warning system against a possible pollution in the Balıklıgol Lakes. For that, a study with a title of "monitoring and dispatching system for the Balıklıgol Lakes in Sanlıurfa by using GIS" has great importance to realize this idea.

References

Altay A, Ozturk M (2012) Halophytic plant diversity of Milleyha (Samandağ-Hatay) freshwater ecosystem, land degradation due to human disturbances & conservation measures. Pak J Bot 44:37–50

Annan JD (2001) Modelling under uncertainty: Monte Carlo methods for temporally varying parameters. Ecol Model 136:297–302

Armagan B, Gok N, Ucar D (2008) Assessment of seasonal variations in surface water quality of the Balıklıgol lakes, Sanlıurfa, Turkey. Fresenius Environ Bull 17(1):79–85

Ashraf S, Abbas F, Ibrahim M, Rashid U, Khalid S, Raza HA, Hakeem KR (2015) Application of Geographical Information System (GIS) for the identification and demarcation of selective heavy metal concentrations in the groundwater of a metropolitan city. J Geogr Sci 25(2):225–235

Baltacı F, Onur AK (2006) Water quality monitoring studies of Turkey with present and probable future constraints and opportunities. 10th International specialised conference on diffuse pollution and sustainable basin management, Istanbul, Turkey

Beck MB (2005) Vulnerability of water quality in intensively developing urban watersheds. Environ Model Software 20(1):381–400

Cetin E, Unlu Y, Ozturk M, Tulu M (2000) Trace element studies on Halilurrahman Lake Sanlıurfa-Turkey. Chemia I Inzynieria Ekologiczna 7(11):1143–1151

Chau KW, Chuntian C, Li CW (2002) Knowledge management system on flow and water quality modelling. Expert Syst Appl 22:321–330

Dogan F, Ozturk M (1991) Studies on the correlations between air and water pollution with the mortality rates in Turkiye. In: Ozturk MA et al. (eds) International urban ecology symposium, Aydın, pp 16–23

Drage BE, Upton JE, Purvis M (1998) On-line monitoring of micropollutants in the river Trent (UK) with respect to drinking water abstraction. Water Sci Technol 38(11):123–130

Eckenfelder WW (2000) Industrial water pollution control. McGraw-Hill, Singapore

Huang GH, Xia J (2001) Barriers to sustainable water quality management. J Environ Manage 61(1):1–23

Infotech (2002) Establishing a pollution control monitoring system for The Golden Horn by using GIS, Metropolitan municipality project Istanbul

Jung PJ, Goh HS, Hyun MH, Lee EJ (2004) Water quality management using WASP5 & WASP builder for a basin of an agricultural reservoir. J KSWQ 20(5):422–431

Karamanis D, Stamoulis K, Ioannides K, Patiris K (2008) Spatial and seasonal trends of natural radioactivity and heavy metals in river waters of Epirus, Macedonia and Thessalia. Desalination 224:250–260

Lindenschmidt KE, Hesser FB, Rode M (2005) Integrating water quality models in the high level architecture (HLA) environment. Adv Geosci 4:51–56

Mujumdar PP, Saxena P (2004) A stochastic dynamic programming model for stream water quality management. Sadhana Acad Proc Eng Sci 29(5):477–497

Nas B, Karabork H, Berktay A, Ekercin S (2006) Application of GIS and geostatistics in assessing water quality in the Beysehir Lake, Turkey. 10th International specialised conference on diffuse pollution and sustainable basin management, Istanbul, Turkey

Ning SK, Chang NB, Yang L (2001) Assessing pollution prevention program by QUAL2E simulation analysis for the Kao-Ping River Basin. Taiwan J Environ Manage 61(1):61–76

Ouyanga Y, Nkedi-Kizzab P, Wuc QT, Shindeb D, Huangd CH (2006) Assessment of seasonal variations in surface water quality. Water Res 40:3800–3810

Ozturk M (ed) (1989) Plants and pollutants in developed and developing countries. Ege University Press, Izmir, 759 pp

Ozturk N (1995) Comparison of biologic, chemical and physical parameters of The Kucukcekmece Lake Water, Master thesis. Istanbul University

Ozturk M, Secmen O (1986) Ecology of wetlands. Nat Man 2:31–32

Ozturk M, Secmen O, Leblebici E (1996) Plants and pollutants in the Eber lake. Ekoloji 20:14–16

Ozturk M, Ozcelik H, Sakcali MS, Guvensen A (2004) Land degradation problems in the Euphrates basin, Turkey. Environews 10(3):7–9

Ozturk M, Alyanak I, Sakcali S, Guvensen A (2005) Multipurpose plant systems for renovation of waste waters. Arab J Sci Eng 30(2C):17–28

Ozturk M, Mermut A, Celik A (2010) Land degradation, urbanisation, land use & environment. NAM Science & Technology, Delhi, 445 pp

Ozturk M, Gucel S, Sakcali S, Guvensen A (2011) An overview of the possibilities for wastewater utilization for agriculture in Turkey. Israel J Plant Sci 59:223–234

Ozturk M, Kebapci U, Gucel S, Cetin E, Altundag E (2012) Biodiversity & land degradation in the lower Euphrates Subregion of Turkey. J Environ Biol (ISSN:0254–8704) 33(Suppl):311–323

Ozturk M, Pirdal M, Mert H, Akgul S (1991) Evaluation of sewage sludge in the plants growth. Ekoloji 1(1):16–21

Quinn NWT (2003) A decision support system for adaptive real-time management of seasonal lakes in California. Environ Model Software 18:503–511

Rauch W, Harremoes P (1999) Genetic algorithms in real time control applied to minimize transient pollution from urban wastewater systems. Water Res 33:1265–1277

Richards K, Sharp M, Arnold N (1996) An integrated approach to modeling hydrology and water quality in glacierized catchments. Hydrol Process 10(44):479–508

Sakcali S, Yilmaz R, Gucel S, Yarci C, Ozturk M (2009) Water pollution studies in the Rivers of Edirne State–Turkey. Aquat Ecosyst Health Manage 12(3):313–319

STM (2005) Standard methods for the examination of water and wastewater, 21th edn. American Public Health Association/American Water Works Association/Water Environment Federation, Washington, DC

Thoms MC, Swirepik J (1998) Environmental flow management in New South Wales, Australia. In: Harding JK, Gilvear DG (eds) Hydrology in a changing environment: ecological and hydrological interactions. Wiley, Chichester, pp 281–288

WHO (1993) The guidelines for drinking water quality, 2nd edn. WHO, Geneva

Yang W, Nan J, Sun D (2008) An online water quality monitoring and management system developed for the Liming River basin in Daqing. China J Environ Manage 88:318–325

Yesilnacar MI, Uyanik S (2005) Investigation of water quality of the world's largest irrigation tunnel system, the Sanliurfa tunnels in Turkey. Fresenius Environ Bull 14(4):300–306

Yucel E, Dogan F, Ozturk M (1995) Heavy metal status of Porsuk stream in relation to public health. Ekoloji 17:29–32

Zacharias I, Dimitriou E, Koussouris T (2005) Integrated water management scenarios for wetland protection: application in Trichonis Lake. Environ Model Software 20(2):177–185

Kinetics and Mechanisms of Biosorption of Copper Ion onto Waste Yeast (*S. cerevisiae*)

M. Sarioglu Cebeci and U.A. Guler

Contents

Abstract Heavy metals are the most common pollutants found in industrial effluents. Several chemical treatment methods are used to remove heavy metals from aquatic solutions. Recently, biosorption process which utilizes various microbial materials (bacteria, fungi, yeasts, algae etc.) has been preferred to remove pollutants. In this study, waste yeast taken from yeast factory in Turkey was used as (working volume of 100 ml) biosorbent in erlenmeyer batch experiments by using temperature controlled shaker and copper (II) ion biosorption was examined. The effects of initial pH (2, 3, 4, 5, 6, 7), initial copper concentrations (25, 50, 75, 100, 150, 200, 250, 300 mg L^{-1}), biosorbent amount (1, 3, 5, 7 10 g L^{-1}), contact time (5, 10, 15, 30, 45, 60, 90, 120, 240, 1440 min.) and temperature (20, 30, 40, 50 °C) parameters on to biosorption process were investigated. Optimum biosorption capacity was found as pH 5, 100 mg L^{-1} of initial copper concentration, 10 g L^{-1} biosorbent amount and 1440 min. of contact time. The experimental equilibrium data fitted to Freundlich and Langmuir adsorption isotherm models. Freundlich models fitted better than Langmuir models. The maximum adsorption capacity of waste yeast was determined as 7.94 mg g^{-1} and Freundlich isotherm values n and k_f were found to be 1.55 and 0.21 respectively. Pseudo-second order kinetic model was suitable for biosorption kinetics. According to calculated thermodynamic parameters (ΔH, ΔG and ΔS), biosorption of copper onto waste yeast was exothermic. As a conclusion; it was found that copper (II) removal by using waste yeast was

M.S. Cebeci (✉) • U.A. Guler
Cumhuriyet University, Department of Environmental Engineering, 56140 Sivas, Turkey
e-mail: sarioglu@cumhuriyet.edu.tr

M. Öztürk et al. (eds.), *Plants, Pollutants and Remediation*,
DOI 10.1007/978-94-017-7194-8_19

low. Activation of biomaterial by pretreatment can be resulted in increasing of removal rate. Use of waste materials in pollution control is important for economic issue.

Keywords Waste yeast • Heavy metal • Biosorption • Batch study

1 Introduction

Heavy metals are the must common pollutants in industrial effluents. Heavy metals are discharged by various industries such as electro plating, metal finishing, textile, storage batteries, mining, ceramic and glass (Panday et al. 1985; Loutseti et al. 2009; Sarioglu et al. 2005, 2009).

Heavy metal pollution is important environmental problem due to toxicity of them to some aquatic life (Aksu 2005). Heavy metal removals are usually studied with conventional physical or chemical treatment processes including coagulation/flocculation, precipitation, ion exchange, membrane technologies and adsorption (Aksu 2005; Kandah 2004; Wang and Chen 2006). Some of these methods have been shown limited usage because of excess amount of chemical usage and huge amount of sludge (Aksu 2005; Wang and Chen 2006; Akar et al. 2008).

Adsorption process is widely used for heavy metal removal from wastewaters. Activated carbon is generally used for heavy metal removal because of high adsorption capacity. Because of quite expensive of activated carbon, alternative low-cost adsorbents were used for heavy metal removal instead of activated carbon. Recently, heavy metal removal has been studied by biosorption which utilizes various dead microorganism including bacteria, fungi, algae and yeast etc. (Aksu 2005; Wang and Chen 2006; Volesky 1990). There are several batch and continous studies for removal of pollutants fom wastewaters by using yeasts like *S. cerevisiae* (Aksu 2005; Wang and Chen 2006).

In this study, copper ion (Cu^{2+}) biosorption onto granule waste yeast taken from Yeast Factory in Turkey was studied under batch experimental conditions. Effects of pH, initial Cu^{2+} ion concentration, contact time, biosorbent amount and temperature onto biosorption efficiency were carried out. Experimental data fitted to equilibrium isotherm models. Biosorption mechanism was determined by calculation of thermodynamic parameters and kinetics.

2 Study Material

Waste yeast (*S. cerevisiae*) used for Cu^{2+} biosorption was taken from İzmir Pakmaya Yeast Factory as granular. It was washed 3 or 4 times with deionizer water before batch studies and dried at 60 °C for 24 h. 1000 ppm of stock Cu^{2+} ion concentration was prepared from $CuSO_4.5H_2O$. HCl was added into stock solution in order to prevent copper precipitation.

3 Measurements

WTW (Inolab) pH meter was used for pH measurement. Samples were taken from determined time intervals from erlenmeyer. It was centrifuged at 4000 rpm for 5 min. and Cu^{2+} ion in supernatant was measured at 217.9 nm by using atomic absorption spectrophotometer (GBC Avanta Σ).

4 Experimental Procedure

Cu^{2+} biosorption onto waste yeast batch experiments were achieved in 250 ml capacity of erlenmeyer with 100 ml working volume at agitation of 120 rpm for several contact times by using temperature controlled shaker. The effects of pH (2, 3, 4, 5, 6, 7), initial copper concentration (25–300 mg L^{-1}), amount of waste yeast (1–10 g L^{-1}), contact time (5–1440 min.) were studied onto biosorption capacity and biosorption efficiency. Lastly, effect of temperature onto biosorption capacity percentage removal was investigated by experimentally and calculation of thermodynamic parameters.

5 Biosorption Isotherm Models, Kinetics and Thermodynamics

Experimental results fitted to Freundlich and Langmuir isotherm models. Langmuir isotherm model explains that monolayer, homogeny adsorption and it is given in the following equation:

$$q_e = q_m bC_e / 1 + bC_e \tag{1}$$

Where 'C_e' is the equilibrium liguid-phase concentration (mg L^{-1}), 'q_e' the equilibrium amount adsorbed (mg g^{-1}), 'q_m' the maximum amount of sorbate per unit sorbent (adsorption capacity) to form a complete monolayer (mg g^{-1}), and b is the Langmuir constant related to the affinity between sorbent and sorbate (mg L^{-1}).

Freundlich model is an ampiric equation for heterogen surfaces (Shahwan and Erten 2002).

$$q_e = k_f C_e^{1/n} \tag{2}$$

Where 'k_f' and 'n' are Freundlich constants that are related to the adsorption capacity and intensity, respectively.

In this study data fitted to pseudo-second order reaction rate equation. Pseudo-second order rate equation is given below (Aksu 2001; Başıbüyük and Forster 2003).

$$t/q_t = \left[1/k_2 q_e^2\right] + t/q_e \tag{3}$$

Where k_2 (mg g^{-1} min.) is the rate constant of the second-order equation, qt (mg g^{-1}) is the amount of biosorption time 't' (min.) and 'qe' is the amount of biosorption equilibrium (mg g^{-1}). The linear plots of t/qt versus 't' for the pseudo-second order model for the biosorption of Cu^{2+} ions onto *S. cerevisiae*.

The initial adsorption rate, h (mg g^{-1} min.), as $t \rightarrow 0$ can be defined as

$$h = k_2 q_e^2 \tag{4}$$

Where the initial adsorption rate (h), the equilibrium adsorption capacity (q_e), and the second-order constants 'k_2' (mg g^{-1} min.) can be determined experimentally from the slope and intercept of plot t/q versus 't'.

Thermodynamic parameters (ΔG: free energy, ΔH: enthalpy and ΔS: entropy) were calculated by using Eqs. (5) and (6) in order to show thermodynamic behaviors of copper biosorption onto waste yeast Dakiky et al. (2002).

$$\Delta G = -RT \ln Kc \tag{5}$$

$$\ln Kc = (\mathbf{DS} - \mathbf{DH}) / R * 1/T \tag{6}$$

6 Findings

In this study; optimum conditions of biosorption were evaluated at pH 5, 10 g L^{-1} of biosorbent amount, 24 h of contact time, 100 mg L^{-1} of initial copper ion concentration (Figs. 1, 2, 3, and 4). As the temperature increased, copper removal percentages increased among studied temperature values (20–30–40–50 °C) (Fig. 5). Experimental results fitted to Langmuir and Freundlich isotherm models (Figs. 6 and 7). Freundlich models fitted better than Langmuir models. The maximum adsorption capacity of waste yeast was determined as 7.94 mg/g and Freundlich isotherm values n and kf were found to be 1.55 and 0.21 respectively. Thermodynamic parameters were calculated as ΔH = −37.92, ΔG = +, ΔS = −0.15 (Fig. 8). According to results of thermodynamic parameters biosorption of copper onto waste yeast was exothermic. Pseudo-second-order kinetic model was suitable for biosorption kinetics (Fig. 9). The model gave good correlation with some heavy metal biosorption studies using *S. cerevisiae* (Wang and Chen 2006).

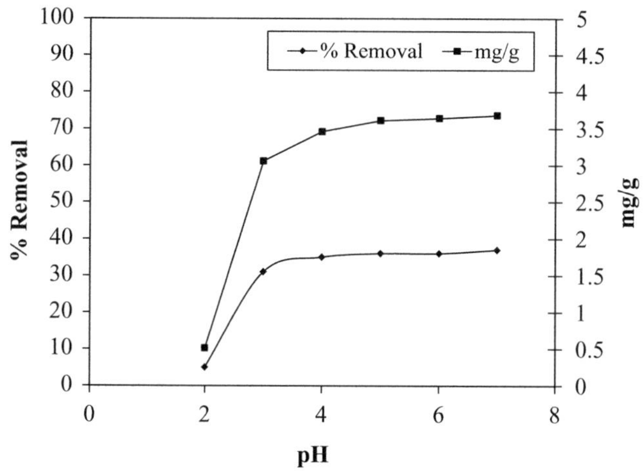

Fig. 1 The effect of pH (Initial Cu^{2+} ion concentration 100 mg L^{-1}, contact time 24 h, biosorbent dosage 10 g L^{-1}, temperature 20 °C)

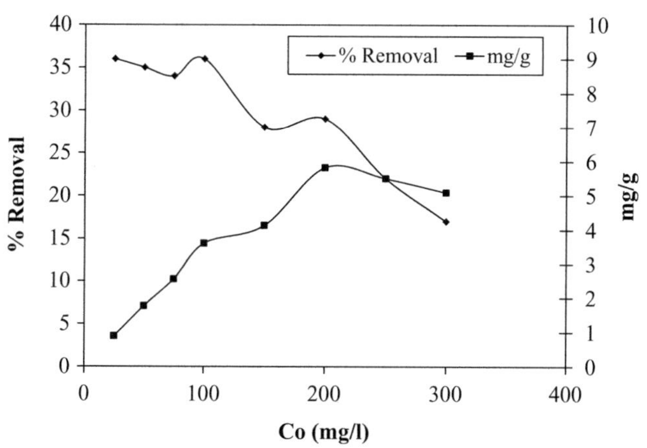

Fig. 2 The effect of initial metal ion concentration (pH 5, temperature 20 °C, biosorbent dosage 10 g L^{-1}, contact time 24 h)

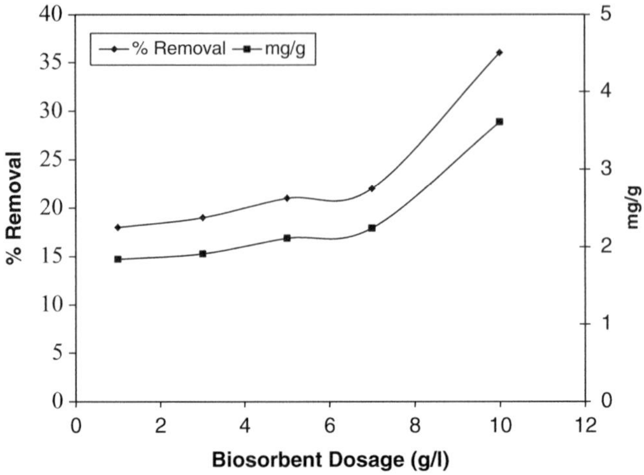

Fig. 3 The effect of biosorbent dosage (pH 5, temperature 20 °C, initial Cu^{2+} ion concentration 100 mg L^{-1}, contact time 24 h)

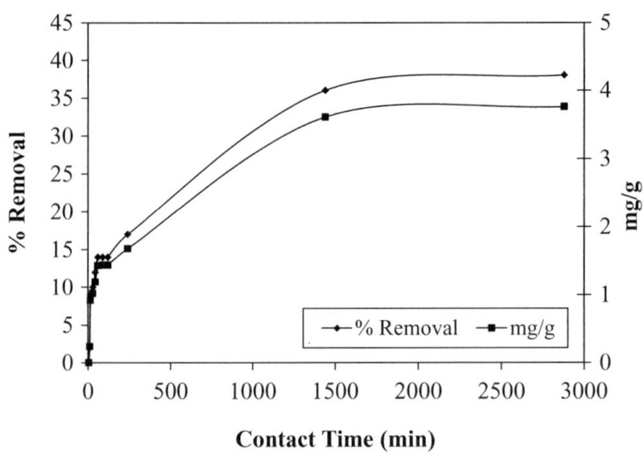

Fig. 4 The effect of contact time (pH 5, temperature 20 °C, biosorbent dosage 10 g L^{-1}, initial Cu^{2+} ion concentrating 100 mg L^{-1})

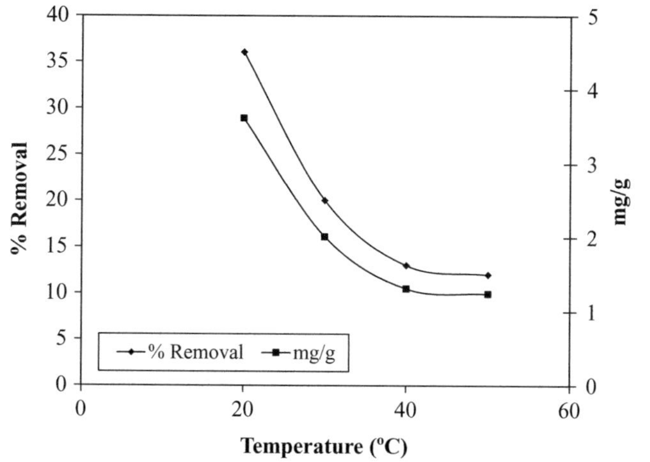

Fig. 5 The effect of temperature (pH 5, initial Cu^{2+} ion concentration 100 mg L^{-1}, contact time 24 h, biosorbent dosage 10 g L^{-1})

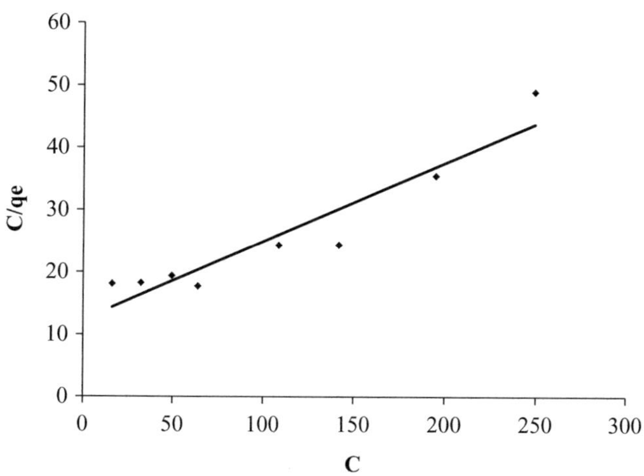

Fig. 6 Langmuir isotherm equilibrium model

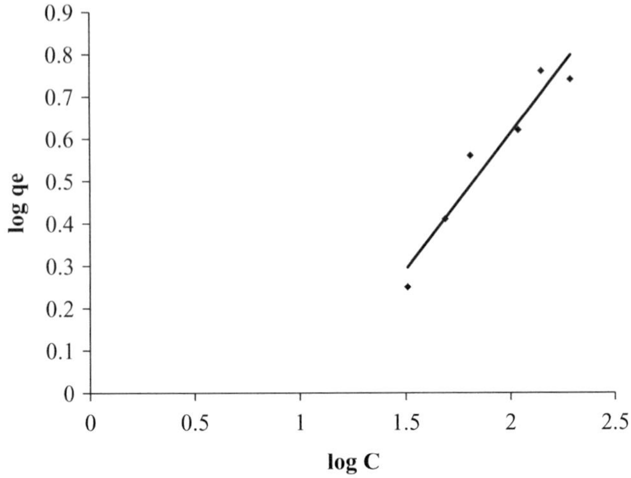

Fig. 7 Freundlich isotherm equilibrium model

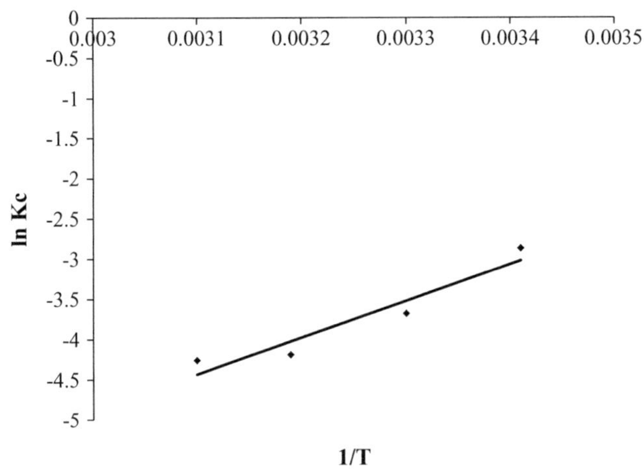

Fig. 8 Plot of lnKc vs. 1/T for the estimation of thermodynamic parameters for biosorption of Cu^{2+} onto *S. cerevisiae* biomass

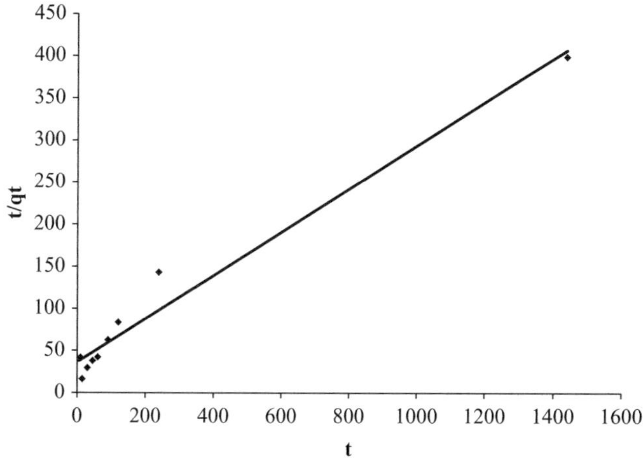

Fig. 9 Pseudo-second-order kinetic plot for the biosorption of Cu^{2+} *S. cerevisiae* biomass

7 Conclusion

Removal of copper ions from aqueous solutions by using waste yeast (*S. cerevisiae*) taken from Yeast Factory in Turkey was investigated. It was found that copper removal percentage was low using natural waste yeast as biosorbent. Activation (treatment) of biomaterial (yeast) physically or chemically need be investigated in detail in order to increase biosorption capacity/removal percentage. As a conclusion, use of pretreated biomaterial efficiently must be considered with regeneration and continuous studies.

Acknowledgement This work was supported by the Scientific Research Project Fund of Cumhuriyet University under the project number M-326.

References

Akar T, Kaynak Z, Ulusoy S, Yuvacı D, Ozsarı G, Akar ST (2008) Enhanced biosorption of nickel(II) ions by silica-gel-immobilized waste biomass: biosorption characteristics in batch and dynamic flow mode. J Hazard Mater 163:1134–1141

Aksu Z (2001) Biosorpsiyon of reactive dyes by dried activated sludge: equilibrium and kinetic modelling. Biochem Eng J 7:79–84

Aksu Z (2005) Application of biosorption for the removal of organic pollutants: a review. Process Biochem 40:997–1026

Başıbüyük M, Forster CF (2003) An examination of adsorption characteristic of a basic dye (Maxilon Red BL-N) and live activated sludge system. Process Biochem 38:1311–1316

Dakiky M, Khamis M, Manassra A, Mer'eb M (2002) Selective adsorption of chromium (VI) in industrial wastewater using low-cost abundantly available adsorbents. Adv Environ Res 6:533–540

Kandah MI (2004) Zinc and cadmium adsorption on low grade phosphate. Sep Purif Technol 35:61–70

Loutseti S, Danielidis DB, Amili AJ, Katsaros C, Santos R, Santos P (2009) The application of a micro-algal/bacterial biofilter for the detoxification of copper and cadmium metal waste. Biosour Technol 100:2090–2105

Panday KK, Prasad G, Singh VN (1985) Copper (II) removal from aqueous solutions by fly-ash. Water Res 19:869–873

Sarioglu M, Atay UA, Cebeci Y (2005) Removal of copper from aqueous solutions by phosphate rock. Desalination 181:303–311

Sarioglu MG, Atay UA, Beyazıt N (2009) Removal of copper from aqueous solutions using biosolids. Desalination 239:167–174

Shahwan T, Erten HN (2002) Thermodynamic parameters of Cs^+ sorption on natural clays. J Radio Anal Nucl Chem 253:115–120

Volesky B (1990) Biosorption and biosorbents. In: Volesky B (ed) Biosorption of heavy metals. CRC press, Boca Raton, pp 3–5

Wang J, Chen C (2006) Biosorption of heavy metals by Saccharomyces cerevisiae: a review. Biotechnol Adv 24:427–451

Effects of Air Pollution on Urban Plants: Nezahat Gökyiğit Botanical Garden

Özge Horasan, Esra Deniz Güner, Gülen Güllü, and Rukiye Tipirdamaz

Contents

Abstract Physiological traits of *Fraxinus angustifolia* Vahl. in response to different levels of polluted air conditions were analysed in Nezahat Gökyiğit Botanical Garden which is located at the intersection of TEM and E5 highways in İstanbul. To determine the level of air pollution at the investigated area, for a 1 year period, monthly passive air pollution (SO_2, NO_2, O_3, VOCs) sampling were conducted at four sampling sites which were chosen as representatives of the whole profile of the botanical garden. Along with air pollution monitoring, some physiological parameters of the leaves of *F. angustifolia*, – total chlorophyll and carotenoid contents, chl a/chl b ratio, net photosynthesis rate, specific leaf dry weight, water content percentage, leaf area and heavy metal concentrations- were analyzed during the growing period. Also PAH, heavy metal and PCB concentrations of the soil were detected in

Ö. Horasan (✉)
Faculty of Science, Department of Biology, Hacettepe University, Ankara, Turkey

E.D. Güner
Faculty of Engineering, Department of Environmental Engineering, Hacettepe University, Ankara, Turkey

G. Güllü
Department of Environmental Engineering, Hacettepe University, 06532 Ankara, Turkey
e-mail: ggullu@hacettepe.edu.tr

R. Tıpırdamaz
Department of Biology, Hacettepe University, 06532 Ankara, Turkey
e-mail: tuz@hacettepe.edu.tr

M. Öztürk et al. (eds.), *Plants, Pollutants and Remediation*,
DOI 10.1007/978-94-017-7194-8_20

the nearby area to see the extend of the pollution on the region. While no SO_2 pollution was observed, NO_2 appeared to exceed the concentration limits that are stated as harmful for vegetation. O_3 concentrations dominated in the area during summer. VOC, PAH and PCB concentrations on the soil were found to be related to the traffic pollution. Excessive levels of Fe, Pb and Cd of the unwashed leaves were determined. Among all the pollutants, only NO_2, O_3, Fe and Cd found to be responsible of plant injury in the area. However the general physiological status of *F. angustifolia* is not seem to be effected from the urban environment conditions of the botanical garden which can be related to its tolerance mechanisms.

Keywords *Fraxinus angustifolia* • Nezahat Gökyiğit botanical garden • Air pollution • Traffic emissions • Heavy metalpollution

1 Introduction

In developing countries, industry and urbanization have resulted in increased concentrations of air pollutants which could cause plant injury symptoms. In Europe, Asia and USA plant injury related to a variety of air pollutants has been well described (Smith 1991; Pandey and Pandey 1994; Dogan and Ozturk 1991, 1994a, b, c; Gratani et al. 2000).

The environment of urban areas usually consists of a mixture of different kinds of air pollutants which are derived from the emissions of traffic sources, industrial processes and domestic heating systems. Mostly encountered phytotoxic air pollutants in urban areas are SO_2, NO_2, O_3 and volatile organic compounds (VOCs and PAHs). Under such urban environment conditions, plant growth, some physiological and biochemical properties of plants are effected of different combinations of these pollutants (Ozturk et al. 1991a, b, c, 1994a, b; Ozturk 1999; Pandey and Pandey 1994). In this instance, it is possible to claim that morphological, physiological and biochemical assays of plants growing in their habitats could be used for monitoring the magnitude of pollution in the area.

Some plant species are characterized with being capable of absorbing, detoxifying and tolerating high levels of pollution (Verma and Singh 2006; Ozturk et al. 2010, 2011; Yilmaz et al. 2011). As a consequence, they show some leaf injury symptoms or have to modify their metabolism for adapting the environment.

In a polluted area, leaves as the synthesizing organs are the firstly effected parts of a plant, considering that they have a close relationship with atmosphere and soil. Some measureable leaf injuries could appear in chlorophyll content, net photosynthesis rate, leaf area and/or accumulation of dry matter (Gratani et al. 2000; Verma and Singh 2006). On the other hand, some adaptation mechanisms to avoid foliar injury could also occur, such as increasing carotenoid and/or relative water contents (Sharma et al. 1980).

Excessive levels of air pollutants are toxic not only for plants but also human health. Therefore, to understand the effects of pollutants on vegetation is important for determining the level and extinguishing the effects of air pollution for the living environment. This investigation that is conducted in Nezahat Gökyiğit Botanical Garden in İstanbul has the aim of determining the air quality in an urban environment with the knowledge of the physiological conditions of plants in the area.

2 Sampling Sites

Nezahat Gökyiğit Botanical Garden, which has been founded in 1996, is located at the intersection of heavy traffic loaded highways, TEM and E5, in İstanbul. In this study, four sampling sites in the garden with different altitudes and distances to the highways were chosen in order to see the variability of the pollutants on the region. Mean values of the analysis data that were obtained from all sampling sites for all sampling periods were calculated to represent the whole profile of the botanical garden (Fig. 1).

3 Air Monitoring

At four sampling sites, every month starting with December 2007 until December 2008 ambient SO_2, NO_2, O_3 concentrations were detected by passive sampling method for weekly period. VOCs were also monitored for 6 months (until May 2008) by the same methodology.

Fig. 1 Location of Nezahat Gökyiğit Botanical Garden and the sampling sites *A*, *B*, *C* and *D* for air monitoring and plant analysis

Passive sampling tubes (Gradko International Co.) were located at four sampling sites in the botanical garden. After a 1 week exposure period they were collected and stored at 0 °C immediately until the analysis. VOCs (benzene, toluene, ethylbenzene and xylene) were analyzed by Thermal Desorber-GC/FID system. Every month, collected SO_2, NO_2, O_3 tubes were send to Gradko International Co. for chemical analysis.

4 Soil Analysis

Soil samples were collected from eight sampling sites in the nearby area on October 2008. Polycyclic aromatic hydrocarbons (PAHs) and trace metals (Mg, Fe, Ni, Pb, Cu, Cd, Zn, Mn, Ca, Na and K) were detected in the soil samples. For PAH determination, collected soil samples were Soxhlet extracted and analysis were performed by using GC-MS similar to literature (Harner and Bidleman 1998; Sofuoglu et al. 2001). While Mg, Fe, Ni, Pb, Cu, Cd, Zn, Mn and Ca were analyzed by atomic absorption spectrophotometer by the extraction method of EPA 3050B for soil, Na and K were extracted by the method of Taleisnik et al. (1997) and detected by flame photometry.

5 Plant Material

Fraxinus angustifolia Vahl., which was found to be mutual at four sampling sites chosen, was used as plant material for the analysis. At all sites, from May to end of September, two trees and then three branches per tree were tagged. Total chlorophyll and carotenoid contents, chl a/chl b ratio, net photosynthesis rate, specific leaf dry weight, water content percentage, leaf area and heavy metal concentrations of the leaves from the same tagged branches were analyzed monthly.

6 Leaf Characteristics

Leaves collected from the tagged branches were immediately weighed and the values were recorded as fresh weight (FW). Then they were dried at 105 °C for 3 days and weighed again which gives the leaf dry weight (DW).

Water content percentage (WC %) of leaves were determined by the formula below:

$$WC\% = [(FW - DW) / FW] \times 100$$

Leaf area was determined by the computer programme SHAPE (Iwata and Ukai 2002) which calculates the area of images based on their contrast with the ground.

For that purpose, leaves were photocopied and the images were transferred to computer by a scanner. Leaf area was determined in cm^2.

Specific dry weight (SDW) (Pandey and Pandey 1994) was calculated by dividing leaf dry weight to leaf area.

Chlorophyll and carotenoid content of leaves were determined in μg/g FW by the method of Lichtenthaler (1987).

Net photosynthesis rate (Pn) was measured by Qubit Infrared Plant CO_2 Analysis Package which analyses CO_2 concentrations both ambient and in its leaf chamber. The difference between the ambient and the leaf chamber CO_2 concentrations gives the amount of CO_2 that the leaf uses for photosynthesis during a particular period of time (μmol CO_2/m^2 sn).

Mg, Fe, Pb, Cu, Cd, Zn, Mn and Ca were analyzed by atomic adsorption spectrophotometer by the extraction method of EPA 3030 E nitric acid digestion for plants. Na and K were extracted by the method of Taleisnik et al. (1997) and detected by flame photometry (μg/g DW).

7 Gaseous Air Pollutants

The annual and seasonal mean of measured air pollutants is given in Table 1. The annual average concentrations of NO_2, SO_2, Ozone, and VOCs (Benzene) were found to be 63, 9.1, 33, 4.58 μg/m^3, respectively at the NG Botanic Garden location. The summary of meteorological conditions during the sampling period were given in Table 2. As you may see from the Table, the temperature was around 11 °C and 21 °C during winter and summer periods respectively. And there were no seasonal difference for the dominant wind direction and speed on the sampling area.

Measured air pollution parameters (NO_2, SO_2, Ozone and VOCs) were compared with the data established for some other cities and also with the national and international standards in order to see the extend of pollution over the area.

Table 1 Annual and seasonal averages of SO_2, NO_2 and O_3 concentrations for this study

	Annual (μg/m^3)	Seasonal (μg/m^3)	
Pollutant	Mean	Winter	Summer
NO_2	63 ± 18.7	66.84	58.3
SO_2	9.1 ± 12.5	11.9	8.5
O_3	33 ± 24.2	20.55	47.9

Table 2 Winter-Summer seasonal ambient weather conditions in İstanbul

	Temperature (°C)	Relative humidity (%)	Mixing height (m)	Wind shifts (m/sn)	Wind directions
Winter	11	72	799.83	2	NE
Summer	21	64	963.31	2.1	NE/ENE

The SO_2 concentrations for NG Botanical Garden were found to be generally low as expected since the main residential heating source on the region is natural gas from the year 1993. The location of NG Botanical Garden generally have lower rates of SO_2 for other cities compared, under the influence of the national as well as international legislations (Table 3).

Observed NO_2 levels were generally high in the region. Although high NO_2 levels were observed in all sites during whole sampling period, as it is expected due to heavy traffic load, winter period have higher NO_2 concentrations rather than in the summer (Table 2). Higher NO_2 levels during winter period could be due to the lower mixing height levels in this period (Schnitzhofer et al. 2008). Annual mean abundances of compounds were assessed with respect to guideline limits and it was found that all the measured parameters were under the legislation thresholds, although some monthly values were exceeded the limits during winter period.

Measured ozone levels in the region were around from 4.8 to 108.6 to $\mu g\ m^{-3}$. Higher ozone concentrations in the summer period were observed in the area which may be attributed to enhancement of photochemical reactions during the summer period. A typical opposing relationship between NO_2 and ozone concentrations was recorded at all sites. Depletion of ozone by NO_2 may have been the cause of the lower ozone values in the areas having high traffic densities than the lower ones (Danalatos and Glavas 1996). The continuous year-long measurement of ozone yielded an annual mean of 33 $\mu g\ m^{-3}$, which was below its EU guideline limit of 120 $\mu g\ m^{-3}$, and thus ozone is currently of little concern at this location (Table 3).

Many volatile organic compounds (VOCs) are both primary and secondary products of vehicle emissions. Other sources of such compounds include industrial

Table 3 Comparison of SO_2, NO_2, ozone and VOCs data (in the units of $\mu g/m^3$ for the year) with the national, international limits and traffic-related exposure studies

Study region	SO_2	NO_2	O_3	VOCs (Benzene)
This study	9.2	63	33	4.6
Kolkata[a] (India)	12.3	32.5		
BuenosAires[b] (Argentina)	69			
Eskişehir[c] (Turkey)	47.13	22.57	46.08	
İstanbul (Turkey)		60[d]	77–76[e]	
Grater Cairo[f] (Egypt)				2.03
Tirol[g] (Austria)		72		3.3
EU guideline limits	122 (24 h)		120 (8 h)	
Turkish limits	150	100	240 (1 h)	5

[a]Gupta et al. (2008)
[b]Fagundez et al. (2001)
[c]Özden et al. (2008)
[d]Topçu et al. (2003)
[e]İm et al. (2006)
[f]Khoder (2007)
[g]Schnitzhofer et al. (2008)

plants (factories, power stations, etc.), household heating and fuel oil combustion engines (Lee et al. 2002; Wallace 1996). Among all of the measured VOCs species, toluene (25.73 μg m^{-3}) was the most abundant compound followed by benzene (4.58 μg m^{-3}), (m,p,o)-xylene (5.86 μg m^{-3}) and etilbenzen (2.42 μg m^{-3}). These results are similar to those which claim that VOCs have high levels in the areas close to the traffic (Khoder 2007; Ras-Mallorqui et al. 2007). Suggesting that the BTEX levels in NGBB results from the traffic density in the surrounding areas. Compared to EU guideline limit values the compounds measured, Benzene was found to be of particular higher at this location (Table 3).

8 Soil-PAH Results

The concentration of the sum of 16 EPA-PAHs (Σ16PAH) in the surface soils were calculated as 147.06 ± 161.81 μg/kg drywt (mean ± SS). It seemed that the levels of Σ 16 EPA-PAHs were nearly 200 μg/kg drywt which is the PAHs limit value of pollution of soil (Tang et al. 2006). There has been a shift from coal to natural gas for domestic heating purposes starting from early 1990s and liquefied petroleum gas (LPG) has been widely used by taxis from the beginning of 1998. The emissions from the using diesel fuel motor, a big portion of PAH are emitted to the atmosphere than using from benzine fuel motor (Khalili et al. 1995; Guangdi et al. 2008). Low quality liquid fuels and long before densely using coal for domesting heating may explain PAH concentrations to reach high levels in this region soil.

9 Plant Analysis

Among all detected air pollutants in the area, only NO_2 was found to be excessing the limit values for vegetation (Adaros et al. 1991), thus may have an effect on plants. Water content percentage of the leaves of *F. angustifolia* increased with the high levels of NO_2 concentrations ($r = 0.73$ $p < 0.05$). As stated by Kammerbauer and Dick (2000), in plants exposed to auto pollution water content may increase. Also NO_2 treated bean plants showed higher water content in their leaves (Srivastava et al. 1994). It is probably because of the osmotic adjustment of the plant to the increasing levels of nitrate in the leaves as a result of inhibited nitrite reduction. Therefore it is possible to claim that increasing water content of the leaves of *F. angustifolia* may be a tolerance mechanism to the ambient conditions with an annual average of 63 ± 18 μg/m^3 of NO_2.

Although O_3 did not seem to exceed the limits with an annual average of 33 ± 24 μg/m^3, it may have a negative effect on specific dry weight of the leaves ($r = -0.55$ $p < 0.05$). It is known that O_3 may cause a reduction in biomass of plants by inhibiting the production of dry matter (Krupa 1997). Also some other relatives of *F. angustifolia* such as *F.americana* and *F.pennysylvanica* was stated as sensitive to O_3

and used as indicator plants (Krupa et al. 1998). As a result, it is possible to assert that *F. angustifolia* may be also stated as sensitive to O_3 similar to some other *Fraxinus* species because of its tendency to have a decrease in specific dry weight when exposed to O_3. On the other hand O_3 has relatively low concentrations in the area besides specific dry weight of the leaves was also found negatively related with the percentage of water content of leaves ($r = -0.56$ $p < 0.05$) which is positively correlated with NO_2. It has been shown that when plants are exposed to several pollutants, it becomes almost impossible to determine their effects particularly (Pandey and Pandey 1994). Also, some plant injury symptoms could occur due to interactive effects of a variety of pollutants at considerably lower concentrations than those required for either gas alone (Bennett et al. 1990) which could explain why O_3 might have an effect on the specific dry weight of leaves of *F. angustifolia*.

In NG Botanical Garden area, despite NO_2 and O_3, no SO_2 pollution was observed. Thus any of the plant parameters analyzed did not seem to have a correlation with SO_2.

In this investigation high concentrations of some trace metals were determined similar to the studies which point out a high metal accumulation in plants growing roadside (Alfani et al. 1996). Results of the analysis showed that unwashed *F. angustifolia* leaves have excessive amounts of Fe, Cd and Pb with the values of 1.95 ± 2.5 mg/g DW; 3.90 ± 1.75 µg/g DW; 70.98 ± 8.78 µg/g DW respectively (Table 4).

Fe was found to be very high in the unwashed leaves of *F.angustifolia* however none of the leaf parameters seemed to be affected. Therefore it is possible to claim that high Fe content of the leaves may not be the result of a cellular contamination but may be due to the dust accumulated on the leaves which could be derived from the high Fe contented soil in the area. Similarly, Pb was also determined as excessive in the leaves although no significant soil contamination was observed in the area. Besides any of the leaf parameters were not found to be negatively correlated with Pb. Pilegaard and Johnsen (1984) concluded that Pb content in plants is correlated with the aerial deposition but not with soil concentrations which is strengthened by the results of this study. Furthermore Prasad (2004) stated that Pb is generally bound to the epicuticular waxes of the leaves rather than being transported into cell thus accumulated on the leaf surface which may also explain the uneffected state of *F.angustifolia* in this study.

Cd was found to be not only overreaching the limit values in the unwashed leaves of *F.angustifolia*, but also may have an effect on chlorophyll and carotenoid con-

Table 4 Comparison of the analysis results and the limit values

	Unwashed leaves of *F.angustifolia*	Limits for plants[a]	Soil	Limits for soil[b]
Fe (mg/g DW)	1.95 ± 2.5	0.14	20.1 ± 7.5	–
Cd (µg/g DW)	3.90 ± 1.75	0.1–2.4	1.07 ± 0.55	1
Pb (µg/g DW)	70.98 ± 8.78	1–13	24.4 ± 2.03	50

[a]Prasad (2004)
[b]Temmerman et al. (1984)

tents. While Chl a (−0.36 $p<0.05$), chl b (−0.62 $p<0.05$) and total chlorophyll (−0.63 $p<0.05$) contents showed negative correlations with Cd, carotenoid (0.85 $p<0.05$) and chl a/b ratio (0.61 $p<0.05$) were determined as positively related with Cd contamination. Chl b content of the leaves was effected more significantly than chl a content was, which resulted in an increasing chl a/b ratio. It is known that chl a is more sensitive to stress conditions than chl b and the high chl a/b ratio is an indicator parameter of stress tolerance in plants (Pandey and Pandey 1994). Carotenoids, which stand as phyto-protectives to stress conditions in the choloroplasts, showed an increase under Cd contamination. Regarding to increasing chl a/b ratio and carotenoid content of the leaves of *F.angustifolia* in response to Cd, those parameters may be pointing out its tolerance mechanisms. However, according to Arduini et al. (2006) *F.angustifolia* is highly sensitive to Cd contamination. Considering that high Fe and Pb concentrations caused no significant effect on leaf parameters and only Cd induced some injury symptoms, it is also possible to assert that *F.angustifolia* is more sensitive to Cd than Fe and Pb.

The overall results of this investigation showed that the area is under the influence of traffic emissions in terms of air quality, and *F.angustifolia* is considerably tolerant to urban conditions considering that the physiological traits of the leaves generally remained uneffected. As stated by Tolunay (2003), plants show higher tolerance to urban conditions when all requirements, such as water, minerals and sun light, are being provided.

References

Adaros G, Weigel H, Jäger H (1991) Single and interactive effects of low levels of O_3, SO_2 and NO_2 on the growth and yield of spring rape. Environ Pollut 72(4):269–286

Alfani A, Bartoli G, Rutigliano FA, Maisto G, Virzo De Santo A (1996) Trace metal biomonitoring in the soil and the leaves of Quercus ilex in the urban area of Naples. Biol Trace Elem Res 51:117–131

Arduini I, Godbold DL, Onnis A (2006) Cadmium and copper uptake and distribution in Mediterranean tree seedlings. Physiol Plant 97(1):111–117

Bennett JH, Lee EH, Heggestad H (1990) Inhibition of photosynthesis and leaf conductance interactions induced by SO_2, NO_2 and SO_2+NO_2. Atmos Environ 24A:557–562

Danalatos D, Glavas S (1996) Diurnal and seasonal variations of surface ozone in a Mediterranean coastal site, Patras, Greece. Sci Total Environ 177:291–301

Dogan F, Ozturk M (1991) Studies on the correlations between air and water pollution with the mortality rates in Turkiye. In: Ozturk MA et al (eds) International urban ecology symposium, Aydın, pp 16–23

Dogan F, Ozturk M (1994a) Air pollution problems in urban areas of developing countries – a case study from Turkiye. Ege Univ Sci Fac J 16:9–17

Dogan F, Ozturk M (1994b) Ecological evaluation of urban problems in Turkey. In: Wang M, Lou YX (eds) Urban ecological development, research and application. China Environmental Science Press, Beijing, pp 198–204

Dogan F, Ozturk M (1994c) Ecological evaluation of urban problems in Turkey. Ege Univ Sci Fac J 16:1–8

Fagundez LA, Fernandez V, Marino L, Martin TH, Persano I, Rivarola Y, Benitez DA (2001) Preliminary air pollution monitoring in San Miguel, Buenos Aires. Environ Monit Assess 71:61–70

Gratani L, Cresente MF, Petruzzi M (2000) Relationship between leaf-life span and photosynthetic activity of Quercus ilex in polluted urban areas (Rome). Environ Pollut 110:19–28

Guangdi W, Qiang Z, Peng M, Jordania R, Howard WM, Chris G, Eric P (2008) Sources and distribution of polycyclic aromatic hydrocarbons in urban soils: case studies of Detroit and New Orleans. Soil Sediment Contam 17:547–562

Gupta AK, Karar KA, John K (2008) Spatio-temporal characteristics of gaseous and particulate pollutants in an urban region of Kolkata, India. Atmos Res 87:103–115

Harner T, Bidleman TF (1998) Octanol–air partition coefficient for describing particle/gas partitioning of aromatic compounds in urban air. Environ Sci Technol 32:1494–1502

İm U, Tayanc M, Yenigu O (2006) Analysis of major photochemical pollutants with meteorological factors for high ozone days in Istanbul, TURKEY. Water Air Soil Pollut 175:335–359

Iwata H, Ukai Y (2002) SHAPE: a computer program package for quantitative evaluation of biological shapes based on elliptic fourier descriptors. J Hered 93(5):384–385

Kammerbauer J, Dick T (2000) Monitoring of urban traffic emissions using some physiological indicators in Ricinus communis L. plants. Arch Environ Contam Toxicol 39:161–166

Khalili NR, Scheff PA, Holsen TM (1995) PAH source fingerprints for coke ovens, diesel and gasoline engines, highway tunnels, and wood combustion emissions. Atmos Environ 29(4):533–542

Khoder MI (2007) Ambient levels of volatile organic compounds in the atmosphere of Greater Cairo. Atmos Environ 41:554–566

Krupa SV (1997) Air pollution, people and plants. American Phytopathological Society, St. Paul

Krupa SV, Tonneijk AE, Manning WJ (1998) Ozone. In: Flagler RB (ed) Recognition of air pollution injury to vegetation: a pictorial atlas. Air & Waste Management Association, Pittsburg, pp 2.1–2.28

Lee SC, Chiu MY, Ho KF, Zou SC, Wang X (2002) Volatile organic compounds (VOCs) in urban atmosphere of Hong Kong. Chemosphere 48:375–382

Lichtenthaler HK (1987) Chlorophylls and caretonoids pigments of photosynthetic biomembranes. Methods Enzymol 148:350–382

Özden Ö, Döğeroğlu T, Kara S (2008) Assessment of ambient air quality in Eskişehir, Turkey. Environ Int 34:678–687

Ozturk M (1999) Urban ecology and land degradation. In: Farina A (ed) Perspectives in ecology. Backhuys Publishers, Leiden, pp 115–120

Ozturk M, Gemici Y, Secmen O, Gork G (1991a) Impact of urbanisation on plants in West Anatolia, Turkiye. In: Öztürk et al (eds) International urban ecology symposium, Aydın, pp 359–363

Ozturk M, Ozcelik H, Ay G (1991b) Effects of anthropogenic factors on the plant cover of Van & its environs. In: Ozturk MA et al (eds) International urban ecology symposium, Aydın, pp 376–385

Ozturk M, Erdem U, Gork G (eds) (1991c) Urban ecology. Ege Univ. Press, Izmir, p 427

Ozturk M, Erdem U, Secmen O, Celik A, Guvensen A (1994a) Urban ecological problems of Kusadası. Ekoloji 12:21–24

Ozturk M, Erdem U, Alparslan E (1994b) Ecological restoration of human disturbed areas – a case study from Gediz Basin-West Anatolia Turkey. In: Wang M, Lou YX (eds) Urban ecological development, research and application. China Environmental Science Press, Beijing, pp 216–223

Ozturk M, Mermut A, Celik A (2010) Land degradation, urbanisation, land use & environment. NAM Science & Technology Centre, Delhi, 445 pp

Ozturk M, Okmen M, Guvensen A, Celik A, Gucel S (2011) Land degradation, urbanisation and biodiversity in the Gediz Basin-Turkiye. In: Ozturk et al (eds) Urbanisation, land use, land degradation and environment. NAM proceedings. Daya Publishing House, Delhi, pp 74–93

Pandey J, Pandey U (1994) Evaluation of air pollution phytotoxicity in a seasonally dry tropical urban environment. Environ Monit Assess 33:195–213

Pilegaard K, Johnsen I (1984) Heavy metal uptake from air and soil by transplanted seed plants of Achillea millefolium and Hordeum vulgare. Ecol Bull 36:97–102

Prasad MN (2004) Heavy metal stress in plants: from biomolecules to ecosystems. Springer, Berlin/New York

Ras-Mallorqui MR, Marce-Recasens RM, Borrull-Ballarin F (2007) Determination of volatile organic compounds in urban and industrial air from Tarragona by thermal desorption and gas chromatography–mass spectrometry. Talanta 72:941–950

Schnitzhofer R, Beauchamp J, Dunkl J, Wisthaler A, Weber A, Hansel A (2008) Long-term measurements of CO, NO, NO_2, benzene, toluene and PM10 at a motorway location in an Austrian valley. Atmos Environ 42:1012–1024

Sharma GK, Chandler C, Salemi L (1980) Environmental pollution and leaf cuticular variation in Kudzu (Pueraria labata Wild). Ann Bot 45:77–80

Smith WH (1991) Air pollution and forest damage. Chem Eng News 11:30–43

Sofuoglu A, Odabasi M, Tasdemir Y, Khalili NR, Holsen TM (2001) Temperature dependence of gas-phase polycyclic aromatic hydrocarbon and organochlorine pesticide concentrations in Chicago air. Atmos Environ 35:6503–6510

Srivastava HS, Omrod DP, Hale BA (1994) Responses of greening bean seedling leaves to nitrogen dioxide and nutrient nitrate supply. Environ Pollut 86:37–42

Taleisnik E, Peyrano G, Arias C (1997) Response of Chloris guyana cultivars to salinity, 1. Germination and early vegetative growth. Trop Grassl 31:232–240

Tang XY, Tang L, Zhu YG, Xing BS, Duan J, Zheng MH (2006) Assessment of the bioaccessibility of polycyclic aromatic hydrocarbons in soils from Beijing using an in vitro test. Environ Pollut 140:279–285

Temmerman LO, Hoening M, Scokart PO (1984) Determination of normal levels and upper limit values of trace elements in soil. Z Pflanzenernahr Bodenkd 147:687–699

Tolunay D (2003) Air pollution effects on annual ring widths of forest tress in mountainous land of İzmir. In: Topçu S, Yardim MF, Incecik S (eds) Proceedings of the 2nd air quality management at urban, Regional and global scales symposium, Istanbul, 25–28 September, pp 118–125

Topçu S, Anteplioglu U, İncecik S (2003) Surface ozone concentrations and its relation to wind field in Istanbul. Water Air Soil Pollut 3:53–64

Verma A, Singh SN (2006) Biochemical and ultrastructural changes in plant foliage exposed to auto-pollution. Environ Monit Assess 120:582–602

Wallace L (1996) Environmental esposure to benzene: an update. Environ Health Perspect 104:1129–1136

Yilmaz R, Sakcali S, Gucel S, Ozturk M (2011) Urban ecology of Edirne-Turkey. In: Ozturk etal (eds) Urbanisation, land use, land degradation and environment. NAM proceedings. Daya Publishing House, Delhi, pp 215–230

ERRATUM

Nickel Metal Uptake and Metal-Specific Stress Alleviation in a Perennial Desert Grass *Cenchrus ciliaris*

Faiz-ul Hassan Nasim, Rabia Khalil, Ayesha Sumreen, Muhammad Shafiq Chaudhry, and Muhammad Ashraf

M. Öztürk et al. (eds.), *Plants, Pollutants and Remediation*,
DOI 10.1007/978-94-017-7194-8

DOI 10.1007/978-94-017-7194-8_21

The publisher regrets that in Ch 5 (page 99) of the print and online versions of this book, the affiliations of one of the following authors are listed incorrectly.

Below is the correct listing of M. Ashraf's affiliation

M. Ashraf: Chemistry Department, The Islamia University of Bahawalpur, Bahawalpur, Pakistan

The online version of the original chapter can be found at
http://dx.doi.org/10.1007/978-94-017-7194-8_5

Index

M. Öztürk et al. (eds.), *Plants, Pollutants and Remediation*,
DOI 10.1007/978-94-017-7194-8

MIX
Papier aus verantwortungsvollen Quellen
Paper from responsible sources
FSC® C105338
FSC
www.fsc.org

If you have any concerns about our products,
you can contact us on
ProductSafety@springernature.com

In case Publisher is established outside the EU,
the EU authorized representative is:
Springer Nature Customer Service Center GmbH
Europaplatz 3, 69115 Heidelberg, Germany

Printed by Libri Plureos GmbH
in Hamburg, Germany